MEDICAL INSTRUMENT DESIGN AND DEVELOPMENT

MEDICAL INSTRUMENT DESIGN AND DEVELOPMENT

FROM REQUIREMENTS TO MARKET PLACEMENTS

Includes a case study in ECG implementation

Claudio Becchetti

Alessandro Neri

A John Wiley & Sons, Ltd., Publication

This edition first published 2013

Registered office

John Wiley & Sons Ltd, The Atrium, Southern Gate, Chichester, West Sussex, PO19 8SQ, United Kingdom

For details of our global editorial offices, for customer services and for information about how to apply for permission to reuse the copyright material in this book please see our website at www.wiley.com.

Library of Congress Cataloging-in-Publication Data

Becchetti, Claudio.
Medical instrument design and development : from requirements to market placements / Claudio Becchetti, Alessandro Neri.
p. ; cm.
Includes bibliographical references and index.
ISBN 978-1-119-95240-4 (cloth)
I. Neri, A. (Alessandro) II. Title.
[DNLM: 1. Electrocardiography–instrumentation. 2. Equipment Design–methods. 3. Electrical Equipment and Supplies. 4. Equipment Safety. 5. Software. WG 140]
R856.A1
610.28–dc23

2013002809

A catalogue record for this book is available from the British Library.

ISBN: 9781119952404

Set in 10/12 pt Times by Thomson Digital, Noida, India

Printed in the UK

To those who share the key to
knowledge and shed light for
those who want to enter

in memory of
My Father and of Italo Corti, MD, Cardiologist

Contents

Foreword

This book fills a gap in the technical literature, because it approaches, as a whole, the design and implementation process necessary to build and to put on the market a piece of medical instrumentation. Even if many good books deal with the design of biomedical equipment, and others discuss the signal processing that is implied or the problems related to certification, no one links all these steps together in the ingenious way that the authors have devised here.

The book presents an integrated view of the whole process involved from the first phases of the conception, through the design, realization and signal processing and on to the hardware realization and certification. This approach is made possible by means of a unique feature of the book: a manufacturer has revealed its know-how and the structure of a commercially available instrument (the Gamma Cardio CG from Gamma Cardio Soft, which complies with the ECG technical standard ANSI AAMI EC 11). Therefore, the integrated view that this book provides is done with reference to the technical details of the design of a marketed instrument; here, concepts and tools find their practical application. This hands-on approach is extremely useful, in that the electrocardiograph is a product that is both sufficiently easy to deal with for a prospective biomedical engineer, but encompasses all the main areas involved, thus helping in understanding medical instrumentation design. In this way, not only will the readers be familiar with the basic concepts, principles and analytical tools needed to design a medical instrument but they will also have the clear idea of how all these concepts and tools are translated into practice.

The field of medical instrumentation shares with industrial product implementation the knowledge of various engineering methodologies and techniques, and these aspects must be combined with specific requirements, such as the medical device certification, a point that is crucial in order to put an instrument onto the market. Moreover, the scenario has evolved, in that an increasing number of medical instruments are being conceived to be closely linked into a hospital information system and/or to be integrated in a distributed management in the context of a virtual (distributed) health record. Part of the book is thus devoted to seeing medical instrumentation in the context of cloud computing, for which the authors introduced the new term of c-Health, thus also helping in understanding future developments.

The book is self-contained because in every chapter it presents the theoretical concepts, on signal and information theory, digital signal processing, electronics, software engineering and digital communications, which remain valid across technological evolutions of medical instrumentation. Moreover, the book explores the additional know-how required for product implementation such as the business context, system design, project management, intellectual property rights and the project life cycle.

Readers can thus use the book in many ways: they can pick up a chapter to refresh some concepts (on electronics, software, algorithms, etc.) or go through the book in order to view the whole design process. Therefore, this book is addressed to upper-level undergraduate and graduate students in engineering, and to engineers interested in medical instrumentation design with a comprehensive and interdisciplinary perspective.

From my experience, I'm convinced that this book will give the reader a clear idea of the whole process involved in the design of medical instrumentation from conception to marketing and beyond.

Tommaso D'Alessio
Professor of Biomedical Engineering
University Roma TRE

Preface

A technology is a true progress when it is in everyone's hands
H. Ford

This book is written for biomedical engineering courses for **undergraduate or graduate level**. The book shows how medical devices and instruments are designed from conception to market placement. As shown in the figure, this implies an **interdisciplinary system approach** where mainstream engineering courses provide the know-how of the associated development stages. More specifically, system engineering in Chapter 1 looks at how product design is performed considering project management, the intellectual property rights and the business context. Chapter 2 deals with the concepts and requirements of bioinstrumentation. The other development stages that have their associated courses are: biomedical design (Chapter 3), signal processing and estimation (Chapter 4), applied electronics (Chapter 5), medical software (Chapter 6), e-Health, telecommunications and medical informatics (Chapter 7) and the certification/approval process (Chapter 8).

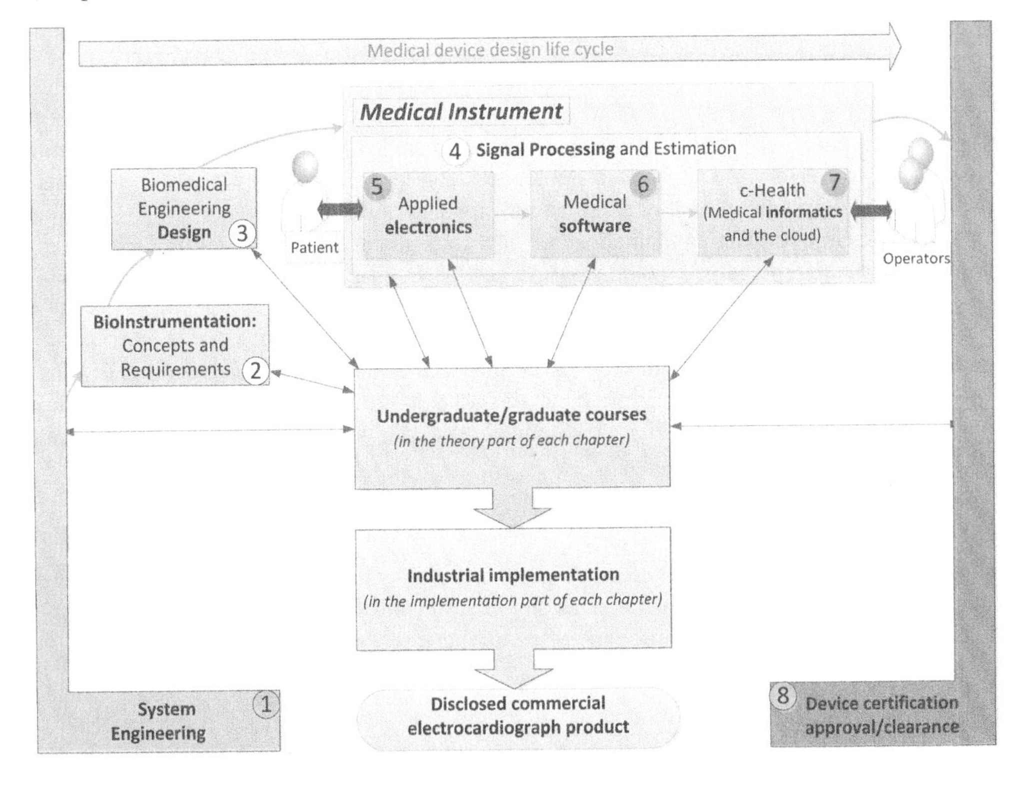

In addition, the book **discloses all the electronic details of a marketed product**, the Gamma Cardio CG, which complies with the ECG technical standard ANSI AAMI EC 11. Each chapter is divided into two parts: **theory** and **implementation**. The implementation part shows how theory is translated into the **ECG device** disclosed by the Gamma Cardio Soft Company.

The ECG is a suitable device because it is sufficiently simple to be addressed in a book at university level, but it is also adequate to show the main components required for any complex medical device design. The ECG also has a crucial impact on society since it is the main prevention and diagnostic tool for cardiovascular diseases, currently the prime cause of death in the world.

The book is also addressed to **practitioners** interested in medical device design, since this text shows the theory and its relationship with the implementation details of a marketed ECG, in terms of requirements, specifications, software, hardware and certification documents. As shown in the previous figure, the book embraces a **medical instrumentation model** that includes an electronic part (discussed in Chapter 5) in contact with the patient, managed by medical firmware and software (Chapter 6) that is integrated into the e-Health medical systems through telecommunication networks (Chapter 7). This product model is applicable to most medical devices and electronic equipment. More recent social needs, and advances in ICT, have fostered the need for alternative, cost-effective solutions for accessing and sharing medical data and services. Cloud computing technology is expected to have a large market share in distributed medical solutions. This issue is addressed in Chapter 7 where we introduce the new term "**c-Health**" referring to "*the set of healthcare services supported by cloud computing*". Medical electronic devices are usually modeled with the signal processing concepts addressed in Chapter 4, and they have to be certified according to applicable regulations and standards. Chapter 8 covers the medical instrumentation approval by regulatory bodies referred to around the world as 'certification', 'approval-clearance', or with other terms according to local regulation.

Pedagogy

Through this book, we emphasize the system-wide technical design approach that encompasses the basic theory, the associated implementation techniques – disclosed for a market product – and the application of regulations and standards.

A good **background** in mathematics and electronics is helpful for using this book, but undergraduates will be helped by the various checklists, recap tables and notes that summarize the main technical background.

The book has 250 figures and 145 tables; most of them offering a schematic representation of the main concepts. Text in bold is used to communicate concepts in speed reading. Figures have been designed with different graphic aspects to help readers' memorization. The implementation part is organized to offer a **didactic perspective**, starting from a general medical device and arriving at the ECG details.

Organization

Chapters are self-contained, addressing the relevant biomedical engineering courses. At the same time, chapters are logically linked as they describe the design process from conception to certification (from Chapter 1 to Chapter 8) with increased technical details.

The sections of each chapter are introduced following the sequential and logical process of the design. For example, in the electronics chapter, the paragraphs are organized according to the flow of the input signal. Each chapter begins with a **conceptual map** showing the relations among the sections and the associated topics.

Regulations, standards and technologies are critical elements of product design. We have addressed the main concepts and principles of these topics avoiding specific regulations, standard versions or technological implementations as possible. Regarding references, we have preferred a selected list of public domain material and tutorials that are helpful to further investigations. References to standards are placed at the end of the book.

Disclaimer

This book contains the details of a certified medical instrument (the Gamma Cardio CG) sold in the European market and released under an open source license as specified below.

The Gamma Cardio CG is an electrocardiograph owned by the Gamma Cardio Soft Company certified under European rules (CE certification), compliant to the US recognized standard for ECG (AAMI EC 11).

The Gamma Cardio Soft Company discloses its product through an open source license to support ECG diffusion worldwide and to open the associated know-how for universities since the ECG is the main prevention diagnostic tool for cardiovascular diseases, which are now the prime cause of death in the world.

The product is composed by a hardware board containing firmware controlled by a software application running over Windows® platforms. The Gamma Cardio CG hardware, firmware and software (see www.gammacardiosoft.it/book) are licensed under a Creative Commons license called Attribution-Non-Commercial-ShareAlike 3.0 Unported available at http://creativecommons.org/licenses/by-nc-sa/3.0/ and the GNU GPL License for software and firmware.

You are free:

1. to share – to copy, distribute and transmit the work
2. to remix – to adapt the work

But this is under the following conditions:

a. Attribution – You must attribute the work in the manner specified as follows (but not in any way that suggests that authors, editors or the manufacturer endorse you or your use of the work): "The whole work (hardware, software, firmware, design, ideas, novelties, projects) is owned by Gamma Cardio Soft S.r.l. – Italy. In no way are you allowed to suggest that Gamma Cardio Soft endorses you or your use of the work."
b. Non-commercial – You may not use this work for commercial purposes.
c. Share Alike – If you alter, transform or build upon this work, you may distribute the resulting work only under the same or similar license to this one.
d. Waiver – Any of the above conditions can be waived if you get permission from the copyright holder (Gamma Cardio Soft).

The Gamma Cardio Soft name is a trademark that can be used only after approval from the Company.

Warning: only approved **medical devices** may be placed in the market and used for medical purposes. The approval procedure must be performed under the regulation applicable in the target market (e.g., Medical Device Directive in the European Union, or under FDA approval-clearance procedures in the USA).

The product is disclosed as it is; Gamma Cardio Soft, the authors, the contributors, the editor, Wiley and anyone involved in the book, can in no way be deemed liable for any damage through the use of the information contained in this book or for the products described or derived from this book.

The Gamma Cardio Soft hardware, software and firmware is provided to you 'as is' and we make no express or implied warranties whatsoever with respect to its functionality, operability or use, including, without limitation, any implied warranties of merchantability, fitness for a particular purpose or infringement. We expressly disclaim any liability whatsoever for any direct, indirect, consequential, incidental or special damages, including, without limitation, lost revenues, lost profits, losses resulting from business interruption, loss of data or any damage to health, regardless of the form of action or legal theory under which the liability may be asserted, even if advised of the possibility or likelihood of such damages.

NO INFORMATION contained in this book or in the associated material SHALL CREATE ANY WARRANTY.

This book does not provide medical advice.
The contents of the book and the accompanying material, such as text, graphics, image and software are for informational purposes only. The content is not intended to be a substitute for professional medical advice, diagnosis or treatment. Always seek the advice of your physician or other qualified health provider with any questions you may have regarding a medical condition. Never disregard professional medical advice or delay in seeking it because of something you have read in this book.

If you think you may have a medical emergency, call your doctor or the emergency telephone number of your area immediately. The authors, the publisher, the Gamma Cardio Soft do not recommend or endorse any specific medical test, procedure, opinion or other medical information that may be mentioned on the book. Reliance on any information provided by the book is solely at your own risk.

By using this book, you agree to the specified terms.

Acknowledgment

The book is the result of 20 years of professional and teaching activity that could never have matured without the work, the suggestions, the constructive criticism and the contributions of many people.

A special mention to Gamma Cardio Soft Company (www.GammaCardioSoft.it) that has disclosed all the hardware and most firmware, software and certification documents of their electrocardiograph product. 'True progress is that which places technology in everyone's hands'. They intend by this action to support an open source hardware–software ECG, increasing device availability. The ECG is the main prevention and diagnostic tool for cardiovascular diseases, currently the prime cause of death in the world. An increased availability of ECGs translates into fewer casualties.

The open knowledge model of this book is derived from a previous book (*Speech Recognition: Theory and C++ Implementation*, 1999) where the approach of showing theory and disclosing all the details of an industrial product was first implemented.

We wish to thank the biomedical engineering department of the University of Roma TRE with Professor Tommaso D'Alessio for the support. We wish to thank the reviewers who have helped improving the book.

Going back ten years, we wish to thank students who started with the first dissertations on medical devices and Professor Fabio Frattale Masciolli for his support of this project. We wish to thank the engineering design laboratory course students who, for four years, have given didactic suggestions translated into this book. The interaction with them helped us to improve the clarity of our lectures and of the general exposition of technical issues.

Some colleagues have contributed to specific sections; we wish to thank them all for their fruitful collaboration and acknowledge chapter contribution shown in the following table:

Chapter	Topic	Contribution
1	System Engineering	Claudio Becchetti
2	Concepts and requirements	Theory: Maurizio Schmidt, Claudio Becchetti Implementation: Claudio Becchetti
3	Biomedical design	Theory: Maurizio Schmidt, Claudio Becchetti Implementation: Claudio Becchetti
4	Signal processing	Alessandro Neri
5	Electronics	Claudio Becchetti
6	Medical software	Angeloluca Barba, Claudio Becchetti
7	Medical informatics	Section 7.1: Alessandro Neri, Marco Leo, Claudio Becchetti Theory: Marco Leo, Claudio Becchetti, Implementation: Stefano Giordano, Claudio Becchetti
8	Certification	Theory: Lorenzo Spinelli, Claudio Becchetti, Implementation: Claudio Becchetti

A special mention to our families: writing a book requires a huge amount of time stolen from our lovely wives and children.

1

System Engineering

Give a man a fish, feed him for a day. Teach a man to fish, feed him for a lifetime.

Chapter Organization

Less than one-third of medical device development projects succeed in getting to the marketing stage (Kelleher, 2003). Table 1.1 adapted from (Kelleher, 2003) summarizes the main **reasons for failure** in medical device development. Certification is also added as a possible critical issue since it is a significant barrier especially for new companies (CB, 2004; Anast, 2001; EEC, 2007).

The application of engineering methods and processes may significantly reduce the risk of failures from the problems in Table 1.1. With this in mind, the **theory part** of this chapter introduces main engineering approaches here defined as '**tools**' described in the sections indicated in the final column of Table 1.1. These widely used tools are described here and applied to the design and development of medical instruments.

The tools and the corresponding paragraphs are introduced in a logical sequence as shown in Figure 1.1, where blocks represent sections indicated in the circled number.

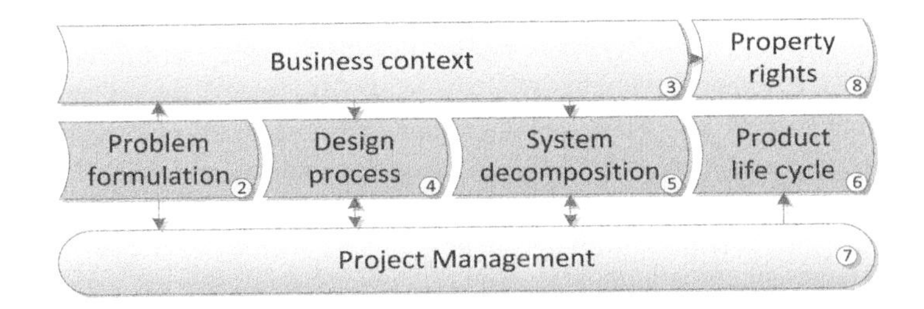

Figure 1.1 Chapter 1 structure.

Medical Instrument Design and Development: from Requirements to Market Placements, First Edition.
Claudio Becchetti and Alessandro Neri.

Table 1.1 Problems, solutions and tools in medical design

Problem	Possible reason	Possible solutions	Section: tools
Specifications	Customer not involved Launching products prematurely	Customer involvement Requirement management	1.2: Problem formulation 1.4: Product design process
Technology	Insufficient experience New technology	Buy in technology through consultancy Risk management	1.7: Project management 1.6: Product life cycle
Manufacturing costs	Unexpected costs in manufacturing	Subsystem design at early stage Design by manufacturing	1.5: System decomposition 1.4: Product design process 1.3: Business context
Intellectual property right **(IPR)**	Recently registered patents	Effective offensive and defensive IPR strategies	1.8: IPR
Financial return	Underestimated development costs	Build reasonable worst-case business plan	1.3: Business context
Certification	Underestimated costs Unexpected problems	Design by certification Hire consultant	1.6: Product life cycle Ch. 8: Certification

The main stream of product design starts from problem formulation (Section 1.2) to the product life cycle (Section 1.6). Any unique temporary activity such as design can be considered as a project that require specific management (Section 1.7). During and after design, intellectual property right (IPR) has to be faced with care to avoid infringing existing patents and to eventually protect the own product innovation as discussed in Section 1.8.

The **implementation part** of this chapter applies theory to a medical instrument example: the electrocardiogram (ECG/EKG). **This book fully discloses a commercially available ECG product**: the Gamma Cardio CG (CGV model) from the Gamma Cardio Soft company. The product's details are used to explain how theoretical concepts are translated into real marketable medical instruments. The implementation describes the '**know-how**' of design and the '**know-why**' of design decisions and rationales. The know-why has a general validity beyond ECG, since the rationale is common to medical devices generally. This design-oriented approach is useful in understanding the main concepts related to physiology, the principles underlying medical instruments and the applications of these instruments.

The **book organization** follows the design process as outlined in Figure 1.2. Chapter 1 addresses the tools required to design electronic products applying the concepts to medical instrumentation. The development process starts with product concepts and requirements (Chapter 2) that feed the design stage (Chapter 3). Signal processing and estimation is pervasive in all the conceptual blocks of a medical instrument. Chapter 4 introduces the

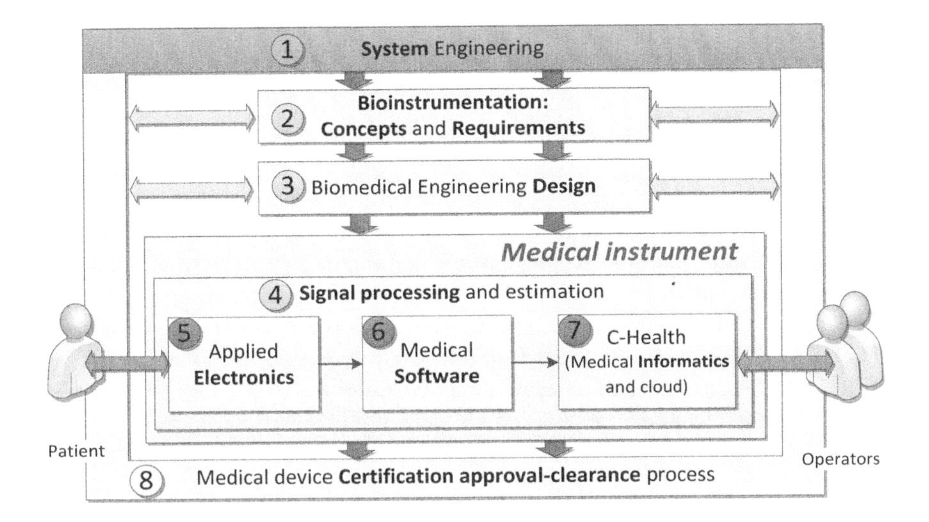

Figure 1.2 Book structure.

mathematical framework required to handle signals and to assess performance through the whole medical system including electronics (Chapter 5), software (Chapter 6) and telecommunications/ICT components such as the e-Health products. Advances in medical device technology are increasing the need for alternative, simpler solutions for accessing and sharing medical data and services. Cloud computing is gaining prominence as a technological solution for distributing medical solutions. This issue is addressed in Chapter 7, where we introduce the new term '**C-Health**' referring to 'the set of healthcare services supported by cloud computing'. Chapter 8 covers medical instrumentation approval by regulatory bodies referred to as certification, clearance or other terms, according to specific authority.

Part I: Theory

1.1 Introduction

This chapter outlines the engineering tools and methodologies that are introduced in a logical order. Referring to Figure 1.1, design is first a problem that has to be formulated (Section 1.2). A systematic approach to **describing a problem** helps greatly in solving the problem itself. The **business context** (Section 1.3) has to be assessed at all the design stages to avoid technology-driven market failures. Design has to be performed in a **systematic process**. In Section 1.4, a simplified model of product design is introduced. This model is similar to the problem solving processes used in many other applications.

Any design of non-trivial devices has a high level of complexity given by the wide set of design details that have to be addressed and solved coherently. Unfortunately, the human mind is only able to cope with a very limited problem size at one time, with limited details. Techniques for reducing and **decomposing complexity** are therefore compulsory for non-trivial product design. So, Section 1.5 introduces the system-subsystem decomposition techniques used in software and in general at system level.

Product development implies a **life cycle** process that encompasses the various steps required for creating new products (plan, analysis, design etc.). These steps can be structured using various approaches to suit the specific constraints, the features of the product and the context, as discussed in Section 1.6. The design of a medical product is a typical project activity and therefore it is worth analyzing major aspects and risks in managing projects (Section 1.7). Finally, during and after design, intellectual property rights (IPR) have to be faced with care to avoid infringing existing patents and to eventually protect the developers' own product innovation (1.8).

1.2 Problem Formulation in Product Design

Product design is first a problem to be defined and solved. It is therefore worth focusing on the main aspects of **problem solving** and in particular on problem classification based on its formulation and solutions (see Figure 1.3). Problem formulation is the set of inputs and constraints available for problem solution. Effective formulation is critical to the outcome, especially considering that problems are usually ill-defined, multi-dimensional and with a wide constraining context. Effective formulation often requires an iterative approach: '*First find out what the question is – then find out what the real question is.*' (Vince Roske)

Problem classification based on the characteristics of the input formulation and on the solution helps in analyzing the problem characteristics and results in more effective problem-solving strategies. Problems may be divided into puzzles, dilemmas and messes (Pidd, 2002). Problems given to students are usually just 'puzzles' because both formulations and solutions are well defined. At the other extreme, product design is usually a 'messy' or 'wicked' problem: there are many different ambiguous solutions based on tradeoffs, and formulation is usually ill-defined. Often, requirements become clear only when the problem (e.g., the design project) is close to being finished.

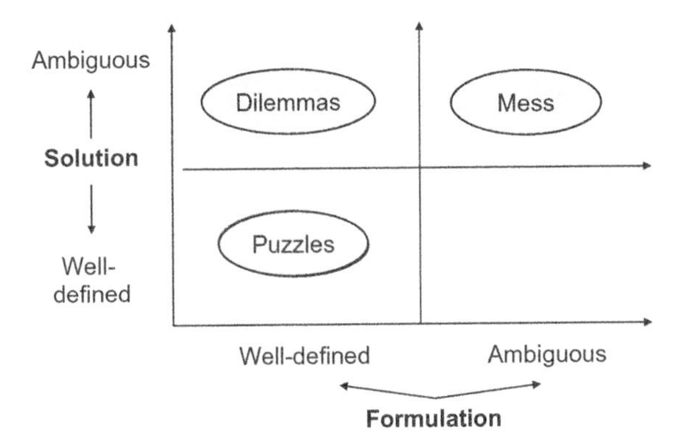

Figure 1.3 Problem classification.

In these **messy problems**, the traditional step-by-step logic useful for solving puzzles may not be sufficient, but here other techniques such as 'abstraction' and 'divide and conquer' may help.

In product design, a simplified model of the product is derived through abstraction. From this first model, other more detailed models are derived, each getting closed to the real system to be implemented. This technique is exploited also for system-subsystem decomposition (Section 1.5) using the 'divide and conquer' technique. In this paradigm, a complex problem can be faced when it is broken down into smaller solvable components.

A systematic approach to **defining a problem** helps greatly in solving the problem itself: *'If I had only one hour to save the world, I would spend fifty-five minutes defining the problem, and only five minutes finding the solution.'* (Albert Einstein).

In a simplified approach, problem management may be schematized as depicted in Figure 1.4. A customer assigns to the executors a problem to be solved with a clear **goal**. The customer then identifies the **output** to be delivered when problem is solved.

The customer itself usually defines the **input** and **constraints** to be considered. Within constraints, the customer may include **time-scheduling**, **milestones**, **budget**, **resources**, **expected quality**, **expected organizational interfaces** and **testing procedures** and **minimum results**. The input may include at first level: customer needs, technological environment and preferred standards. The executor has to provide the output, given the input and constraints. During the execution of the task, the executor generates the assumptions required to solve the problem and the possible risks. The problem-solving activity in more complex cases will be performed through a project using techniques encompassed in the project management area (Section 1.7).

All the previous elements should be agreed between the customers and the executors before the project initialization phase so that all the stakeholders are committed to solving the problem with the same elements. The executor has to define the problem formulation in as much detail as possible before starting, to avoid misunderstandings during the task execution.

Complex problems usually involve social and psychological issues. The definition may not be simple and **negotiation** on the various constraints probably has to take place with all the stakeholders, who are conditioned by their own background. If problem definition does not take into account all the stakeholders' perspectives, people involved in the task may

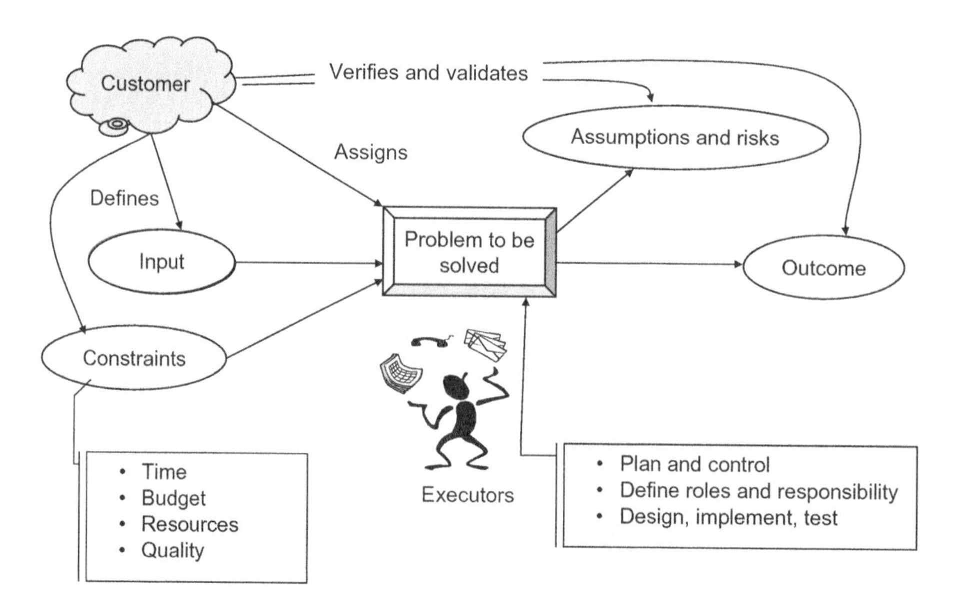

Figure 1.4 The problem framework.

Table 1.2 Problem formulation checklist ☑

1.	Real problems, such as product design, are messy, so adopt proper strategies.
2.	Take proper time to define the problem, specifying the goal, outcome, input, assumptions, risks and constraints (scheduling, budget, resources, quality).
3.	Negotiate the problem definition with the customer to obtain a contract.
4.	Promises have to be kept.

possibly not cooperate (Pidd, 2002). In product design, a company may outsource the development of a new device. The main negotiation may start with budget, time and performance. The customer will tend to reduce budget and time and try to obtain the maximum quality for the outcome. Executors have the opposite goal. The customer will also impose other constraints for problem-solving. Negotiation will end in some sort of compromise, frozen in a formal agreement. Errors in this step will result in project failure, with extra cost, delays or lack of quality.

A mistake in problem formulation will probably end in project failure. In Table 1.2, a simple **checklist** outlines the main steps of problem formulation.

1.3 The Business Context for Design

Scientific literature has systematically investigated the reasons for failure and the best practices to facilitate the success of new product development. There is quite good agreement in identifying two main reasons for failure (De Brentani, 1989; Timmons, 2004):

1. product development outside market needs
2. product development too late to exploit windows of opportunity.

Regarding the first error, there are two points of view. On one hand, it is recognized that products should be market-driven. On the other hand, literature recognizes from the empirical evidence that technology-driven products prevail (Brown, 1991). This probably suggests that technology-driven approaches should be the impulse for a new product development but it must then be market-validated and market-designed (Johnson, 1997). Business literature generally agrees that some sort of primary data are compulsory for designing and validating products (CB, 2004). The design of the marketing is also another issue that is quite commonly thought to be a critical success factor in literature (Bailetti and Litva, 1995).

This section aims to show how business and marketing considerations usually included in the product business plans impact on product design. A deeper coverage of **business plans** may be found in (Bygrave, 2004). For our scope, a business plan is mainly a document explaining:

1. the business goals for the specific products (sales, revenues, profits etc.)
2. the way to achieve those goals
3. the arguments that explain the feasibility of goal achievement through a specific set of actions.

A first question that the business plan must respond to is

Why should customers buy our product and not those of our competitors?

In marketing terms, this is referred to as '**differential competitive advantage**' (Kotler, 2002); that is, the customers' perceived advantage of our product over our competitors' products.

Assuming a marketplace where only product features determine buyers' behavior, the two main product parameters become price and 'performance' where the latter encompasses many factors (implemented functions, quality, robustness, aesthetic design and service).

In this context, the business plan must define the required set of product performance and features to achieve the proper sales target.

More specifically, the business plan must include market research of the target customers that identifies the **perceived performances** and **features** that customers are willing to pay for and that they prefer when choosing one specific product instead of another. This is translated into a set of mandatory and suggested features, which are included in the business plan.

It is market analysis that often identifies critical success factors (Johnson, 1997), and *this may be far from what is expected by engineers involved in product design*. The key element is the 'perceived performance'. Let us consider, for example, a doctor buying a medical instrument. This doctor does not usually have the proper know-how to assess technical performances although the performance is normally clearly evidenced in the technical specifications of the product. For example, a cardiologist may have no knowledge of sample rate, resolution or 'common mode rejection rate', which are main factors for assessing real ECG performance. In this context, doctors will use proxies to assess product performance (Kotler, 2002): brand, colleagues' suggestions, expected service or even the esthetic design, on the instinctive assumption that 'the quality of the packaging relates to the quality of the product itself'.

For many products, it would be expected that price and technical performance should rise together in a competitive product. Unfortunately, for medical products this is not always true due to the *limited customer capability of technically evaluating products.*

The market product analysis will then contain a set of features that are perceived by potential customers to differentiate the product. These features are an input for the product design problem.

Performance, cost and price are strictly related. In a simplified framework, each performance has an associated '**product marginal cost**', that is, an increase of the production cost of the equipment due to the implementation of the specific feature. This is true especially for features implemented in hardware. The implemented features also have an associated '**feature investment**' that is the additional investment required to develop the feature.

Let us suppose that our product will implement M features; the **total cost** to produce n items of equipment has a $Cost_{Tot}(n)$ which can be expressed as:

$$Cost_{Tot}(n) = n \times \sum_{i=1}^{M} PMC_i + \sum_{i=1}^{M} PI_i \tag{1.1}$$

where PMC_i is the product marginal cost to produce the i-th feature (e.g., the cost of raw materials, labor and sales commissions) and PI_i is the product investment required to develop the i-th feature.

Revenues from selling n units are:

$$Revenue_{Tot}(n) = n \times Sales\ Price \tag{1.2}$$

Finally, the profit is given by the difference between revenues, production costs and investment:

$$Profit(n) = Revenue_{Tot}(n) - Cost_{Tot}(n) = n \times \left(Sales\ Price - \sum_{i=1}^{M} PMC_i \right) - \sum_{i=1}^{M} PI_i \tag{1.3}$$

The **profit** graph in Figure 1.5 is interesting, because it shows that when a certain quantity of product has been sold (i.e., the break-even point), the business becomes profitable. In this simplified framework, the fixed costs are neglected. Fixed costs are those costs that do not change with volume of production (facility, equipment, salaries) and are apportioned to the product. Those fixed costs may be simply added to the investment cost $\sum_{i=1}^{M} PMC_i$ in the previous formulas.

It is easily shown that the threshold quantity referred to as the **break-even point** is then equal to:

$$break\text{-}even\ point = \frac{\sum_{i=1}^{M} PI_i}{\left(Sales\ Price - \sum_{i=1}^{M} PMC_i - Fixed\ Cost \right)} \tag{1.4}$$

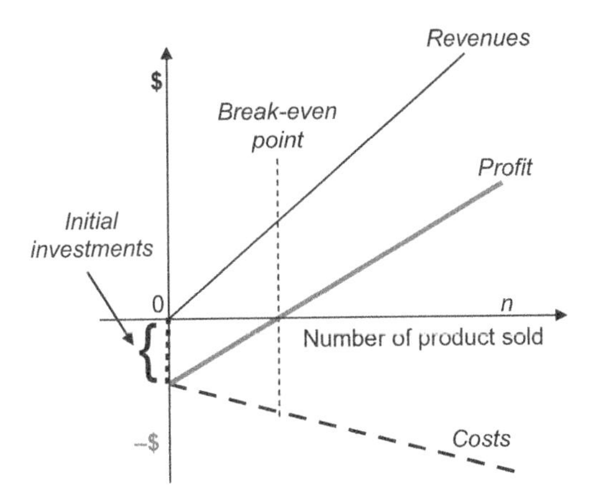

Figure 1.5 Gross profit versus number of products sold.

Business profitability is achieved when the product quantity sold is greater than the break-even value. When sales forecast cannot achieve break-even, the previous formula suggests approaches to ensure business sustainability (i.e. quantity sold > break-even value):

- Reduce the implemented features.
  This will reduce both the product costs $\sum_{i=1}^{M} PMC_i$ and the investment $\sum_{i=1}^{M} PI_i$, but may have a negative impact on sales.
- Increase the sales price, again with a negative impact on sales.
- Reduce fixed costs.

The first bullet is of particular interest for product design and development. Often, the majority of users use only a minority of the product's features. Therefore, the selection of the features to be implemented has to be particularly accurate to avoid including features that are expensive in implementation/production but of less interest to the target market.

In the previous formulas, the **markup** value can be derived as the difference between the selling price and the cost:

$$Markup = Sales\ Price - \sum_{i=1}^{M} PMC_i \tag{1.5}$$

Markup evaluation is critical because the higher the price, the faster the break-even. Unfortunately, markup generally has an inverse effect on sales because customers are generally less willing to buy more expensive products. This concept is explained in economics with the so-called '**price elasticity of demand**' (Kotler, 2002). For some goods, demand and price are related by an inverse proportionality (perfectly elastic curve). For medical instruments, this is less true due to the nature of the product. Physicians use medical instrumentation to decide on patient health, and therefore price considerations seem less relevant. Also, physicians only give considered attention to innovations after extensive

Table 1.3 Business context for design checklist ☑

1.	Define and select perceived features that guarantee product competitive advantage, using data either from the customer or from market research.
2.	For such distinctive features evaluate the marginal production cost and the investment required for development.
3.	With the customer, perform a benefit-analysis for the implementation of such features.
4.	Reduce features to be implemented basing on the previous item.

clinical trials that guarantee effectiveness. In addition, since price is a proxy for quality, too low a price may even reduce sales because product is then perceived to be of poor quality.

The previous discussion is somewhat simplified with respect to real market dynamics, where other factors have to be considered (Kotler, 2002). In any case, it allows for drawing some useful **considerations** on product design:

1. product developers may focus on technical excellence
2. *but* . . . product technical superiority and innovation do not yield a clear and automatic market advantage
3. *and* . . . feature implementations increase costs and development time
4. *therefore* . . . selection of features to be included in a product should be carried out after market research and after cost/technical considerations have been used so as to reduce costs/risks and increase the probability of market success

Considering also that '*what is not implemented will not cost anything and go wrong*', product designers should carefully reduce performance to a minimum set of 'must have' features to obtain a successful market product. Such performance should be implemented with investment cost that is not higher than that included in the business plan. In addition, during design, final expected production cost has to be monitored during each stage because exceeding production marginal cost or investment cost may make the business plan unprofitable. For this purpose, program management techniques should be carried out in order to avoid such risk. Table 1.3 shows some main points related to this paragraph.

1.4 The Engineering Product Design Process

Engineering design is the process that includes activities required to place products on the market. The scientific industrial literature offers many different models for product design since different models can be used according to the nature of the product and the context. In this section, a simplified model will be introduced with the aim of highlighting the rationale for development process activities. In this way, readers may find it easier to adapt models to their specific development.

The general process of product design is similar to procedures for systematic problem-solving used in many other fields of application. This may be explained since any solution obtained through a systematic rational process requires answers to the general questions indicated in the first column of Table 1.4.

Table 1.4 General activities in product development

General questions for product development	Corresponding activity name in the proposed model
1. Why the product?	Mission-concept
2. What is the plan?	Project planning
3. What product is needed according to the customer/ beneficiary?	Requirement analysis
4. How will the product implement these requirements?	Design specification
5. How should the product be implemented and integrated?	Implementation and system integration
6. How can it be tested according to the criteria:	Test
a. Has the right thing been done?	Validation
b. Have the things been done right? (i.e. Does it work?)	Verification

In the second column of Table 1.4, there is a possible definition adapted from the **MIL-STD 498 standard** (MIL, 1994). This standard although substituted by newer standards is still often preferred due to its advantages: free availability of templates and guides, flexibility in terms of development strategy, methods and tools and system-orientation that encompasses hardware–software integration.

A more frequently quoted source of errors in project management is that of embracing a **process standard** as is, without adequate analysis of the real needs of the product development. Every item of the standard (e.g., documentation) requires extra time, and some may be an unnecessary waste of time. The MIL-STD 498 standard is helpful here since it contains most of the 'components' required for usual projects and it is conceived so as to be easily tailored for specific goals.

The process model in Table 1.4 is still quite simple for handling real products and therefore it should be completed, introducing stages that address the most recurring problems of development. Product developments are characterized by many errors in every step of Table 1.4, but errors at different stages have different costs for fixing. Figure 1.6 summarizes some studies on errors in software projects (adapted from Davis, 1993).

The cost of fixing errors grows exponentially with the time interval from the start of the project and the time when the error is fixed, as shown in Figure 1.6. Boehm (1981) identifies a cost escalation in the range from 40 to 1000 times for fixing an error when the software is delivered, compared to fixing it at requirement stage. Errors at requirement or design specification stages are only 'paper errors', in that they just need time to analyse and correct the changes to the documentation. At the implementation stage, software errors have to be solved through debugging and fixing that requires much more time than at the previous stage. After formal tests, things get even harder because software has to be generally retested from scratch and a formal procedure has to be performed (e.g., problem report, change request). After certification, apart from administrative procedures, the main risk is of product withdrawal and damage compensation.

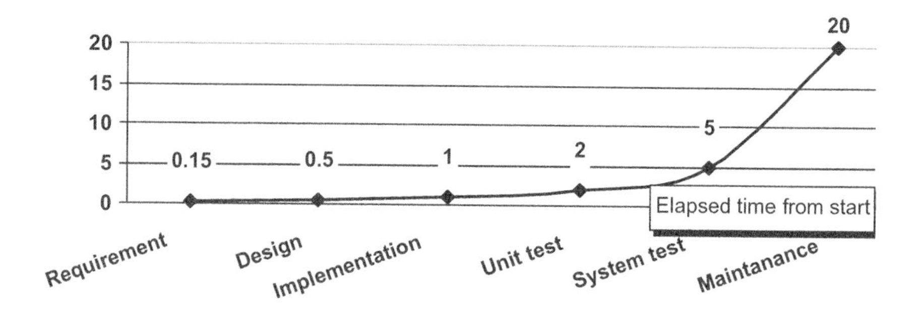

Figure 1.6 Cost of repairing errors at different design stages.

Even hardware has similar cost escalations. According to a NASA study (NASA, 2010), hardware fixing has on the whole an exponential behavior similar to that of software, with specific differences at specific stages. In medical devices, as well as in all safety-critical fields where rigorous product certification holds, the cost may be even higher considering also the consequences of the errors.

The obvious conclusion is that *'the project design process should include activities to detect errors as early as possible'*.

A frequent strategy consists of **estimating performance** of the product at the early stages. This can be done through three approaches:

1. analytical models
2. low fidelity simulations (e.g., not using the final software)
3. high fidelity simulations (e.g., using the final software in a simulated environment).

Analytical models are powerful and rapid for estimating performance over many parameters. A product model may be derived using formulas in the context of signal-information theory, signal processing, digital signal processing, circuit theory, electronics and software engineering. The product performance may be also simulated through proper tools that handle hardware, signals and software behavior at various levels of complexity. This topic is now extremely relevant for product development; performance estimation will be outlined in Chapter 4.

It is therefore suggested to add a stage of performance evaluation to the activities included in product development. Figure 1.7 depicts a possible **product development life cycle**.

The first step is the product **concept** that generally is included in the **plan**. The concept should outline briefly the scope and the content of the project. Then a plan is recommended

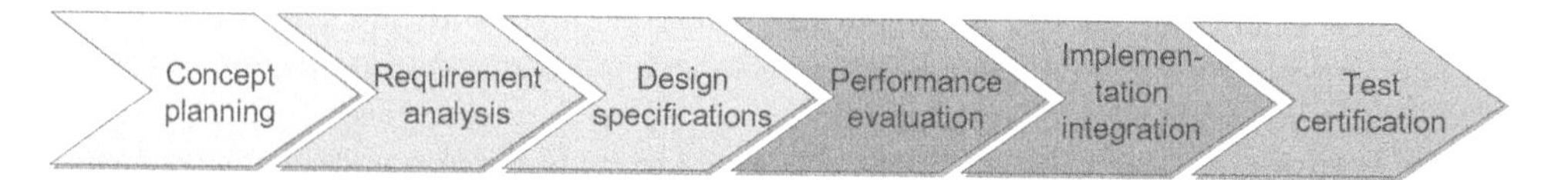

Figure 1.7 Product development life cycle.

as discussed in Section 1.7. **Requirement analysis** defines what the product must do according to customer/buyer needs.

Requirements contents are usually generated by practitioners in the field. Analysts and marketing experts usually generate these requirements from physicians, nurses and other stakeholders working where the product will be used. The output of this process is a set of requirements: characteristics that the product must possess in order to be acceptable to the customer.

A requirement should have these attributes:

1. It is uniquely identified by an identity code.
2. It is feasible and it can be verified through inspection, demonstration, test or analysis. (i.e., a test can be written to ensure that the requirement is met).
3. It must not conflict with other requirements.
4. It can be translated into a technically and economically feasible design specification.
5. It is mandatory and it identifies a real business need/opportunity.
6. It is unambiguous and complete and it addresses only one specific feature.

An example of requirements on an ECG may be (adapted from AAMI, 1991):

Req 1.1 *The ECG shall be able to display differential voltages of* $\pm 5\,mV$.
Req 1.2 *A* $20\,\mu V$ *peak-to-peak sinusoidal signal shall yield a visible ECG trace.*

All the previous conditions are satisfied by these two requirements: the business need is related to the fact that cardiologists must have such performance to detect specific pathologies.

In a more focused and efficient approach, requirement analysis should limit strictly to what the customer considers as a condition for product acceptance (MIL 498 Guide, 1994). To this end, company internal marketing should help product development. The analysts should bear clearly in mind which are the technology and cost constraints, because, at this stage, it is common to introduce requirements that are expensive or even unfeasible. Every requirement normally determines an increase in both product marginal cost and product development cost that cannot be estimated clearly at this early stage. Therefore, in the requirement review, we *must take into consideration if each requirement is really necessary*. Additional requirements will increase the cost, the development time and the risks considering the trivial rule that: *'What does not exist cannot fail'*.

To simplify the requirement analysis, **use cases** are usually employed. These are a descriptive method of defining the behavior of a system when achieving the business goals required by customers. A use case describes informally how users will practically handle the system to achieve their intended goals in different operative scenarios. The use case will contain the set of actions that the product will perform in response to users' input. The use case will then include the corresponding expected output of the product *ignoring the real product implementation*. From a set of use cases, a coherent set of requirements may be defined.

Every requirement must be translated into at least one design specification. So a **traceability matrix** is used where the first column contains all the requirements, one per row, and the second column includes a reference to the associated specifications introduced to implement the specific requirement.

The **design step** is the process of developing a set of specifications that explains how the product is implemented in response to requirements. The design specifications are performed by developers, who are expert in implementation also referred to as the 'solutions domain'. The analyst, in turn, should be expert in the domain of the problems, although competences are required in implementation to avoid unfeasible requirements.

A typical specification in response to the previous requirements is

Req 1.1 *The ECG shall be able to display differential voltages of* $\pm 5\,mV$.
Req 1.2 *a* $20\,\mu V$ *peak-to-peak sinusoidal signal shall yield a visible ECG trace.*

> Spec. 1 the ECG A/D converter shall have at least 10-bit real resolution, assuming that 0.01 mV is the minimum vertical excursion that produces a vertical modification on the display.

This is because the minimum resolution should allow us to detect less than $0.01\,mV$ *to display a* $0.02\,mV = 20\,\mu V$ *peak-to-peak signal and therefore we need at minimum* $(5 + 5\,mV)/0.01 =$ *sample intervals* $= 1000 < 1024 = 2^{10}$ *corresponding to a 10-bit resolution.*

Spec 1 obviously does not cover completely the two corresponding requirements since other specifications are required on all components that handle the signal (i.e., hardware, firmware and software). The complete design will also include what is required to obtain the product, such as software diagrams, hardware circuits diagrams, and mechanical components drawings.

After this stage, performance evaluation may test the requirements and specifications. Analytical models may validate some target performance such as: sensitivity, frequency response, resolution, power requirement, cost and so on.

Hardware design is performed using proper tools (e.g., CAD – computer aided design) that allow the evaluation of electronic, mechanical, thermal and radiation behavior through high quality simulations. This evaluation can be assessed at the **performance evaluation** stage. These evaluations may be helpful also for the risk analysis and certification stage. Other tools may validate signal processing or the software algorithms.

After performance evaluation, the **implementation** can take place at various levels of complexity using demonstrators, prototypes and the final sample products. All the components (software, hardware, mechanical objects, sensors) are then integrated within the system. Manufacturing issues have to be considered at earlier stages to avoid expensive reworking.

The final stage of this simplified life cycle is the **test** that will include validation of all requirements, and verifications of the functions of the product with clinical trials usually being required by **certification** (EEC, 2007). This testing stage is critical because, after certification, error fixing becomes much more expensive. Errors are managed through proper processes as explained in Chapter 6 and, in safety-critical cases, they require proper notifications to the regulatory bodies.

The proposed cascade model of Figure 1.7, usually referred to as 'waterfall model', is still too simple for a real environment. There are two further topics to be addressed:

1. the lack of system decomposition
2. the limits of the 'waterfall' approach life cycle.

Table 1.5 Engineering design process checklist ☑

1.	Apply an engineering product development process (basic steps: planning, requirement analysis, design specification, implementation, testing).
2.	Since the cost of fixing errors increases exponentially toward the end of the project, define a strategy to discover errors as early as possible.
3.	To do so, consider performance evaluation that allows assessing product performance at earlier stage.
4.	Also consider system design decomposition (Section 1.5), and product life cycle consideration (Section 1.6).

System decomposition is required because the system itself is generally too complex to be handled as a single unit. The *divide et impera* (divide and conquer) approach is a major strategy for handling the product complexity by decomposition into smaller units as explained in Section 1.5.

The **life cycle waterfall** model considers a sequential development strategy where each stage starts at the end of the previous: first planning, then requirement analysis, design and so on. Unfortunately, this model is critical because in innovative projects like new product design, some details only become clear at the later stages of implementation. This may invalidate the previous development stages that then have to be reworked (Parnas *et al.*, 1986). This topic will be addressed in Section 1.6. Table 1.5 shows a simple checklist including the suggestions discussed in this section.

1.5 System-subsystem Decomposition

Divide and conquer

Any design of non-trivial products has a high level of complexity given by the set of the design details that have to be addressed and solved coherently. Unfortunately, the human mind is only able to deal with a very limited problem size at one time with restricted details. The problem refers to when Romans were facing enemies but they had insufficient forces. The Roman strategy may be summarized by the phrase 'divide and conquer', that is, to try to divide the enemy forces into small components until the component size could be faced alone by the Romans' forces, as shown in Figure 1.8. The components should be as 'independent' as possible, so that the Romans could fight against one component, hoping that the other components would not arrive and change the balance of power.

The strategy then consists in fighting one component at time (from one to six in Figure 1.8), hoping that the other components do not join in the battle.This strategy is effective for design problems as well. Project design forces have limited capacity to solve simultaneously all the design details (the enemy). The strategy then consists of breaking the system down into smaller subsystems until subsystem sizes can be faced by the project designers' capability. In this section, the term 'subsystem' will be used. When dealing with decomposition, other equivalent terms are also used according to the context: modules, components, packages, configuration items, classes, boards and so on.

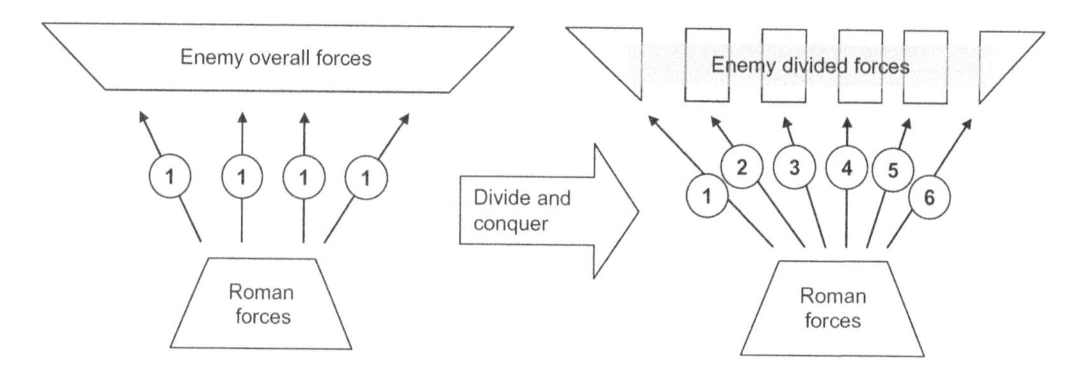

Figure 1.8 Divide and conquer strategy.

The subsystems should be as 'independent' as possible so that each subsystem can be designed without considering side effects on and from the other subsystems. In the end, the design of all subsystems must cover all of the system design, since missing design elements will be probably be a source of errors in the product implementation.

The strategy can be implemented as follows. The first step consists of defining level one (L1), that is, the system itself with its external interacting entities (usually termed 'actors') and the interfaces between the system and the actors (see Figure 1.9). An interface may be defined as the point of interaction between subsystems where the method of interaction is defined.

The system is then decomposed into subsystems at L2 level that inherits L1 interfaces and introduces other L2 interfaces (I2.1 and I2.2 in Figure 1.9). Each L2 subsystem in turn has its interfaces, and can be decomposed into subsystems that constitute the L3 level. Other levels can be defined until the appropriate level of detail is reached. However, all the L2 subsystems must partition L1 completely, that is, the set of subsystem interfaces at any level decomposes exactly each upper level.

The final level should contain the appropriate degree of complexity and design detail for the subsystems and interfaces. Figure 1.10 shows the first level diagram related to a biopotential-based medical instrument.

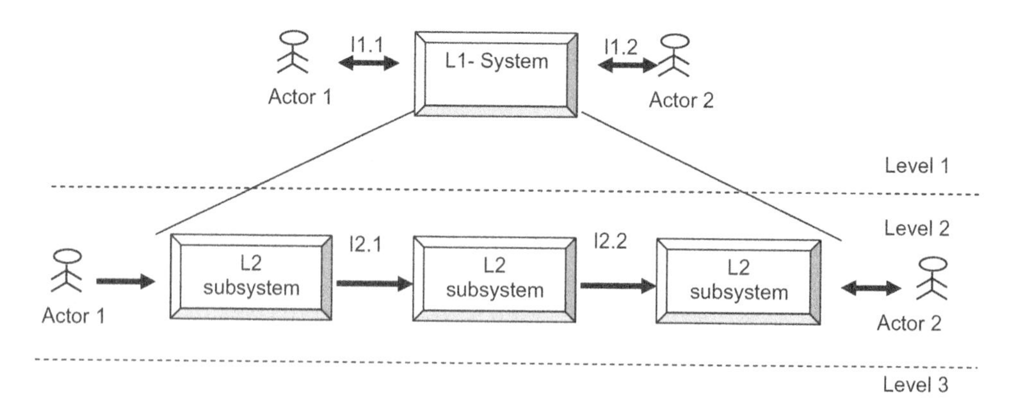

Figure 1.9 System-subsystem decomposition.

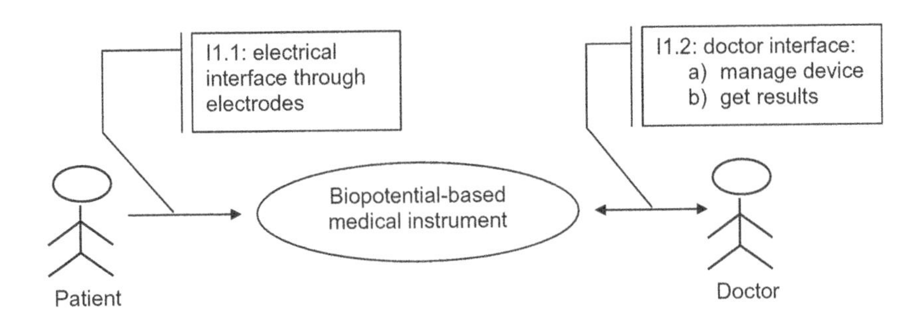

Figure 1.10 Medical instrument system diagram.

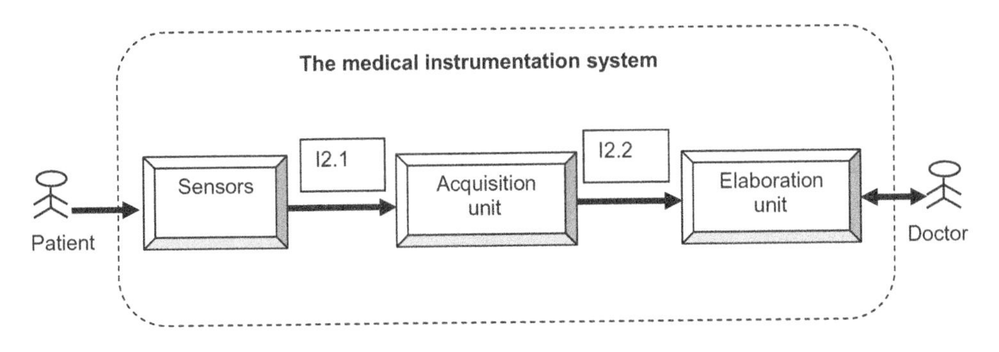

Figure 1.11 Medical instrument subsystem decomposition.

This level is complete in terms of interfaces (I1.1, I1.2) and actors, although it is rather generic. The subsequent levels will specify the design in more detail. The system of Figure 1.10 can then be decomposed into a second level that includes the decomposition of the elements of the first level as shown in Figure 1.11.

This hierarchical approach is also suited for structuring the overall design job so that different subsystems can be worked on in parallel. Figure 1.11 can be easily organized into a **work breakdown structure** (WBS) that is required to organize tasks for the working team in the project plan. The levels of the WBS (Figure 1.12) mainly coincide with system-subsystem decomposition.

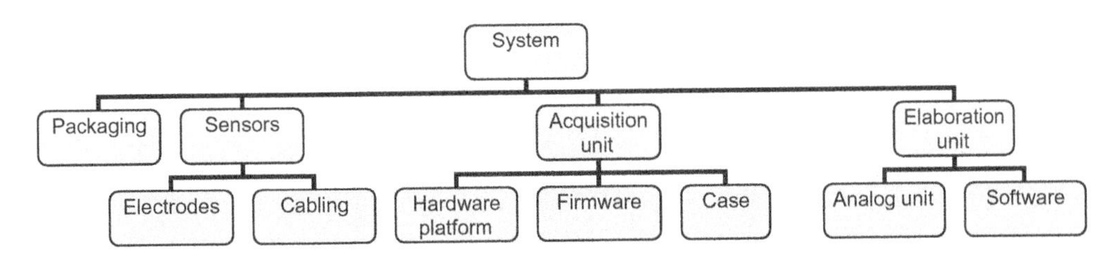

Figure 1.12 Product breakdown structure for a medical instrument.

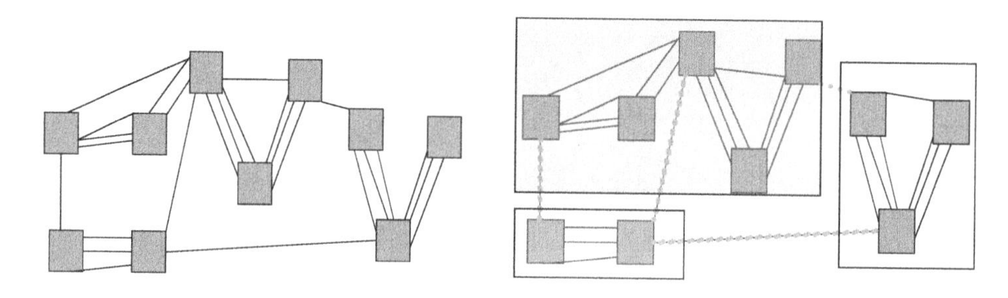

Figure 1.13 System example (left) and its subsystem decomposition (right).

The project team can then address in parallel the activities and, at the end of subsystem implementation, an integration/testing step will yield the overall product.

A product breakdown structure is also easily derived using Figure 1.12. This is helpful for product implementation and production.

Given that some decomposition is required, the question of how to **select a suitable subsystem decomposition** arises. Many different solutions are available for system-subsystem decomposition; that is, many system partitions are possible and some criteria are needed to select a good partition.

The left-hand side Figure 1.13 shows a possible system, where lines denote interactions between functions (the boxes). A subsystem decomposition consists of aggregating functions into different subsystems. An effective decomposition criterion may consist of minimizing the complexity of subsystems. Complexity mainly arises because a modification in one subsystem causes effects on other subsystems and this effect may even be cyclical. This means that a single subsystem cannot generally be analyzed without taking into account the effects on/of other subsystems.

In Figure 1.13 the decomposition is performed minimizing interactions among subsystems: this yields a second level with three subsystems and four interfaces evidenced with green dotted lines. The functional decomposition of Figure 1.13 is roughly obtained by minimizing the 'coupling' between subsystems and aggregating the functions that are strongly related in the same subsystem. This approach is referred to as 'cohesion'. The overall paradigm is usually referred as **maximal cohesion and minimal coupling** and although it is a rather old idea (Stevens, 1974), it still remains the major guideline for defining software and hardware architectures.

The **degree of cohesion** of a subsystem may be expressed in how much the functions are related within a specific subsystem. Cohesion may be qualitatively measured using concepts of the upper part of Figure 1.14 (adapted from Stevens, 1974).

The lowest degree of cohesion is achieved when functions are casually grouped within a subsystem. Figure 1.14 shows other intermediate steps toward high cohesion: functions may be grouped according to better criteria (logically similar, executed at a similar time etc.).

Figure 1.14 bottom shows also a metric for measuring coupling. **Coupling** may be defined as the level of interdependency between subsystems. Low coupling is a preferred approach since it reduces system complexity and therefore it improves subsystem design test reusability, as well as system integration and maintenance.

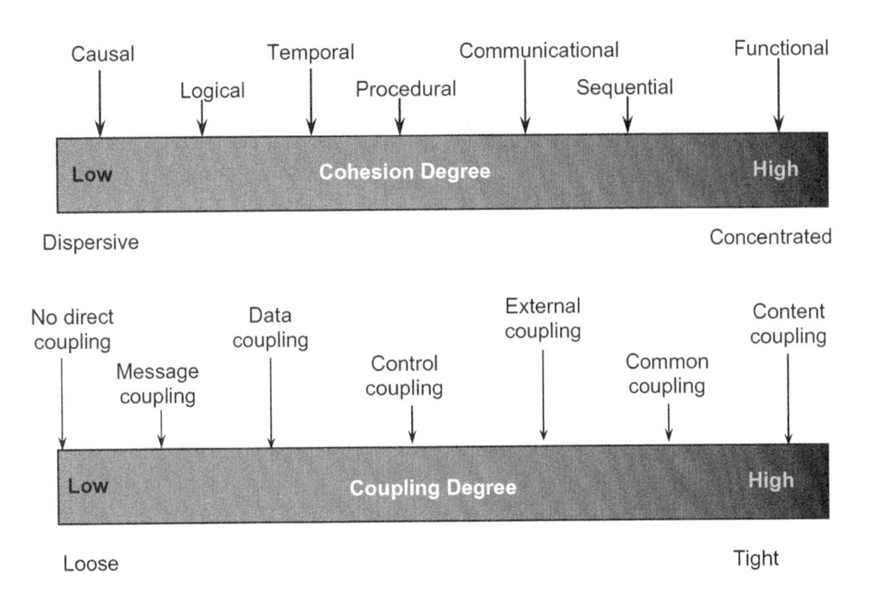

Figure 1.14 Metrics for coupling and cohesion.

The best approach is 'no coupling' at all. In this way, subsystems may be handled independently. A good approach is **message coupling** where a subsystem offers its services with message exchanges through a defined public interface. In this case, there is no direct dependency between subsystems. In **data coupling**, subsystems share some data, and data modifications in one subsystem may affect the others. In **control coupling**, the control flow of one subsystem is handled by another module. **External coupling** occurs when subsystems depend on external entities.

In common coupling, subsystems share global data. The worst coupling (content coupling) applies when one subsystem accesses and modifies internal elements (data, flow...) of other subsystems that heavily affect their behavior.

The approach for **maximal cohesion and minimal coupling** is a design guideline to be preferred. If this guideline is ignored, the following **disadvantages** are to be expected:

- Subsystems will be harder to design and test because of side effects in the subsystems.
- System integration will be more time-consuming since the behavior of the system is affected by the behavior of the subsystems and their more complex interactions.
- Subsystems will be less reusable since they are strongly connected to other subsystems.

It is worth noting that the criterion of cohesion and coupling also has some **disadvantages**:

- Extra effort is required to design and develop proper subsystems.
- Since proper architecture may not be clear at beginning, design reworking is to be expected.
- Efficiency may be reduced due to the introduction of extra interfaces.

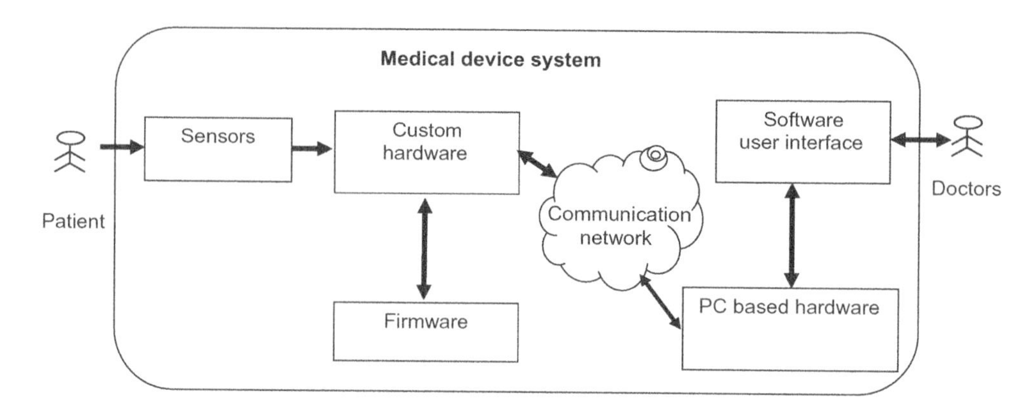

Figure 1.15 General product architecture.

A good architecture solution may imply the design of subsystems that offer well-defined services through simple defined standard interfaces. In hardware, plug-in modules using standard buses are examples. In software, similar concepts may be applicable using virtual software bus architectures or CORBA (common object request broker architecture). Even more, in software, many new development environments embed service-oriented plug-in components that can be easily combined to obtain the required application with minimal coding (Becchetti, 2001). The system architecture shown in Figure 1.15 is always more considered in electronic and medical product design because

1. PC based hardware is highly efficient and low cost due to wide distribution (economies of scale),
2. the software marginal cost is almost zero,
3. software designed for PC based platforms has a much lower development cost with respect to firmware owing to better tools and know-how and the wide availability of software components, and
4. by contrast, custom hardware is expensive to develop and maintain.

Costs, risks and development time may be reduced by reducing custom hardware to those functions that cannot be performed with PC based platforms for various reasons (insufficient computational power, stringent real-time requirements, certification requirements, hardware-specific functions). Functions associated to custom hardware could include signal acquisition, high data rate signal processing, electrical isolation and specific actuators. It should be noted that in medical devices, PC based platforms are often used but masked to users that do not feel to be in front of usual PCs. The wide use of PCs also has some disadvantages: operating systems and the hardware itself change more rapidly than custom hardware. This means that more upgrading has to be performed within the product life cycle. System subsystem decomposition is a powerful tool for any designer whatever the non-trivial problem. Table 1.6 below summarizes the main concepts.

Table 1.6 System-subsystem decomposition ☑

1. System design complexity and its associated costs, risks and development time may be reduced by applying system-subsystem decomposition process as follows:
   a. at the first level define actors, interfaces and system with limited detail level
   b. create a second level by decomposing the system into subsystems and interfaces
   c. create additional levels iteratively by decomposing subsystems from the previous level
   d. define other levels until the level of detail is satisfactory
   e. check coherence among levels, analyzing the corresponding interfaces and subsystems.
2. Define subsystems minimizing coupling between subsystems and maximizing cohesion within each subsystem.
3. Consider the generic product architecture (see Figure 1.15) as a possible system decomposition template to reduce cost, risks and development time.
4. At the end of the process, there should be a hierarchical structure of the subsystems and interfaces useful also for a work breakdown structure or a product breakdown structure.

1.6 The Product Development Life Cycle

The product development life cycle is a process that encompasses the various steps required for deriving a new product (plan, analysis, design etc.). These steps can be organized in various approaches to suit the specific constraints and the features of the specific product development.

In the **waterfall model**, shown in Figure 1.16, the various steps are performed sequentially and each step has to be concluded before going on with the following one. This model is suitable for some fields of application (e.g., the construction sector) where development processes are standardized, and requirements are not expected to change significantly throughout the project.

Unfortunately, in hardware–software product development, requirements may change considerably due to the use of unknown technology or simply because the product has never been implemented before. Therefore specifications and performance become clear only during the project life. In addition, the waterfall model suffers from one main risk. Results are available only at the very end of the project when tests are performed, as shown in Figure 1.17 (left); this increases the risk for time delays and cost escalation.

Other approaches are based on a different consideration: to reduce risks, the overall performance of the product may be achieved incrementally through successive steps of

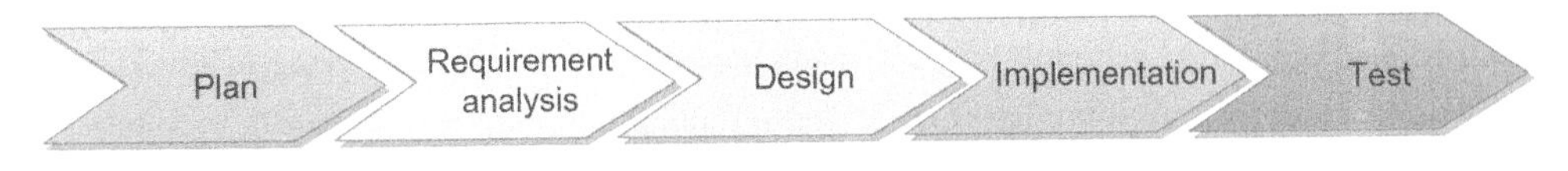

Figure 1.16 The waterfall design process.

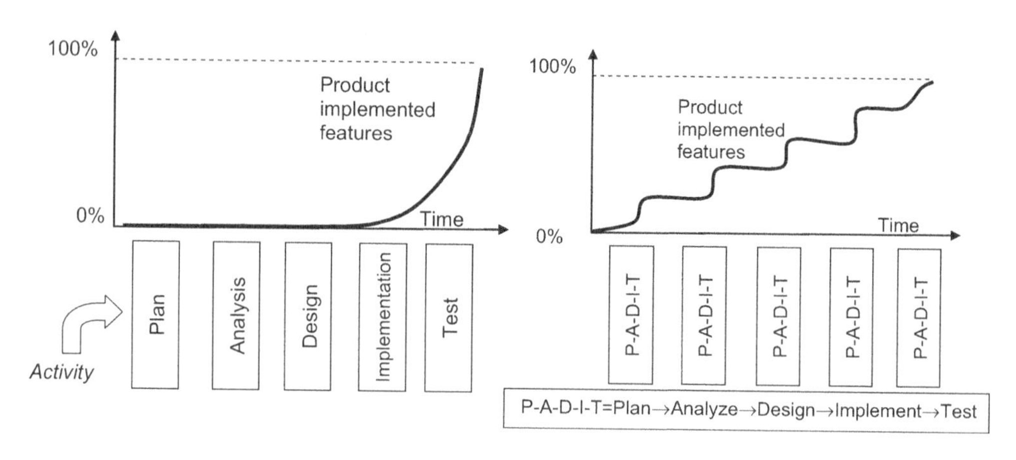

Figure 1.17 Comparison between waterfall and evolutionary life cycles.

smaller time length. Each step produces intermediate products, and includes all the basic development stages (plan, analysis, design, implementation, test) referred to as P-A-D-I-T in Figure 1.17. Furthermore, at the end of each step, the available result is increased in terms of implemented features and it is always closer to the final product.

In these '**evolutionary**' **models**, the intermediate results available during the project make allowance for tuning requirements, specifications and the previous intermediate results. Customers may also be included in the loop. They may validate/revise intermediate results that can even be used as a preliminary product. This is particularly valuable when development time is limited, and the customer may be temporarily satisfied by a product with limited functionality. This approach is also well suited to innovative products where requirements cannot be determined at startup. Through this approach, requirements may be easily changed during the project.

In hardware development, the process may include **intermediate results** such as: hardware or software simulators, hardware emulators, prototypes or first production samples. These intermediate results may implement an increasing number of functions so that different prototypes are released with a growing level of requirement coverage. In software, many models are based on obtaining the final product through intermediate results (iterative and incremental models, spiral models, extreme programming, Agile models, rational unified process – Larman and Basili, 2003). Models vary on the organization of the intermediate steps that are tailored to the specific project needs. For example, Agile Programming (Beck, 2001) and in particular Extreme Programming are well suited when customer requirements are expected to change significantly. In this case, there are frequent releases over very short time intervals that have to be validated by the customers.

In recent years, the availability of **pre-implemented components** in software, firmware and hardware is becoming more common. There are also many tools that allow code generation from draft design. In some cases, coarse implementation may be performed, graphically combining available functional blocks.

In addition, public domain open source libraries, or even commercial software components, may greatly reduce the development time, avoiding 'reinventing the wheel'. In hardware, there are many tools for simulation/emulation, and signal processing analysis may also be performed by powerful tools. Microprocessors and associated firmware may be emulated, and prototyping boards may be a convenient solution for rapid prototyping.

In (Wiki, 2011) there is an exhaustive list of **tools that allow for fast generation of software** in various fields. These tools perform fast development of prototypes, and therefore they are a valuable support for evolutionary life cycle models. The corresponding life cycle model is referred to as **rapid application development** (RAD, Wiki, 2012) and it is focused on rapid prototyping.

The development is obtained iteratively; at the analysis stage, processes and data model are detailed; design and implementation are performed mostly on the tool itself. Testing is reduced since most of the code is already implemented within the tool.

These tools may foster software implementation and testing because of the software already implemented, but it should be highlighted that the overall development performance cannot be improved by orders of magnitudes (Brooks, 1987). Some other warnings should be considered: due to the relative ease of prototype development, the designer may neglect the basic design steps such as software architecture design. This could yield a set of prototypes that can hardly become a real software product.

In general, the **strategy to reuse components** in product development may reduce the product development life cycle. Some care should be taken with licensing; the selected components might not be permitted to be employed in commercial products, as discussed in Section 1.7. As Software Of Unknown Pedigree (SOUP), they may require special care to be used in medical devices as discussed in Chapter 6. In addition, as suggested by Figure 1.18, when a component has to be substantially modified in order to be suitable for the required product, the cost of development may be much higher (Figure 1.18 right) than developing the component 'from scratch' (Figure 1.18 left).

The product life cycle has to be selected with care since an unsuitable model may have a considerable negative impact. Table 1.7 outlines some basic steps on this issue.

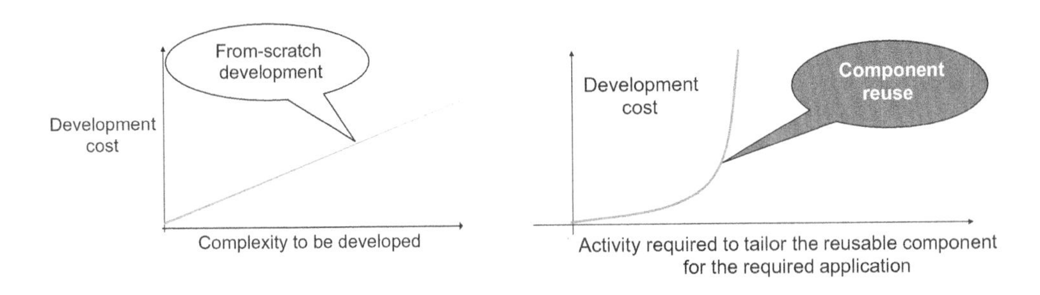

Figure 1.18 Reuse versus from-scratch development.

Table 1.7 Product lifecycle checklist ☑

1.	Assess organization, risks, customer expectations and requirement stability.
2.	Select a proper development life cycle that suits your project characteristics.
3.	Reduce risks by anticipating project results especially through performance estimation, simulations and prototypes.
4.	Involve potential customers in the development loop to validate intermediate results.
5.	Pursue a component reuse strategy in software, firmware and hardware, taking into account licensing, safety risks and cost of adaptation.

1.7 Project Management in Product Design

The design of a medical product is a typical project activity and therefore it is worth analyzing the major aspects and risks in managing projects.

In general, 'a project is a temporary endeavor undertaken to create a unique product or service' (PMI, 1996). The temporary and unique nature of the project has a deep impact on the probability of success of the project itself. 'Anything that can go wrong, will [probably] go wrong' (Murphy's law), especially when there is a lack of experience or with new product design. Referring to IT, according to KPMG's Canada Survey (1997) over 61% of the projects analyzed were considered by respondents to be a failure. More than three-quarters were out of schedule by 30% or more, and more than half required significant extra budget.

A discipline (project management) has been developed, in order to define rules, processes and best practices required to increase the probability of success.

The success of a project has to be assessed over three axes: time, cost and performance. According to IT managers (Bull Survey, 1998), **IT projects fail** because of

- delay (75%)
- excess costs (55%)
- non-achievement of planned performance with proper level of quality (37%).

In product design, delay affects time to market. For some products, arriving too late translates into a market failure because competitors' products have already reached most of target customers.

Extra costs may negatively affect the business plan. Business becomes unprofitable over a certain level of cost increase as shown in Section 1.3. In product design, the lack of planned performance or poor quality results in reduced sales and therefore again in possible business failure.

We now consider the main project management issues that affect project design. Figure 1.19 shows a typical **project management cycle**.

At project start, a first **development plan** is conceived. The plan should include at least 'who does what, when, how and why'. More formally, the plan should set goals (the why) of the project and define tasks (the what) required to achieve the goals. Tasks are generally defined in a hierarchical form under a work breakdown structure (WBS).

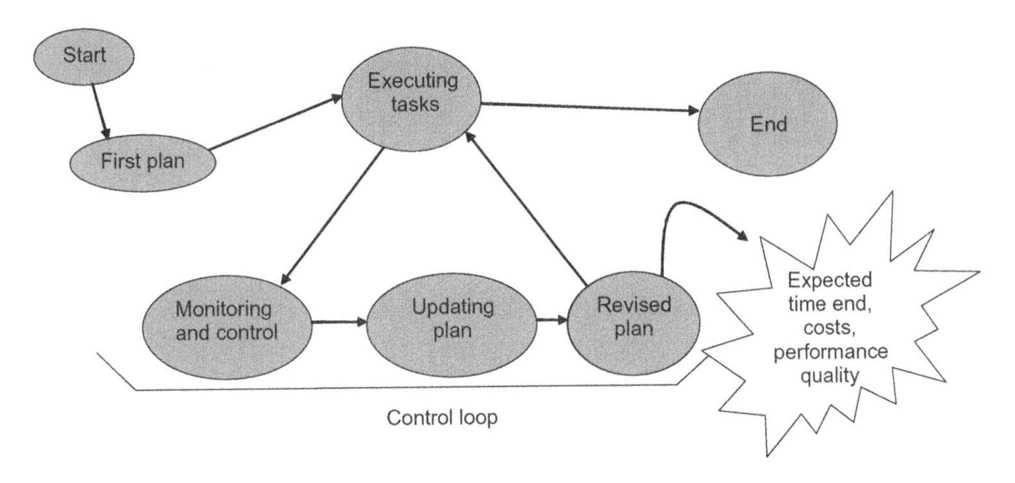

Figure 1.19 Project management life cycle.

Tasks are organized through time scheduling (usually a Gantt chart – Meredith, 2009), showing the temporal sequence of the tasks (when). The project deliveries and the associated milestones also have to be specified. Finally, the plan should include the procedures (how) and the human resources (who) involved in the project.

There are valuable template plans for specific fields of application. For example, in software projects, the military standard MIL-STD 498 (MIL, 1994) contains free templates to develop systems with lower risk. These templates can be used as a checklist to ensure that all applicable tasks are addressed. The software development plan template of the MIL standard, although complex at first glance, is worth investigation, since it is conceived to address effectively the topics that may jeopardize projects if neglected.

Risk management is a critical aspect that is sometimes underestimated in projects. In medical design, risk management is compulsory to ensure patients' and users' safety. Unexpected risk occurrence is a major reason for project failure. Since product development projects aim for innovative outcomes, project risks may occur more frequently. Detailed coverage of risk management is addressed in the international standard ISO/DIS 31000 (2009) and, for medical devices, in ISO/EN14971, discussed in Chapter 8. We focus here on project risks and, for our purpose, at minimum it is important that

1. risks are identified and monitored throughout the project
2. the probability (P_i) of each risks is assessed
3. the economic cost impact C_i and technical consequence of the risk is evaluated
4. an estimated extra cost C_{Tot} due to the risks is assessed. for example as:

$$C_{Tot} = \sum_i P_i \times C_i \tag{1.6}$$

5. actions are taken to reduce most impacting risks
6. an extra budget for risk management (e.g., Extra Budget > C_{Tot}) must be set aside.

Table 1.8 The risk register

id	Description	Probability (%)	Cost Impact	Estimated cost (Cost × Probabililty)
1	Hardware component (instrumentation amplifier) in phase out	20%	100k €	20 k€

id	Description	Mitigation	Contingency Action	Action By	Contingency	Action When
1	Hardware component (instrumentation amplifier) in phase out	Check for replacement from supplier catalogue	Change PCB layout and certification documents. Recertification tests may be required	HW project manager	30 k€	Before final design delivered

In more complex projects, **contingency** actions are planned to address risk occurrences. Positive risks also have to be taken into account, that is, risks that opportunities happen instead of threats. In non-trivial projects, a **risk register** should be updated regularly. A typical risk register is shown in Table 1.8.

Each row relates to a specific risk. The risk reported in Table 1.8 indicates that some hardware components may be in 'phase-out' (no longer produced) before finishing the design. The probability of such risk and the cost in euros are estimated in the additional columns. The risk mitigation consists of assessing regularly the phase-out plan of the supplier and evaluating component substitutes. If the risk occurs, the contingency action has to be implemented. So it is decided that extra budget (contingency) should be set aside to manage the risk occurrence. During the projects, decisions are also taken based on assumptions, that is propositions that are taken for granted or accepted as true. Some **assumptions** may turn out to be wrong during the project and this may require additional cost to overcome the associated wrong decisions. Assumptions are managed similarly to risks. In general, assumptions are recorded in a separate register during a project and they are managed periodically. An assumption register is similar to a risk register since the effect of a wrong assumption is similar to a risk occurrence.

A **monitoring and control** plan is a critical activity that, when neglected, increases the probability of project failure. It is important to introduce quantitative indicators to assess project progress (MIL 498 Guide 1994). Performance indicators, when periodically updated, may suggest confident estimates of project end and budget to finish the project.

Typical performance indicators are some 'activity performed' values. In software/hardware design projects, the activity indicator may be related to the number of successful tests over the total number of tests performed. When all the requirements are validated through

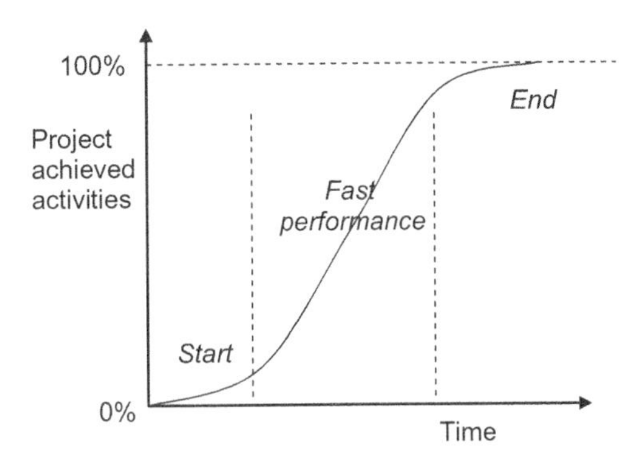

Figure 1.20 Performance behavior vs. time.

tests, the correct implementation of the design is verified by additional tests and the final qualification test is passed, the project is then considered complete.

The curve of achieved performance vs. time or sustained cost vs. time has generally the typical S-shaped behavior depicted in Figure 1.20 (Meredith, 2009). At the beginning of the project, the results cannot be achieved quickly because some time is required for project team to be effective – typically, the team needs time to form itself. At the beginning, due to human factors, some 'storming' time is required in order to arrive at an equilibrium where processes, roles and responsibilities are accepted in all the team components. Such a stage, also called 'norming', is a prerequisite to start to be 'performing'. A comprehensive overview of group dynamics may be found in (Handy, 1998).

Additional time is required by the team to achieve the background know-how. This time lag is generally inversely proportional to the experience of the task. The so-called '**learning curve**' shown Figure 1.21 is usually depicted as taking into account this effect. The effort required to become expert in a specific type of project/technology is inversely proportional to the experience already gained in similar projects/technologies.

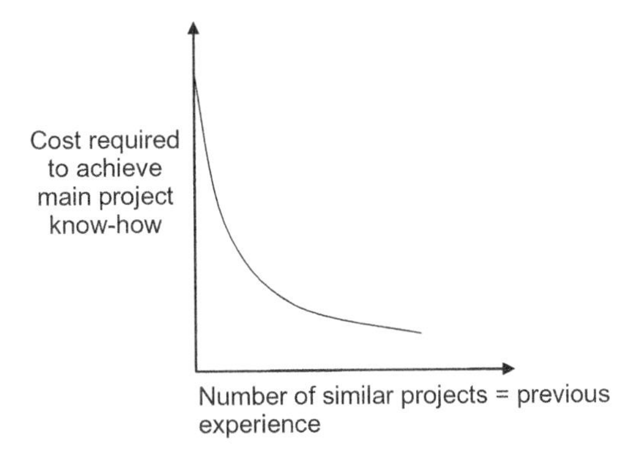

Figure 1.21 Learning curve.

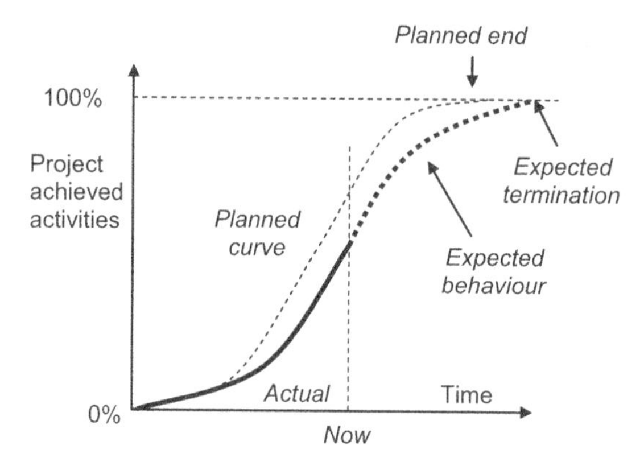

Figure 1.22 Estimation of project termination.

After a first stage, the project usually exhibits a 'fast performance', where the number of achieved results per period rises quickly as shown in the last part of Figure 1.20. The final period of a project is generally smoother when performance is S-shaped. The termination can be roughly estimated by comparing planned and actual performance over time. As shown in Figure 1.22, the plots of the planned and expected behavior may suggest project termination. At the beginning of product design project, it is critical that an adequate design effort is planned for requirement analysis and design specification. In software design projects, the lack of requirement analysis and proper design often produces the shape depicted in Figure 1.23. The project is planned to produce results quickly. At the beginning, some results seem to be achieved but, due the lack of careful design, such results are unsatisfactory, and reworking is often required. Poorly designed and implemented software may in turn generate large numbers of bugs, which, in the worst case, may require a complete rewrite of the software.

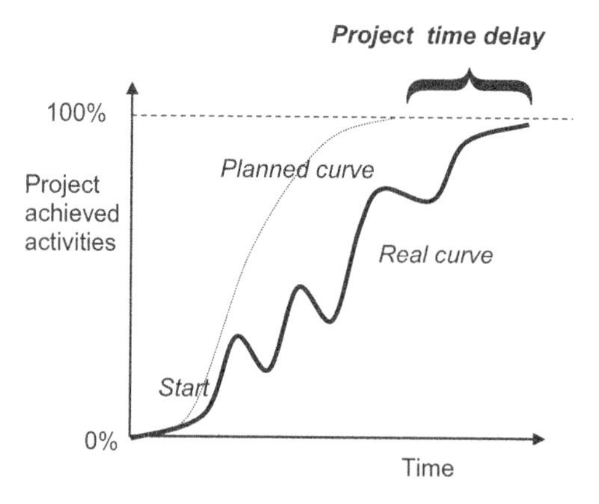

Figure 1.23 Performance with limited design.

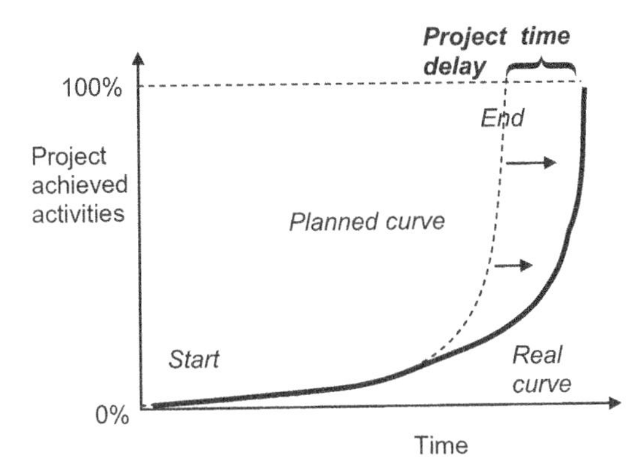

Figure 1.24 Example of risky project performance curve.

Other approaches to project management are possible. For example, Figure 1.24 shows a performance curve where results are concentrated at the end of the project. This approach is very risky because deviations from the planned performance behavior are discovered only in the final stages and therefore little time is available for recovery actions. Such an approach is in general discouraged in product design.

In case of **project delays**, some **recovery actions** may be considered based on the relations between cost, performance and time. Reduction of project scope is a main option when accepted by the customer. In design projects, cost is often underestimated. According to the Standish Group (1995) survey, in software design, 52.7% of projects cost over 189% of their original budget. In some cases, performance cannot be even implemented due to technological barriers, and projects have to be cancelled if the reduction of project scope is not accepted.

Introducing extra resources (i.e., extra costs) is another strategy to reduce project delay. Such an option creates extra effort in project coordination that may in turn reduce the probability of success. In addition, in some fields such as software, the amount of resources is not always linearly proportional to productivity and there were even projects where the introduction of extra resources has triggered definite project failure.

In brain-intensive activities, such as design projects, **human factors** have to be managed with care because resources have a significant impact on productivity and they may possibly not be replaceable in specific cases. In software, for example, the difference in productivity between programers may be as high as a 25× factor, while in more general activities the differential in productivity does not usually go over 3 or 4. This shows that extra care should be considered in resource selection to assess the specific resource of productivity. In general, we may say that human productivity depends on available **time, capability**/skill and **motivation**. Motivation needs to be continuously assessed by the project manager, taking into account that in such projects resources cannot easily be replaced. Conflicts and negotiations are topics to be well monitored by project managers; indeed, such topics have a significant space in project management books.

Project management is a complex discipline that aims to fulfill project outcomes within the planned quality, time and costs. With this in mind, a simple **checklist** of activities to be performed before starting a project is outlined in Table 1.9.

Table 1.9 Project management checklist ☑

1.	Produce a project development plan (who does what, when, how and why).
2.	Check if a project plan template exists for the specific type of project (e.g., a software development plan) to ensure that all the critical issues have been addressed.
3.	At the beginning address what could go wrong (risks) with risk mitigation and contingency actions; allocate extra budget to address risk consequences.
4.	Manage assumptions as for risks.
5.	Monitor and control the project regularly in terms of cost, time and performance.
6.	Remember the human factors (choose appropriate resources and monitor motivation).

1.8 Intellectual Property Rights and Reuse

Dwarfs standing on the shoulders of giants

Standing on the shoulders of giants in product development means that the available know-how is used to reduce cost, risk and development time of the product to be designed. The know-how may concern hardware schemes, algorithms, industrial design, man–machine interfaces, trademarks, copyrighted material or software/firmware code. Some of these elements may be patented, others may be used only under specific conditions expressed in the associated licenses. The reuse of such know-how is a great opportunity to hide strong risks.

As shown in Table 1.1, *late discovery of patents is one of the main reasons for project failure in medical devices.* A patent is a set of exclusive rights granted by some countries to protect an invention, that is, something new, useful and non-obvious. In product development, **patent analysis** is the first step in designing a product for technical and commercial reasons. This analysis is compulsory to understand what *must not be done*, in order to avoid infringement of patents. On the other hand, patent analysis is a primary source to discover the state of the art on the specific products to be designed. This allows working beyond the state of art of product design and avoiding reinventing the wheel, or, worse, performing applications research on solutions already available that cannot be used. Patent investigation is nowadays quite simple using Internet.

Main patents are: 1) international patents, 2) US patents, 3) EU patents.

1. The **international patent**, called PCT (Patent Cooperation Treaty), permits 'reserving' of patents almost worldwide. International patents are managed by an international office that first carries out a novelty research. Later, the invention will enter the national or regional phases, requiring every country to examine it and grant patent. From then on, the patent will be divided into many national patents, every one of which will have its independent progress. The organization where the international patent applications are filed is the World Intellectual Property Organization (WIPO).
2. In the **USA, patents** are managed through the United States Patent and Trademark Office. Note that there are some subtle differences between US and international patent legislation.

3. The **European patent** allows for the application and possession of a patent in several European states, through a single procedure. The patent application granted by the European Patent Office is valid not only within EU countries but also in the other countries that have joined the agreement.

International, US and EU patents can now be easily searched through the Internet sites for their respective office. The **US patent** office (http://www.uspto.gov/) offers, on the net, the Web Patent Full-Text Database (PatFT) that contains the full text of patents from 1976 to the present. It also provides links to the Web Patent Full-Page Images Database (PatImg). More recently, US patents may be searched easily through **Google Patent** (http://www.google.com/patents) that allows access to over 8 million patents and 3 million patent applications (2011). Further information is available at http://www.google.com/patents/. **International patents** can be searched at the Wipo website (http://www.wipo.int). **EU patents** can be searched at http://www.espacenet.com/. In the implementation section (Section 1.18), gives examples of ECG patents that will be given in the implementation section.

Patents are also a strategy to give value to R&D expenditure and to obtain a competitive advantage on one's own product since, through patents, the owner has the right to exclude others from making, selling or using the claimed invention (and the associated product) in the countries where patents are valid. At the beginning of the project, it is therefore worth considering whether patents may be obtained to avoid that innovative product design being copied and thereby obtaining a better sales performance.

For **software and firmware**, legal copyright is generally applicable in Europe, while in other countries patent may be granted. In any case, reusable components have associated **licenses** that must be read carefully to determine whether commercial use is feasible.

Open source software is also a major resource when its license allows commercial reuse. For example, the **GNU** General Public License **GPL** (GNU, 2011) allows for using software in commercial products but, in this case, the software that includes open source material must itself be open source. Since commercial software is generally an asset for a company, the GNU GPL open software cannot generally be used in commercial applications. By contrast, the **LGPL** (GNU Lesser General Public License) is less restrictive. Proprietary software which does not contain part of the LGPL but it is designed to work with LGPL (i.e., is compiled or linked with LGPL) does not itself have to be open source (GNU, 2011). The LGPL open source library is therefore generally used in commercial applications. The checklist in Table 1.10 summarizes the main suggested actions related to IPR and product design.

Table 1.10 Intellectual property right checklist ☑

1.	Perform a deep patent-license analysis on all components of the product to be designed (hardware, software, algorithms, design . . .).
2.	Extract state-of-the-art information for the specific product.
3.	Determine the feasibility of reuse based on patents/licenses, contents and technical pros and cons.
4.	Consider issuing a patent application to protect innovative design.

Part II: Implementation

1.11 The ECG: Introduction

The implementation parts of this book show how the theory sections are translated into a real medical instrument placed on the market. To this end, the implementation sections have the same unit number as their counterparts in the theory section.

The implementation parts show the design of an electrocardiograph (ECG or EKG) manufactured by Gamma Cardio Soft (www.gammacardiosoft.it). Gamma Cardio Soft has the vision for supplying low-cost high-performance devices, disclosing the devices' implementation details. The company has the goal to increase the world diffusion of ECGs and other medical devices with the final aim of reducing deaths due to cardiovascular disease which is the leading cause of death and disabilities in the world (WHO, 2011a).

According to the FDA classification, an ECG is

> *'a device used to process the electrical signal transmitted through two or more electrocardiograph electrodes and to produce a visual display of the electrical signal produced by the heart. The ECG is used to perform the electrocardiogram test. ECG is an electrical recording of the heart used in the investigation of heart diseases. The electrical signal recording shows the electrical heart activity and is used to diagnose various cardiovascular diseases: myocardial infarction, arrhythmia.'*
>
> (FDA, 2005)

By the end of the book, readers will understand **the design and the implementation of a diagnostic 12-lead resting PC ECG** (the Gamma Cardio CG product by Gamma Cardio Soft). A PC ECG is an ECG composed of a hardware acquisition unit connected to a PC supplied with dedicated application software. The PC and the hardware unit might also be contained in the same package, so that users do not perceive the presence of a conventional PC platform. The proposed PC ECG, already available on the market, is developed following the theory and implementation concepts included in this book. The implementation sections take into account errors discovered during the real development, and they propose improved solutions and processes. A professional PC ECG may be considered sufficiently simple and complete to be a good example to explain medical instrumentation design in practice. The PC ECG elements (hardware, software, firmware, certification etc.) and the development process are similar to any medical device development. Therefore, the process and the concepts included in the implementation section may be considered as exhaustive examples to describe the common features of medical instrumentation device development. Such concepts are general and they can be considered exhaustive also for designing electronic products.

1.11.1 The ECG's diagnostic relevance

The ECG is a primary instrument for preventing cardiovascular diseases (CVDs), which are the leading cause of death and disability in the world. In 2008, it was estimated that around 17.3 million deaths were due to CVDs (WHO, 2011a). It is estimated that this number will rise to 23 million within the 2030. CVDs account for 23.6% of all deaths in the world, as

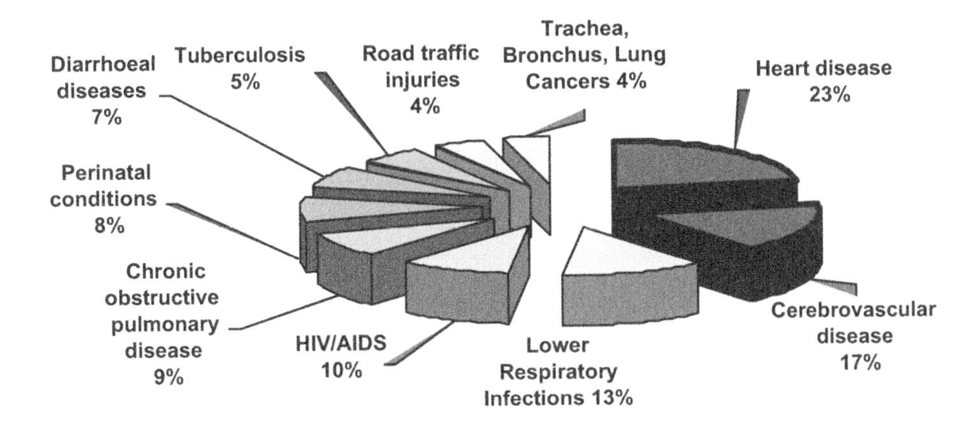

Figure 1.25 The ten leading causes of death (WHO, 2011b).

shown in Figure 1.25, where the ten leading causes of death are depicted according to a World Health Organization report (WHO, 2011b).

The cost of CVDs is a significant part of all national health budgets. In the USA, CVDs are responsible for 17% of national health expenditure. These costs are estimated to increase significantly due to aging. By 2030, it is expected that 40.5% of the US population will have some form of CVD (AHA, 2011).

Although a significant number of CVDs are preventable, there is continuous increase in CVDs due to insufficient preventive measures. Cardiac monitoring devices are a main contributor to the decline in mortality rate: a timely diagnosis of cardiovascular diseases may significantly increase the life expectancy of patients. The ECG is the primary test for assessing CVD, and it is suggested as a routine examination in many cases, such as for patients over 40 years of age (Keller, 2005).

1.11.2 ECG Types

There are three main types of ECGs: Holter ECGs, event monitoring ECG systems and diagnostic ECGs.

Holters are used to record long-term heart activity (e.g. 24-hour monitoring). They have the poorest signal resolution (technically 40 Hz signal bandwidth) and are used to discover specific diseases that do not need a high signal quality but require hours of recording for diagnostic purposes (e.g., arithmias).

Event monitoring ECGs have a better signal resolution (100 Hz signal bandwidth) and are usually used in hospitals to continuously monitor critical patients and to raise the alarm in case of patient problems.

Diagnostic ECG are used to discover a wider range of cardiovascular diseases since they provide better signal resolution (150 Hz signal bandwidth). *This book will focus on the development of this last type.*

Diagnostic devices may perform resting ECG or optionally exercise ECG (stress test). A resting ECG takes approximately 3–4 minutes. An exercise ECG will take longer since it includes several minutes to half an hour of exercise (depending on what can be tolerated), as well as some monitoring after exercise.

The users of this product are doctors, usually specialists, who are responsible for performing the examination and giving an interpretation of its results.

The main advantages of ECGs for patients are the absence of annoying preliminary procedures and the limited time necessary to perform the examinations.

The main advantages for doctors are the wide range of diseases that can be diagnosed and the wide and consolidated clinical experience. Actually, test conditions have not basically changed in the past 70 years.

In addition, the test quality does not depend on the skills of the operator as is the case, for example, with ultrasound diagnoses. The test output data are standard. This means that the test may be carried out and the ECG interpretation made in different places/time and by different operators. So nurses or trained practitioners are now usually employed to physically conduct the test on patients, delivering the test results to the doctors for interpretation.

The diagnostic ECG (from now on, simply ECG) is physically composed of a device that is connected to the patient's body with 10 electrodes. The device shows the heart's electrical activity through the plotting of 12 signals – also called 'leads' – that are obtained as a combination of the 10 input signals. This justifies the term 12-lead diagnostic ECG.

1.12 The ECG Design Problem Formulation

The design of a medical instrument can be considered as a 'messy' type problem as discussed in Section 1.2. The problem formulation will probably be ambiguous because, at the beginning, the input and outcomes are not defined.

The ambiguity is due to the fact that in a new design many decisions and inputs will not be supported by previous experience while solutions belong to an unknown domain. This increases the risk of errors, such as poor performance, increased cost and delays that cannot be tolerated by the customer.

This strengthens the need for an accurate problem definition and negotiation with customer before starting all the activities of the actual product design.

Problem definition may, at minimum, include jointly agreed: 1) goals, 2) outcomes, 3) input, 4) assumptions, 5) risks, 6) constraints (scheduling, budget, resources, quality, requirements).

In Table 1.11, these items are outlined. They can be the basis for a customer–supplier contract. The **goal** of the activity consists of performing all the tasks to design, develop and implement the **diagnostic ECG for a PC platform**, that is produced by Gamma Cardio Soft. The device, also named Gamma Cardio CG, is suited for usual medical practice and it can be produced through conventional industrial processes. This has been proved because the **outcomes** of the design project will include samples developed through the final manufacturing process. Samples must comply with medical device regulations, and project documentation shall be complete for manufacturing and certification. The first **input** is the requirements included in customer documents (e.g., a business plan) which are probably incomplete at the beginning of the project. These requirements should define features that are perceived to differentiate the product from the competition (see Section 1.3). This should suggest that a proper life cycle has to be selected to create and validate requirements and to avoid the risks of incomplete requirements, as discussed in Section 1.6. The customer requires a double **market target**: the usual ECG target (cardiologist, ambulatory and hospital) and non-conventional market (non-cardiologist, nurse for a telemedicine approach trained personnel,

Table 1.11 ECG design problem definition

Goal	Design and implement a professional PC ECG
Outcome	1. First samples already developed through final industrial process, ready for the market (software, hardware, firmware, sensors) 2. Medical Device European Certification, compliant to US FDA approval requirement 3. Complete manufacturing and design documentation
Input/main requirements	Business plan requirement (partially defined) Target market: hospital, cardiologist, ambulance (non-cardiologist version to be developed for nurse in telemedicine approach, medical students, trained personnel) Suitable for harsh environment, third-world medical environment
Technical constraints	**Medical device regulation:** EU Directive EEC (2007) **Electromagnetic compatibility:** IEC/EN 60601-1-2 **Security certification:** IEC/EN 60601-1, IEC/EN 60601-2-25 standards
Mandatory requirement	**Performance standard:** AAMI EC 11 standard **93/42/EEC Annex IX classification:** class II B **Electrical Isolation IEC/EN 60601-1:** class II CF **Target manufacturing cost:** €250, all included
Assumptions	Involvement of customer on main steps for validation of requirements and milestone deliveries
Risks	Commercial, scheduling
Scheduling	One year completion
Budget	€2 million
Required resources	• cardiologists for requirement validations • analysts in medical device instrumentation • developers in safety-critical PC-based software and firmware • medical device system engineers for system integration and testing • consultants in medical device certification • instruments for software, firmware and hardware development and test

third-world harsh medical environment). This last requirement creates some challenges in both technical and marketing areas because the product should include extra technical features, and the marketing proposal should not confuse potential customers. Products that are suitable for a different type of customers may confuse potential customers creating a negative impact on sales.

Constraints are on the compliance of the product and the producing and selling company with compulsory regulations for medical device (Table 1.11: Technical constraints). Such items will be discussed in more detail in the Chapter 8 on certification.

The customer also imposes **mandatory requirements** based on its market investigation. The customer believes that the highest performances on technical and security specifications will be a perceived differentiator with respect to competitors' products.

From a technical point of view, the mandatory requirement for an AAMI EC 11 standard compliance imposes technical performance over the product. In Europe, it is sufficient that technical performances are validated comparing performance of the medical instrument to be developed with a certified one over a significant set of trials. Certification through compliance over a standard that specifies performance is a plus. AAMI EC 11 (AAMI, 1991) is a recognized standard for ECG performance that specifies all essential performance to be achieved. For safety. European medical devices are classified into four classes: I, IIa, IIb, III. The higher the class, the higher has to be the safety required for the product in design and manufacturing. In the case of the Gamma Cardio CG, the customer required a class IIb device, while most ECGs are in class IIa.

According to the EEC (1993) Annex IX rule 10, class IIb are devices 'intended to allow direct diagnosis or monitoring of vital physiological processes, unless they are specifically intended for monitoring of vital physiological parameters, where the nature of variations is such that it could result in immediate danger to the patient, for instance variations in cardiac performance, respiration, activity of CNS'. If the previous is not verified for the device, the instrument is class IIa.

The customer also asks for the highest protection against electric shock. ECG devices may be in class BF or CF but, according to IEC/EN 60601-1, CF type will 'provide a higher degree of protection against electric shock than that provided by type BF'. A detailed discussion on these certification topics is included in Chapter 8.

The final item included in the checklist of the corresponding theory section (1.2) is: '*Promises have to be kept*'. The project is quite challenging since outcomes are difficult to obtain considering the available budget, timeframe, requirements and constraints.

Extensive use of assumptions and contractual terms and conditions have to be set up by the supplier in the agreement with the customer in order to avoid project failure.

1.13 The ECG Business Plan

Product design may be commissioned either by a customer external to the company (outsourced development) or by the internal customer. For example, the commercial department of the company may need a new product. In both cases, **market research**, usually included in a business plan, has to be supplied before developing the product. The market research process is generally divided into four categories: definition of objectives, research plan development for collecting data, research plan implementation with data collection/analysis and data interpretation (Kotler, 2002).

The first set of objectives includes the assessment of

- the business opportunity for designing and producing a PC ECG
- the market size and trends
- the legal, regulation, certification and other existing barriers
- the financial evaluation of competitors and competitive prices
- the distinctive features of similar existing products
- the market demand for each selected target.

The second set of objectives includes the assessment of

- the client's expectations of the product in terms of mandatory and optional features
- the price target that customers are willing to pay
- the preferred channel for purchase
- the expected method of promotion.

A complete coverage of these topics is out of the scope of this book. In the following, an example of how these data are obtained is proposed for the market size and trend (adapted from Becchetti, 2005).

1.13.1 Market Size and Trend

The medical device market projections include the cardiovascular monitoring equipment market that embraces (Frost & Sullivan, 2003):

- resting ECG/EKG
- ECG Data Management Systems
- stress testing
- Holter monitoring
- event monitoring.

Resting ECG is the target market since a 12-lead resting ECG is developed here. Cardiac devices do usually experience growth even in recession periods.

The ECG global market is expected to grow to USD 4.1 billion by 2015 with a compound annual growth rate (CAGR) of 9.7% in the 2009–2015 period (Axel, 2011).

We may then assume for a rough estimation that country market share is similar for the cardiovascular monitoring market and for ECG (Becchetti, 2005). Regarding country market share, the USA is then at around 40%, Europe 25% and Japan 12%. The market trend for 2009–2015 is assessed in Table 1.12 using the previous data.

A business plan must then specify what and how a market share can be targeted.

A business plan may be concerned with resources involved in the development project because it clarifies the perceived distinctive features which guarantee the product's competitive differential advantage (i.e., the features that make buyers prefer a specific product).

Table 1.12 Medical device and ECG market size/trends

	Country market share%	ECG market ($ billion) (2009)	ECG market ($ billion) (2011)	ECG market ($ billion) (2015)
USA	40%	0.9	1.1	1.6
EU	25%	0.6	0.7	1.0
Japan	10%	0.2	0.3	0.4
others	25%	0.6	0.7	1.0

1.13.2 Core and Distinctive Features

The product features may be divided into the minimum core features, also referred to as 'threshold product features' and distinctive features also called 'critical success factors' (Johnson and Scholes 1997).

Customers take for granted that any ECG product of any manufacturer has at least the core features. A complete survey of ECG products and associated features is found in the Healthcare Product Comparison System (HPCS) published by the ECRI Institute (ECRI, 2012). Distinct features act as motivators for customers and differentiate between product classes and manufacturers.

The following items are possible minimum **core features**:

- quality and reliability of recorded data
- safety for patients and users
- adequate mean time before failures (MTBF)
- user-friendly interface.

Quality and reliability consists of recording all the required signals (12 tracks or leads) with no error on the track plot that might lead to a wrong diagnosis. This feature may be guaranteed by a comparative validation using certified ECGs or through performance standards (AAMI, 1991).

Safety is guaranteed by complying with compulsory standards of safety detailed in Chapter 8 on certification (see also Table 1.11).

MTBF is related to the fact that an ECG is a device that is mission critical, and often safety critical. Failure in performing examinations may have an impact on patients' health and may cause loss in terms of doctors' image and money.

The lack of a **user-friendly interface** is often a main cause of errors and injuries (Sawyer, 2005). The layout and design of the graphic user interface greatly affect the user's ability to successfully perform functions and extract information during the operation of the device, especially during critical situations. In addition, doctors may not have a technical attitude, so a complex interaction with medical devices may impede ECG usage.

The following **distinctive features** are examples of differentiating ECG performance:

1. capability to perform simultaneous lead recording (1, 3, 6 or 12 simultaneous signals)
2. interpretation capability
3. display for recording preview
4. networking PC connection, telemedicine, healthcare standards.

The benefit of these additional capabilities is briefly explained below. It should be considered that in past years ECG devices did not usually have such capabilities.

Simultaneous lead registration may be required to diagnose more easily some very specific diseases. This feature is not included in all ECGs; it also allows reducing time to record the ECG. ECG is usually recorded for 10 seconds for each signal. The patient must be at rest when the ECG is recording, otherwise artifacts may be displayed on the recorded signal. If the device is capable of recording simultaneously all the 12 leads (= signals) the test is performed in 10 seconds. In a single-channel ECG where only one lead is recorded at once,

$10 \times 12 = 120$ seconds are required, and some patients may find it difficult to stay at rest for 120 seconds.

Interpretation capability is one of the most innovative capabilities in recent years. To make a diagnosis, doctors have to compare the current record with typical signal patterns of diseases and to evaluate numerical parameters in comparison with standard values.

All patterns and standard parameter data must be clearly presented to the doctor who should be able to compare current and typical patterns and find most similar ones.

In the past, this procedure was easy only for cardiologists. Current interpretation programs may perform the task automatically producing a diagnosis whose quality is claimed to be comparable to that of cardiologists (Willems, 1991).

This issue should also be considered in the light of doctors' responsibility. Incorrect cardiovascular diagnosis may cause death from non-discovered diseases. Doctors feel safer if their diagnosis is backed up by computer programs. On the other hand, current interpretative programs may show errors and many cardiologists do no use these programs.

Many ECGs are supplied with an LCD **display**. The doctors' main benefit consists of having a result preview, thus avoiding useless printouts. This is reasonable since:

1. ECG paper is usually specific and expensive
2. printing is also time-expensive since ECG printers are usually slow
3. recordings are affected by artifacts that make diagnosis difficult and therefore, repetitive recording and printing are often required to have good results.

The display allows printing only when the signal is adequate for diagnosis, thus saving money and time.

Network/PC connection provides potential extra benefits for the doctor and in general for hospitals. The ECG may implement a *healthcare standard* for communication and storage so that patient information can be managed in an interoperable and consistent approach (e.g., see the HL7 standards – HL7, 2011). Examinations may be transmitted and saved in a permanent archive. The network connection also enables *telemedicine* functions. For example, nurses or non-specialist doctors may perform the ECG examination directly, where the patient lives. The results can then be transmitted to a specialist for interpretation.

Once distinctive and core features are identified, technicians have to assess the associated marginal production cost and investment for development. Investment cost may be linked directly to labor hours, materials, consultant subcontracts and new instrumentation. Assumptions and implementation strategy have to be considered in order to derive at least a draft cost estimate.

In Table 1.13 a draft analysis is outlined assuming that a 12-lead diagnostic PC ECG is developed. For some features, evidenced with the (*) indication, no extra hardware is required since it uses the hardware already available on PC platforms. Marginal production costs are in this case associated only to hardware improvements.

Having defined the distinctive features and evaluated the associated costs, a technical market analysis is required to select features to be developed and implemented. The reduction of implemented features will decrease cost, risk and time, but may lead to a reduction in sales. The proper tradeoff has to be chosen, evaluating the gross profit graph (Figure 1.5) for various implementation scenarios.

Table 1.13 Distinctive ECG features and associated costs

Feature	Extra activity required	Extra investment (€×1000)	Delta on Marginal production cost
Simultaneous recording of all leads	Hardware has to be redesigned in comparison to single-channel recording Specific software development	500	€20 (additional cost of hardware)
Interpretation capability	Intensive medical investigation on protocols to be implemented Heavy software development	400	0
Healthcare standards	Investigation on standards Software development*	100	0
Simultaneous printing of all leads	Software development*	100	0
Real-time display of track recording	Software development*	50	0
Network connection	Software development*	100	0
Telemedicine	Software development*	100	0

Based on the previous considerations and taking into account the customer suggestions outlined in Table 1.11, the PC ECG product will fully implement the last four items of Table 1.13. Healthcare standards will be partially implemented. Instead of a full implementation of the interpretation capability, tools for assisted interpretation will be included. Simultaneous recording will be postponed to a future implementation.

1.14 The ECG Design Process

The European Medical Devices Directive EEC 93/42 and successive amendment 2007/47 (EEC, 2007), requires that medical devices are compliant to the '**essential requirements**' contained in Annex I of the Directive. Such requirements impact on functional features of the product and on the design process (risk analysis, specification etc.). Conformity to essential requirements may be demonstrated through compliance over **harmonized standards** (EEC, 2009).

The Table 1.14 contains the selected **Medical Device European standards/regulations** for the certification of the PC ECG in the European market. Similar standards are applicable in other countries (e.g., USA (FDA, 1998; FDA, 2005)).

Compliance with Table 1.14 standards guarantees the safety usage of the medical device, that is, the 'Freedom from unacceptable Risk' (ISO 14971, IEC/EN 60601-1). A comprehensive risk management is set in action to ensure that a safe device is obtained. In addition, standards in Table 1.14 impose requirements that are specific for ECG, medical electrical systems, electromagnetic compatibility and programmable electrical medical systems.

Table 1.14 ECG design applied standards

Standard Number	Topic	Title
ISO/EN 14971	Risk management	Medical devices – Application of risk management to medical device
EN 1041	Information from supplier	Information supplied by the manufacturer of medical devicc
EN 980	Device labeling	Symbols for use in the labeling of medical devices
ISO/EN 10993-1	Biological evaluation	Biological evaluation of medical devices – Part 1: Evaluation and testing
IEC/EN 60601-1	Safety general requirement	Medical electrical equipment – Part 1: General requirements for basic safety and essential performance
IEC/EN 60601-2-25	ECG safety requirements	Medical electrical equipment – Part 2-25: particular requirements for the safety of electrocardiographs
IEC/EN 60601-1-2	Electromagnetic compatibility requirements and test	Medical electrical equipment. General requirements for safety. Collateral standard. Electromagnetic compatibility. Requirements and test
AAMI EC 11	Diagnostic electrocardiographic devices requirements	Diagnostic electrocardiographic devices, ANSI/AAMI EC11-1991

Most of the previous standards are focused on safety and impose functional requirements on the product. Regarding life cycle stages, the standard implies: system-subsystem requirement specification, system-subsystem architecture, detailed design and implementation, including software development, test specification, verification and validation (plan and result), modification change procedures and a final assessment report (see IEC/EN 60601-1).

The Table 1.14 includes some standards related to ECG medical devices to guarantee compliance to essential requirements of medical device EU Directive (93/42, 47/2007). Other standards may be applicable: for example, Table 1.15 shows additional harmonized standards.

Such standards are not mandatory, and other standards may be used. The use of harmonized standards is, however, the more direct approach to complying to the EU medical directive.

In addition, the designer should consider

1. usual methodologies for hardware/software design of mission- or safety-critical systems
2. typical countermeasures to face new product design (uncertain functional market requirements, technology risks etc.).

Regarding the first item, MIL-STD-498 (MIL, 1994) or similar newer standards are a good candidate for software development and hardware software integration as outlined in Section 1.7.

Table 1.15 Other harmonized standards for ECG

Standard Number	Topic	Title
IEC/EN 60601-2-27	Monitoring ECG essential performance	Medical electrical equipment – Part 2-27: Particular requirements for the safety, including essential performance, of electrocardiographic monitoring equipment
IEC/EN 60601-2-47	Ambulatory ECG essential performance	Medical electrical equipment – Part 2-47: Particular requirements for the safety, including essential performance, of ambulatory electrocardiographic systems
IEC/EN 60601-2-51	Recording and analyzing ECGs	Medical electrical equipment – Part 2-51: Particular requirements for safety, including essential performance, of recording and analyzing single channel and multichannel electrocardiographs
IEC/EN 62304	Medical device software	Medical device software – Software life cycle processes

For the second item, a proper life cycle should be chosen so that those market/technology risks are minimized. An iterative and incremental life cycle with intermediate results may be a proper choice.

The previous constraints and considerations allow for defining a first system design process as shown in Figure 1.26. A further refinement will be obtained in the ECG product life

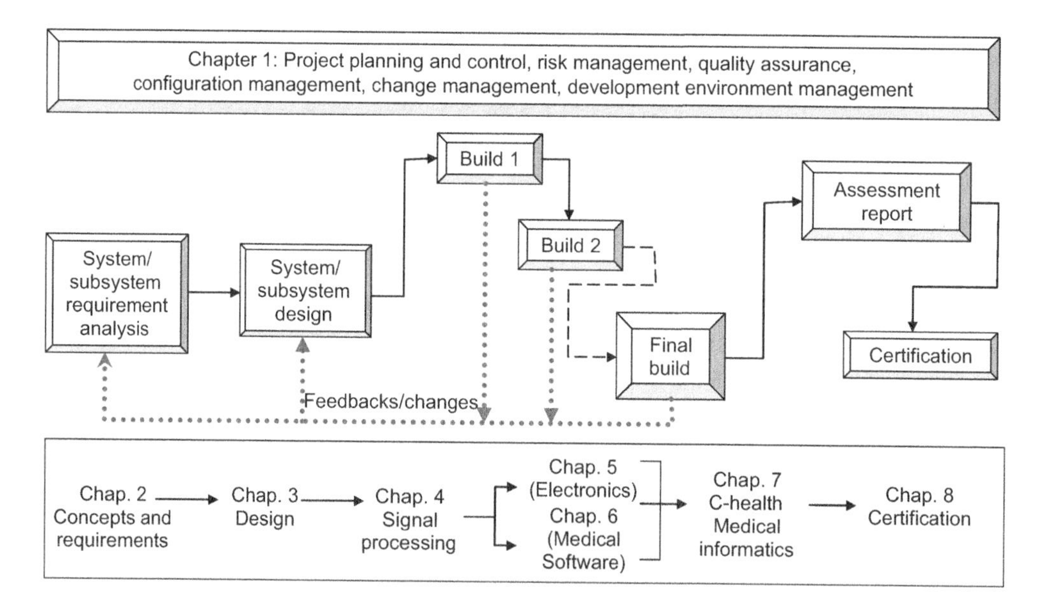

Figure 1.26 ECG process design.

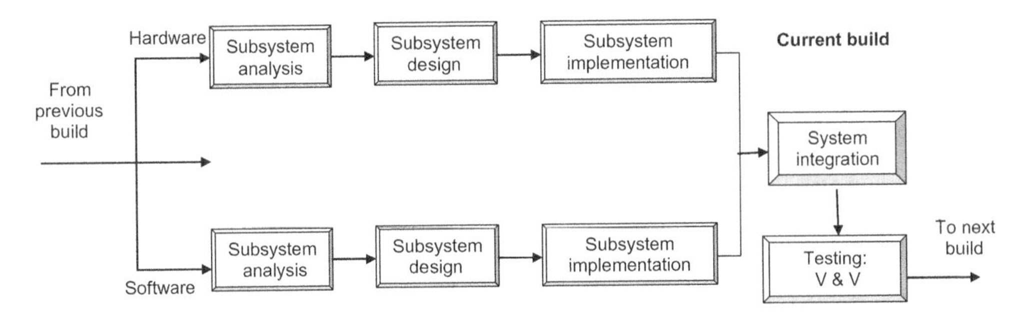

Figure 1.27 Builds within the ECG development.

cycle (Section 1.16) after a draft system-subsystem decomposition is performed (Section 1.15). Figure 1.26 shows the main activities ordered on a time basis. In the central part of the figure, the **core activities** are outlined. Such activities are performed sequentially. The box on the bottom indicates chapters where these activities are discussed. The correspondence among the build and the chapters are better evidenced in Figure 1.27. The top box of Figure 1.26 includes **transverse activities** that are performed throughout the entire design project. These activities are outlined below.

1.14.1 Transverse Activities of the ECG Design Process

Risk management refers mainly to the activities prescribed in IEC 14971 (see Tables 1.14 and 1.15). *Quality assurance* implies the respect for company quality assurance procedures throughout the entire projects. *Configuration management* is focused on establishing and maintaining consistency among all deliveries of the system (documents, software, hardware schemes, requirements, specifications, operational information) throughout its life. 'The configuration management effort includes identifying, documenting and verifying the functional and physical characteristics of an item; recording the configuration of an item; and controlling changes to an item and its documentation' (MIL, 2001). Configuration management allows for managing the versions of each item of the project recording, the changing history and the currently valid version of all items. Configuration management will be further examined in Chapter 6.

When *change management* is in place in a project, changes require a defined process before being formally approved. Typical change management procedures define how to change software/hardware in case of change requests or problem reports. The *development environment* includes all the 'facilities, hardware, software, firmware, procedures, and documentation needed to perform software engineering. Elements may include but are not limited to computer-aided software engineering (CASE) tools, compilers, assemblers, linkers, loaders, operating systems, debuggers, simulators, emulators, documentation tools, and database management systems' (MIL-STD-498, 1994). Elements also include tools for hardware design, development and testing, and a library. The *development environment management* implies a process for setting all items of the development environment under control.

1.14.2 Core Activities of the ECG Design Process

Referring to Figure 1.26, the first activity of the design process is a system-subsystem requirement analysis. Requirements will be given by risk analysis, applicable standards and functional and market requirements. System interfaces are also defined. Before considering this activity, a concept of operations and a business analysis is performed as outlined in Sections 1.12 and 1.13. These latter activities are outside of the design process. The *system-subsystem requirement analysis* also includes a first subsystem decomposition based on requirements and the considerations from the problem domain. The system-subsystem decomposition related to the design is performed in the next activity (system-subsystem design). In this step the subsystems and their interfaces are identified and the associated specifications are outlined.

The next steps are called ***builds***, and they contain activities as shown in Figure 1.27. Builds are versions of the ECG system that meet a specified subset of the requirements that the completed system will meet (adapted from MIL-STD-498, 1994). These builds yield intermediate results that allow the tuning of product requirements and specifications. So in Figure 1.27, each build provides feedback for an possible change of the system-subsystem requirements or the design specifications with an update of the related documents. Within a build, for each subsystem, a complete development cycle is performed over a subset of requirements. The subsystems are the hardware platform, the firmware embedded in the hardware and the software running on the PC. Finally, an integration step is performed among the three subsystems to be developed and the other available subsystems: sensors cabling and network telecommunications. Finally, the appropriate testing is completed with the proper verification and validation. Verification implies that the system works correctly according to specifications. Validation implies a compliance check over the implementation of the requirements or, more simply, that the system is in line with the intended purpose. Verifications and validations have to be performed particularly over safety requirements identified in the risk analysis. Testing activity includes a test plan, test descriptions and results.

In the final build, the product implements all the requirements and is fully tested. Therefore, a final assessment report is produced before certification is obtained. The two previous schemes in Figures 1.26 and 1.27 are still generic. More detailed activities can be obtained when a system decomposition and the product life cycle are available (Sections 1.15 and 1.16).

1.15 ECG System–subsystem Decomposition

The functional scheme of the PC-EC may be derived from the general scheme outlined in the Section 1.5 (see Figure 1.11). This newer scheme (Figure 1.28) shows functions, physical interfaces (continuous lines) and logical interfaces (dotted lines). The patient is connected through electrodes and cabling to a **hardware platform** which converts and translates the patient signals into digital signals. The hardware platform contains **firmware** that manages the board and the message exchange with the PC. The hardware platform is digitally connected to a PC where the specific ECG **software application** is running. The software application performs most of the ECG functions and handles the protocol message exchange with the firmware. The PC is also physically connected with the Internet for telemedicine services that may be remotely accessed by other healthcare operators.

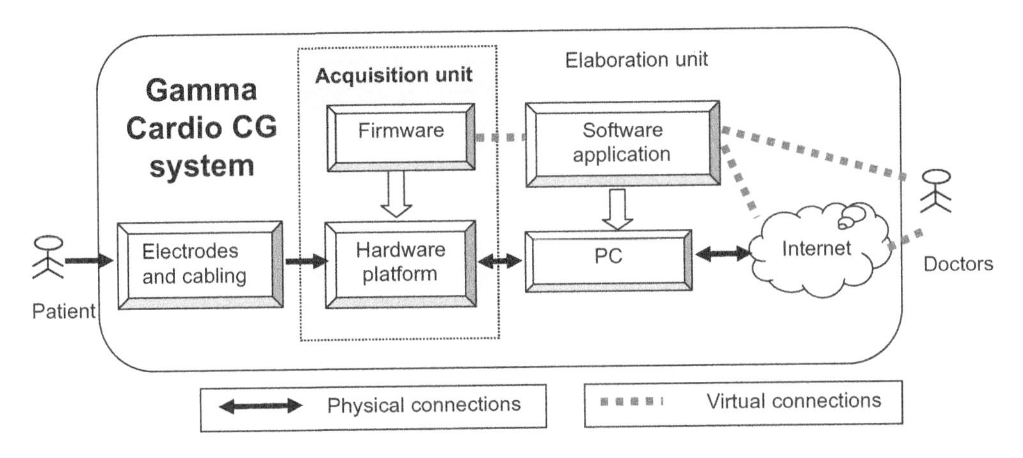

Figure 1.28 Gamma Cardio CG functional scheme.

The hardware platform, the firmware and the software application are the subsystems that must be developed. These subsystems have to be integrated with sensors (electrodes and cabling) and the telecommunication network. In the following, the hardware platform and the software application of the Gamma Cardio CG are decomposed into subsystems.

1.15.1 Hardware Platform Decomposition

The hardware platform can be further decomposed into analog and digital unit within the board. The **analog unit** handles the analog signals coming from patients. These signals are digitized by the **digital unit** that contains the analog to digital converters (ADC). The digital unit also has a microprocessor that controls the board and exchange messages with the PC through the firmware. This decomposition obeys the principle of *maximize cohesion and minimize coupling*. Analog and digital units are connected through a simple interface, mainly constituted by the ADC. The functions of these two units have a self-contained structure. This decomposition allows for better parallelization of the activities, integration and testing. The analog unit can be developed quite separately with respect to the digital one. Chapter 5 describes this design schemes in detail.

1.15.2 Software Application Decomposition

The software application can be decomposed, taking into account its main functions. The software application exchanges messages with the firmware to manage the hardware board and to perform digital acquisition of the patient's electrocardiogram. This signal has to be stored in the PC memory. The signal is then processed and adapted according to the visualization media features and to doctor needs. For example, different vertical and horizontal scaling will be needed for PC screens, printers or remote visualizations.

A **graphic user interface** (GUI) has to handle all the commands coming from the users. The GUI will also have the task of showing data and messages received from the other modules.

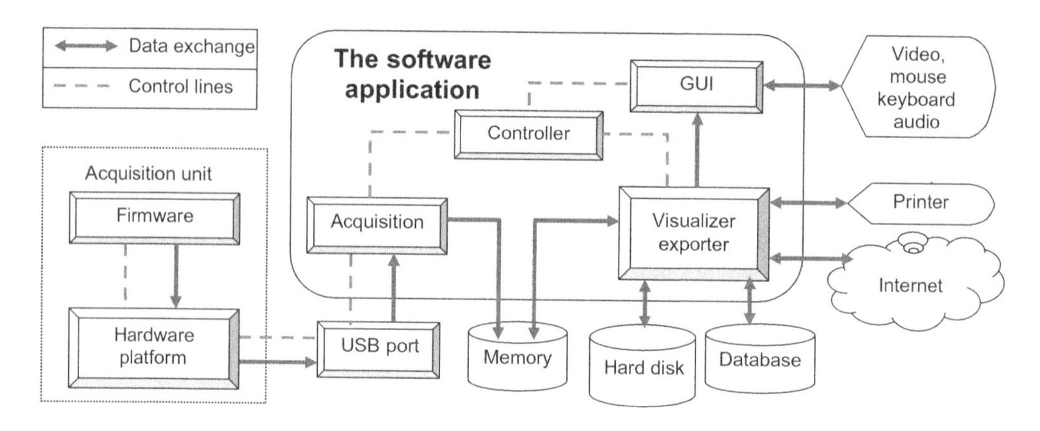

Figure 1.29 The Software application subsystem decomposition.

A **controller module** will monitor and control all the ECG activities. The scheme in Figure 1.29 shows the corresponding software application modules and the associated controls flows (dotted lines) and data exchange (continuous lines).

The **acquisition module** is in charge of managing the acquisition unit and exchanging control and data with firmware. Patient data are stored in the internal memory under the supervision of the controller.

The '**visualizer exporter**' module processes the patient's raw signal to obtain proper visualization features and to store all patient data on databases or files. This module has the task of sending ECGs to the printers, to the network and to the GUI module, which in turn is in charge of receiving the user's commands and showing data and messages coming from the controller. The GUI module interfaces with video, mouse, keyboard and the audio system.

The software decomposition is also in line with the *maximum cohesion minimum coupling* principle. The modules perform well-defined tasks and are loosely coupled. Interfaces are also well-defined.

In the next section, system-subsystem decomposition will be used to define a life cycle design process.

1.16 ECG Product Life Cycle

This section addresses the problem of defining an advantageous life cycle approach for products that integrate hardware, software and firmware. This discussion will be helpful for any medical or electronic product with these specific features. At the beginning of the project, it is worth investing time in order to study and define the most valuable strategy for the life cycle (for example, to reduce late-discovered errors). The cost of repairing errors at different design stages is exponentially proportional to the time passed from the beginning of the project (see Figure 1.6). This means that an inappropriate life cycle may yield additional effort, possibly leading to project failure.

In medical device design projects, errors may also jeopardize safety. The selected life cycle must therefore reduce risk related to reduced safety. The life cycle selection strategy must also take into account many factors, such as requirement stability, technical knowledge, clarity of project objective, project size, project management team experience, customer behavior and degree of formality (CMS, 2008).

A possible approach to deriving a life cycle process consists of focusing on possible risks that may occur during the project and then defining the associated strategy to overcome those risks.

The main risks for ECG products are

1. distorted visualization of the ECG signal
2. expensive error-fixing in the overall system integration
3. requirement validation and instability
4. law/license infringement in component reuse or patents.

Law/license infringement risks may be reduced by a proper patent investigation before starting the design. Software and hardware licenses of components to be reused may be verified before usage. Strategies to reduce the other risks will be outlined in the following.

1.16.1 Overcoming Risk of Inadequate Visualization of ECG Signal

A wrongly reproduced output measure is the main risk for any medical instrument. From a false output, doctors may make an erroneous diagnosis. For example, a distorted ECG signal may hide a myocardial infarction.

The distorted signal may be caused by an error in specifications (the selection of an incorrect algorithm or filter specification) or by a problem in software, firmware or hardware.

A problem of this kind is not easy to discover. Most of the ECG modules manage the ECG signal; modules are also coupled, that is, a perturbation in one module produces effects on many other modules. This means that the origin of the problem may be far away from the module where the effect appears. When specification deficiencies give a contribution to the problem, the error detection may be cumbersome. Worse, the errors may even not appear in the testing steps.

The **approach** to overcome such risks may be based on a '**divide and conquer**' approach: divide the possible error sources and test them individually, eliminating side effects from other modules. First, errors in specifications have to be tested. This is achieved through both low and high fidelity simulations to be discussed in Chapter 4.

For example, the module and phase response of the filters may be theoretically analyzed and then the system response may be simulated through appropriate tools at different levels of simulation. For analog filters, tools may compute the approximate filter response based on ideal values of resistors and capacitors. Then, the effect of commercially available real values and the result of the component tolerances may be analyzed. Finally, circuit simulator tools such as SPICE (see wiki for details) may plot the expected output with respect to test input signals (sinusoids, square waveforms etc.). These **simulators** may also estimate the effect of parasitic resistances and capacitances that are usually present in the final printed circuit board (PCB).

Similar considerations hold for digital filters, where available tools allow for design and performance evaluation. Such tools may also simulate the effect of real filter values that differ from ideal ones due to the available precision of microprocessors.

System specifications	Validation (theoretical evaluation and simulations)	Subsystem decomposition and specifications	Validation at subsystem level	Unit implementation	Unit testing	Unit integration

Figure 1.30 Product development life cycle.

These simulations may validate specifications at unit level (i.e., testing each module alone) and at system level. When specifications are reasonably tested with simulations, the implementation step may start testing each single module alone. The interface of the module under test with the other ones is simulated through 'dummy modules'.

Taking into account the previous considerations, the resulting life cycle is outlined in Figure 1.30. The steps before system specifications (requirement analysis etc.) are omitted.

Reduction of risks and development time may be further obtained when modules are developed and tested, following a proper temporal sequence.

The risk of signal distortion may be reduced by testing all the modules that manage the ECG signal, starting from the simplest and the least coupled (i.e., with fewer connections to other modules).

Referring to Figure 1.29, the GUI should be tested beforehand, developing a dummy software module for testing. This module will be a software signal generator whose output is visualized through the GUI. This test will prove that the GUI module is able to show typical ECG signals without perceived distortion. For example, a square wave may be used to evaluate phase distortion (see Figure 1.31). Other signals may be used (McSharry *et al.*, 2003, 2011; IEC, 1999) to evaluate typical visualization problems such as incorrect scaling or aliasing. Aliasing is due to undersampling signals which cause distortion or artifacts (see Chapter 4 and 5).

Referring to Figure 1.29, the subsequent module in the signal chain is the '**visualizer exporter**'. This module adapts input signals to the requirements of the target (GUI, files, printer, Internet), modifying scaling, sampling rate and data format and applying proper filtering. The signal generator developed in the previous step is useful also for testing the visualizer exporter module functions. Template files from different standards may be used for testing. Filtering should be tested using digital signal processing knowledge. For example, a unit response signal may evidence problems in the filter impulse response. Frequency domain analysis is helpful for validating system requirements that contain constraints on the maximum frequency bandwidth. Chapters 4 and 5 discuss these topics in more detail.

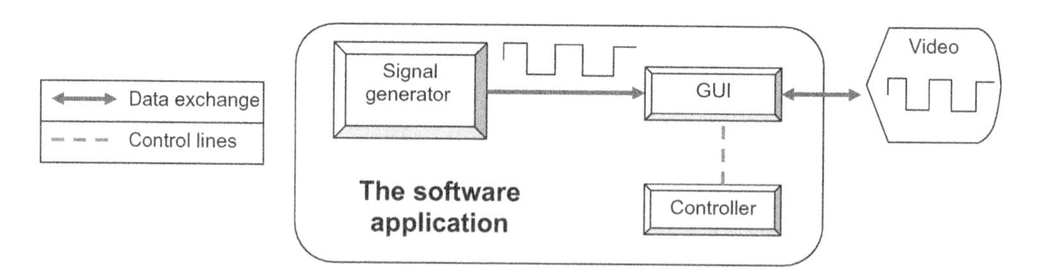

Figure 1.31 GUI module testing.

Assuming that the 'memory block' in Figure 1.29 is a simple storage function, either in a shared memory area or on disk, the next module in the signal management chain is the **acquisition** module.

This module handles data messages coming from firmware, which are contained in a proper protocol, and then extracts ECG signals from data messages. Artifacts and signal distortion may originate from modules that manage the signal before the acquisition module. For example, problems may raise in the acquisition hardware unit and, in particular, in hardware transmission components that are not suited for the specified throughput. Again, validation through theoretical analysis and simulations has to be performed. For example, given the throughput between the PC and the acquisition board, the hardware components that handle the signal must be suitable for the throughput. Theoretical analysis and simulations using SPICE may verify design choices.

In addition to the previous problems, the acquisition module itself may introduce signal modifications when the protocol is not correctly designed or implemented in the software application. In testing the 'acquisition module', we again have more factors that may result in signal modifications and the divide and conquer approach may help.

The acquisition module should first be tested with a dummy module that simulates the firmware of the acquisition unit. The first step consists of testing the protocol exchange between the dummy module (i.e., the firmware simulator) and the acquisition module.

The **firmware simulator** is also helpful because, once tested, it may become the core process of the firmware that has to be developed for the acquisition unit. Then the firmware simulator may be connected to a signal generator to test if the acquisition module is able to handle any type of signals. Finally the acquisition module may be tested with the set of modules already tested, as shown in Figure 1.32.

Modules configured as in Figure 1.32 may also be helpful for performing demonstrations or for training operators. Operators may not have real patients for training ECG acquisition, so if the signal generator can itself create ECG signals, non-expert users can experience training on the system.

With this final step, the software application is tested module by module and the ECG signal should not experience distortion or artifacts.

The hardware acquisition unit can be tested with a similar approach through module testing and then system testing. The hardware may embed a simple signal generator so that hardware can be tested with limited external instrumentations. In addition, an automatic testing function can be implemented and used to test the system during upgrades. This

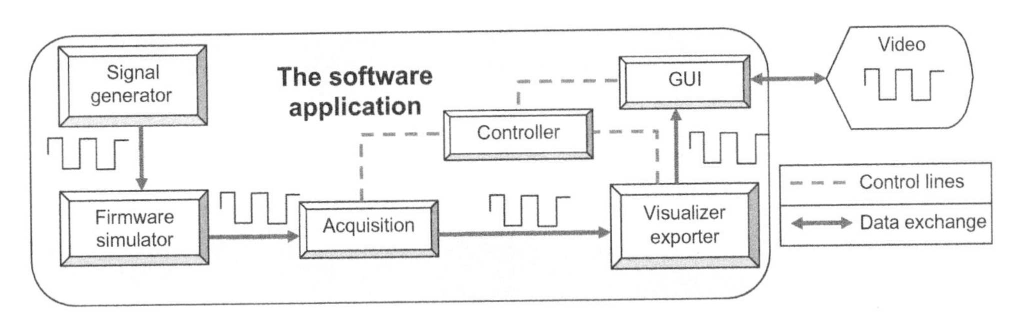

Figure 1.32 Acquisition module testing.

function is also helpful to reduce human intervention during test in production. Most of the hardware testing can be performed without human intervention, and this will reduce labor cost for producing the hardware, and therefore the marginal production cost.

1.16.2 Overcoming Risk of Error Fixing in System Integration

Error fixing risk may be reduced using the strategy pursued in the previous paragraph. Modules are coupled and then the module where the error happens may be far away from the problem where the error is generated.

For example, a random artifact on an ECG visualization on the GUI could come from interference generated by fluorescent lights at mains frequency or in a higher frequency spectrum. This may influence non-shielded digital circuits (e.g., photodiodes) that in turn may corrupt data messages. Since software engineers may not consider hardware problems at first time, defect removal may be particularly time consuming.

The strategy outlined in the previous subsection is as follows:

1. Although module coupling is minimized in design, modules maintain some coupling: perturbation in one module can create effects that are distant in space and time.
2. The search for causes of defects may be complex and time consuming.
3. Unit testing with dummy modules that emulate the module interface will increase implementation/testing time but it will greatly reduce debugging time.
4. Dummy modules may in turn be helpful for adding extra functions to the product or may reduce time spent in the development of other modules.

1.16.3 Overcoming Risks for Non-stable/Unfeasible Requirements

Requirements are generated from applicable medical device directives or standards, from marketing research and from technical considerations. Requirements may be unfeasible with the technology chosen for the product, and the cost of redesign will increase exponentially with the time passed from the beginning of the project to the time when the reworking is needed (see Figure 1.6). This suggests testing requirement feasibility at the earliest stages of development and to continuing the tests throughout all the following stages, using theoretical analysis and simulations.

For example, the overall minimum system noise specified by the AAMII ECG standard (AAMI, 1991) may be estimated analytically from component specifications and circuit diagrams. This will help avoid feasibility problems before starting the implementation stage.

The main impacting requirements have to be continuously evaluated as an action to reduce risks. For example, the following requirements are suggested to be monitored throughout the project:

1. signal quality (frequency response etc.)
2. electrical safety performances (risk current etc.)
3. electromagnetic compatibility (EMC) performance
4. MTBF (Medium Time Before Failures)
5. power consumption

6. production cost
7. computer/board hardware resource utilization by software and firmware (memory, computational power etc.).

A failure to achieve previous requirements may have a deep impact on product development cost or time.

Signal quality can be tested as outlined in the previous subsection. The other parameters can be evaluated through a theoretical analysis, hardware simulations or through prototypes.

Production cost is estimated considering the costs of

1. materials (electronic/mechanic components, PCB, packaging accessories)
2. direct man-hours for production (soldering through-hole technology (THT) components, firmware download, mechanical assembly, testing, packaging)
3. services required for production (surface mount assembly, logistics, purchasing etc.).

Materials cost is easily estimated through on-line catalogs. Big manufacturing companies may have lower prices using component distributors. This analysis is also useful for detecting components that are going to be no longer produced. Man-hours are estimated considering for example the number of THT components, the number of testing procedures and so on.

The previous two cost types may be greatly decreased by a proper design. By contrast, service costs are less influenced by design.

In addition to the previous list, requirements from market research are also essential to achieve forecast product sales. These requirements are usually focused on perceived performance that changes buying behavior of the target market. The fast development of a demonstrator using a GUI module may help in measuring customer satisfaction before going into implementation.

This section highlights that, for some design projects, it is more advisable to use an iterative and incremental life cycle driven by risk reduction/avoidance. Risks are mainly due to complexity in implementing project requirements that are given by ECG standards, market analysis and technical constraints. For this reason, risks has to be monitored continuously throughout the project.

1.17 The ECG Development Plan and Project Management

The first complex issue in a real product development is the compilation of the plan document – the system development plan (SDP). The SDP includes the developer's strategy and the implemented approaches for managing the development effort. The plan must include processes, tools and methods to be used, and it must also include the approach to be followed for the activities, project schedules, the organization and resources (MIL 498, 1994).

A document template for an SDP may be used so that all these issues are addressed before starting the design project. A template such as (DI-IPSC-81427A, 2000), although conceived for software, may be adapted for this purpose. In this section, the more general and applicable items of the template will be outlined for the ECG development to show a real example. For the sake of simplicity, simple references to sections in this book will be given when the SDP contents are already covered here. The sections of DI-IPSC-81427A will be referenced as: 'SDP XX. Title'.

SDP 1. Scope

Scope and system overview are outlined in Table 1.11.

SDP 2. Referenced documents

EU directive (EEC 2007) for European certification and selected applicable standards included in Table 1.14.

SDP 3. Overview of required work

This is outlined in Section 1.12. Requirements and constraints of the system are included in the general requirements outlined in Table 1.11, medical device standards (Table 1.14) and marketing research (Table 1.13). Requirements and constraints on project documentation stem from the medical device certification needs (EEC 2007).

SDP 4. Plans for performing general development activities

SDP 4.1 System development process

An iterative and incremental development process driven by risk reduction/avoidance is selected. The process is described in Section 1.16. After the system-subsystem specification and design description (see Figure 1.26), the builds included in Table 1.16 will be performed. Risks of the previous activities have been discussed in the previous section.

SDP 4.2 General plans for system development

SDP 4.2.1 System development methods

An iterative and incremental development method will be used, as shown in Figures 1.26 and 1.27. Project Planning and Control will be performed through a project management and

Table 1.16 Builds

Num. Build	Builds	Objective	Development activity
1	GUI validation	Validate and test operator interaction through a GUI simulator	GUI simulator, ECG signal generator see Figure 1.31
2	Software application first build	Validate and test first build of the overall software application	Develop modules (controller, acquisition, firmware simulator, visualize export modules) as in Figure 1.32
3	Digital hardware first build and preliminary integration	Validate and test demo board hardware, firmware and communication among demo board and software application	Develop firmware demo board with signal simulator and integrate hardware and software application
4	Analog hardware development on demo board	Validate and test analog frontend	Develop analog interface
5	Hardware on target platform	Validate and test the overall system	Develop target board

collaboration tool. Configuration management will be performed through a software configuration management tool for all the deliverables (documents and software). This tool will also help for the workflow of the change management process and the development environment management. Risk management will be performed according to IEC/EN 14971. Quality assurance will be performed according to the company's own processes. Further details are included in Section 1.13.1.

SDP 4.2.2 Standards for software-hardware products

Requirement management will be performed through a specific tool. The unified modeling language (UML) standard will be used for analysis and design. The usual guidelines for proper software coding will be used. Some components will be designed using RAD tools for software and firmware development. Demo boards will be used for rapid prototyping.

SDP 4.2.3 Reusable software products

Reusable software libraries will be selected, considering that the overall software is mission critical. The following criteria will be used for selection:

1. license for commercial use (see Section 1.8)
2. availability of source code
3. maturity of the software
4. fit with requirements, considering reuse problems outlined in Figure 1.18 and Section 1.6
5. software performance and resource utilization.

Candidate software must be evaluated through specific testing to evaluate performances and safety impact before usage.

SDP 4.2.4 Handling critical requirements

Critical requirements are mainly the 'essential requirements' contained in Annex I of the directive plus the requirements contained in the selected harmonized standards (see Section 1.14 and Table 1.14). These requirements will be recorded in the medical device's 'design dossier' as specified in the Medical Device Directive (EEC, 2007) Annex II.3.2 (c) and 4.2, Annex III.3, Annex VII.3, and Annex VIII.3.1 and 3.2. Validation and verification will also be included in the design dossier.

SDP 4.2.5 Computer hardware resource utilization

The hardware board has constrained availability of memory, processor power and resources. Over-utilization may negatively impact performance such as the time required to avoid ECG signal distortion.

Software running on a PC may also experience problems due to excessive use or misuse of resources (memory, CPU objects, I/O etc.) with critical side effects. The previous hardware resources will therefore be monitored at any release of software or hardware.

SDP 4.2.6 Recording rationale

The rationale for the 'key decisions' will be recorded in an electronic log file and included in specific documents when applicable.

SDP 5. Plans for performing detailed development activities

SDP 5.1 Project planning and oversight

The sequence of activities has been outlined in Figure 1.26. The detailed plan can easily be obtained, considering also the sequence of the specific tasks outlined in Section 1.16 to reduce the overall risks. At the end of each activity, a review is organized to accept the associated deliverables and to update the project plan.

SDP 5.2 Establishing a system development environment
The development environment includes

- hardware design tools: CAD for hardware design and simulation
- firmware design tools: integrated design environment IDE with compilers, operating systems and libraries etc.
- software design tools: IDE, design tools, requirement tools, operating systems, libraries etc.
- software for instrumentation: oscilloscopes, signal generators etc.

All the software and firmware required to design the product must be under configuration control, and relevant testing documents must include versions of the software used.

SDP 5.3 System requirements analysis

SDP 5.3.1 Analysis of user input
User input is included in Table 1.11.

SDP 5.3.2 Operational concept
Operational concepts are outlined in Section 1.11.

SDP 5.3.3 System requirements
System requirements derive from AAMI EC 11 standard and from applicable standards in Table 1.14. System requirements will be discussed in more detail in Chapter 2.

SDP 5.4 System design
System-wide decisions and system architectural design were outlined in Section 1.15.

SDP 5.5 Software requirements analysis
Software requirements will derive from the decomposition of system requirements into subsystem requirements, as outlined in Section 1.5.

SDP 5.6 Software design: computer software configuration item (CSCI)
Software architectural design and wide decisions are outlined in Section 1.15. The approach for software design is outlined in Section 1.5.

SDP 5.7–5.19
The sections from 5.7 to 5.19 of SDP are mainly related to testing, integration, qualification evaluation of software and system. Such paragraphs of DI-IPSC-81427A (2000) are beyond the scope of this book.

SDP 6. Schedules and activity network
This section contains the list of milestones, deliverables and their relationship. Readers may derive such elements from previous sections as an exercise.

SDP 7. Project organization and resources
This section includes the detailed organization of the project in terms of employed resources, roles and responsibilities and management of the project. Again, readers may derive such elements from previous sections as an exercise.

1.18 IPR and Reuse Strategy for the ECG

Patents may be powerful methods of obtaining a sustainable competitive advantage since they give the owner the right to exclude others from making, selling or using the claimed invention in those countries where the patent is valid. When developing a new product, it is therefore necessary to perform a thorough search for existing patents that could impede product sales. In addition, the patent analysis may reveal that the product under design has innovative aspects that are worth patenting.

Depending on the nature of the invention, there may be

1. 'utility patents' including inventions that operate in a new and useful manner
2. 'design patents' that relate to the invention's unique ornamental and aesthetic properties.

Medical devices and, in particular, the cardiovascular monitoring sector, use mainly the first type of patent. Some companies have also reverted to the second one: HP, for example, use a novel aesthetic in the ornamental design of a medical device (HP, 1995). This type of patent can hardly be considered a barrier for new entrants, since new entrants may easily change a device's design. By contrast, utility patents may represent potential barriers for new entrants.

Utility patents must have all these essential features. Patents must be new, non-obvious and useful or industrially applicable. An insight into such features is helpful to assess patents that may have an impact on an ECG design:

1. **Novelty requirement:** the invention must be original or must include something that has never been distributed before in the world. This means that if an invention has been published or sold in a specific country it can no longer be patented in any other. If the invention has been patented only for a specific country, it can be used but not patented in countries not covered by the original patent.
2. **Non-obvious requirement:** the invention must not be obvious to the people skilled in the area of the invention. The 'intrinsic' innovation, also called the 'inventive step', concerns the internal aspects of the invention, which must be non-obvious in order to be patented; it must represent progress, a step forward in the present state of the art. This issue is sometimes hard to define and is generally a source of litigation.
3. **Industrialization requirement:** only inventions that do not rely on personal skills but can be reproduced through industrial processes may be patented. More specifically, the following items may be patented:
   i. *processes* including conventional processes (e.g., the method for making plastics) and software processes

   ii. *machines* including conventional machines, those with moving parts (e.g., a telephone) and software machines
   iii. *manufactured products* inventions with non-moving parts (e.g., books)
   iv. *compositions of matter* (e.g., chemicals, alloys and pharmaceuticals)
   v. new uses of any of the above.
4. **Limited time validity requirement:** patents of this type expire within 20 years.

ECG is a well-consolidated technology whose first experiments date back to the 19th century, and its core technology has been available since 1942. Therefore, no patent technology is possible on the core technology due to the limited time validity requirement (20 years). Technical implementation schemes have also been published for more than 20 years (Webster, 1978, 2009). This means that patents on technical implementations of the core technology, if existing, will already have expired. For ECG development, patents may only concern specific technological improvements for solving problems where non-patentable solutions already exist.

In any case, care must be taken, since ECG patent analysis is quite complex. ECG/EKG is present in over 26,000 patents as shown by Google patent searches only into US patents (http://www.google.com/patents). The patent number reduces to 2000 when considering patents associated with using a PC. The 'PC ECG' related patents are only 26. The analysis of the abovementioned patent database confirms that there are no patents preventing the development and sale of the ECG or PC ECG when based on standard implementations. Obviously, the implementation must not contain any patented scheme. Since patents must fulfill novelty requirements, published material cannot be patented. Therefore, specific patents related to implementations may be replaced by non-patentable published implementations. In addition, patents older the 20 years counting from the filing date can be used (WTO, 1995).

Regarding open-source software, there are libraries that can be used in an ECG product; this is more true if they contain source code, so that integration is easier (see Chapter 6 for additional details). These libraries may be helpful for signal processing and data storage management. Other libraries may be used for general purpose functions (data compression, reporting etc.). Licenses have to be read carefully to avoid infringements.

A typical free software copyright is depicted below (see also GNU, 2007):

> 'This library is free software; you can redistribute it and/or modify it under the terms of the GNU Library General Public License as published by the Free Software Foundation; either version 2 of the License, or any later version. This library is distributed in the hope that it will be useful, but WITHOUT ANY WARRANTY; without even the implied warranty of MERCHANTABILITY or FITNESS FOR A PARTICULAR PURPOSE. See the GNU Library General Public License for more details'.

The use of free software may be sometime critical since proper product quality and safety may not be achieved and integration time may be considerable. A non-free library may also be a cost-effective alternative. Figure 1.18 outlines other technical concerns: when libraries do not fit the required functions properly, the required integration time may encourage developing completely new software.

References

AHA (2011), American Heart Association, Forecasting the Future of Cardiovascular Disease in the United States.

AAMI (1991), Association for the Advancement of Medical Instrumentation, American National Standard, Diagnostic Electrocardiographic Devices, ANSI/AAMI EC11-1991.

Anast D. G. (2001), Marketing and regulatory requirements in some emerging medical device markets, *Biomedical Market Newsletter*.

Bailetti A. J. and Litva P. F. (1995), Integrating customer requirements into product designs, Journal of Product Innovation Management, vol. 12, no. 1, pp. 3–15 (13), January 1995.

Becchetti C. (2005), A marketing plan for a start-up company selling a low-cost, hi-tech, high diffusion biomedical device, MBA Thesis, Warwick Business School.

Becchetti C. (2001), What are the most urgent research problems of component-based software engineering for resource-constraint systems? IEEE Proceedings of Sixth International Workshop on Object-Oriented Real-Time Dependable Systems (WORDS '01).

Beck K. *et al.* (2001), Manifesto for Agile Software Development, Agile Alliance, http://agilemanifesto.org/ Accessed May 2011.

Boehm B. W. (1981), Software Engineering Economics, Prentice-Hall, Englewood Cliffs, NJ.

Brooks F. (1987), No Silver Bullet – Essence and Accidents of Software Engineering, IEEE Computer.

Brown R. (1991), Managing the S curves of innovation, Journal of Marketing Management.

Bull Survey (1998), Spikes Cavell & Co.

Bygrave W. (2004), The Entrepreneurial process, in The Portable MBA in Entrepreneurship, Bygrave W. editor, John Wiley and Sons, 2004.

CB (2004), Centre for the Promotion of Imports from developing countries (CB), Medical Devices and Medical Disposables, EU Market Survey 2004.

CMS (2008), Center for Medicare and Medicaid Services, selecting a development approach, 2008.

Davis, A. M. (1993), Software Requirements: Objects, Functions, and States. Prentice-Hall, Englewood Cliffs, NJ.

De Brentani U. (1989), Success and Failure in New Industrial Services, Journal of Product Innovation Management, no. 6, pp. 239–258.

DI-IPSC-81427a (2000), Data Item Description: Software Development Plan (SDP), US Department of Defense.

ECRI (2012), Electrocardiographs, Multichannel; Interpretive; Single-Channel, Healthcare Product Comparison System (HPCS) ECRI Institute.

EEC (2007), Council Directive 93/42/EEC concerning medical devices amended by 2007/47/EC.

EEC (2009), 2009/C 293/03: Commission communication in the framework of the implementation of Council Directive 93/42/EEC of 14 June (1993) concerning medical devices.

FDA (1998), Diagnostic ECG Guidance (Including Non-Alarming ST Segment Measurement), US Department of Health and Human Services Food and Drug Administration Center for Devices and Radiological Health. Guidance for Industry.

FDA (2005), Guidance for the Content of premarket submission for Software Contained in Medical device, Guidance for Industry and FDA Staff, US Department of Health and Human Services FDA.

Frost and Sullivan (2003), Report on World Cardiovascular Monitoring Equipment Market.

GNU (2007), GNU General Public License: http://www.gnu.org/copyleft/gpl.html

GNU (2011), GNU operating system licenses: http://www.gnu.org/licenses/

Handy C. (1998), Understanding Organizations, Penguin Business Library.

HP (1995), Hewlett Packard, The ornamental design for a cardiograph, US Patent DES 359,651, 1995, available at US Patent and Trademark Office: http://www.uspto.gov/.

IEC 60601-3-2, ed 1 (1999) Medical electrical equipment – Part 3-2: Particular requirements for the essential performance of recording and analysing single channel and multichannel electrocardiograph.

Johnson G. and Scholes K (1997), Exploring corporate strategy, Prentice Hall, 1997.

Kelleher B. (2003), The seven deadly sins of medical device development. Medical Device & Industry Magazine 2003: http://www.devicelink.com/mddi/archive/01/09/002.html

Keller S. (2005), US National Library of Medicine Medline, Medical Encyclopaedia, ECG.

Kotler P. *et al.* (2002) Principles of Marketing, Prentice Hall, 2002.

KPMG, Canada Survey, 1997, http://www.parthenon.uk.com/project-failure-kpmg.htm, accessed January 2013.

Larman C. and Basili V. R. (2003), Iterative and Incremental Development: A Brief History. Computer **36** (6): 47–56.

McSharry P. E., Clifford G. D. and Tarassenko L. (2003), A Dynamical Model for Generating Synthetic Electrocardiogram Signals, IEEE Trans. Biomed. Eng.

McSharry P. E. *et al.* (2011), ECGSYN – A Realistic ECG Waveform Generator, http://www.physionet.org/physiotools/ecgsyn/, accessed 2012.

Meredith J. R. *et al.* (2009), Project Management: A Managerial Approach, John Wiley & Sons.

MIL, 498 Guide (1994), MIL-STD-498 Overview and Tailoring Guidebook.

MIL-HDBK-61A (2001), Configuration Management Guidance 7, DoD USA National Information Systems Security Glossary, ANSI/EIA-649-1.

NASA (2010), Error Cost Escalation Through the Project Life Cycle, NASA Johnson Space Center.

Parnas D. *et al.* (1986), A rational design process: How and why to fake it, IEEE Trans. Software Engineering.

Pidd M. (2002), Tools for Thinking – Modelling in Management Science John Wiley & Sons.

PMI, Standards Committee (1996), A Guide to the Project Management Body of Knowledge.

Sawyer D. (2005), An Introduction to Human Factors in Medical Devices, US Dept. of Health and Human Services Public Health Service FDA.

Standish Group (1995), The Chaos Report.

Stevens G. Myers, L. (1974), Constantine, Structured Design, IBM Systems Journal, **13** (2), 115–139, 1974.

Webster J. G. (1978), Medical instrumentation: application and design, Boston: Houghton Mifflin.

Webster J. G. (2009), Medical instrumentation: application and design, John Wiley & Sons.

WHO (2011a), World Health Organization, Global atlas on cardiovascular disease prevention and control.

WHO (2011b), World Health Organization, The top 10 causes of death, Fact sheet N°310 (2008) updated June 2011.

Wiki (2012): List of Rapid Application Development tools, 2012.

WTO (1995) Trips agreement on trade-related aspects of intellectual property rights, World Trade Organization.

2

Concepts and Requirements

Incomplete requirements is the main reason cited for failure

Chapter Organization

Medical instruments rely on specific physiological properties to infer the health condition of specific parts of the human body. The first design step is therefore focused on understanding on which **concept** medical instruments rely on to assess health.

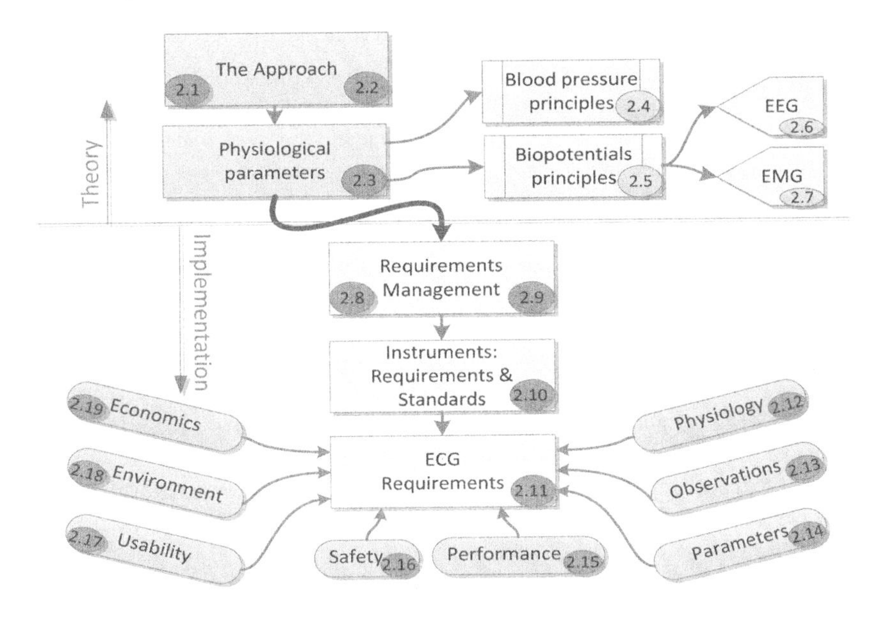

Figure 2.1 Chapter 2 structure.

Medical Instrument Design and Development: from Requirements to Market Placements, First Edition.
Claudio Becchetti and Alessandro Neri.

The second step implies the definition of **what** the medical instrument has to do so that all the manufactured products are *safe* and *effective*. The '**whats**' that a product is intended to do are formalized through **requirements**. This chapter addresses these two steps: concepts and requirements.

The chapter will start from a generic instrumentation model to derive the specific equipment. Referring to Figure 2.1, the concepts for a generic device are outlined in Sections 2.1, 2.2 and 2.3, where medical instruments are introduced, with a simplified model and the associated physiological output parameters used to infer health condition. Sections 2.4 and 2.5 focus these concepts for two main areas: blood pressure and biopotential based devices. In particular, Sections 2.6 and 2.7 will discuss electroencephalography (EEG) and electromyography (EMG), devices that measure one-dimensional (1-D) signals over time. These devices are also a reference for more complex 2-D, 3-D, 4-D examples, such as the imaging devices.

These concepts will be translated into requirements in the **implementation** section, where the ElectroCardioGram (ECG or EKG) will be addressed. The ECG is a good textbook example since it encompasses the main design components of most of the medical devices. In addition, the ECG is also sufficiently simple to be completely described in a book. Sections 2.8 and 2.9 introduce **requirements** management which is one of the most critical design step. While an error in requirements can be fixed easily when products are at requirement stage, if such errors are discovered at later stages they require exponentially more effort to be corrected; that is, the cost to fix requirement errors is roughly exponential to the time passed from the project design start. This explains why poorly defined requirements and specifications are a main reason for medical device development project failures. *Inadequate requirements and specifications account for more than two-thirds of overall failures* (Kelleher, 2003).

Section 2.11 will discuss the ECG requirements referring to a specific marketed product: the **Gamma Cardio CG**. The owner (the company *Gamma Cardio Soft* www.GammaCardioSoft.it) has disclosed the Gamma Cardio CG details in this book with the aim of increasing the diffusion of ECG know-how, raising the worldwide number of devices and reducing the associated pathological impact. Cardiovascular diseases are the prime cause of death worldwide, and the ECG is the main medical instrument for diagnostic practice.

Sections 2.12–2.19 go into detail, describing the ECG requirements associated to the eight areas shown in Figure 2.1. Through the comparison of theory and implementation, readers will be able to see how concept models are implemented in one of the most important area (cardiology) in terms of deaths and specifically in one of the more significant instruments (the ECG).

At the end of this chapter readers will be able to:

- outline the link: pathology → measurement → instrumentation → method for the medical devices, the pathologies and the medical instruments
- describe the general framework and particularize the scheme for a major medical device
- appreciate the essential role of requirements through a real marketed example.

We will show that requirements are not just technical, but that many other areas impact the design of a medical instrument.

Medical instruments require some background to understand their use and the associated detected pathologies. In this chapter, we will outline this background especially in the implementation part for the ECG. For further investigation, the reader may consider a manual of medicine such as (Harrison, 2011).

Part I: Theory

2.1 Introduction

The process of healthcare delivery includes prevention, diagnosis, therapy and rehabilitation. According to (Grimes, 2004), there are four distinct ways through which **technology** takes place in **healthcare delivery**:

- At *level 0*, both diagnosis and therapy are virtually unassisted by technology: the diagnosis is drawn by direct observation and the therapy may perhaps be provided manually.
- At *level 1*, technology assists in the diagnosis, that is, it provides clinically relevant parameters to help the practitioner determine the presence of a pathology. In addition, technology may help treatment, for example, by improving the effectiveness of the practitioner's therapeutic intervention.
- At *level 2*, technology intervenes at the decision/interpretation level, that is, by proposing a possible diagnosis that can be used by the practitioner. This can be done not only through the parameters collected in the diagnosis stage, but also through other patient-related information collected and interpreted by the information processing system.
- At *level 3*, the technology integrates all the information gathered from the patient, it interprets this information (e.g., through expert systems) and delivers the therapy based on the results of the interpretation. At this level, we may say that biomedical technology is the master in the healthcare delivery process.

It is worth noting that some current technologies already operate at level 3; for example, implantable cardioverter defibrillators (ICD) can automatically perform defibrillation procedures according to the heart condition. The ICD monitors heart conditions, and when irregular rhythms are detected, electrical pulses are generated to revert the heart to a normal rhythm. The device is able to increase pulse energy to fulfill the defibrillation effect.

Independently from the level at which a healthcare service is delivered, **medical devices** may be organized into three different **groups**:

- ***diagnostic equipment***, where information flows from the patient to the technology
- ***therapeutic equipment***, where the flows is the other way around – the treatment is performed by the machine on the patient (this includes rehabilitation equipment)
- ***information processing systems***, where the information gathered is interpreted to assist in the clinical decision process.

These groups are becoming less distinct by the year. For example, most therapeutic equipment contains decision support sections, and diagnostic elements as well, and many therapeutic appliances modulate the treatment based on information gathered from the patient as in the implantable cardioverter defibrillators.

In this book we will focus on diagnostic equipment that is represented as a **device model** that performs the following tasks:

1. sense the patient information
2. transform this information into an electrical signal
3. remove artifacts, and sources not meaningful for diagnostic purposes, from the input signal
4. extract parameters and indexes that have a physiological meaning and can be profitably interpreted by the clinician practitioners to infer health condition.

Medical practice shows that instrumentation products that did not take into account the '**doctor–patient in the loop**' often end up as a market failure. So, policies are recommending that innovative products increasingly be patient/user centered and demand driven, focusing on healthcare needs (CEU, 2011). Medical device requirements must encompass how the instrumentation is used by doctors and technicians, and perceived by patients, because doctors, technicians and patients are essential components of the diagnostic process.

A reasonable approach is therefore

1. define the medically relevant problem
2. define the measures to detect the problem
3. define the instrument to detect the problem
4. define how the healthcare service that uses the instrument is organized, taking into account the patient/doctor in the loop.

This approach will be further addressed in the implementation part within the Gamma Cardio CG requirements.

2.2 The Medical Instrumentation Approach

When dealing with generic diagnostic equipment, the ultimate goal is to reveal and quantify the extent of a pathology based on the flow of information that comes from the patient, and it is gathered via the instrumentation at hand (Von Maltzahn, 2006).

Note that the information flows from the patient, either as the response of the body to an excitation source coming from an external device, or as the result of an internal source of information autonomously generated by the body. In measurement theory, this source is generally called the ***measurand***, that is the information that needs to be captured and isolated by the instrumentation with the final goal of diagnosing the presence or absence of a pathology, and quantifying its extent.

As shown in Figure 2.2, this type of information can originate from a specific organ or anatomical area, or it can be the result of a function linking a number of different organs. It can also be accessible directly on the body surface (such as in the case of biopotentials recorded non-invasively), it can be captured by energy emanating from the body area (as in the case of medical imaging), picked up internally (as in the case of e.g. invasive blood pressure sensors) or originate from a tissue specimen extricated from the body (e.g., biopsies).

The measurand is usually immersed in a rather **noisy** and **hostile environment**, comprising a number of unwanted sources that are either directly associated with the non-ideality of

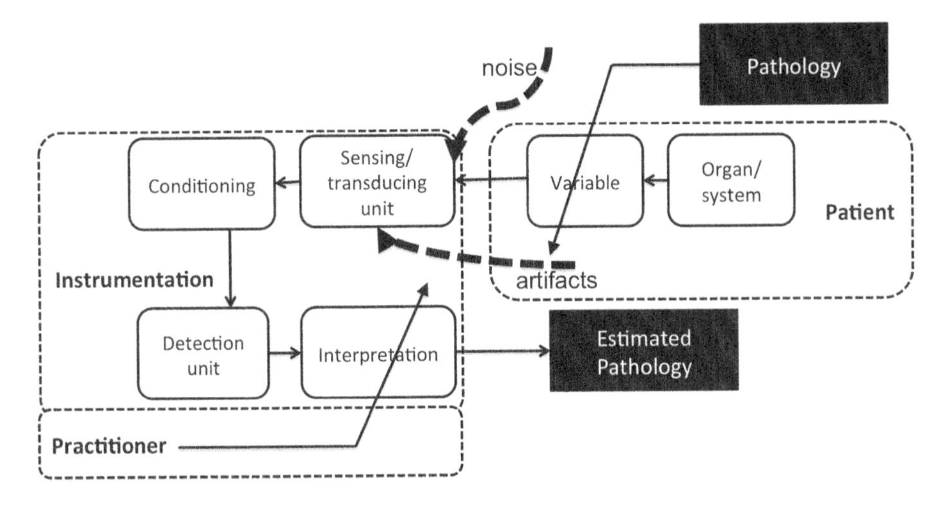

Figure 2.2 Information flow from patient to instrument.

the sensing area itself, or related to the body, which can generate a number of artifacts that can be unintentionally gathered by the instrument. For example, in the ECG, the skeletal muscles produce measurable electricity that is added as an artifact to the wanted signal (i.e., heart activity). In addition, noise coming from the environment may be even 3-4 times greater than the signal to be detected as in the ECG case. As a result, the instrument will include various hardware, firmware and software components to minimize unwanted signals in the decision process. At the end of this process, a diagnostic instrument will solve a detection theory problem, where the signal under investigation can be expressed either in the form of a binary variable corresponding to the presence or absence of a certain pathology/event, or as a numeric outcome associated with the extent of that pathology/event.

The process can then be summarized as follows:

1. The presence of a **pathology** determines a change in the function or the anatomy of an organ in a specific domain (e.g., size, voltage amplitude, shape, pressure).
2. The instrument measures a variable in a different domain.
3. The modifications to this variable are projected into this new domain, where variables can be corrupted by the presence of noise and artifacts.
4. Through the help of hardware devices and software solutions this representation domain helps clinicians answer the questions of whether the pathology is present and to what extent.

In the **design of a biomedical instrument**, the process can basically be reversed, in the sense that the input comes from a clinical requirement. Once the pathology has been defined, its manifestation in the domain at either the anatomical or the physiological level may correspond to a change that needs to be captured. The biomedical instrument will therefore be designed including a number of variables at least equal to or better than what is required to detect the change with sufficient accuracy.

Table 2.1 EEG recommendation adapted from ECRI

Basic EEG monitors	Specifications
Applications	OR, CCU, epilepsy monitoring, diagnostic EEG
Number of channels	8
EEG data resolution, bits	12
Sampling rate, Hz	>200
Freeze memory	2 hours
Electrode-impedance check	automatic
Noise, μV	according to standards
Common mode rejection ratio (CMRR – see Chapter 5 for details), dB	100 @ 60 Hz
Preamplifier input impedance, MΩ	10
Evoked potentials	according to standards
Spike and seizure detection	preferred
Artifact detection	yes

In the case of a biomedical instrument corresponding to an established way of capturing data, these requirements may be determined by the clinical specialist and their recommendations. For instance, most of the medical devices grouped according to the Global Medical Device Nomenclature (GMDN) refer to recommendations given by the ECRI Institute (ECRI, 2012a). Table 2.1 shows an example of EEG recommendations.

These requirements have been devised by the ECRI Institute, which is an independent not-for-profit organization designated as a Collaborating Center of the World Health Organization (WHO).

These requirements usually entail a group of technical specifications that can be summarized as follows:

- specifications related to the quality of captured data (e.g., sensitivity, frequency range, resolution, intrinsic noise, linearity, reliability)
- specifications related to the processing and visualization of the data
- specifications related to the usability of the system, and to the range of choices that the operator can handle to optimize the recording in the specific case
- general specifications, such as size, weight, power requirements.

When, instead, the medical instrument to be designed is not included in the range of consolidated state-of-the-art devices, specifications needs to be heuristically determined by the designer and then validated according to the applicable regulatory framework. In this case, it must be considered that each source of **uncertainty** originating from the instrument will be propagated through all sections of the medical system itself in a way that needs to be kept under control, because it could result in a decrease in the diagnostic performance of the instrument itself. For example, as discussed in Chapter 5, the amount of amplification that is needed to allow a viable representation of a biopotential, is generally split into a number of different cascaded amplifiers, each of them adding an inherent amount of noise, expressed through the noise factor parameter. The amount of noise that is generated in the very first amplifier will be amplified by all the others amplifiers, thus resulting in a propagation of the uncertainty, and this needs to be kept under control, by choosing an appropriately low-noise amplifier for the first stage.

The **diagnostic performance** can be estimated by specifically determining two factors extracted from the parameter (or set of parameters) that the instrument makes available to distinguish between the presence or absence of a pathology:

1. the ***sensitivity***, cited as *SENS*, and defined as the probability of detecting an event $\Pr\{Ev\_det/Ev^{-}\}$ with the instrument (in this specific case, a pathology), when the pathology is present
2. the ***specificity***, cited as *SPEC*, and defined as the probability of revealing the absence of an event, when it is not present

$$\begin{aligned} SENS &= \Pr\{Ev\_det/Ev^{+}\} \\ SPEC &= 1 - \Pr\{Ev\_det/Ev^{-}\} \end{aligned} \tag{2.1}$$

We recall that $\Pr\{Ev\_det/Ev^{-}\}$ is the probability of detecting an event conditioned that the event is present. Considering the Bayes's Theorem (see eq. 2.5), such probability can be expressed as:

$$\Pr\{Ev\_det/Ev^{-}\} = \Pr\{Ev\_det,\ Ev^{-}\}/\Pr\{Ev^{-}\} \tag{2.2}$$

where $\Pr\{Ev\_det,\ Ev^{-}\}$ is the joint probability that the event occurs and that it is detected, and $\Pr\{Ev\_\}$ is the probability that the event *Ev* occurs.

In order to estimate these two parameters, a decision criterion needs to be put into practice. This usually implies the determination of a cutoff thresold value *th* that separates the supposed presence of a pathology from its absence. By changing this value, both the sensitivity and the specificity change. The function described by the point ($x = 1 - SPEC(th)$, $y = SENS(th)$) when varying the cutoff value *th*, graphically describes of the diagnostic power of the instrument.

In detection theory, this function is referred as the **Receiver Operating Characteristic** (ROC). The ROC of an ideal instrument is represented by the point $x = 0$, $y = 1$. Figure 2.3 shows an example of an ROC function for a parameter extracted with a medical device where ($1 - SPEC(th)$, $SENS(th)$) are reported on the (x,y) axes varying the value of *th*.

More specifically, an overall measure of the diagnostic performance can be determined by calculating the area under the ROC curve. If the cutoff value is set, a point measure of diagnostic ability can be obtained by calculating the positive likelihood ratio, LR^{+}, that is the

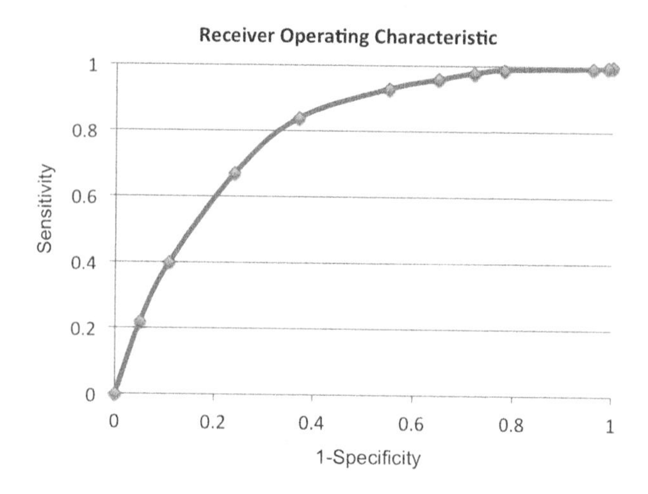

Figure 2.3 Typical receiver operating characteristic (ROC) curve.

ratio between the sensitivity of the system and the false positive rate for that specific cutoff value ($LR+ = \infty$ for ideal instruments):

$$LR^+ = \frac{SENS}{1 - SPEC} = \frac{\Pr\{Ev\_det/Ev^+\}}{1 - \Pr\{Ev\_det/Ev^-\}} \tag{2.3}$$

In **probability theory**, *SENS* corresponds to the conditional probability of revealing a specific pathology (Ev_det) given that the pathology is present (Ev^+), whereas $1 - SPEC$ corresponds to the conditional probability of detecting the same pathology provided that the pathology is not present (Ev^-).

Then, in order to completely characterize the diagnostic power of an instrument, it may be worthwhile combining these parameters with the information coming from the prevalence (*PREV*) of that specific pathology (that is, the probability that the pathology occurs):

$$PREV = \Pr\{Ev^+\} \tag{2.4}$$

Introducing Bayes' Theorem:

$$\Pr\{Ev^+/Ev\_det\} = \frac{\Pr\{Ev\_det/Ev^+\} \times \Pr\{Ev^+\}}{\Pr\{Ev\_det\}} \tag{2.5}$$

we can estimate the **instrument diagnostic power** $\Pr\{Ev^+/Ev\_det\}$ by calculating the posterior probability, or post-test probability, which is the probability of having that specific pathology, given the detection of this pathology by the instrumentation. This can be calculated once the values of *SENS*, *SPEC* and *PREV* are known.

By substituting into the denominator, according to the total probability theorem, we have the following $\Pr\{Ev\_det\} = \Pr\{Ev\_det, Ev^+\} + \Pr\{Ev\_det, Ev^-\}$:

$$\Pr\{Ev^+/Ev\_det\} = \frac{\Pr\{Ev\_det/Ev^+\} \times \Pr\{Ev^+\}}{\Pr\{Ev\_det/Ev^+\} \times \Pr\{Ev^+\} + \Pr\{Ev\_det/Ev^-\} \times \Pr\{Ev^-\}} \quad (2.6)$$

$$\Pr\{Ev^+/Ev\_det\} = \frac{SENS \times PREV}{SENS \times PREV + (1 - SENS)(1 - PREV)} \quad (2.7)$$

Equation (2.7) links the diagnostic power of an instrument/system/service, once we know the sensitivity and specificity of the instrument in detecting the pathology under examination, and the prevalence of the pathology. Note that these values can be estimated analytically if the distribution of the parameter values for the population under examinations are known Alternatively, these values can be experimentally evaluated on a subsample of the population that undergoes the diagnostic process with the instrument under test, with information regarding the presence or absence of the pathology that is available to the examiners (Campbell, 2007).

Given the experimental nature of these tests, it is common to have a heuristic more than an analytical approach, based on the results coming from a population sample. Guidelines for the assessment of the diagnostic accuracy of biomedical instrumentation are reported in different branches of medicine, including systems for biopotential recording, medical imaging and laboratory equipment.

The **diagnostic power** defined in equation (2.7) is one of the specifications that need to be addressed to determine if a medical device is appropriate for the detection of a specific pathology. Other elements may include the added risk in performing the examination, the required test time and the personnel training for the use of the specific device. It is nonetheless assumed that, *by using the statistical measures reported in this paragraph, one can assess the feasibility of using a specific medical instrumentation to detect a pathology* (or class of pathologies), as compared to the established means.

2.3 Extraction of Physiological Parameters

In general terms, we may assume that the presence of pathology may result in a modification of the function or the anatomy of an organ that can be observed or quantified in some way. This is the underlying hypothesis that allows us to diagnose a specific pathology using a medical instrument that is suitable for detecting the modifications associated with the desease. This modification can either be directly observed through the use of the technology at hand, or indirectly estimated from a number of measurable units and with the availability of some modeling tools that take into account the relation between the observable variable (and its modifications), and the inner variable whose modifications are directly related to the pathology. For example, if one refers to an anatomical modification of the lungs, associated with the presence of a malignancy in the lung, this can be directly observed by using specific diagnostic imaging devices, from an X-ray scan to more sophisticated systems. Alternatively, the modifications of the respiratory functions driven by the malignancy can be indirectly estimated through the use of a pulmonary function analyzer, able to gather waveforms associated with the inspiration and expiration of the breath.

The quantification of these modifications is the main objective of a medical instrument. Note that this transformation may involve

- point parameters (e.g., blood pH)
- 1-D waveforms, as a function of space or time (e.g., EEG, ECG)
- 2-D waveforms, as a function of 2-D space, or changes over time of a 1-D space variable (e.g., an X-ray image)
- 3-D waveforms, which depict the behavior of a variable in 3-D space, or the changes over time of a 2-D projection (e.g., computer tomography)
- 4-D waveforms, which describe the behavior over time of a variable represented in 3-D space (e.g., time-resolved cardiac magnetic resonant imaging).

Regardless of the dimensionality of the variable to be measured, the relation with the presence or absence of a pathology is usually associated with a set of parameters. Such parameters are extracted

- from the measured variables: relevant points in the waveforms, such as in the case of ECG waveforms, where amplitudes, distances over time and ranges are used to determine the presence of a pathology,
- from ensemble parameters, extracted with the use of some statistical method, and represented in the domains of interest, such as in the case of quantitative EEG analysis, where second-order moments are calculated from the raw EEG data and classified in the frequency domain, and
- from some simplified function calculated from the raw data, when it is deemed easier to gather data in the simplified domain than in the original one, such as in the case of the quantification of apnea based on the bispectral index (BIS), calculated to enlighten phase couplings in EEG rhythms.

All these physiological parameters range in a variety of possible values, and the medical instruments have to be designed so as to optimally capture variations within these ranges. A non-exhaustive list of amplitude and frequency ranges of the 1-D waveform variables that are used to capture the physiological parameters is given in Table 2.2.

Most variables are physiological representations of phenomena that are captured also in non-medical contexts (e.g., pressures, flows, potentials, concentrations). Apart from regulatory issues, the main difference with variables captured in medical applications resides in the rather low amplitude ranges, and generally, but not always, smaller frequency ranges usually encompassing DC and close-to-DC components. The rather low values in the amplitude ranges, combined with the difficult accessibility of the parameters to be monitored and the inherent presence of noise and physiological artifacts, make the design of medical instrumentation a rather challenging task. Moreover, the amount of predictability and determinism in physiological waveforms varies depending on the application; the presence of statistical variability makes it necessary to use some statistical approaches.

Table 2.2 Physiological 1-D parameters characteristics

Physiologic parameter/method	Amplitude range (unity values)	Maximum observable frequency (Hz)
Blood flow	0–300 (ml/s)	20
Blood pressure, invasive, arterial circulation	10–400 (mm Hg)	50
Blood pressure, non-invasive, arterial circulation	25–400 (mm Hg)	50
Blood pressure, non-invasive, venous circulation	0–80 (mm Hg)	50
Cardiac output (flow)	2–30 (l/min)	20
Electrocardiography (ECG)	0.1–5.0 (mV)	300
Electrocorticography (ECoG)	0.01–3.0 (mV)	150
Electroencephalography (EEG)	5–300 (μV)	120
Electromyography, invasive (EMG)	0.1–5.0 (mV)	15000
Electromyography, non-invasive (surface EMG)	0.005–2.0 (mV)	500
Electrooculography (EOG)	0.02–2.0 (mV)	50
Electroretinography (ERG)	0.01–1.0 (V)	50
Membrane action potential, intracellular recordings (AP)	0.01–100 (mV)	30000
Nerve conduction studies (NC)	0.01–10 (mV)	10000
Respiratory function, rate	1–60 (breaths/min)	10

When dealing with medical instruments for diagnostic purposes, there are different **classifications**:

- Each medical instrument can be classified according to the **physical quantity** that is measured. In such cases, we will distinguish between the instrumentation that senses electrical potentials, and devices that convert into potentials a kinematic or dynamic property, such as displacement, flow, force or pressure – or even temperature or concentration.
- Alternatively, medical instruments can be divided according to the '**district**' (i.e., the body area) that is monitored at different scale levels (e.g., for cardiovascular systems: organ level, muscle level, coronary level, cellular level). At organ level, we may consider the nervous system, the digestive system and the musculoskeletal system. The classification may also be based on the **medical specialty**: we will have medical instruments for cardiology, for nephrology, for diagnostic imaging, for rheumatology and so on. Apart from a limited number of specialties that by definition include monitoring multiple organs, this classification is similar to the one based on the district to be monitored.

We will use the first classification based on physical quantity when we focus on the principle of operation of each instrument (such as in the case of this book), since it makes sense that systems that monitor the same quantity in different districts share the same functional design. We will instead use the second classification if we are interested in describing the variables that can be measured from each district: in this case, instruments accessing the same organ will share the same objective, that is the monitoring of some function, by way of a number of physiological parameters.

Based on the physical quantity to be measured, diagnostic medical instruments may grouped as those for

- transduction of forces, flows and pressures
- extracting parameters from biopotentials
- extracting concentrations and other chemical properties
- recording medical images.

In the following we will focus on the first two items since these 1-D instruments are representative of the more complex 2-D and 3-D devices. The implementation part of this chapter will then focus on a specific instance of the 1-D group – the ECG.

2.4 Pressure and Flow

The human body is able to generate transport phenomena, which concern the exchange of mass, energy or momentum between systems. In biomedical engineering, the most studied transport phenomena are thermoregulation, ventilation, perfusion and microfluidics.

With specific reference to perfusion, when the heart exerts its pumping action on the blood, this allows the transfer of delivery of nutritive arterial blood to a capillary bed in the biological tissue. Measuring the amount of blood flow is thus necessary to determine if and to what extent this transport phenomenon is within the correct physiological range, or deviates from it as the result of a number of different anomalies. This transport phenomenon is triggered by the alternating pumping action of the heart, which is also responsible for the variations of pressure that are exerted on the blood vessel walls, and that propagate along the peripheral vascular system. The waveform associated with the pressure changes is generally referred to as **blood pressure**; this term usually refers to the arterial pressure of the systemic circulation.

The measurement of blood pressure is one of the most common medical procedures in both clinical practice and surgical environments. This procedure is executed to assist the physician in determining the functionality of the cardiovascular system.

From an simplified engineering approach, the core of the **cardiovascular system** is the heart, which can be considered as a double blood pump composed of two different stations, each one composed of two chambers, synchronized by a 'conduction control system'.

For each pumping station, the two upper chambers (right and left atria) are synchronized, and their contraction is almost simultaneous. The two lower chambers (right and left ventricles) have a similar behavior. Figure 2.4 shows that the pumping action of the heart is 'double' since there is first an atrial contraction and then a ventricular contraction.

The **right atrium** receives the deoxygenated blood from the body through the two venae cavae and pumps it to the right ventricle through the tricuspid valve during the atrial

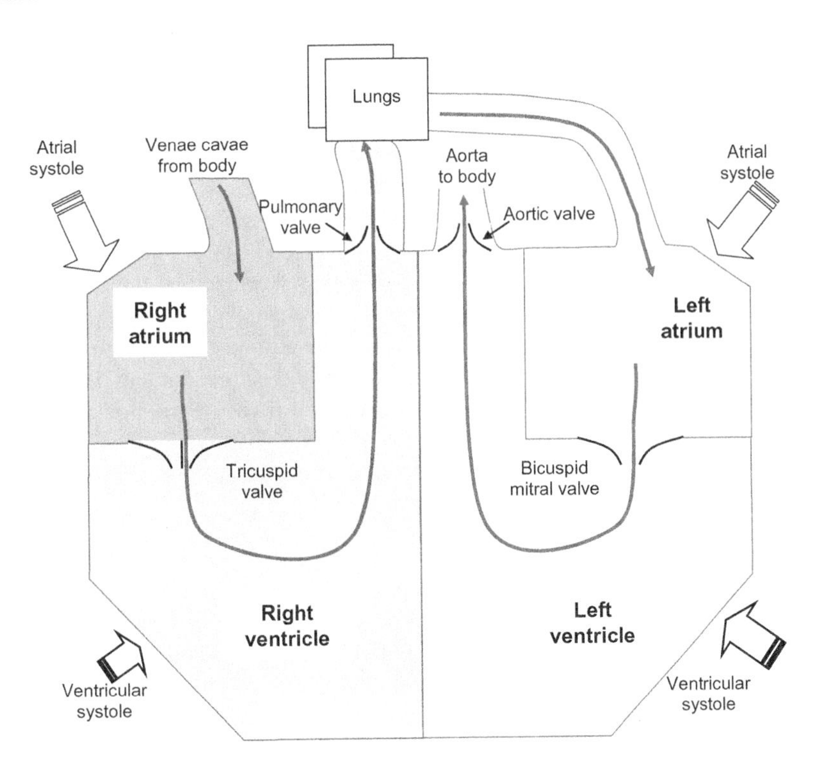

Figure 2.4 Schematic representation of the heart blood circulation.

contraction (i.e., atrial systole). The **right ventricle** contraction (i.e., ventricular systole) flushes blood forward into the lungs. The two unidirectional valves of the right heart part avoid inverting circulation during contraction. Blood is then oxygenated in the **lungs** and it is collected in the **left atrium.** The atrium's contraction moves blood into **left ventricle**. Again valves impede blood circulation in the reverse sense. Finally, the ventricular contraction pulls oxygenated blood into the body through the aorta. The blood then travels through arteries, arterioles and capillaries, thus allowing nutritive material to be transferred to the peripheral organs. The venous blood will then travel through venules and veins coming to the two venae cavae, which merge on entering the right atrium through the pulmonary valve.

Even from this simplified scheme, some possible failures can be envisaged. For instance, the reduced contractility of the heart will result in a reduced pumping power; problems in valve functionality may determine pathological changes of the cardiac output. The circulatory system connected to the heart may also have an impact on pumping efficiency: the vessel lumen reduction determines an increased systemic resistance to fluid circulation. Increased blood viscosity may cause impairments in flow; changes in vessel elasticity may lead to changes of circulation efficiency. These conditions may require extra pumping power and a raised pressure level in the circulatory circuit that in turn may affect heart functionality in the long term.

As pointed out before, the structure of each station is composed of two different chambers: the atria and the ventricles. The atrium accepts the returning blood, and it may be considered as a filling chamber, whose limited contraction provides the power needed to transfer the blood mass to the ventricle. This chamber provides the blood with acceleration as a result of a strong contraction; this acceleration allows the pumped blood to reach the peripheral organs. The difference in contractility between the atria and the ventricles is the result of different muscular thickness and physical characteristics.

This different role for the four chambers suggests other possible insights. For example, the system may work (in degraded conditions) even when the atria are not working properly. This may happen when atrial contraction is no longer synchronized with the ventricles (e.g., when the bundle of His, pertaining to the heart control system, is completely interrupted). In this last case, the ventricles start beating at their own rate allowing for blood circulation.

In the following, we will refer first to blood pressure indicators that can be used to assess the functionality of the circulatory system. Then, we will focus on the flow indicators. The list of indicators is not exhaustive; we will use the definition of some of these indicators as a paradigm to represent the connection between a dysfunction and the ability of diagnostic equipment to capture and quantify the extent of this dysfunction.

2.4.1 Blood Pressure

From a clinical perspective, it is clear that one of the variables of interest in regards to the circulatory system is the pressure. The blood pressure is defined as the pressure exerted by the circulating blood on the vessel walls of the entire system, and it is clinically considered as one of the main significant vital signs.

Without losing generality, when using the term blood pressure with no additional specification, we usually refer to the arterial pressure that is exerted onto the arteries in the systemic circulation.

Since the **pumping action** of the heart is an alternating mechanism that interleaves contractions with silent periods, the value of the blood pressure is not constant over time, but follows a pseudo-periodic behavior with alternations between different values. The silent interleaving period is called diastole while the contraction is called systole.

As can be seen in Figure 2.5, the pressure exerted on the aorta follows an alternating pattern. The pressure waveform that is generated alternates between the maximum value measured during the systole (P_s), and a minimum value exerted at the end of the

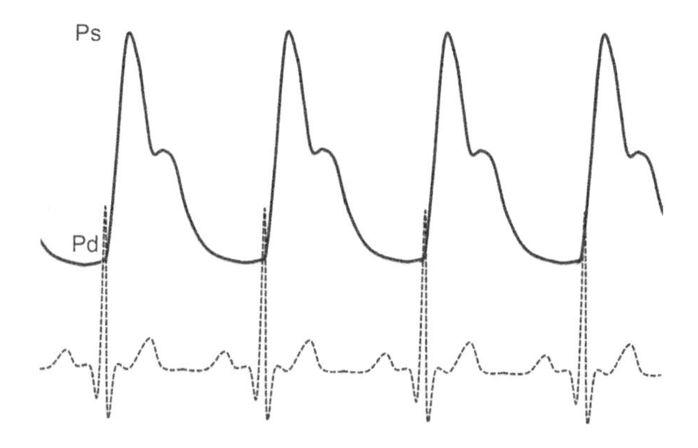

Figure 2.5 Pressure and ECG timings.

diastole (P_d). Although the pressure felt by the ventricle and the aorta is similar during systole, at the end of the systole and during all of the diastolic period they diverge. The ventricle wall pressure tends to zero right after the systole, as the result of an empty chamber, while the aortic pressure is always positive, as the result of blood being always present in the systemic circulation and the aortic valve.

In clinical practice, pressure is usually expressed in mm Hg, or millimeters of mercury, which is a unit not included in the International System of Units (SI). The unit is defined as the pressure exerted by a millimeter of mercury that is equal to 1 torr, roughly corresponding to 133.3 Pa. Common point parameters that are extracted from the waveform pressure at the aortic level are its maximum (i.e., systolic) value P_s, and its minimum (i.e., diastolic) value P_d.

Other parameters that can be estimated from the blood pressure are

- the **mean arterial pressure** (MAP), that is the average value of the blood pressure over a cardiac cycle. Its value can be determined experimentally also by multiplying the value of the cardiac output CO by the systemic vascular resistance R_{VS}, and then summing up the right atrium pressure P_{RA}. Alternatively, a MAP estimate $\hat{MAP}$ can be evaluated from the point parameters P_s and P_d:

$$MAP = (R_{VS} \times CO) + P_{RA}, \qquad \hat{MAP} = P_d + \left(\frac{P_s - P_d}{3}\right) \tag{2.8}$$

This estimated value $\hat{MAP}$ comes from approximating the blood pressure waveform to a trapezium as shown in Figure 2.6.

- the **systemic pulse pressure** P_{pulse}, that is the difference between the maximum and the minimum value recorded in the systemic circulation. Since P_{pulse} is defined as the difference from the plateau values of blood pressure, it can be considered as the minimum pressure required to feel the pulse. The value is generally evaluated at the systemic level, even if it can be measured also in the pulmonary circuit:

$$P_{pulse} = P_s - P_d \tag{2.9}$$

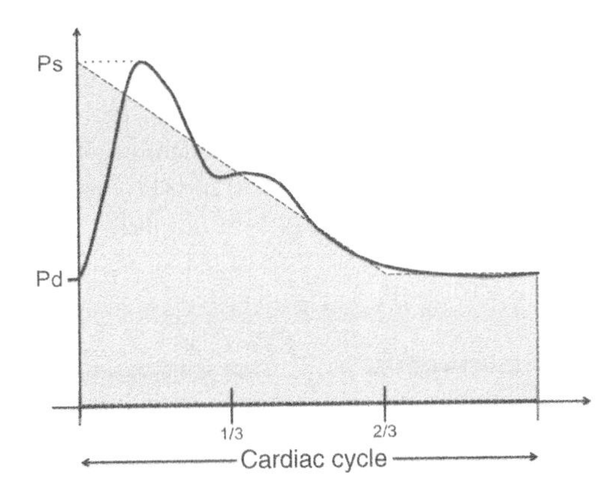

Figure 2.6 Trapezium approximation.

Table 2.3 Indicative adult normal values for blood pressure parameters

Feature	Indicative value
P_s	90–119 mm Hg, 120–139 denotes pre-hypertension
P_d	60–79 mm Hg, 80–89 denotes pre-hypertension
MAP	70–110 mm Hg
P_{pulse}	$0.25 \times P_s - 60$ mm Hg for systemic circulation, lower values indicate drop in stroke volume, higher values at rest may be signs of increased artery stiffness.

Table 2.3 shows the range values for the four pressure parameters, in normal adults. According to the International Classification of Diseases (WHO, 2012), the additional ranges of Table 2.3 denote the presence of an altered condition from normality that cannot be diagnosed as hypertension.

In regards to the mean arterial pressure values in Table 2.3, the lower range are considered as signs of reduced perfusion that is the delivery of blood to a specific organ. At the same time, when systemic pulse pressure falls below the lower range, it may be associated with a decrease of stroke volume that is the blood volume pumped out from the hearth.

2.4.2 *Blood Flow and Hemodynamics*

Another relevant physical quantity associated with circulation is the blood flow. Blood flow is often monitored to estimate the effectiveness of the pumping action of the heart by means of the cardiac output. A reduced (or even abnormal increase of) cardiac output may be a sign of heart failure. Blood flow is also monitored to determine the perfusion of the heart.

In this context, hemodynamics refers to the physical factors governing the flow of blood. As in every fluid, these are based on the relation that links the pressure difference (ΔP) at the ends of a vessel, with the flow (Q) of that fluid, which is characterized by a resistance to the flow (R), according to the equation:

$$\Delta P = QR \tag{2.10}$$

The volumetric flow rate SI unit is ($m^3\ s^{-1}$), even if the clinical practice unit measure is ml/min, where 1 ml/min $= 6 \times 10^{-5}\ m^3\ s^{-1}$. When fluids flow through pipes such as the blood vessels, there are two possible types of flow: turbulent and laminar flow, as shown in Figure 2.7.

Laminar flows tend to be experienced at lower fluid speed. Turbulent flows occur at higher speed and they are less orderly flow regimes. In **laminar flow**, the fluid always moves in parallel layers within a vessel and the layers close to the vessel walls experience a reduced speed due to friction with the vessel walls.

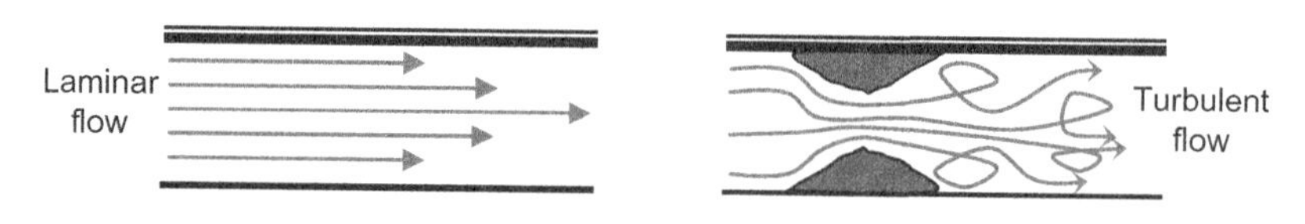

Figure 2.7 Laminar and turbulent flow.

When experiencing laminar flow in a vessel of length L, and radius r, the resistance offered by the system can be expressed as

$$R = \frac{8\eta L}{\pi r^4} \tag{2.11}$$

where η is the fluid viscosity, determined as the ratio between the shear stress τ and the shear rate dv/dt. For **a Newtonian fluid**, the viscosity, by definition, depends only on temperature and pressure (and also the chemical composition of the fluid if the fluid is not a pure substance), not on the forces applied to the fluid (i.e., the exerted shear rate). In the specific case of blood, this independence is not valid, since blood has a pseudo-plastic behavior, that is its viscosity decreases with the increase of shear stress. This process needs to be taken into account when considering the different resistance that can be obtained as a result of different exerted pressures. At the same time, blood has no memory, that is its behavior, in the presence of a constant shear stress, does not vary over time.

For **non-Newtonian fluids** like blood, shear stress τ will depend on shear rate dv/dt according to the following formula:

$$\tau = \tau_0 + K(dv/dt)^n \tag{2.12}$$

where τ_0 is the shear at zero rate, and K and n are two fluid-specific constants. In the case of pseudo-plastic fluids, such as blood, $\tau_0 = 0$ and $n < 1$.

Now, combining together equations (2.10) and (2.11) we obtain the Hagen–Poiseuille equation for the **pressure difference** ΔP:

$$\Delta P = \frac{8\eta L Q}{\pi r^4} \tag{2.13}$$

where η is again the fluid viscosity, Q the flow and L, r are the length and the lumen of the vessel respectively.

This equation allows us to calculate the pressure difference due to the presence of a viscous fluid in a vessel. The pressure difference is inversely proportional to the fourth power of the vessel lumen r. This essentially means that even *small differences in the caliber of the vessel (e.g., caused by occlusions) involve a large change of the needed pressure* to maintain the same amount of flow. For example, a change of pressure difference of 20 mm Hg with respect to a normal value of 120 mm Hg can be caused by an occlusion of the vessel caliber of less than 4% of its original value.

We recall that equation (2.11) holds if the blood flow is laminar, that is its velocity profile through the vessel section is parabolic, reducing to zero at the vessel walls, and increasing to its maximum near the center of the section. The laminarity of blood flow depends on a number of factors, including blood's velocity, viscosity, density, and the characteristics of the vessel section. In general terms, turbulence occurs when the flow speed is high enough that the inertial components of exerted force prevail over the viscous ones. In order to determine if the flow is laminar or turbulent, the dimensionless **Reynolds number** Re is introduced, and defined as follows:

$$\mathrm{Re} = \frac{\rho Q}{2\pi\eta r} \tag{2.14}$$

where ρ is the fluid density. In general, the threshold value of 2040 determines the turbulent or laminar nature within a circular vessel. This threshold holds for straight vessels. In systemic circulation, with its branches and curves, blood may experience turbulent flow even at rather lower values, around 200–400. Moreover, since blood is a pseudo-plastic fluid, its viscosity varies with flow and the Reynolds number expressed by (2.14) needs to be adjusted as follows:

$$\mathrm{Re} = \frac{\rho Q}{2\pi r K[(3n+1)/(4n)]^n (8V/D)^{n-1}} \tag{2.15}$$

where D is the diameter and V is the mean velocity. For a Newtonian fluid, $n = 1$, $K = \eta$ and equation (2.15) simplifies to equation (2.14). The Reynolds number is usually assessed to evaluate whether turbulent flow occurs.

In the specific case of blood, because it is composed of different components, including water, cells, proteins and inorganic ions, the approximation to a Newtonian fluid may be an oversimplification. The approximation it is still valid as far as the plasma is concerned (i.e., without considering the effect of cell presence), but it fails when considering the effect of the cells. It is to be remembered that plasma constitutes as much as 50% of blood, while the remaining part is composed of red blood cells, white blood cells and platelets. The approximation of blood being Newtonian holds when dealing with vessels whose diameter is higher than a few hundred μm. As far as the laminarity of blood flow is concerned, it is agreed that this assumption is valid for the entire arterial circulation, excluding the most proximal arteries, including the aorta, and the main pulmonary arteria, where signs of turbulence cannot be ruled out.

As for the characteristics over time of this flow, this is, as predictable, highly pulsating in the proximal vessels, both for systemic circulation and for pulmonary circulation. In order to quantify the amount of **pulsatility**, the dimensionless Womersley Number α is introduced as follows:

$$\alpha = r\sqrt{\frac{\omega}{\eta/\rho}} \tag{2.16}$$

where ω represents the angular frequency (in rad/s), ρ is the fluid density and η is the fluid viscosity as before. The degree of pulsatility thus increases with the size of the vessel, and with the (square root of the) heart rate. In general terms, α can be considered as a ratio between the strength of the inertial forces and the viscous forces: if α is less than 1, the rate of the flow pulse is slow enough to let the pulse profile decrease to its rest state between successive pulse time instances. This corresponds to a nearly parabolic profile of the flow rate, while when this value increases above 10, the pulse profile is nearly flat and the inertial forces prevail. In the blood vessel system, this number greatly varies: in the range of 15–20 in the aorta, and less than 0.05 in capillaries and microcirculation.

With low values of α, the behavior of the vessel is almost resistive, with pressure and flow in phase, whereas when α is higher, flow lags behind the pressure gradient, and its behavior may be considered as capacitive, with the flow rate going in quadrature with respect to the pressure gradient. One of the interesting effects associated with this parameter is that it

Table 2.4 Indicative normal values for blood flow parameters in adults

Vessel	Velocity (cm/s)	α
Abdominal aorta	80–130	11–15
Common carotid	65–95	4.5–5.5
Brachial	40–75	2.4–3.6

regulates the relationship between pressure gradient and flow rate: if we hypothesize that a sinusoidally varying pressure gradient is present, with angular frequency ω_0, the corresponding flow rate is a scaled (and possibly shifted) replica of the pressure gradient, with the same angular frequency. In the case of $\alpha > 1$, if we suppose that the pressure gradient angular frequency changes from ω_0 to $k \times \omega_0$, with $k > 1$, the corresponding flow rate not only changes its angular frequency, but also its amplitude, which decreases as k increases. The behaviour is similar to a signal passing through a low pass filter. This can be explained by considering that, due to its viscous properties, the blood cannot 'keep up' with the fast reversals in pressure gradient, and as a result its flow diminishes in amplitude.

In Table 2.4 the average values of both the Womersley number, and the maximum of the velocity profile in the main branches of the systemic circulation is reported.

In hospital settings, **parameters extracted from blood flow** rate are either directly related to the blood flow rate of the main vessels of systemic circulation, or to its behavior in those small vessels that directly supply the heart with blood: the coronaries. In the first case, there are two main indexes evaluated from a combination of blood pressure and blood flow:

- The total peripheral resistance, ***TPR***:
  TPR is the physiological replica of the resistance R that was introduced in the first part of this section, and it is obtained as the ratio between the observed flow rate and the observed blood pressure. This is generally calculated as an average over the entire cardiac cycle, according to the following formula:

$$TPR = \frac{Q_m}{MAP} \tag{2.17}$$

where Q_m is the average flow rate, which corresponds to the clinical term cardiac output, *CO* and MAP is the mean arterial pressure. *TPR*'s SI unit is (Pa·s·m^{-3}), and for ease of use, this can be replaced by the Woods Units, (mm Hg · l/min), where one Wood Unit corresponds to $8 \cdot 10^6$ Pa m^3/s. Physiological values at rest lie in the range 11–17 Woods Units for the systemic circulation (systemic vascular resistance, SVR), and 1.9–3.0 for the pulmonary circulation (pulmonary vascular resistance, PVR).

- The total arterial compliance ***C***:
  C is defined as the ratio between the change in arterial blood volume (ΔV) due to a given change in arterial blood pressure (ΔP):

$$C = \frac{\Delta V}{\Delta P} \tag{2.18}$$

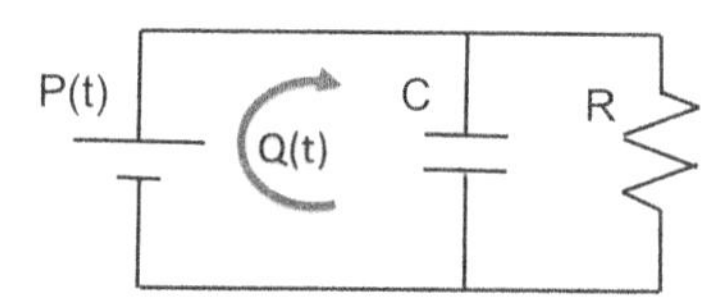

Figure 2.8 Windkessel model (two elements).

C is a measure of the capacitive power of the vascular system, and its SI unit is (m^3/Pa); alternatively, it can also be expressed in terms of (ml/mm Hg). Normal values for the adult population are in the range 1.0–2.1 ml/mm Hg.

Both these parameters are the quantitative representations of the **Windkessel model** represented in Figure 2.8, which is a simplified circuit for blood circulation. The equation that governs this two-element lumped parameter circuit is

$$C \cdot \frac{\partial P(t)}{\partial t} + \frac{P(t)}{R} = Q(t) \tag{2.19}$$

Equation (2.19) represents a first-order model that incorporates the dependence of the flow from pressure in static conditions (i.e. the resistance R), and its dependence on its variations over time (i.e., the compliance C).

More complex models have been proposed, in order to also encompass other effects: that is, the local inertia and the local compliance of the very proximal ascending aorta, that can be represented by a characteristic resistance in series with the original two-element circuit. For this, the 'inertance' is introduced to measure the fluid pressure gradient required to cause a change in the flow rate. The inertance accounts for the inertial effect of the overall vascular system and can be represented by an inductance in parallel with the characteristic resistance.

As far as the blood flow of the coronary system is concerned, a number of other interesting indexes are generally being used in clinical practice. Many of them are directly calculated from the coronary pressure-flow relation, and are thus similar to the ones that have been introduced above.

If we refer to the simplified model represented in Figure 2.9, the section of clinical relevance is represented by the branches representing two coronaries. P_a represents the aortic pressure and P_v is the venous pressure. Q_N and Q_S are the blood flows along the two branches, and R_N and R_S are the resistances offered by the two capillary branches. In the absence of lumen decrease for the S branch (i.e., in physiological conditions), we have $P_d = P_a$, $R_N = R_S$, $\rightarrow Q_N = Q_S \rightarrow Q_C = 0$. This means that, even in the presence of a collateral branch linking the two branches, the absence of pressure difference between the branches makes the collateral branch virtually unused.

A blood vessel may experience a **stenosis**, that is a pathological restriction of vessel section. Stenosis may be due to many causes, such as atherosclerosis, diabetes, infection ischemia or neoplasm. In the presence of a stenosis, the lumen of the S branch decreases, and it determines a change of the pressure, with P_d lower than P_a. With this modification, the flow Q in the stenotic branch also decreases, even if it gathers blood from both the

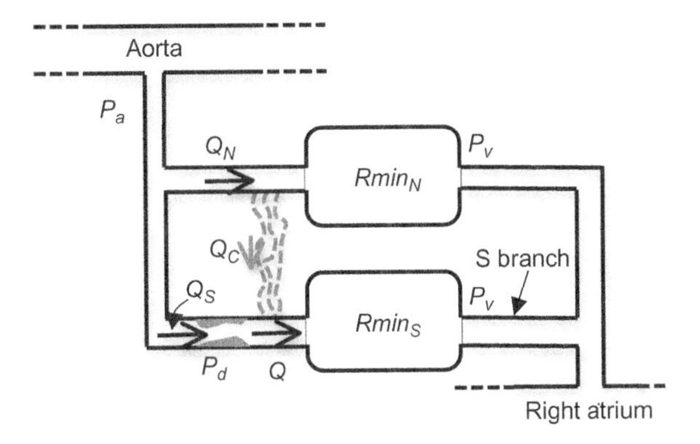

Figure 2.9 Model of the coronary circulation.

stenotic section Q_S, and from the collateral branch Q_C. **Myocardial fractional flow reserve** (FFR), is a clinically relevant index, which takes this model into account, and is defined as

$$FFR = \frac{Q}{Q_N} \tag{2.20}$$

In physiological conditions, this index yields 1; FFR <0.75 corresponds to inducible ischemia and it is advised to intervene to dilate the lumen.

If we consider that, for each branch:

$$Q_N = \frac{P_a - P_v}{R_N}, \quad Q = \frac{P_d - P_v}{R_S} \tag{2.21}$$

we have:

$$FFR = \frac{P_d - P_v}{P_a - P_v} \cdot \frac{R_N}{R_S} \tag{2.22}$$

under the hypothesis that the two resistances R_N and R_S offered by the two capillary branches are equal:

$$FFR = \frac{P_d - P_v}{P_a - P_v} \tag{2.23}$$

In terms of medical requirements, the cutoff point is thus defined at around 0.75, that is a value lower than 0.75 implies a need for a surgical revascularization procedure, whereas values higher than 0.80 are associated with no need for intervention.

2.5 Biopotential Recording

The recording of biopotentials has its origin in the presence of voltage differences across the cell walls. This difference comes from the specific characteristics of the cell membrane,

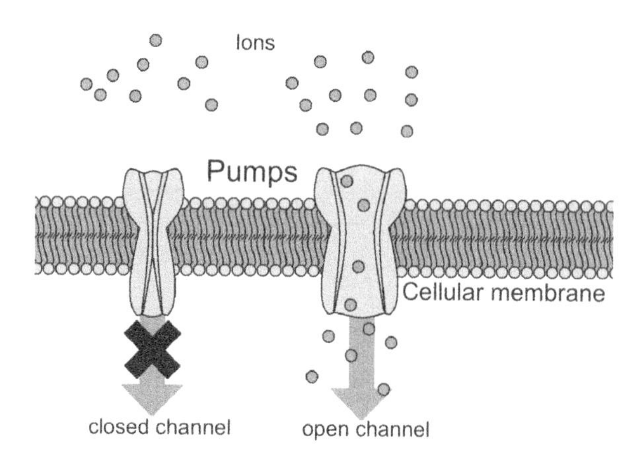

Figure 2.10 Cell membrane representation.

which represents a discontinuity between the intracellular volume and the extracellular volume.

In general terms, the **cell membrane** depicted in Figure 2.10 is a very thin structure made of an electrically inert material (a phospholipid bi-layer).

The cell membrane displays a high electrical capacitance, accompanied by a relative resistance to ion flux across it. In order to make the membrane selective to specific ions, the bi-layer is often interleaved by protein constructions that cross the layer multiple times, making it possible to let specific ions cross the membrane.

These protein constructions are generally called **ion-specific channels**. Some of these channels are active, that is they not only allow the passage of ions according to the concentration gradient, but they can also expend energy in making some ions move against this process, that is from lower concentration areas, to higher concentration ones. In this specific case, we call these proteins 'pumps'. In both cases, however, these channels are open or closed, based on a number of factors, or sources that make them function, such as an electrical source, or an electro-chemical source that can be either generated from an external source, or as the result of an internal propagation along the cell.

It is acknowledged that, in order to get insights on the cell function, an intra-cellular recording is required, that is a means to directly record the voltage difference between the intra-cellular medium (possibly in the proximity of the cell membrane) and the extracellular medium (also in proximity of the cell); therefore, we will focus in this section of the chapter on the systems able to gather information on the propagation of these electrical phenomena across multiple cells, and on their macroscopic effect at the organ level. For this reason, we need extracellular recordings, where voltage difference is picked up between two different points in the extracellular medium. In this way, we are not able to directly capture the electrical polarization and depolarization of each cell, but rather we will indirectly get information on the propagation of this effect along a number of cells under examination.

By using the time course of this voltage difference – which we will call **biopotential** in the following – we will extract a number of relevant parameters able to quantify the degree of functionality of the organ where these electrical phenomena occur.

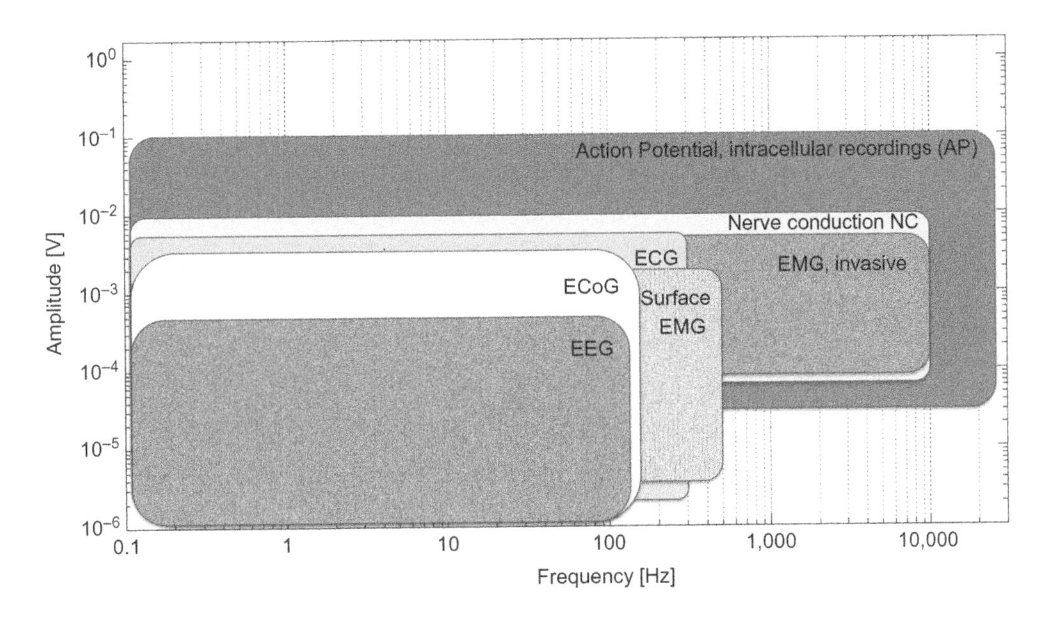

Figure 2.11 Biopotentials: typical values.

This happens since the flow of current generated by cell polarization and depolarization (and their propagation) does not follow a specific path, but it radiates throughout the whole surrounding conducting volume (the volume conductor), which may be a section of an organ, an organ itself or the entire body. The biopotential associated with the current flow through the volume conductor is usually tiny, in the order of a few mV at most, depending on the placement of the two pick-up points where the voltage is recorded.

In the case of biopotentials, the structure of the temporal waveforms can be substantially different, in amplitude, frequency characteristics and shape, depending on the particular phenomenon associated with the biopotential under analysis. In Figure 2.11, the characteristics of both amplitude and frequency ranges of most biopotentials of interest in clinical practice is reported.

As we may notice from Figure 2.11 derived from Table 2.2, biopotentials span a very large range of values in both amplitude and frequency characteristics. In the following, we will focus on two relevant biopotentials – the electroencephalogram (EEG) and the electromyogram (EMG) – leaving the most popular one, the electrocardiogram (ECG), to the implementation section. While it is acknowledged that each biopotential has its own generation process and its own relation to the physiological source, we will consider this triplet as the one that globally represents the general characteristics of most biopotentials.

2.6 Electroencephalography

The **electroencephalogram** (EEG) is a biopotential that appears as the result of the electrical potentials associated with brain activity. In the case of transcutaneous recordings, the biopotential is obtained as the difference of voltage between an electrode placed on the scalp surface and a reference electrode where no significant activity can be picked up.

Brain activity, at the macroscopic level, can also be recorded by using electrodes that are inserted in direct contact with the cortex, and in that case we will call the corresponding biopotential the **electrocorticogram** (ECoG) – or even within the skull, with thin electrodes. In that case, we will call the biopotential **depth recording**. In the specific case of the surface recording, the biopotential comes out as the weighted combination of the action potentials that emerge from a number of excitable cells, the neurons, present in the brain. In other words, EEG records electrical potentials at some distance from their source; this is described as the volume conduction, which models the effects coming from a non-negligible medium between the source and the sensor. **Volume conduction** is usually referred to as the volume space where a transmission of electric or magnetic fields takes place, generated by current sources. Excitable tissues when activated become current sources, generating fields in the surrounding passive conductive biological tissue.

Volume conduction plays an important role in almost every biopotential recording. In the case of EEG recorded on the scalp surface, this is very important. In general terms, we will differentiate between near-field potential, when the recorded potential is close to the generating source, and far-field potential otherwise. A near-field potential waveform changes significantly even with small changes in the position of the electrode, while this happens only to a minor effect in case of far-field potentials.

In order to determine what happens at a distance from the generating source, we may consider the relationship between what happens across the membrane (i.e., intracellular recordings as measures of the action potential) and what happens in proximity to the cell membrane itself (i.e., extra-cellular recordings, as the components of, for example, EEG signals). This relation is not simple, as it involves the medium between the source and the pick-up site, and this medium is generally traversed by currents coming from different sources.

Without going into too much detail on the analytical relation between intracellular and extracellular recordings, we may summarize by saying that the potential associated with the extracellular potential is a weighted sum of the effects of each membrane potential, where the weighting coefficient depends on the distance of the pick-up site projection on the membrane cell from the axial position. This clearly holds if we hypothesize that the extracellular medium is uniformly characterized by a conductivity σ_e.

In the specific case of EEG, one may consider that we are in far-field, where the distance from the generating source is higher than a few length constants. In the case of neurons in the brain, this value – which roughly corresponds to the ratio between the membrane resistance and the sum of the *resistance per unit length* in both extracellular and intracellular medium – is usually in the order of a few millimeters at most. As a consequence, the EEG, as recorded by a number of electrodes on the scalp surface, is primarily generated by the post-synaptic potentials (either excitatory EPSP or inhibitory IPSP) of cortical nerve cells, more than from the propagation of their action potential. This is because the effect that the discharge of a single neuron or single nerve fiber in the brain generates in terms of potential field to be recorded at a distance does not significantly appear at the pick-up sites. When, instead, hundreds of thousands of neurons discharge synchronously, the sum of these effects appears even at a distance.

Thus, **the EEG does not necessarily record the *amount* of electrical activity, rather the amount of synchronicity among nerve cell firings.** Another important aspect to be kept in mind concerns the kind of nerve cells whose synchronous activity is recorded. As a consequence of the anatomy of the brain, only the activity of certain nerve cells can be

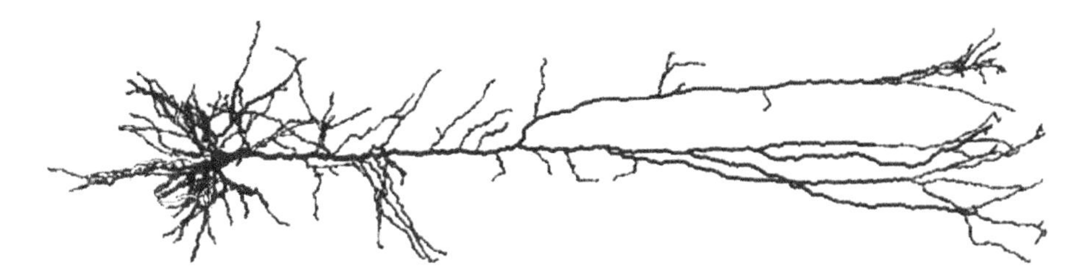

Figure 2.12 Representation of an apical dendrite.

recorded. As a matter of fact, EEG will measure the post-synaptic potentials of those nerve cells that are distributed in organized layers within the brain, and whose apical dendrites are aligned normally to the scalp surface, as represented in Figure 2.12. In this way, the post-synaptic potential comes out as the result of a current flow that mainly appears in the same direction, and can thus be picked up efficiently. The major role in EEG generation is thus played by the large cortical pyramidal neurons present in deep cortical layers.

As a result of this process, EEG displays characteristics that are peculiar both in terms of magnitude and in terms of frequency content. More specifically, the amplitude of the EEG signals varies from a few hundreds of nanovolts (nanovolt $= 10^{-9}$), up to a few hundreds of μV, whereas the frequency band is generally lower than 50 Hz. Owing to the generation process, which involves a large number of sources, the EEG signal deterministic component is almost negligible, that is it is practically impossible to link each characteristic of the oscillation present in the EEG to a peculiar physiological source (as will be the case with ECG). However, the amount of stochasticity is pretty high, and thus most parameters that can be extracted from EEG signals (the so-called **quantitative EEG** – qEEG), generally refer to ensemble measures, such as the variance and the higher order statistics, such as the kurtosis and the skewness. To be more specific, qEEG refers to the processing of EEG data to highlight specific waveform components, possibly transforming the data into a domain that better elucidates relevant information (e.g., Fourier transform to represent the data in the frequency domain as shown in Ch. 4 and 5), and extract numerical results for review or comparison (Nuwer, 1997).

Historically, the frequency components of the EEG signal have been associated with different physiological, or in some cases pathological, conditions. As a result, the EEG signal is generally analyzed in different bandwidths, to separate the different *rhythms*, which are defined in Table 2.5, together with the different conditions they are associated to. The data in

Table 2.5 EEG rhythms

Rhythm	Frequency (Hz)	Amplitude (μV)	Normal conditions
Alpha	8–13	20–200	Relaxation, eyes closed
Beta	13–30	5–10	Activity, concentration
Gamma	30–100		Multimodal sensory processing
Delta	<4	20–200	Slow wave (non-REM) sleep
Theta	4–8	5–20	Drowsiness, idling

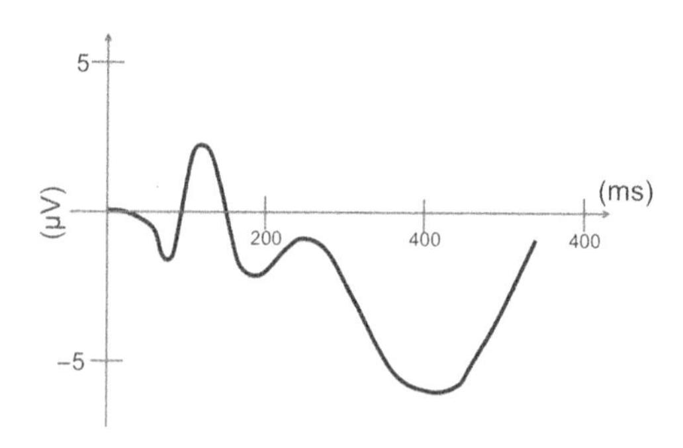

Figure 2.13 ERP example with stimulus at time 0.

Table 2.5 show that, generally, the higher the frequency associated with a rhythm, the lower is the amplitude with which this rhythm appears.

Notwithstanding the highly stochastic nature of the EEG, some inferences can be collected on specific components of the EEG, which appear as a response to a particular environmental excitation, usually known as a stimulus. We will call these responses **event-related potentials** (ERP).

ERPs appear with a specific latency with respect to the stimulus, with a deterministic shape, that is they consistently appear with the same shape with a specific delay from the stimulus. In the case of ERP analysis, we will extract a number of time and amplitude parameters, namely the *latency*, and the different amplitudes and timings of successive peaks (both minima and maxima) in response to the stimulus. Figure 2.13 shows an example of an event-related potential, together with the peaks that are generally recorded and extracted from it. Timings and amplitudes may vary depending on the characteristics of the stimulus.

In terms of the clinical significance of the qEEG parameters, we will review them according to the classification based on the time domain, the frequency domain and the statistical properties (i.e., the probability distribution). We are not going to list all the parameters, since it is beyond the scope of this chapter. We will just highlight some of them, based on the specific application where they are used in clinical environments.

In the first group (**time domain parameters**), if we refer to rest EEG (where no event-related potentials are voluntarily studied), the parameters are related to specific disorders or events:

- When in anesthesia, for instance, **total EEG power** is somewhat associated with the *degree of anesthesia*. The total power, usually calculated as an average over a few seconds, is then a simple time domain parameter that can capture this information.
- **Zero-crossing frequency**, calculated by detecting the number of times per second the EEG pattern crosses the zero-value, was historically used to capture information on the status of a patient. This parameter has been replaced by parameters evaluated in the frequency domain.
- In the case of deep anesthesia, the presence of a specific EEG pattern in the time domain, called *burst suppression*, can be gathered through a specific time domain parameter, called **burst suppression ratio** (BSR). This is calculated by detecting when the EEG activity

does not exceed ±5.0 μV for more than 500 ms. BSR is then calculated by dividing the amount of time where burst suppression is present by the total monitoring duration.

In the case of **frequency domain**, multiple parameters are generally used to assess the status of a patient. In the specific case of brain ischemia, differences were noted in terms of frequency distributions among the different bands and rhythms. A number of clinically relevant parameters are thus used to assess the extent of brain functionality in this sense. The first group is composed of parameters representing the amount of each rhythm with respect to the overall power:

- relative power of delta rhythm (range 1–3 Hz, RDP)
- relative power of delta and theta band (1–7 Hz, RDTP)
- relative power of alpha rhythm (range 8–13 Hz, RAP)
- relative power of $beta_1$ rhythm (range 14–20 Hz, RB_1P).

From the powers calculated for each rhythm and band, we can also calculate some ratios between them. Most significant, in clinical terms, are the following:

- delta/alpha ratio (DAR), as the ratio between the power associated with the delta rhythm and that associated with the alpha rhythm
- (delta + theta)/alpha ratio (DTAR), where the amount of power in the delta and theta band is divided by that associated with the alpha rhythm
- (delta + theta)/(alpha + $beta_1$) (DTABR), where power in the delta + theta band is divided by the amount of power obtained by adding power in the alpha and $beta_1$ bands.

Finally, with specific reference to the **probability distribution**, the amount of 'predictability' of the EEG pattern is a phenomenon that needs to be monitored in specific clinical applications. For example, in order to monitor the amount of anesthesia, since unpredictability of an EEG signal decreases with increasing brain levels of anesthetic drugs, a usual parameter is based on the calculation of Shannon's entropy, extracted as the integral over a specified bandwidth, of the product between the power spectral density and the natural logarithm of its inverse. In the case of deterministic patterns, this value reaches its minimum, and it increases with the amount of irregularity (i.e., unpredictability).

With regards to the minimum requirements for the design of monitoring EEG, ECRI (2012a) considers a minimum sampling rate of 200 samples/s, and it does not provide restrictions to the maximum amount of noise when recording the EEG; most of the EEGs that are on the market refer to noise RMS values up to a few μV.

2.7 Electromyography

An electromyogram (EMG) is a signal recorded from a biopotential generated by skeletal muscles. More particularly, we usually refer to the **motor unit** (MU), as the complex that includes the motor neuron that descends from the spinal cord and the muscle that this motor neuron innervates. Electromyography is the technique that is used to capture the electrical sources generated from the motor units. Note that muscles are innervated by a multitude of MUs, ranging from some tens for a small muscle located in the hand, to more than a thousand for the bigger ones. An MU receives a stimulus from the upper neurons, and transmits it to the muscle that, as a result, will transform this depolarization stimulus into a muscular contraction that produces force. As a consequence, this will generate movement. While the process of depolarization and propagation of the stimulus is somewhat similar to that reported in the case of electrocardiography, differences may arise in the analysis and interpretation of the data. The amount of **force** that is **generated** by the muscles is directly related to two distinct factors:

1. the number of MUs that are active (the so-called MU recruitment),
2. MU status, i.e. the number of times per second they 'fire' (the so-called MU firing rate that is, the frequency with which the MU are activated by nerves impulse or, equivalently, the number of depolarizations per second of an MU).

Three types of MUs are identified, with different characteristics:

1. S MU (also called type I), which are characterized by a Slow firing rate
2. FR MU (also called IIa), which are Fast, and Resistant to fatigue
3. FF MU(also called IIb), which are Fast (with high firing rates) and Fatigable (i.e., they cannot maintain the same force production for long).

Type I fibers are generally recruited when low force levels are needed, while for increasing force levels type IIa, and then IIb, come into play.

With regards to the recording method, we need to distinguish between

- **intramuscular recordings**, when thin wires or needles are placed in the muscle
- **surface recordings**, when electrodes are placed on the body surface in proximity to the muscle.

In this latter case, the technique is sometimes referred to as **surface electromyography** (or sEMG), to distinguish it from invasive EMG.

In the first case, recording aims to study the physiology of the MU, and to identify possible alterations associated with pathologies: for instance, how the MU varies its behavior when there is loss of nerve supply to the muscle. In surface recordings, the focus is instead on the muscle behavior as a whole. When used as a diagnostic tool, it supplies information on macroscopic anomalies of muscle behavior: for instance, the ability of a muscle to produce force, when needed, and with the amount needed. We will here focus on some parameters of clinical relevance in the case of intramuscular recordings, given that this list is not exhaustive of all the clinical signs that can be captured through EMG.

When recording the electrical activity of one MU, we will call this biopotential the **motor unit activation potential** (MUAP). Some information can be extracted directly from the shape of the MUAP; they are usually ensemble parameters calculated from the shape: the number of peaks, turns and zero-crossings and the presence of possible satellite potentials generally used to characterize the MU. In Figure 2.14, an example of different MUAP trains is displayed, together with the points of interest from which the main shape-related parameters can be extracted.

Another clinically relevant parameter in MUAP analysis is the firing rate. This value is not constant over time, and a number of parameters can be extracted from the instantaneous

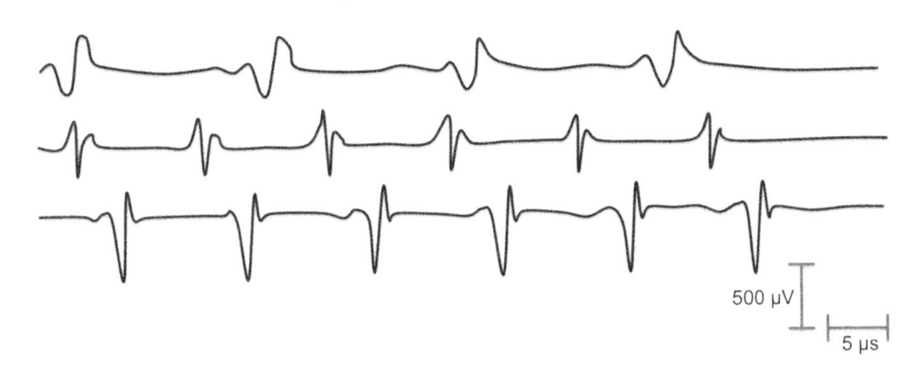

Figure 2.14 Typical MUAP waveforms.

firing rate over time, that is the inverse of the time elapsed between two successive depolarizations. For instance,

- in motor neuron diseases, MUAPs display higher firing rates, higher amplitudes and higher firing rate variability
- in myopathies, the behavior is opposite, and in some cases MU recruitment is substantially different (i.e., higher)
- in multiple sclerosis patients, while the shape of MUAP does not significantly differ from control population, firing rates are lower, and firing rate variability is higher
- in patients suffering from amyotrophic lateral sclerosis (ALS), MUAPs look larger, with polyphasic shape, as compared to a control population.

An interesting and clinically significant parameter is obtained when analyzing multiple MUs: given a pair of MUs, it is of interest to gather information on the shared behavior between them, by looking at the cross-correlation between the MUAP firing rates. This is called **common drive**, and its value depends on a number of factors, including the effect of ageing on descending command from the central nervous system.

EMG are often clinically used to determine the **nerve conduction velocity** (NCV), which is a measure of the rate at which a nerve can carry information from a stimulus location to the muscle it innervates. This measure can be made for both motor nerves and sensorial nerves. In the latter case, the response of the nerve to the stimulus is called **sensory nerve action potential** (SNAP), and its amplitude is in the order of 5–10 μV. While it is acknowledged that NCV is influenced by the amplitude and characteristics of the stimulus, and it depends also on the specific nerve. Normal reference values lie in the range 40–70 m/s in adults, and reduce in presence of neuropathies (cutoff values are generally 30–35 m/s for diagnosis of demyelination in arm nerves).

Regarding the recommended specifications of the EMG that has to record the previous parameters, ECRI advises on the need for an overall noise <2 μV (p-p), a CMRR higher than 100 dB at 60 Hz, and an input impedance higher than 100 MΩ. In terms of frequency range, ECRI reports that, for invasive recordings, a frequency range spanning up to 35,000 Hz should be considered (ECRI, 2012a). These performance terms (CMRR, frequency range etc.) are explained in Chapter 5.

Part II: Implementation

2.8 Introduction

This book fully discloses a marketed PC-based ECG product with the associated know-how to design it and the 'know-why' of design decisions. Such know-why, with the associated design methods, is generally applicable to other medical devices.

The implementation parts of the chapters of this book describe first a proposed approach to address the specific chapter problem. Again, this approach is generally valid for any product when properly tailored. Then, the approach is applied to the real case of a commercial ECG, the **Gamma Cardio CG** of the Gamma Cardio Soft. The decision rationale is also outlined to allow readers an easier application to other device types. This design-oriented method is also useful for understanding the main concepts related to pathology with the associated diagnostic instrument.

Regarding this chapter, *inadequate requirements are a prime reason for failure* in medical device design (Kelleher, 2003) in software design (Standish Group, 1995) and probably in any complex human venture. Errors in requirements incur a cost for fixing that is hundreds of time greater when discovered at the later development stages. This is discussed in Chapter 1, but it is also suggested by common sense.

It is then worth focusing on the **product requirement generation**, on the physiology of the part of the body to be diagnosed and on the measurement method. The implementation part of this chapter will therefore explain the main principles of heart physiology and the associated ECG usage.

Section 2.9 introduces main concepts of requirements. In Section 2.10, a general method for requirement generation is proposed. This method is then tailored to medical instruments and specifically to the Gamma Cardio CG in Section 2.11. The use of standards and their benefits in requirement generation is also covered in 2.10. The proposed approach implies focusing on the physiologic mechanisms of the part to be analyzed (i. e. the *heart* in Section 2.12), the corresponding method of observation (Section 2.13), the ECG with its principles and applications (Section 2.14). The ECG performances are addressed in Section 2.15. ECG is a simple reference for any medical instrument; the design approach has a general validity, which holds for requirements related to safety (Section 2.16), usability, marketing (Section 2.17) and environment (Section 2.18) since they all derive from the need to ensure the safety and effectiveness of medical device products.

The economic constraints are then discussed in Section 2.19. They are a relevant part since they often contrast with the other requirements and some tradeoff has to be considered in real product developments.

At the end of this chapter, readers will be able to understand requirements generation and the associated rationale so as to tailor the process to define different requirements for medical devices. The reader will also be able to understand ECG requirements, the principles of heart physiology, the ECG itself and its applications.

2.9 Requirements Management

Before going into the design specifications that state how the system will be realized, the design team must define what the system must do and how well it must performs its functions. These concepts are organized in functional requirements (i.e., the 'what' done by the product) and non-functional requirements focused on performance.

A single **requirement** may be defined as a textual description related to system functions/performances. It must satisfy the following conditions:

C1. It is uniquely identified by a code.
C2. It can be verified through inspection, demonstration, test or analysis; that is, a procedure can be written to ensure that the requirement is met.
C3. It must not be in conflict with other requirements.
C4. It can be translated into a technically and economically feasible design specification.
C5. It is mandatory and it identifies a real business need/opportunity.
C6. It is unambiguous and complete and it addresses only one specific feature.

For each requirement a qualification method has to be specified so that the requirement can be validated. According to (SSS, 1994) **qualification methods** may include

1. demonstration: observable functional operation not requiring the use of instrumentation, special test equipment or subsequent analysis
2. test: using instrumentation or other special test equipment to collect data for later analysis
3. analysis: the processing of accumulated data obtained from other qualification methods
4. inspection: visual examination of system components, documentation, etc.
5. special qualification methods: any special qualification methods for the system, such as
   - special tools, techniques, procedures, acceptance limits, use of standard
   - samples, preproduction or production samples, pilot models or pilot lots.

The qualification methods are the same as included in the example shown in the sub-requirements associated to AAMI EC 11 standard (Table 2.10, Table 2.11 and Table 2.12).

Requirements for medical devices come from a variety of **requirement sources**. Some requirements are generated by people in the medical field. Analyst and marketing company resources derive these requirements from physicians, nurses and operators located where the product will be used. Market requirements can also be derived from company sources, marketing intelligence, competitors, domain experts and prospective users. Applicable standards are another source of requirements.

Focusing on medical instruments, Figure 2.15 outlines the process of requirement generation. The previous sources are the input of the process while the output is the set of engineered requirements that can be grouped into signal, safety, usability, environmental and economic factors requirements (adapted from Webster, 2009).

The **requirement process** may be quite complex due to the amount of input. A systematic and rationalizing process is suggested to collect data and then to engineer and reduce the overall requirements. The process must guarantee that conflicting and redundant

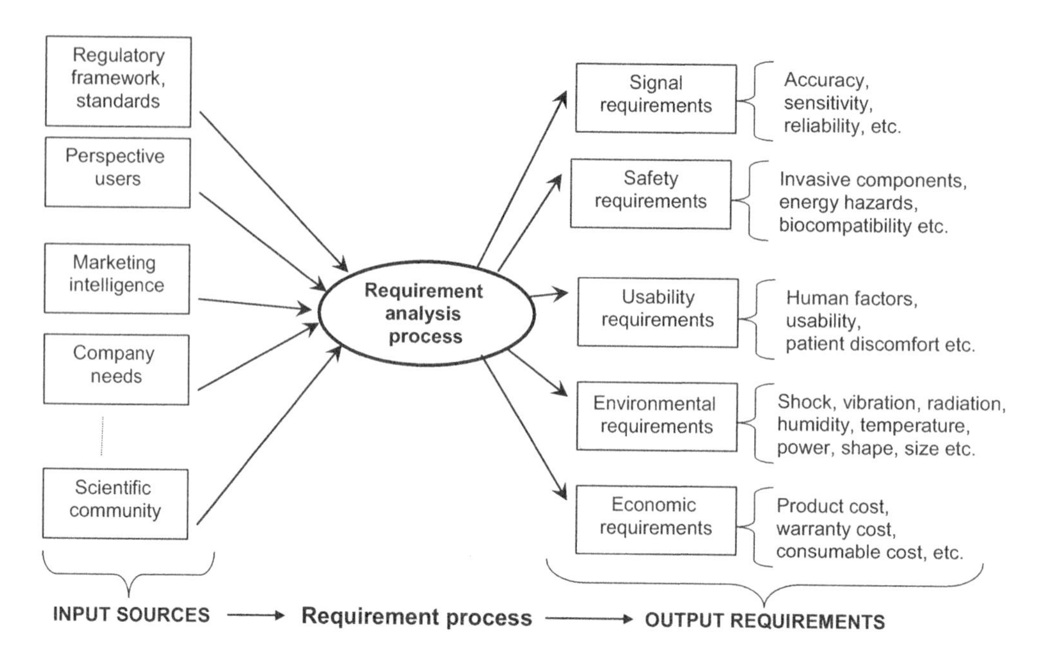

Figure 2.15 Requirement process model.

requirements are removed. In addition, an analysis is needed to evaluate and eliminate unfeasible requirements due to chosen technology, cost constraints or market factors. Every additional constrait usually determines an increase in both product marginal cost and product development cost that cannot be estimated clearly before design (see Chapter 1). Therefore, in the requirement review, care must be taken to consider whether what specified is really compulsory. *The more the requirements, the more the cost, development time and risks.*

The requirement management is a continuous process throughout the project since requirements are validated (i.e., if they are right), verified (i.e., if their implementation is correct) and updated during the various steps of product development.

A **requirement management tool** is often necessary to speed up this process and help team collaboration among designers. In addition, a tool can track each single requirement to the specific design specifications that implement the requirement. The tool helps also in the activity of validation, verification and collection of the proper documentation needed.

The **requirement analysis is an iterative and incremental process** that involves the review of various options in order to obtain the candidate set. The first step of the process includes the definition of functional signal requirements of Figure 2.14 based on standards and clinical research. Then this first set has to be checked with respect to safety. Usability, economic and environment-related requirements conclude the first iteration. Other iterations are usually performed to obtain a proper quality level.

The following section outlines the relation between scientific investigation, standards and R&D efforts in order to generate the more relevant functional requirements.

2.10 Medical Instruments Requirements and Standards

Figure 2.16 shows a possible model for generating requirements for a new product. The process includes the following tasks:

1. select the **patient component** that must be addressed (e.g., the heart and the associated health condition)
2. determine a **method to 'observe'** the patient component
3. define observation parameters that are clinically relevant (i.e., evidencing health condition)
4. evaluate the performance of selected parameters (e.g., bandwidth, accuracy etc.) and the associated requirements
5. generate **safety** requirements related to the patients and operators involved
6. define **usability and marketing** requirements to improve patients' and users' experience
7. define and check **environmental** requirements
8. define and check **economic** requirements
9. iterate the previous eight steps until all requirements are satisfied.

In **step 1**, we focus on parts/features of the body that can be analyzed through the medical instrument that has to be designed. After some iterations of the previous process, the focus of the medical instrument may change to obtain a set of feasible requirements.

Step 2 is clarified by Figure 2.16. For some instruments such as those using X-rays, a probing signal (i.e., the radiation) is delivered to the patient. For an ECG, an output electrical signal is connected to the patient to improve signal accuracy. These probing signals have an impact on patient safety. The signal observations measured from the patient are then recorded. Among the set of observations $\{O_{\mathrm{T}}\}$, the subset of clinically significant

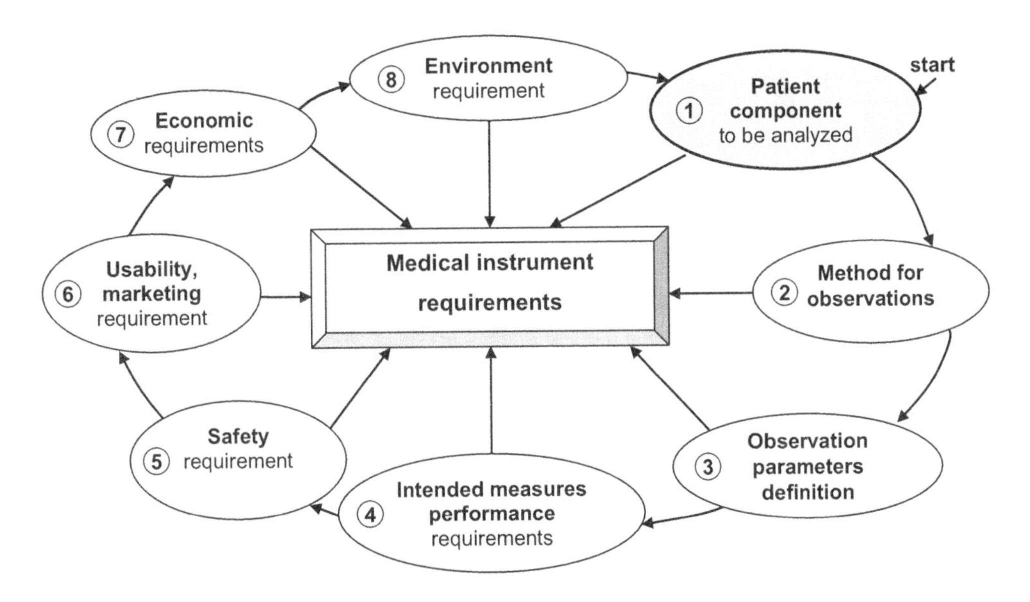

Figure 2.16 Requirement generation scheme.

observations $\{O_C\}$ has to be isolated. This subset will contain clinical information related to the state of the component of the patient to be analyzed. In the **step 3**, this procedure implies a clinical research to compare observations with the patients' condition.

In the **step 4**, the relevant features of the subset $\{O_C\}$ are defined. An example of the features of the subset may be for an ECG: type of signal = unidimensional, bandwidth = [0.05–150 Hz].

These features are used to determine the performance requirements related to the measuring function.

Note that steps 2, 3 and 4 have been particularized for measuring devices. For generic medical devices, items 2, 3 and 4 can be restated to address the performance of the specific device with respect to the intended purpose.

The next step **(step 5)** takes into account patient - operator safety that may be impacted by the device. The risk management process must evidence possible unacceptable risks due to the use of the equipment. Requirements have to be set to reduce the residual risk to an acceptable level. The residual risk is defined as '*the risk remaining after risk control measures have been taken*'. This definition is included in the reference standard used for risk management: 'Application of risk management to medical devices' (IEC/EN 14971, 2009). Other characteristics that impact safety may be evidenced by answering specific questions such as the one included in Appendix C of (IEC/EN 14971). These questions allow the outlining of a complete picture of possible risk, considering energy/substances transferred to the patient, biological materials, sterilization and so on. This topics will be discussed in more detail in Chapter 8.

For example, risks due to hazards derived from transferred energy are those due to current leakage and radiations. Biological or chemical threats may be represented by viruses. Operational failures may be generated by incorrect measurements. Misleading information is usually associated with inadequate labeling or instructions. Table E.1 of IEC/EN 14971 may help in defining other medical device hazards.

Other safety requirements are generally derived from the IEC/EN 60601 family of standards outlined in Table 2.6 that addresses electrical medical products. This family contains also 'vertical' standards related to specific devices such as X-ray equipment or ultrasonic defibrillation equipment.

In **step 6**, it is considered that **usability** requirements are to be introduced to improve usage by doctors and to reduce discomfort to patients. Improvements in usability are addressed to reduce users' training and avoid incidents, improving the overall safety. Medical instrument misuses due to poor interface design is one of the common sources of incidents (Fairbanks, 2004; Kohn, 2000). Some possible errors are due to simple reasons such as: connection swap, misperceived or non-perceived information from the device even due to an unclear display or lack of feedback from the device in critical events. Specific standards such as IEC/EN 60601-1-6 have been introduced to reduce such risks.

Effective usability requirements may also be a critical success factor from the business point of view due to the increased user satisfaction, reduced cost in customer training and improved perceived quality of the product. So usability requirements are also associated to marketing requirements aimed to obtain a more competitive product.

In **step 7**, the requirements related to **environmental** constraints are assessed. Environmental requirements may come from needs related to the conditions where the device is intended to operate (e.g., humidity, temperature, space), to be transported or stored.

Table 2.6 IEC/EN 60601-X main standards

Standard Number	Topic	Title
IEC/EN 60601-1	Safety **general requirement**	Medical electrical equipment – Part 1: General requirements for basic safety and essential performance
IEC/EN 60601-1-2	**Electromagnetic compatibility** requirements and test	Medical electrical equipment. General requirements for safety. Collateral standard. Electromagnetic compatibility. Requirements and test
IEC/EN 60601-1-6	**Usability** requirements	Medical electrical equipment – Part 1-6: general requirements for safety – collateral standard: usability
IEC/EN 60601-1-8	**Tests** and guidance for alarm systems requirements	Medical electrical equipment – Part 1-8: general requirements for safety – collateral standard: general requirements, tests and guidance for alarm systems in medical electrical equipment and medical electrical systems

Finally, in **step 8**, the economic constraints are assessed through the set of proposed requirements and a tradeoff is selected so that the cost target can be met, reducing the economic impact of some requirements. The **economic** target figures are derived from the marketing survey on the prices of competing devices and the associated maintenance and accessory prices as suggested in Chapter 1.

This eight-step iterative procedure allows us to define a set of coherent requirements for a medical device. During design, requirement feasibility is continuously evaluated in order to quickly evidence critical requirements. As discussed in Chapter 1, it is recommended that a specific set of critical requirements has to be continuously monitored throughout the project:

1. signal performances (e.g., frequency response)
2. electrical safety performances (e.g., leakage current, harmful energies)
3. electromagnetic compatibility (EMC) performances
4. mean time between failures (MTBF)
5. production cost.

The **use of standards** have a positive impact on requirement management since the associated requirements are already assessed in terms of feasibility and effectiveness. Standards compliance is not compulsory in medical device design. On the other hand, the use of standards recognized by national/regional authorities usually gives the presumption of conformity with the essential requirements that would otherwise have to be proved. See for example the Medical Device Directive for the European area, (EEC, 2007 in the Summary of

Regulations and Standards at the end of Chapter 8). Standard application implies other positive effects. Standards:

1. are appealing from a customer marketing point of view
2. guarantee product quality for the customer
3. create a product competitive advantage
4. allow for saving in product industrial research
5. reduce risks.

Regarding the last two items, requirement definition implies that a strong effort in product research and requirement feasibility is not guaranteed without standards. Standards reduce such effort, since requirements are derived directly from the standards themselves. In this chapter, standards have been introduced for safety (Table 2.6). For some more common medical devices, there are also standards that define the minimum essential performance requirements.

These **performance standards** may greatly reduce the research effort. Referring to Figure 2.17, since observed signals have to be associated with the medical patient condition with a significant level of correlation, an exhaustive set of patients has to be selected for clinical validation, and valid performance requirements will be obtained only after intensive medical research. Standards such as AAMI EC 11 (AAMI, 1991) for ECG may greatly simplify steps 2–4 of the previous procedure since feasible requirements are clearly defined with associated qualification methods and rationale. In this case, the manufacturer has to design product only to

1. introduce the applicable requirements of the standards
2. justify the rationale for non-applying certain requirements
3. test the device using the provided test methods and procedures or other equivalent tests.

The following sections will show the application of the concepts discussed in this section on an ECG device placed on the market.

2.11 ECG Requirements

The following sections will detail the ECG system requirements based on the model outlined in the previous section. As discussed in Chapter 1, only those characteristics

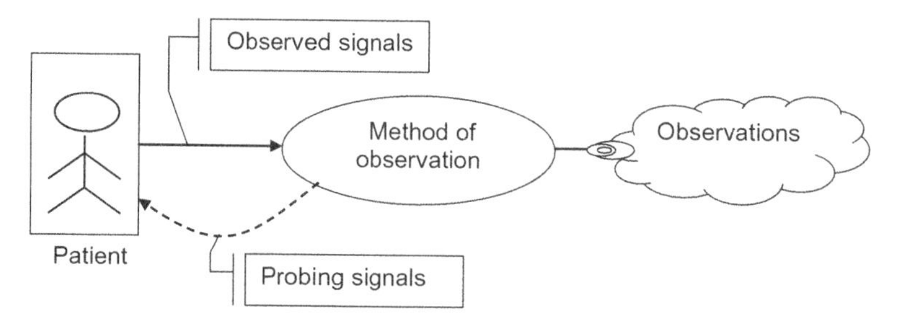

Figure 2.17 Method for observation.

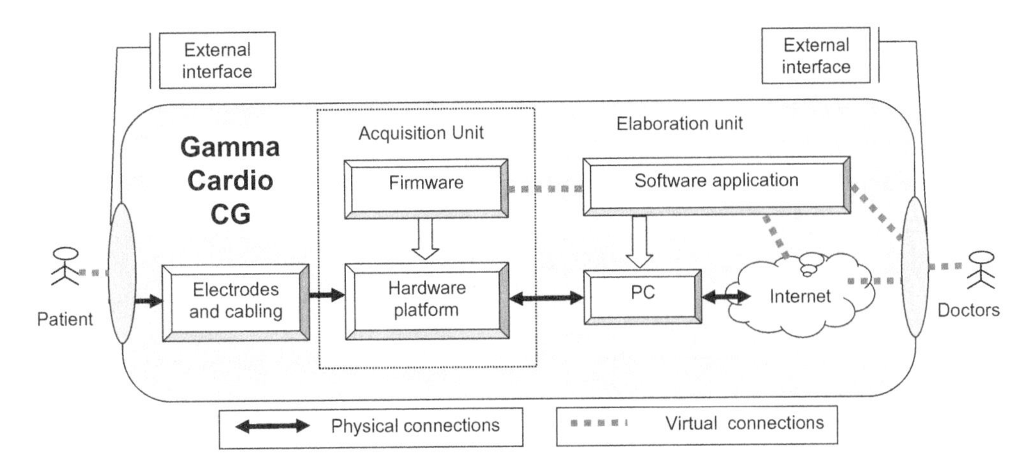

Figure 2.18 Gamma Cardio CG system-subsystem scheme.

that are necessary conditions for system acceptance should be included within the requirements set.

System requirements may be organized in a corresponding document template such as the MIL 498 (MIL, 1994) System/Subsystem Specification (SSS, 1994). The adoption of templates makes it easier to include all the relevant requirements or information that, if missed, might cause extra costs in the following product implementation. For sake of simplicity, a specific template format such as (SSS, 1994) will not be followed; in any case, relevant SSS data items are included in the following sections. The requirements will be generated following the eight-step procedure introduced in the previous section that is summarized in Figure 2.16.

A real case of a medical instrument placed on the market may help in understanding requirement management. As outlined in the previous chapter, the Gamma Cardio CG, a CE certified PC ECG system sold on the EU market and compliant to US ECG standard (AAMI, 1991) will be fully disclosed with the associated requirements.

The following sections will focus on requirements of the Gamma Cardio CG whose conceptual scheme is shown in Figure 2.18. Such figure summarizes also the first requirement (R1). Requirements will be identified with the label 'RX', where X is an increasing numeric index. The following R1 and R2 are the first main requirements:

R1 Product composition:

the Gamma Cardio CG system will have two external interfaces (patient and doctor) and will be composed of:

1) the acquisition unit (firmware and hardware)
2) the 10 reusable electrodes
3) the patient cable
4) the PC cable connection

5) software application running on the PC
6) the physical support where software is memorized
7) the instruction manual
8) the packaging.

R2 Medical Device Directive compliance:

the Gamma Cardio CG system will be compliant to EU Medical Device Directive (see EEC 2007 at the end of Chapter 8).

2.12 The Patient Component

The first step for defining requirements of a medical instrument consists of detailing the component of the body that has to be analyzed. The ECG aims to analyze electrical heart activity and the associated pathologies. In order to understand the requirements and measures obtained from the ECG, a simplified description of heart functions is outlined in the following.

2.12.1 The Heart's Pumping Function and the Circulatory System

From a simplified engineering approach, the heart can be considered as a **double pump** of four chambers (see Figure 2.19 adapted from Figure 2.4) synchronized by a **conduction 'control system'**.

The two upper chambers (right and left atria) are synchronized and their contraction is almost simultaneously. The two lower chambers (right and left ventricles) have a similar behavior. Heart blood circulation has been already addressed in Section 2.4 and it will be here only outlined. In this paragraph, the conduction 'control system' will be also addressed. The pumping action is double since there is first an atrial contraction and then the ventricular contraction.

Referring to Figure 2.19, the **right atrium** receives the blood from the body ① and pumps it to the right ventricle during the atrium's contraction (i.e., atrial systole ②). The **right ventricle** contraction (ventricular systole) flushes blood forward into the lungs ③. Blood is oxygenated in the **lungs** ④ and then it is collected in the **left atrium** ⑤. The atrium's contraction moves blood into **left ventricle** ⑥. Finally, the ventricle's contraction pulls oxygenated blood into the body through the aorta ⑦.

Even in this simplified scheme, some possible failures can be envisaged. The different roles for the four chambers suggest possible insights. For example, the system may work (in degraded condition) even when atria are not working properly. This may happen when atrial contraction is no longer synchronized with ventricles (e.g., when the His bundle, pertaining to the heart control system is blocked as explained in the following). Ventricles start beating at their own rate, allowing for blood circulation, thanks to a secondary pacemaking activity at the lower chamber level: this medical condition, which leads to bradycardia, hypotension and hemodynamic instability, can be detected from the ECG, since two independent rhythms can be noted: one corresponding to the pacemaking activity of the **sino-atrial (SA) node**, and one associated with the secondary pacemaking activity.

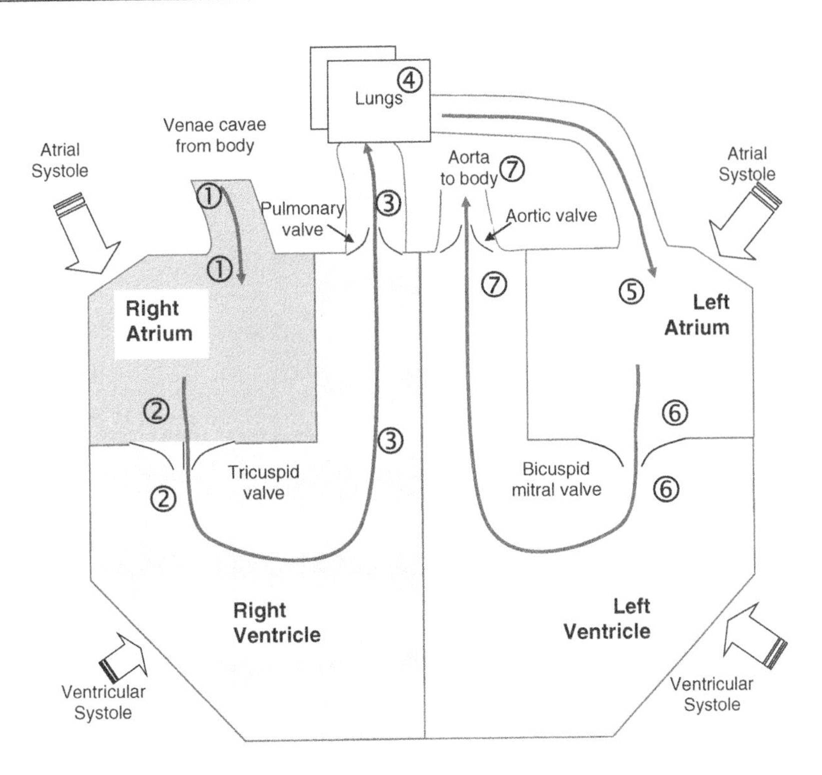

Figure 2.19 Heart blood circulation.

2.12.2 Heart Conduction 'Control' System

The conduction control system of the heart comprises a set of specialized cells that synchronize atrial and ventriclar contraction. The contraction is triggered by an electrical signal originated by the SA node located close to the two vanae cavae in the upper area of the heart. The trigger signal is periodically and spontaneously generated by the SA node (the natural pacemaker) at a frequency rate that is normally between 60 and 100 beats per minute (bpm). The frequency adapts according to body conditions.

The electrical signal is transmitted within both the right atrium and the left atrium through the **Bachmann's bundle** as shown in Figure 2.20.

The electrical signal induces a 'depolarization' of atria cells that, in turn, causes the atrial contraction (also called atrial **systole**). Note that the heart cells, like most other cells, exhibit at rest a voltage potential in the order of tens of mV between the inner and outer parts. In the resting state, the cells are then said to be polarized. When stimulated by an external electric, or electrochemical source over a certain threshold, the cells start a depolarization process. The electric signal is then propagated to the neighborhood cells since heart cells are interconnected through low resistance conduction paths. The cell conduction system can be considered as an active element (with negligible attenuation) with respect to the propagated signal that has an intrinsically digital nature – an all-or-nothing behavior. The excitation signal induces a depolarization that in turn stimulates contraction in muscle type cells. This step is then followed by a repolarization phenomenon that leads cells again to the resting state.

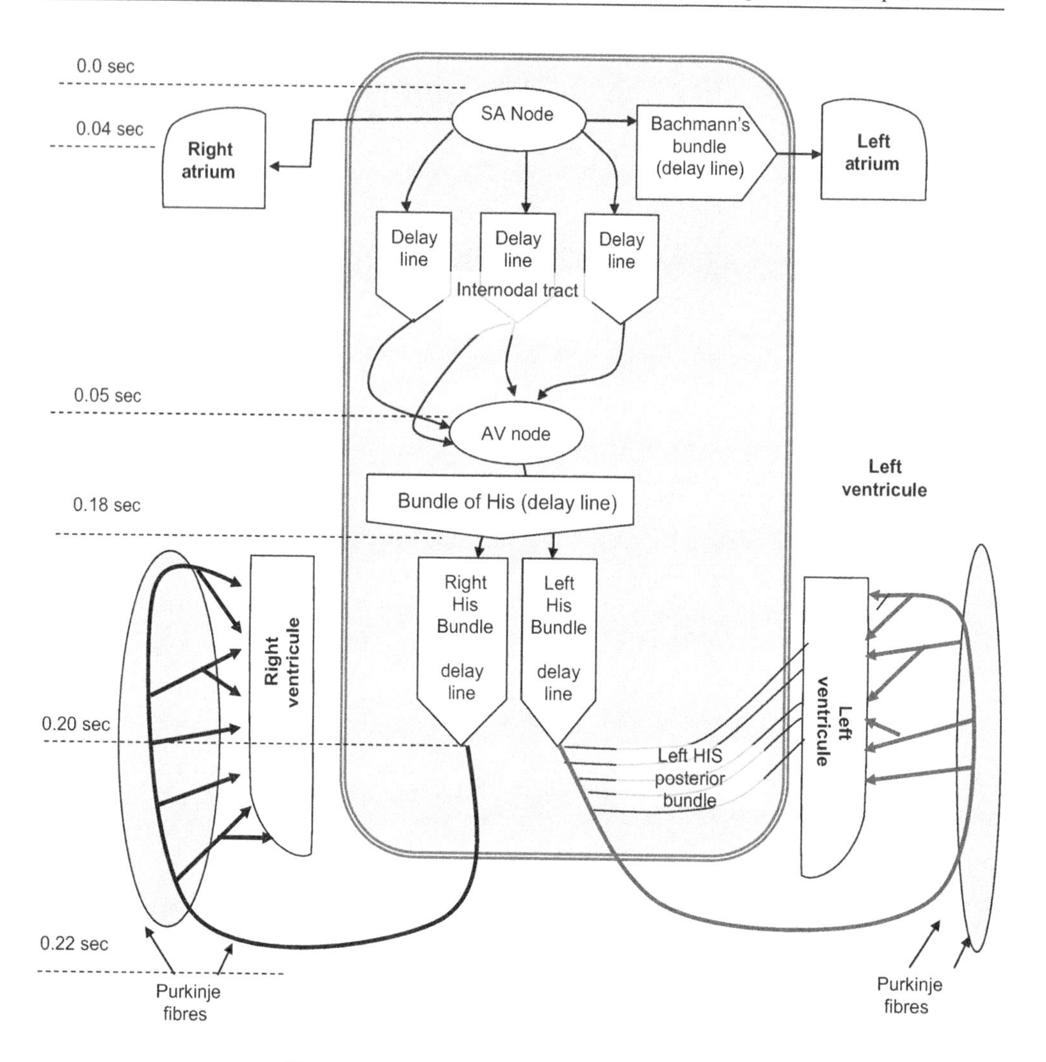

Figure 2.20 Heart 'conduction control system'.

Depolarization and repolarization are purely electric phenomena of heart cells due to chemical factors (sodium and potassium ion exchange) that are followed by mechanical changes (i.e., contraction, relaxation) (Dubin, 2000). During atrial contraction, the electrical signal is transmitted through the internodal tract to the **atrium ventricular node** (AV) located between the right atrium and the right ventricle. Note that between ventricles and atria there is an electrically insulated surface that inhibits premature ventricular contraction triggered by the electrical signal generated by the right atrium polarization.

The atria pump the blood to the ventricles. After atrial contraction, some delay is therefore needed so that all fluid contained in the atria to be transferred to the ventricles before ventricular contraction. On the left side of Figure 2.20, an approximated propagation time of the electrical signal within the heart is shown.

The most of the delay is experienced within the AV node and the following **His bundle** where the electrical control signal is transmitted to the ventricles. The His bundle splits into three branches: the right, the left anterior and the left posterior bundle that transmits the signal to the ventricles. The signal is then transmitted to the ventricular walls by the Purkinje fibers so that ventricles start contraction. After ventricular contraction, the atria and ventricles start repolarization to go into a relaxing resting state (also called **diastole**). The complete process of a single heartbeat requires a time given by 60/(beats per minute abbreviated as bpm) seconds that is in the range of 0.6 to 1 second for a normal beat rate.

The previous explanation, even if oversimplified, may already suggest possible failures. For example, the time of arrival of the electrical signals along atria and ventricles is tuned to fulfill the mechanical features of the circulation systems and the heart pump. Frequency rates far different from the normal range at rest (60–100 bpm) may indicate some critical condition and in turn they may not allow correct mechanical fluid circulation and atrium/ventricle filling. Another possible problem arises when the conduction system is partly or totally interrupted, as for example for a His bundle complete block. In this case, ventricles will beat independently with respect to atria contraction. This is due to cardiac muscular cell features that may start pacemaking activity autonomously. These cells have a non-stable membrane potential that tends to rise to positive values until the positive excitation threshold is reached. At that time, a contraction is obtained. Other heart pathologies will be outlined in Section 2.14.

2.13 The ECG Method for Observation

During the process of requirement definition, we intend to outline also the basic principles underlying the body component to be analyzed. In the previous section, the heart functions were described. In this section, the physical principles underlying the measurement of the heart's electrical signals are first described (Section 2.13.1) to allow readers a better understanding of the selected method of observation.

As described in Section 2.11, in the medical instrument requirement process, a method of observation has to be defined and then it has to be improved, analyzing the clinical relevance of the observed signals.

For ECGs, this process has been in place for more than a century and the observation method is now consolidated in clinical practice as explained in 2.13.2. This method is described in Section 2.13.3.

2.13.1 Recording the Heart's Electrical Signals

The electrical depolarization and the repolarization of heart cells is related to the contraction and relaxation of atria and ventricles, and therefore the measure of the electrical variations of polarization may indicate when and how contractions of heart chambers take place.

The cardiac muscle performs a contraction due to the effect of the propagation of the depolarization along a specific direction. The whole heart system does not simultaneously depolarize or repolarize. The depolarization moves from the inner surface of the ventricular walls to the outer surface and it carries on from top of the heart to the ventricles' bases. This means that there is a different distribution of negative and positive charges that changes with time. An ionic current is generated that, in turn, produces a different value of the electric potential between two areas of the body where the current flows. The different charge

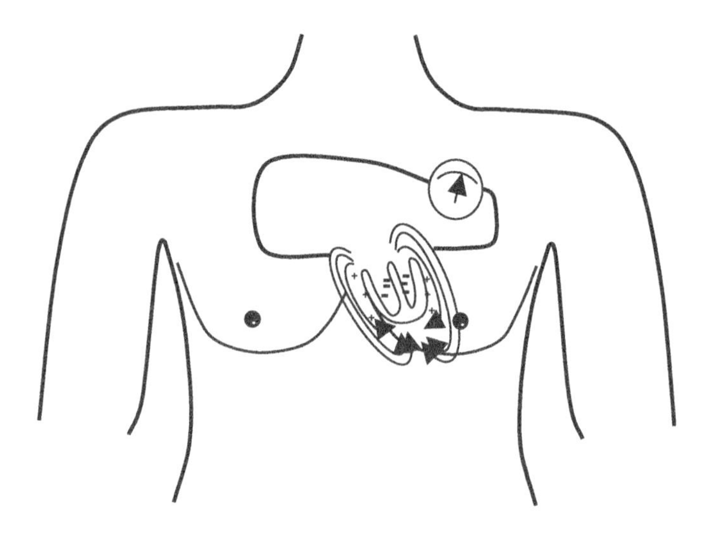

Figure 2.21 The measure of heart potential.

distribution generates an electrical field that changes in intensity and in the direction of the equipotential lines. This potential may be measured as shown in Figure 2.21.

The heart is contained in the thorax, which can be considered roughly as a saline solution with isotropic behavior from an electrical conductivity point of view. This rough and oversimplified assumption allows for deriving a simple heart model.

The electric potential E generated by an electric charge q at a distance r may be expressed by:

$$E(r) = \frac{q}{4\pi\varepsilon r} \tag{2.24}$$

where ε is the dielectric constant.

It is worth noting that a set of electric charges distributed in a specific area has an overall resulting effect of a single charge of appropriate value when measured at a distance that is much greater than the distances between the charges themselves. This allows representing the overall set of positive charges as a unique charge whose value corresponds to the sum of the existing charges. The position of the center of the charges is located at some point inside heart. The same applies for negative charges whose center of charge is normally non-coincident with the center of the positive charges in the depolarization/repolarization steps.

As shown in Figure 2.22, this enables us to define an electric dipole that is characterized by

- the value of the electric overall charge q,
- the distance of the two charges δ,
- the axis of dipole (i.e. direction given by the line connecting the two charges.

We can now consider measuring the heart's electric field at the skin surface through electrodes with the assumptions to be listed in Figure 2.23.

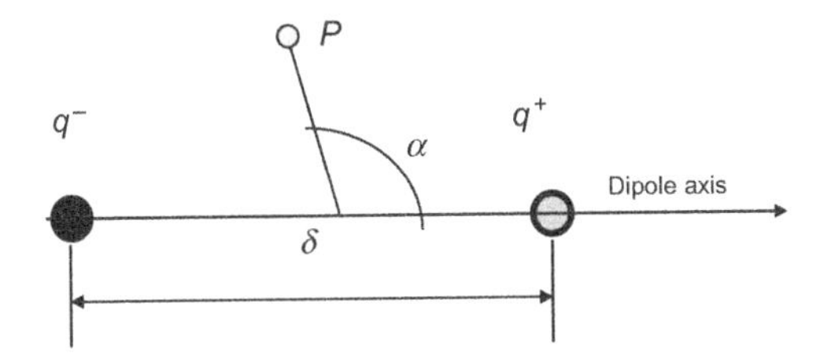

Figure 2.22 Schematic of an electric dipole.

1. the heart is included in a homogenous body conductor
2. the heart generator is a single dipole whose position is at the center of the equilateral triangle (Figure 2.24)
3. the heart is located far from the points of measurement
4. the thorax's limited dimensions do not influence the heart potential
5. the skin electrodes do not introduce interferences in the measurements

Figure 2.23 Assumptions related to the single dipole model.

The measurement points P may be set at the apices of an equilateral triangle. The center of the positive and negative charges produced by the heart generator is located on a single line at distance ($\delta/2$) from the triangle's center, as shown in Figure 2.24, where α is the angle between the dipole axis and the x axis.

The magnitude M of the heart generator dipole $\vec{H}$ is $M = q\delta$; the axis coincides with the direction of the line connecting the two charges. The electric potential $E(P_{\mathrm{LA}})$ at the point P_{LA} of Figure 2.24 is given by:

$$E(P_{\mathrm{LA}}) = \frac{1}{4\pi\varepsilon}\left(\frac{q^{+}}{r_2} + \frac{-q^{-}}{r_1}\right) = \frac{q}{4\pi\varepsilon}\left(\frac{r_1 - r_2}{r_2 \cdot r_1}\right) \tag{2.25}$$

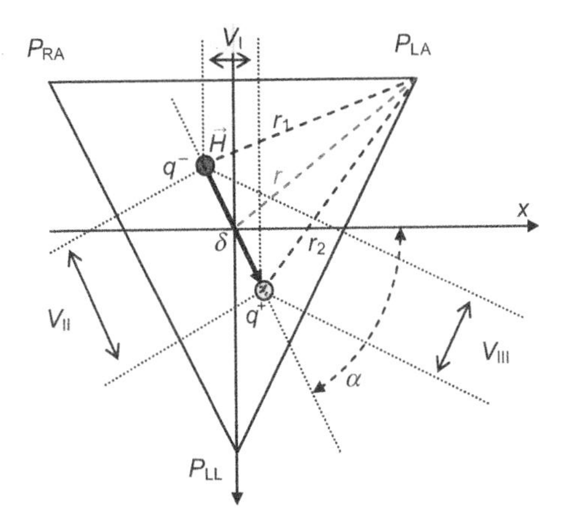

Figure 2.24 Measurements from the ECG triangle.

Considering the simplified assumption that measures are far away from the heart (i.e., $\delta \ll r$), the following holds: $r_1 \cdot r_2 \cong r^2$, r_1, r_2 are approximately parallel and therefore after some trigonometric considerations:

$$r_1 - r_2 \cong \delta \cdot \cos(150^\circ - \alpha) \tag{2.26}$$

where $\cos(\cdot)$ is the cosine expressed in degrees. The voltage potential in the point P_{LA} simplifies to

$$E(P_{LA}) = \frac{-q\delta}{4\pi\varepsilon r^2}\cos(150^\circ - \alpha) \tag{2.27}$$

Considering that the potential in P_{LA} is equal to the potential in P_{RA} when the $\vec{H}$ dipole axis is rotated by 120° anticlockwise we have

$$E(P_{RA}) = \frac{-q\delta}{4\pi\varepsilon r^2}\cos(30^\circ - \alpha) \tag{2.28}$$

and since

$$\cos(30^\circ) = -\cos(150^\circ) = \sqrt{3}/2, \cos(\beta - \alpha) = \cos(\beta)\cos(\alpha) - \sin(\beta)\sin(\alpha)$$

the voltage difference measured between the two points is

$$V_I = E(P_{LA}) - E(P_{RA}) = \left(\frac{\sqrt{3}}{4\pi\varepsilon r^2}\right)M\cos(\alpha) \tag{2.29}$$

where $M = q\delta$ is the module of the dipole vector. The term V_I in equation (2.29) is equal to the projection of the dipole on the P_{LA}– P_{RA} line. Equation (2.29) proves that with the simplified assumptions of Figure 2.23, the voltage differential of any couple of apices P_x– P_y is equal to the projection of the $\vec{H}$ dipole on the line connecting the two apices.

The voltage given by the difference of the other combinations of the three apices shown in Figure 2.24 (i.e. $V_{II}V_{III}$) are easily obtained through other calculations.

Under the previous assumptions, the heart potentials may be measured at the apices of the equilateral triangle. It is then considered that the triangle apices may be located at the left and right shoulders and the left hip, trivially assuming that these body parts are far away from the heart and that the heart is at the center of the triangle. Since it is often uncomfortable to locate electrodes at these positions, the electrodes are located at the left and right wrists and at the left ankle, assuming that limbs are ideal conductors.

The $V_I V_{II} V_{III}$ are defined as **leads** and they are expressed by

$$V_I = E(P_{LA}) - E(P_{RA}), V_{II} = E(P_{LL}) - E(P_{RA}), V_{III} = E(P_{LL}) - E(P_{LA}) \tag{2.30}$$

Where the letters LL, RA, LA indicate the points of measurement, that is

LL = left leg ≅ left hip,
RA = right arm ≅ right wrist
LA = left arm ≅ left wrist

These three leads introduced by Einthoven as in (2.30) are the first three leads of the 12 standard ECG lead that will be introduced in Section 2.13.3. Figure 2.24 is referred to as the Einthoven triangle and the mathematical relations among V_{I}, V_{II} and V_{III}:

$$V_{\mathrm{II}} = V_{\mathrm{I}} + V_{\mathrm{III}} \tag{2.31}$$

is cited as the Einthoven Law.

The previous measures derive from a model whose assumptions are quite critical. Other models have been derived with possibly increased *diagnostic performance* (e.g., see (Malmivuo, 1995) for a discussion), but the Einthoven's leads still remain the most relevant in medical practice because of the clinical validation performed for more than a century. More simply, taking into account Figure 2.17, there is such a consolidated clinical experience with Einthoven's leads and the 12-lead standard method of observation that other methods are usually neglected for conventional medical usage.

The Einthoven model demonstration has been introduced in this book because of its simplicity. The real value of this oversimplified model is the huge clinical experience that makes this model more acceptable for standard clinical usage with respect to other even more effective methods. The next few sections will introduce the other leads that make up the 12-lead standard ECG.

2.13.2 ECG Definition and History

According to the FDA classification (CFR, 2012), an electrocardiograph (abbreviated as ECG/EKG) is a '*device used to process the electrical signal transmitted through two or more electrocardiograph electrodes and to produce a visual display of the electrical signal produced by the heart*'. The ECG has a long history, since first experiments on heart electrical activity go way back to 1870. In 1906 there is the first publication showing normal and abnormal ECG recordings due to Einthoven (Einthoven, 1906). The first method for observation with standard leads $\mathrm{V_I}$ $\mathrm{V_{II}}$ $\mathrm{V_{III}}$ (also simply referred as I, II, III) is reported in (Einthoven, 1912). The standard 12-lead ECG was completed in 1938 for chest leads $\mathrm{V_1}$–$\mathrm{V_6}$ (Barnes, 1938) and in 1942 for augmented limb leads $\mathrm{aV_R}$, $\mathrm{aV_L}$ and $\mathrm{aV_F}$ due to Goldberger (Fye, 1994). These additional leads will be illustrated in the next section. The clinical usage of this device is then particularly consolidated. Despite its age, more than 100 years, ECG is still a research field useful for diagnosing significant new pathologies. For example, in 1992 a new disease ('Brugada Syndrome') was discovered. Brugada Syndrome accounts for 'the commonest cause of sudden cardiac death in individuals aged under 50 years in South Asia' (Brugada, 1992).

2.13.3 ECG Standard Method of Observation

From the previous sections, it is clear that different positions of electrodes lead to different results. It is therefore compulsory to have a standard electrode positioning in order to compare results and obtain the clinical significance.

The 12-lead standard ECG is composed of

- Einthoven leads I, II, III (or $\mathrm{V_I}$ $\mathrm{V_{II}}$ $\mathrm{V_{III}}$)
- augmented Goldberger limb leads $\mathrm{aV_R}$, $\mathrm{aV_L}$ and $\mathrm{aV_F}$
- six chest leads $\mathrm{V_1}$–$\mathrm{V_6}$.

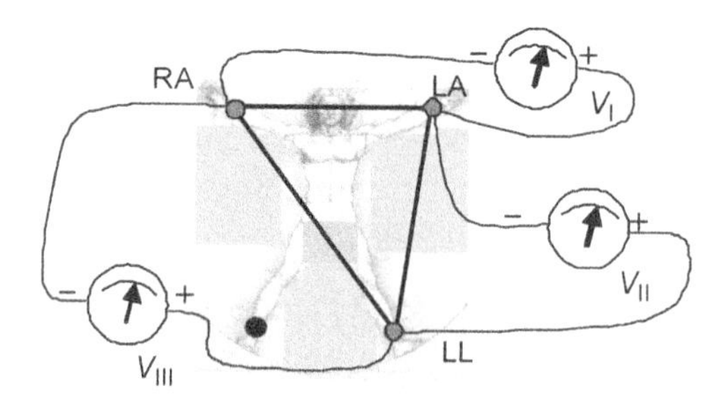

Figure 2.25 The Einthoven leads.

Note that $\mathrm{V_x}$ indicates the x lead while V_x with V in italic, denotes voltage corresponding to the x lead. **Einthoven leads** have been introduced in the previous sections. In Einthoven's model, the body is a homogenous conductor with much larger volume with respect to the heart's electrical generator. The electrodes are then positioned at the apices of an equilateral triangle as depicted in Figure 2.24 and the heart is positioned at the center.

The obtained measures show the projection of the cardiac vector $\vec{H}$ along the three directions of the '**frontal plane**' parallel to the sides of the triangle of Figure 2.24. The 'frontal plane' is the plane parallel to the front containing the shoulders.

Taking into account equation (2.30) and defining V_{LA}, V_{RA}, V_{LL} as the potentials measured at the left arm (i.e., left wrist), right arm (i.e., right wrist) and left leg (i.e., left hip) respectively, the Einthoven leads are given by Figure 2.25 as

$$\text{Lead I}: \mathrm{V_I} = \mathrm{V_{LA}} - \mathrm{V_{RA}},\ \text{Lead II}: \mathrm{V_{II}} = \mathrm{V_{LL}} - \mathrm{V_{RA}},\ \text{Lead III}: \mathrm{V_{III}} = \mathrm{V_{LL}} - \mathrm{V_{LA}} \tag{2.32}$$

These three leads are also referred to as bipolar leads since they measure the difference between two poles. The other leads are instead **unipolar** since they measure the difference of one pole with respect to a reference point obtained as an average of the signals measured at the electrode positions LL, LA and RA.

The so called '**Wilson reference point**' is obtained by connecting together all the three previous electrodes through three identical resistances with a suitable value R.

If the electrocardiograph has an high input impedance with respect to R then the current flow toward the measuring instrument is negligible and the Wilson reference point signal V_{W} is equal to the average of the three signals:

$$V_{\mathrm{W}} = \frac{V_{\mathrm{LA}} + V_{\mathrm{RA}} + V_{\mathrm{LL}}}{3} \tag{2.33}$$

Referring to Figure 2.26, equation (2.33) can be demonstrated assuming that one of the three voltages is non-null at time among V_{LA}, V_{RA}, V_{LL} (e.g., $V_{\mathrm{LA}} \neq 0$ and V_{RA}, $V_{\mathrm{LL}} = 0$) and then superimposing the effects of each voltage generator V_{LA}, V_{RA}, V_{LL}. The voltage at Wilson reference point due to V_{LA} with V_{RA}, $V_{\mathrm{LL}} = 0$ is given by

$$V_{\mathrm{W}}(V_{\mathrm{LA}}) = I_{V_{LA}} \cdot R/2 = \frac{V_{\mathrm{LA}}}{R + R/2} \cdot R/2 = \frac{V_{\mathrm{LA}}}{3} \tag{2.34}$$

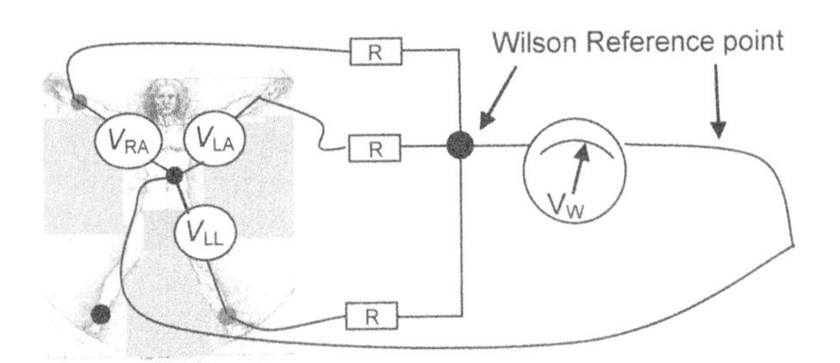

Figure 2.26 Wilson central reference terminal.

The same applies for V_{RA}, V_{LL}. Equation (2.33) is then obtained summing the potentials of the three voltage generator V_{LA}, V_{RA}, V_{LL} at the Wilson reference point.

This justifies the nature of the reference voltage terminal of the Wilson point since it is somehow independent of one particular lead. The Einthoven leads I, II, III measure the components of the cardiac vector $\vec{H}$ along the 'frontal plane' and from a diagnostic point of view do not reveal all the information that is present in the cardiac vector on the other planes. The measures over the projections of the cardiac vector $\vec{H}$ along the '**transversal plane**' (i.e., the orthogonal plane with respect to the head–feet axis) contain the rest of the information required to explain most of the heart's electric activity.

The six standard leads V_1–V_6 over the transversal plane are obtained as the difference between the Wilson central terminal and the signal obtained by an electrode positioned on the specific chest positions indicated in Figure 2.27 and defined in the following (AAMI, 1993):

V_1: 4th intercostal space at the right of the sternum
V_2: 4th intercostal space at the left of the sternum
V_3: between leads V_2 and V_4
V_4: 5th intercostal space at the mid-clavicular line;

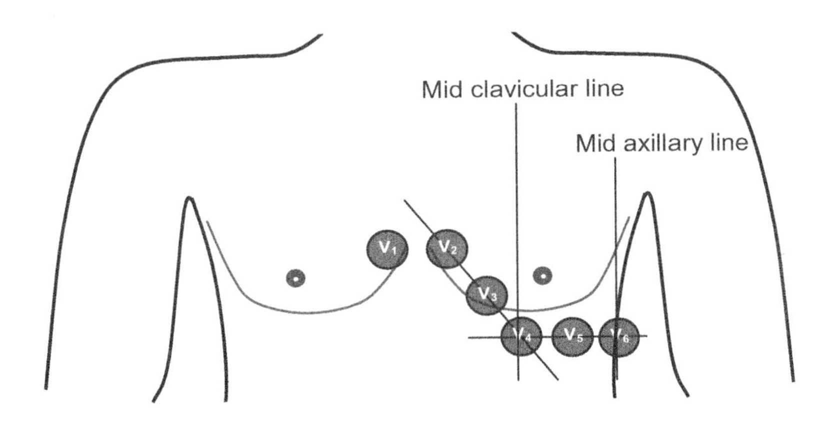

Figure 2.27 Precordial leads.

V_5: at the same horizontal level as V_4, vertically to the left anterior axillary line (the anterior axillary line is the imaginary line that runs down from the point midway between the middle of the clavicle and the lateral end of the clavicle; the lateral end of the collarbone is the end closer to the arm)

V_6: at the same horizontal level as V_4 and V_5, vertically in the mid-axillary line. (The mid-axillary line is the imaginary line that extends down from the middle of the patient's armpit).

The other three leads in the frontal plane that complete the 12-lead standard ECG are also unipolar since they measure the difference of one pole with respect to a reference point. For example, for the left leg V_F we have

$$V_F = V_{LL} - V_W = \frac{2 \cdot V_{LL} - V_{LA} - V_{RA}}{3} \tag{2.35}$$

It was noted by Goldberg (Malmivuo, 1995) that omitting the signal to be measured (e.g., V_{LL}) from the Wilson reference point, the overall voltage aV_F is 3/2 of the previous voltage V_F:

$$aV_F = V_{LL} - \frac{V_{LA} + V_{RA}}{2} = \frac{2 \cdot V_{LL} - V_{LA} - V_{RA}}{2} \tag{2.36}$$

The resulting lead aV_F is called the augmented lead. The **augmented leads** (aV_R, aV_L and aV_F shown in the equation (2.37)) are the last three leads included in the 12-lead standard ECG:

$$aV_L = V_{LA} - \frac{V_{LL} + V_{RA}}{2}, aV_R = V_{RA} - \frac{V_{LL} + V_{LA}}{2} \tag{2.37}$$

Referring to Figure 2.28, these leads are still unipolar since the reference point is obtained from the average signal of the remaining two electrodes with respect to the electrode where the signal is measured. The augmented leads reveal other diagnostic information since it can be demonstrated that they show the projection of the cardiac dipole

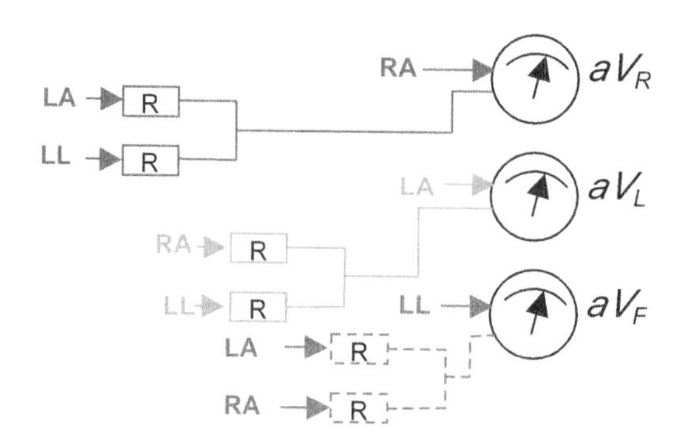

Figure 2.28 Augmented leads.

rotated by 30° anticlockwise with respect to the Einthoven triangle. The demonstration is left to the reader. This new information is obtained through a linear combination of the already available signals obtained by the same electrodes RA, LA and LL.

Taking into account equations (2.36) and (2.37), the augmented leads are linearly related to the Einthoven leads as follows:

$$aV_R = -\frac{1}{2}\cdot(V_I + V_{II}), aV_L = \frac{1}{2}\cdot(V_I - V_{III}), aV_F = \frac{1}{2}\cdot(V_{II} + V_{III}) \tag{2.38}$$

Using trigonometric computation, equations (2.38) can be obtained also through the projection of the cardiac vector $\vec{H}$ over an equilateral triangle rotated by 30° anticlockwise with respect to the Einthoven triangle considering that aV_R is multiplied by -1. More specifically, let us assume that $\vec{H}$ is a vector with Magnitude M and direction defined by an angle α. Defining with "angle(V_x)" the angle associated to the projection V_x, the Einthoven leads are given by:

$$V_x = M\cos(\alpha + \text{angle}(V_x)) \tag{2.38a}$$

that is:

$$V_I = M\cos(\alpha + 0°), V_{II} = M\cos(\alpha + 60°), V_{III} = M\cos(\alpha + 120°) \tag{2.38b}$$

It can be proven that for the augmented leads the previous formulas become:

$$aV_x = \left(\sqrt{3}/2\right)M\cos(\alpha + \text{angle}(aV_x))$$

that is:

$$aV_R = \left(\sqrt{3}/2\right)M\cos(\alpha + 90°), aV_L = \left(\sqrt{3}/2\right)M\cos(\alpha - 30°) \tag{2.38c}$$

$$aV_F = \left(\sqrt{3}/2\right)M\cos(\alpha - 150°) \tag{2.38d}$$

The previous (2.38) equations may be used to estimate the axis from the ECG signal using a simple linear system. It is now possible to define leads over the frontal plane as shown in left side of Figure 2.29. Figure 2.29 shows also the transversal plane given by the position of the electrodes.

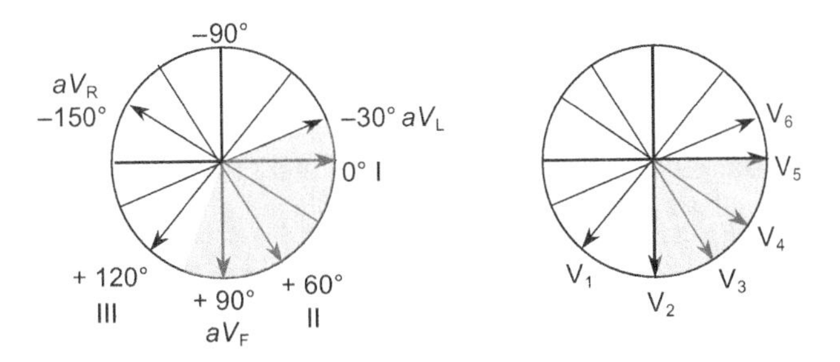

Figure 2.29 Frontal plane and transversal plane.

Some considerations are useful for implementations. According to (Geselowitz, 1964) 'over 90% of the heart's electric activity can be explained with a dipole source model'. From a theoretical point of view, three independent leads are sufficient to describe the dipole under the assumptions included in Figure 2.23. For example, the leads I, aV_F and V_2 may be considered suitable since they are orthogonal. However, the chest leads V_1–V_6 also have non-dipolar components, with additional diagnostic significance because of their proximity to the frontal part of the heart (Malmivuo & Plonsey 1995). The independent minimal leads are therefore the six precordial leads plus two of the frontal plane leads (e.g., I and aV_F). The other six frontal plane leads can be derived by equations (2.31) and (2.38). This means that the 12-lead standard ECG can also be obtained through the acquisition of just eight leads where the other four leads can be computed from the original eight leads. This consideration allows the saving of electronic circuitry, since four leads can be computed via software. From an implementation point of view there may be some problems, for example: in case of limited resolution of the available signals, the derived leads may contain considerable numerical errors.

2.14 Features of the Observations

Referring to the requirements generation approach outlined in Figure 2.16, in this section, we introduce the features of the observation signal. Such features will be used to derive requirements for the ECG signal processing modules.

The following steps are needed to obtain such features:

1. analyze the features of the observed signals (Section 2.14.1)
2. define the set of clinically relevant signals (Section 2.14.2)

This allows us to derive the overall signal requirements discussed in Section 2.15. It is also important to highlight that **artifacts** may arise from the method of observation. Such artifacts that are superimposed on the clinically relevant signal may suggest wrong diagnoses and therefore they are to be taken into account in order to evaluate possible countermeasures in the requirements.

The main ECG artifacts are generated by

- electric power system (Section 2.14.3)
- muscle tremor (Section 2.14.5)
- electrode contacts and other causes that determine baseline wandering (Section 2.14.4).

2.14.1 ECG Signal

In Figure 2.30, a typical ECG waveform of a single heart cycle is shown. The horizontal axis refers to time while the vertical axis corresponds to voltage. The ECG waveforms are usually printed on paper with a background pattern of 1 mm squares and bold vertical and horizontal lines printed at 5 mm intervals, as shown in Figure 2.30. Such paper is referred to as **pre-ruled paper**. The vertical and horizontal scales can be changed within predefined values.

The dimensions of the waveforms in Figure 2.30 are real with vertical scale of (20 mm/mV) and horizontal scale of (50 mm/sec). Each bigger square has a 5 mm size that corresponds to a vertical size of 0.25 mV and a horizontal size of 0.1 sec. This scale is twice

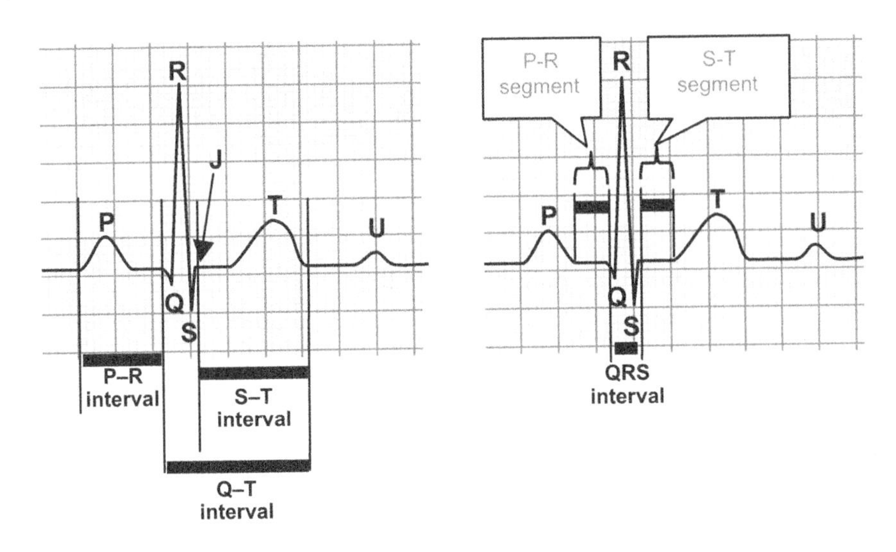

Figure 2.30 Single cycle typical ECG (20 mm/mV and 50 mm/sec).

as large as the standard scale (10 mm/mV and 25 mm/sec) to allow a clearer indication of intervals and segments. The represented waveform shows the systole and diastole of heart atria and ventricles. The waveform deflections are labeled with letters (P to U). The signal is constituted by waves (P, T, U) and the QRS complex delimited by the QRS deflections and the **isoelectric** (i.e., the zero potential) segments.

The first deflection, defined as the **P waveform**, is due to the depolarization of atria. Referring to Figure 2.20, the electric signal is propagating from the SA node towards the AV node, and from the right atrium to the left atrium.

The **P–R segment** is related to the propagation of the electrical signal from the AV node to the bundle of His and to the bundle branches up to the Purkinje fibers. In this interval, the ventricles are filled with blood.

Since the conduction system electrical activity is too limited, such activity is not recorded by non-invasive ECG that is able to record only the higher electrical levels of depolarizations/repolarizations generated by atrial and ventricular systole/diastole. In intracardiac electrocardiograms, where electrodes are positioned directly on the heart, the depolarization of the bundle of His is also visible through a specific **H deflection**.

The **QRS complex** is due to the rapid depolarization of ventricles. Since ventricles perform pumping into the external circulation system, they require more pumping capacity. This performance is obtained through a different thickness of the muscle surrounding ventricles. This, in turn, produces a larger depolarization/repolarization and a consequently larger electric field that is recorded in higher amplitudes of the QRS complexes compared to the P waves.

The Q point is defined as the first negative deflection, R is the successive positive deflection and S is the following negative deflection.

The **J point** is defined as the point where the QRS complex and S–T segment connect. This point is clinically useful for measuring the S–T segment elevation or depression, indicating possible infarction.

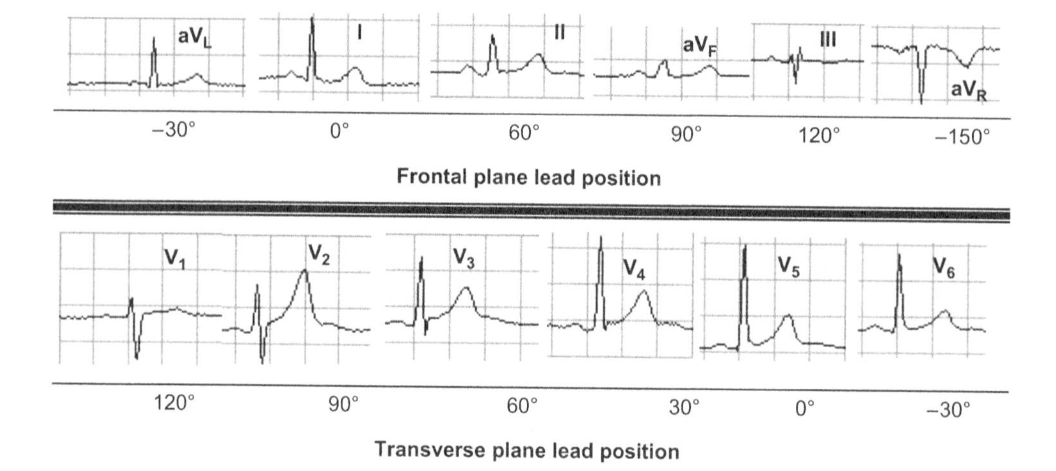

Figure 2.31 Typical 12-lead ECG.

The **S–T segment** is the period where ventricles are in a depolarized status (ventricular contraction).

Ventricular repolarization generates the **T wave** that is sometimes followed by a U wave that may be due to the repolarization of the interventricular septum. The **U wave** is normally low amplitude, or often is absent.

Note that atrial repolarization is of low amplitude and is masked by the QRS complex and therefore it cannot be clearly identified in the ECG signal (Dubin, 2000).

The isoelectric line is normally estimated between the end of the T wave and the beginning of the next P wave.

A 12-lead standard ECG consists of the 12 leads discussed in the previous sections (three Einthoven's leads (I, II, III), augmented leads (aV_F, aV_L, aV_R) and the chest leads ($V_1 - V_6$), as shown in Figure 2.31.

Figure 2.31 also shows the lead recording with respect to the frontal and the transversal plane as in Figure 2.29. These traces show that leads can be considered as projections of the heart dipole on the two planes. More specifically, in the frontal plane, the **dipole axis** is oriented at around 30° (between lead I and II) since the maximum positive of the QRS complex is in $-aV_R$ that is positioned at 30° (=180° − 150°). In Lead III at 120° (=30° + 90°) the QRS complex is almost isoelectric (i.e. positive and negative deflections are similar in amplitude this confirms that the dipole (at 30°) is perpendicular (+90°) to this lead (120°). For the transversal plane the dipole is at around 0° and consequently, at 90° the QRS deflections are almost equal.

2.14.2 *Clinically Significant Signal*

In this section, the clinically relevant information embedded in the ECG signal is introduced. This allows us to obtain the features of the clinically relevant signal required to dimension the signal-processing modules of the instrument.

In an ECG signal, the components that are relevant for a physician are

- the waves' morphology (i.e., pattern shapes)
- the amplitude of some waves
- the length of some intervals
- the axis of the dipole vector for specific waves/complexes.

From such ECG elements, physicians deduce the possible heart pathology, reported in the **ECG interpretation**.

The clinically significant wave patterns are related to the specific shapes. The amplitudes and the intervals are measured on the pre-ruled ECG paper considering that the vertical and horizontal ECG scales are set to 10 mm/mV and 25 mm/s (each horizontal small box = 0.04 sec).

Referring to Figure 2.29 and Figure 2.31, the **axis** of the QRS complex in the frontal plane may be evaluated individuating the lead where the R deflection height is closer to the S height. In this case, the QRS is said to be 'isoelectric' and the axis is perpendicular to the specific lead. To individuate the quadrant (i.e., the 90° wide sector where the dipole lies), if the QRS is positive (i.e. mainly above the isoelectric line) in lead I, then the axis is directed to the right quadrants, and if the QRS is positive in lead aV_F, then the axis is directed to the lower quadrants.

Some normal indicative values for an ECG are shown in Table 2.7 extracted from various sources including (Harrison, 2011) and (Dubin, 2000). Note that normal values may differ in various literature sources. The normal values also vary according to patient features such as age and gender (Philips, 2003). Modern electrocardiographs offer assisted tools to physicians

Table 2.7 Indicative normal values for ECG of adult patients

Feature	Indicative value
Frequency	60–100 bpm, corresponding to RR interval 1–0.6 seconds
Rhythm	Sinus rhythm: every P wave followed by a QRS complex, every QRS complex is preceded by a P wave, P interval $l \geq 120$ ms, P wave positive in I, II, III leads
PR segment	Width 120–200 ms
QRS complex	Width 60–100 ms
QRS axis	−30° to 100° on the frontal plane and isoelectric on the transversal plane as shown in Figure 2.29 by the gray areas
QT interval	< 430 ms, $< 50\%$ RR interval
P wave	– Monophasic behavior (i.e. the morphology does not cross isoelectric line), – height ≤ 2.5 mm positive in I, II, aV_F, V4, V5 and V6
Q wave	– maximum height 25% of the following QRS height in I, II, aVF, precordials and aVL (except when QRS axis > 75°) – length ≤ 40 milliseconds (except for aV_R and III)
ST segment	maximum 1 mm deviation from the isoelectric line

for both measurements and suggested interpretations. The associated algorithms for computer assisted ECG analysis show some interesting insights from an engineering point of view on the algorithms to derive an ECG interpretation (Philips, 2003).

Note that normal values do not automatically determine a non-pathological heart, and similarly *a computer-interpreted ECG with a non-pathological report does not automatically imply a healthy patient*. The analysis from a qualified physician is therefore essential for a confirmed interpretation of the ECG. Such interpretation takes into account other knowledge about the patient and possibly other findings/results from other clinical examinations.

The evaluation of the normal values, as well as the computer-assisted ECG interpretation, is simply a support to identify possible problem areas.

A first rough consideration, based on the topics discussed in this section, suggests that the ECG device must not change morphology and point of measures to avoid misleading ECG interpretation. It is also important to take into consideration pathological ECGs in order to define requirements that do not hide such abnormal features.

In the following, the approach used by physicians in the ECG interpretation is outlined. This is useful for device designers in order to understand the user's approach and improve usability. Approaches to analyzing ECGs are similar. For example referring to (Dubin, 2000; Harrison, 2011), the interpretation of an ECG includes the following steps:

1. frequency rate evaluation;
2. rhythm analysis;
3. axis analysis;
4. hypertrophy evaluation; and
5. infarction.

The **ventricular frequency rate** indicating the number of heart cycles per minute is the simpler parameter to assess. Physicians evaluate it by considering that when the speed of the paper is set at 25 mm per second, each millimeter (i.e., the smallest squares in Figure 2.32) represents a 0.04 sec segment and the larger squares, 5 mm, are equivalent to 0.2 sec. The distance between QRS complexes reveals the beat length in seconds. Beats per minute (bpm) and length in second are related by

$$\text{BPM} = 60/(\text{beat length in seconds}) \tag{2.39}$$

In general, the new generation ECGs automatically evaluate the beats per minute parameter. Figure 2.32 shows an example related to the Gamma Cardio CG. Rulers (i.e.

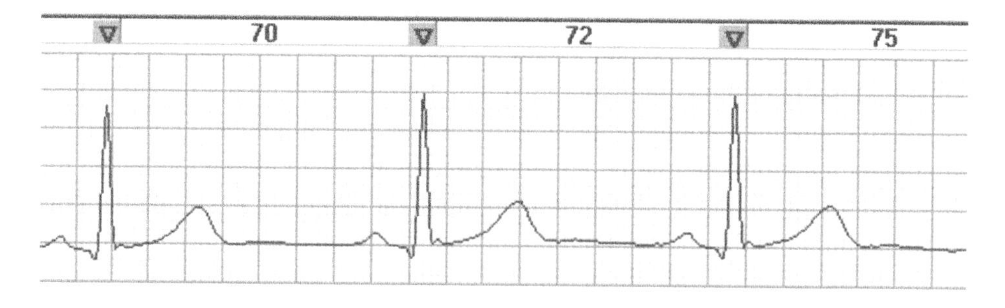

Figure 2.32 ECG frequency rate evaluation (50 mm/sec, 20 mm/mV).

black upside-down triangles) indicate R Peaks and instantaneous BPM, shown between rulers, is evaluated using eq. 2.39 with the time interval between two QRS complexes.

The frequency rate is considered usually normal in the range of 60–100 for adults. Normal pediatric value ranges differ. Higher rates are cited as **Tachycardia** while lower rates are defined as **Bradycardia**.

In normal ECGs, every P wave is followed by a QRS complex and vice versa. This is a necessary condition for the Normal (sinus) rhythm. This reflects the fact that after atria contraction there is a synchronized ventricle contraction. The ECG waveform may show that P waveforms are at rates different from the QRS complex, that is, atrial and ventricular contractions are independent. This suggests problems in the electric conduction system outlined in Figure 2.20.

Arrhythmias (from Greek: absence of rhythm) are a set of conditions with abnormal heart electrical activity that generates too fast, too slow or irregular rates. Abnormal **rhythms** are evidenced by specific wave patterns, abnormal PR, QRS intervals and axis deviations. Abnormal rhythms may indicate cardiac pathologies (Harrison, 2011). Some examples are given below to show readers examples of the specific morphologies.

This first abnormal heart activity may be due to **ectopic** (from Greek: out of place) beats that may be triggered when ventricular/atrial cells or specialized conduction cells discharge regardless of the normal timing generated by the sino-atrial node. This produces an extra contraction also called **extrasystole** that alters the normal rhythm, as shown in Figure 2.33 for ventricle generated extrasystoles.

Since normal timing is altered, contractions of the four heart chambers are less coordinated and the heart's pumping is less efficient.

Abnormal impulses may arise also when **reentrant mechanisms** takes place. In this case, electrical impulses have a circular propagation path that creates an auto trigger effect. Reentrant mechanisms may generate paroxysmal tachycardia (transient sudden accelerated ventricular rate), atrial/ventricular flutter, atrial/ventricular fibrillation.

The flutter is generated by a single ectopic pacemaker (also called a focus) producing periodic impulses at high rates due to the reentrant mechanism.

The fibrillation occurs when an entire chamber experiences the activation of multiple micro-reentry ectopic pacemakers with uncoordinated contractions. In **atrial flutter**, regular atrial movements are generated at a rate rising to 250–350 bpm as shown by the P waves in Figure 2.34. Atrial flutter is characterized by the specific saw-tooth morphology. The regularity is due to the fact that the pacemaker is unique and it beats at quasi-periodic rhythms. In **atrial fibrillation**, numerous ectopic pacemakers generate uncorrelated impulses in the atria

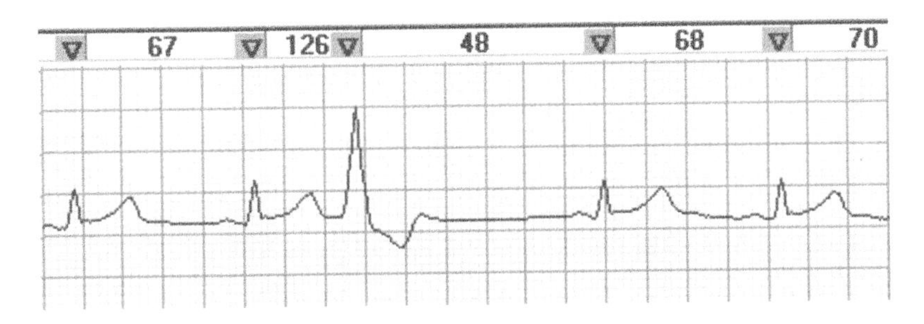

Figure 2.33 Extrasystoles among normal cycles (25 mm/sec, 10 mm/mV).

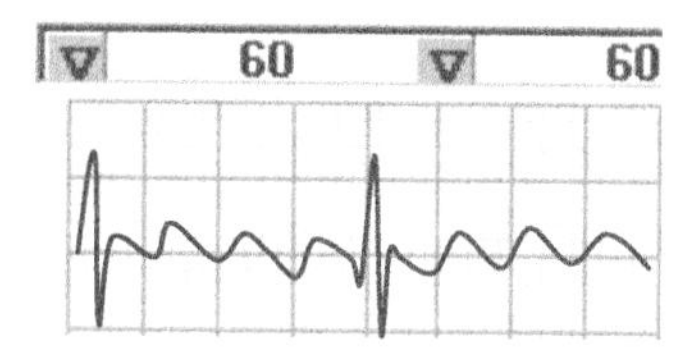

Figure 2.34 Atrial flutter.

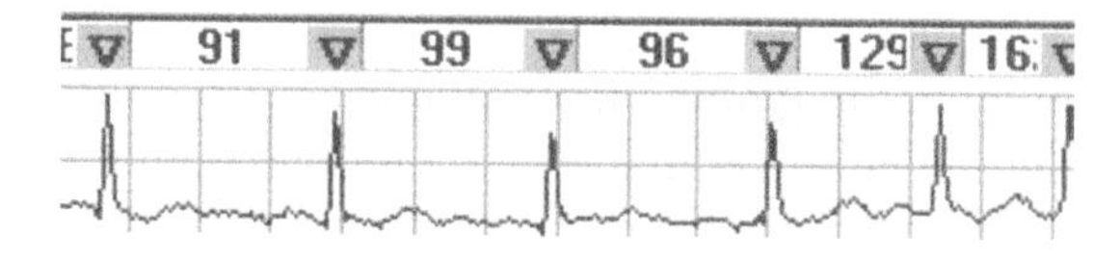

Figure 2.35 Atrial fibrillation.

and therefore P waves are replaced by low irregular waves. The pumping function is in any case performed by the ventricles which maintain an effective pumping rhythm, as shown by the QRS complexes in Figure 2.35.

Ventricular flutter is also generated by a single ventricular ectopic pacemaker that produces impulses at a rate of 200–300 bpm. The corresponding typical sinusoidal morphology is shown in Figure 2.36. In ventricular flutter, ventricular contractions may exceed 300 bpm. At this rate, blood is not able to flow at proper speed due to its viscosity and the ventricles are not properly filled. In order to balance the reduced blood circulation, more ventricular ectopic pacemakers activate uncoordinatedly and **ventricular fibrillation** takes place. As a consequence, the ECG morphology is completely irregular as shown in Figure 2.37. In fibrillation, ventricles do not achieve contraction movements but perform chaotic vibrations. In

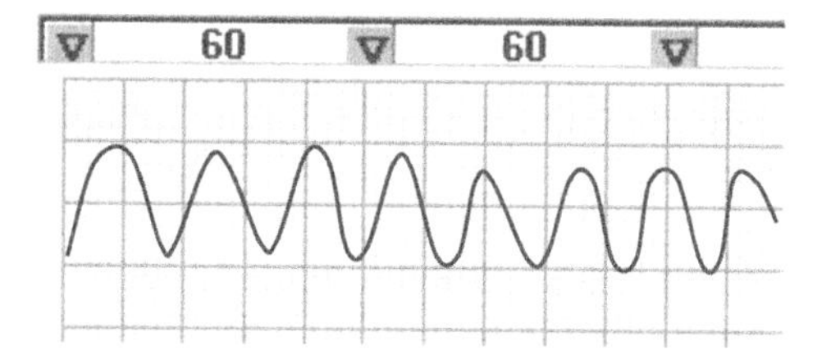

Figure 2.36 Ventricular flutter.

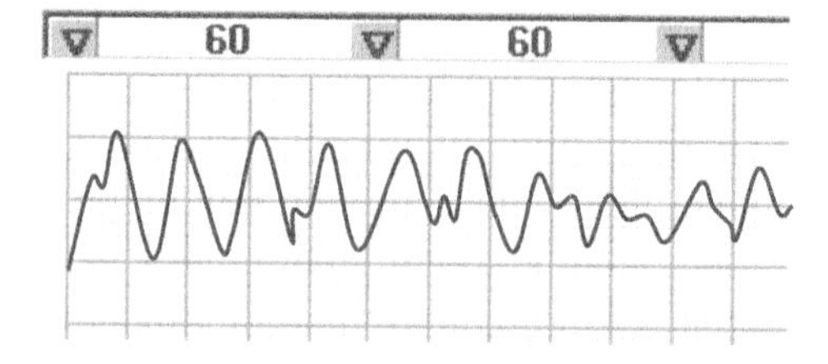

Figure 2.37 Ventricular fibrillation.

Table 2.8 Hypertrophy indications

Interested chamber	ECG indication
Right atrium	– P wave ≥ 2.5 mm in lead II.
Left atrium	– P biphasic (first positive, then negative) in V_1, P biphasic wave – with terminal negative semi-wave wider than 0.04 sec.
Right ventricle	– R amplitude > S amplitude in V_1 – R amplitude in $V_1 > 5$ mm; – deep S in V_6; – right-axis deviation ($90° \leq$ QRS axis $\leq 180°$).
Left ventricle	– S amplitude in V_1 plus R amplitude in V_5 or $V_6 \geq 35$ mm *or* R amplitude in $aV_L > 11$ mm.

this case, there is an immediate risk of death because the pumping function does not take place anymore (Dubin, 2000).

After rhythm evaluation, **analysis of the axis** is generally performed in ECG interpretation. The axis is evaluated as explained at the beginning of this section. The abnormal axis deviation may indicate problems such as diffuse left ventricular disease or infarction (axis $< -30°$) or right ventricular hypertrophy with (R height > S height in V_1) and left posterior hemiblock (with small Q and tall R in leads II, III and aVF, and axis $> 90°$) (Harrison, 2011).

The **hypertrophy** refers to the heart (and more specifically to the ventricle) having an abnormal mass increase owing to an augmented ventricle wall thickness.

Heart hypertrophy may be due to non-pathological needs of the body to increase pumping capacity (athletes, pregnancy) or to other pathological conditions (hypertension, infarction etc.).

Hypertrophy appears differently in ECG according to the interested heart chamber as outlined in the Table 2.8 (Harrison, 2011):

Ischemia and infarction may also be evidenced by the ECG. **Ischemia** is a reduction of blood supply that may happen in the heart. Infarction is the consequence of the ischemia that leads to tissue death (i.e., necrosis).

As shown in Figure 2.38, the ischemia may be individuated by the T wave inversion; dotted pattern indicates normal ECG. The **infarction** may be evidenced by abnormal Q waves (with length ≥ 0.04 sec and height $\geq 33\%$ of total QRS height) in some specific leads according to the location of the infarction or by an ST segment elevation or depression with respect to the isoelectric line.

The waveforms in Figure 2.38 are drawn at standard scales (25 mm/sec, 10 mm/mV). It is clear that any unwanted, even slight, modification of the morphology may induce an incorrect interpretation. For example, the difference between abnormal Q and a normal pattern appear minimal. In the abnormal case, the Q wave is 5 mm height (=0.5 mV) with respect to the overall QRS complex that is 10 mm height. In the normal case, the Q height is

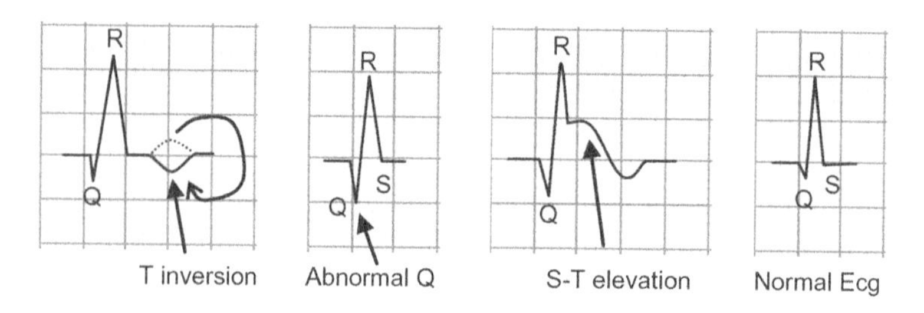

Figure 2.38 Ischemia and infarction waveforms (25 mm/sec, 10 mm/mV).

2 mm. A difference in 2–3 millimeters has to be clearly maintained at minimum by any ECG device.

Some general considerations can be drawn:

1. The amount of clinically relevant morphologies is really high, and the determination of features of the observations is critical without an extensive clinical study.
2. Even a slight modification of the morphology by the ECG device may lead to an incorrect ECG interpretation.

This suggests adopting a performance standard such as AAMI EC 11 (AAMI, 1991), that defines the features of the clinically relevant observations and the minimum requirements for the ECG signal processing modules. This standard prescribes requirements and tests that guarantee that possible wave modifications do not lead to wrong interpretations.

More formally, the European Medical Devices Directive (MDD) (EEC, 2007) prescribes the requirements labeled as 10.1, 10.2 10.3 for devices with a measuring function. Such requirements are reported below in italics with the corresponding Gamma Cardio CG requirement:

MDD 10.1. Devices with a measuring function must be designed and manufactured in such a way as to provide ***sufficient accuracy and stability*** *within appropriate limits of accuracy and taking account of the intended purpose of the device. The limits of accuracy must be indicated by the manufacturer.*

R 3	**AAMI EC 11 compliance:** The Gamma Cardio CG will be compliant to AAMI EC 11 for the applicable parts of the standard.

MDD 10.2. The measurement, monitoring and display scale must be designed in line with ***ergonomic principles****, taking account of the intended purpose of the device.*

MDD 10.3. The measurements made by devices with a measuring function must be expressed in legal units conforming to the provisions of Council Directive 80/181/EEC.

R 4	**User interface and measuring units:** The Gamma Cardio CG user interface will reproduce the control panel of a standalone conventional ECG device. The measuring units will be standard (mm/mV and mm/ms).

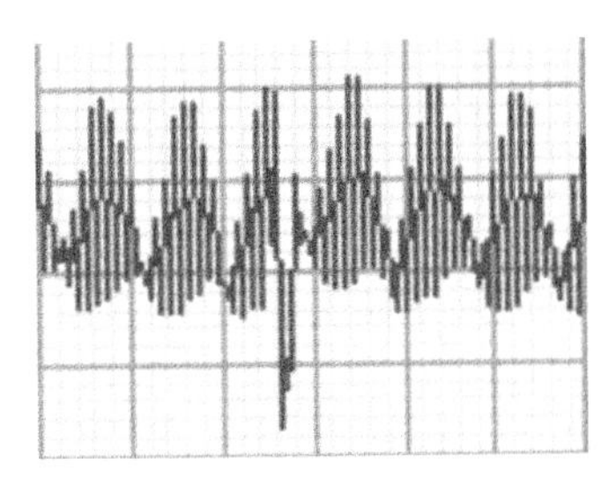

Figure 2.39 Trace with power line noise.

The following sections will introduce other elements that may lead to misinterpretation. These elements are related to artifacts that are generated by sources that are external to the ECG device. The ECG device must therefore implement measures to overcome such artifacts that are generated by: power line interference, isoelectric line instability and muscle noise superimposed signals.

2.14.3 Power Line Noise

Power line noise is an artifact that may be superimposed on the ECG waveform as shown in Figure 2.39. This artifact is generated by the electric power system whose electric fields are captured by the body, the patient cable or by the device itself. Chapter 5 discusses such problem in detail. The artifact is simply a sinusoidal-like periodic waveform with a frequency equal to the power line frequency, that is at 50 or 60 Hz according to the specific country. Sinusoids at frequency multiple of the line frequency (i.e. higher harmonics) may also be present.

Possible sources of this artifact are

- high electromagnetic fields generated for example by other medical devices or by neon lights,
- high voltage electric cables close to the patient,
- poor electrode–skin connection,
- faulty patient cable, and
- poor connection between electrodes and patient cable.

Only some of the previous reasons may be eliminated: some sources of net power interference may be moved away from patient; the electrode–skin connection may be improved by using special gel. But since some sources of noise cannot be eliminated, the following requirement is introduced:

R 5	**Noise net filter (power line interference filter):** The Gamma Cardio CG will be equipped with a filter for net noise removal that can be activated/deactivated, even on an already recorded ECG. The noise net filter will be able to eliminate even power net artifacts that completely hide the ECG trace.

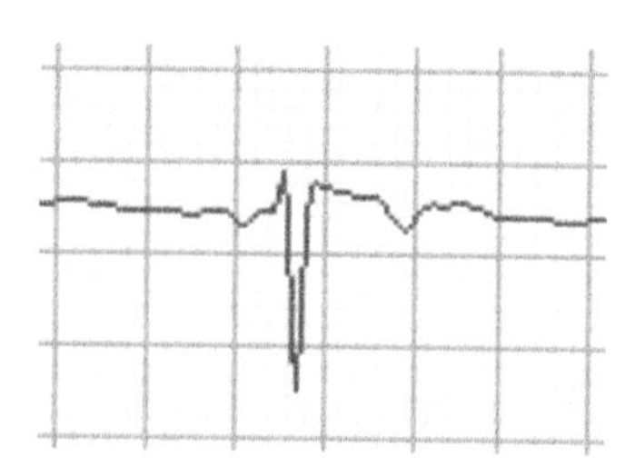

Figure 2.40 Result of denoising the trace in Figure 2.39.

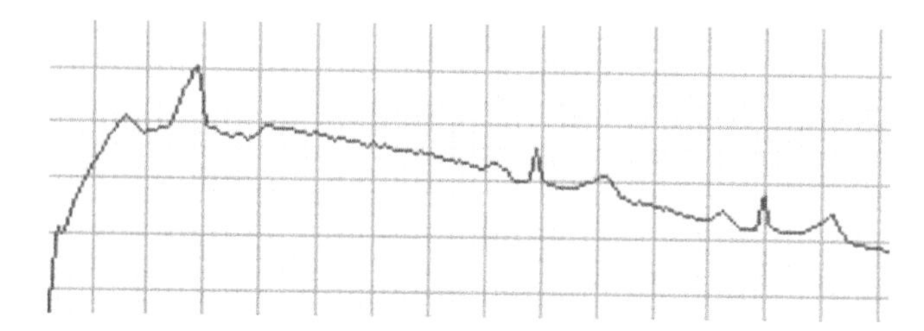

Figure 2.41 Waveform with isoelectric line instability.

The previous requirement prescribes that the filter can be activated/deactivated at any time, allowing physicians to perceive the effect of filter application.

For example, after the insertion of the net filter of the Gamma Cardio CG, the waveform shown in Figure 2.39 appears as in Figure 2.40.

2.14.4 Isoelectric Line Instability

The **isoelectric line** is the line obtained when electrical activity is not present. Such line should be horizontal, but in some cases, it may become unstable with a so-called 'wandering' behavior as shown in Figure 2.41.

Possible reasons are

1. problems in electrodes, electrode–patient cable connections, skin electrode contacts
2. electrodes polarization
3. patient movement.

In the case of a wandering line, the isoelectric line changes position, making interpretation critical. Some reasons for this problem may be overcome; in any case, the following requirement is introduced:

> **R 6 Isoelectric line stability filter (baseline wander filter):**
>
> The Gamma Cardio CG will be equipped with a filter to overcome isoelectric line instability. The filter will be able to be activated/ deactivated even on an already recorded ECG.

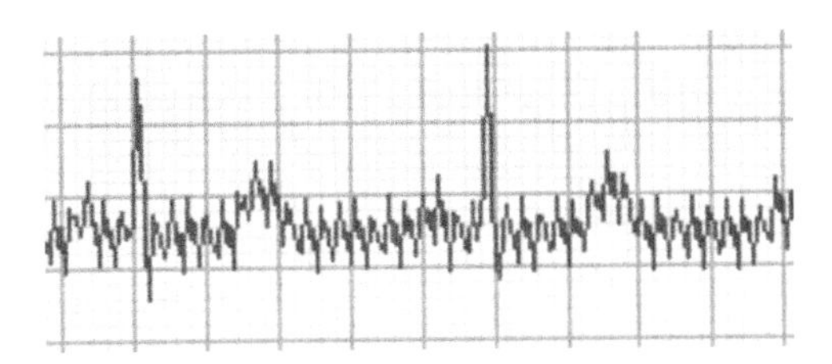

Figure 2.42 ECG with muscle tremor noise.

2.14.5 Muscle Artifacts

When the patient is affected by muscle tremors during ECG recording, the signal may show a pattern as depicted in Figure 2.42. The ECG heart electrical activity is superimposed on a saw-tooth-like pattern.

Muscle tremor is generated by unintentional patient activity possibly due to

1. perceived cold environment
2. irritability
3. too narrow examination table
4. patient disease (e.g., Parkinson's).

Since some causes cannot be removed, the following requirement is introduced:

R 7 Low-pass filter (muscle noise filter):

The Gamma Cardio CG will be equipped with a low-pass filter (25 and 35 Hz) to overcome muscle tremor line instability. The filter will be able to be activated/ deactivated even on an already recorded ECG.

Note that muscle filter usage has a possible side effect since it rounds the signal and it reduces higher frequency components of the ECG waveform such as the QRS complexes. This must be indicated to physicians who perform ECG interpretation, to avoid errors.

2.15 Requirements Related to Measurements

For the implementation part of this book, an ECG has been selected because it is a complete equipment that encompasses the most of the problems that a general medical instrument may face in design. At the same time, the ECG is simple enough to be addressed with sufficient detail in a book that can be used for biomedical engineering courses. This aspect is also evidenced in the analysis of the measurement performance requirements, since ECG encompasses most of the features required by a measuring instrument. This is clear from Figure 2.43.

The generic instrument functions may be modeled by the following three functions: front-end, signal processing, presentation module.

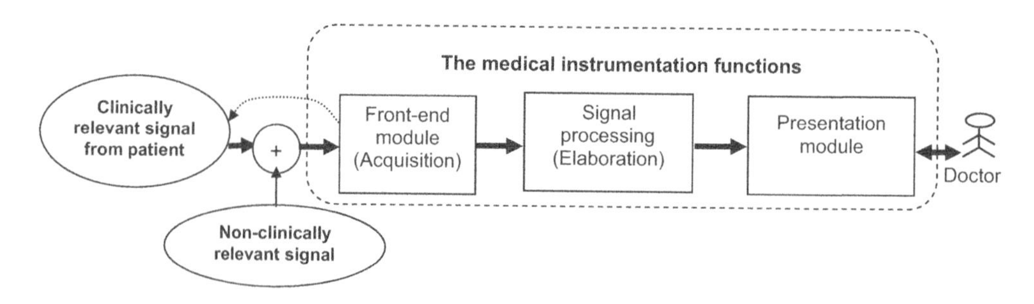

Figure 2.43 Instrument model and associated measurement requirements.

The instrument collects the signal from the patient that can be modeled as a source of clinically relevant signals plus unwanted signals (e.g., external captured noise, non-clinically relevant body signals).

The overall signal collected from the patient is managed by the **front-end** module that encompasses

1. the method of observation
2. any signals that may need to be delivered to the patient to perform measures
3. the functions required to eliminate some unwanted signals.

The **signal processing** has to apply appropriate processing to the clinically relevant signal with functions (e.g. scaling, filtering) and performance to be implemented (accuracy, noise figure, etc.).

The **presentation module** contains functions to adapt the signal for the final output that must be suitable for the physician.

The requirements associated with these modules must be defined following clinical experience or, alternatively, referring to a standard such as the AAMI EC 11 (AAMI, 1991) where available clinical studies have been translated into requirements and associated qualification tests.

Table 2.9 shows the requirements that have to be defined for each of the three modules. Such performances are substantially valid for many different medical instruments. The associated system requirement is indicated in parenthesis. Tables 2.10–2.12 contain the requirements associated with each AAMI EC 11 section.

Note that the standard also contains other requirements related to safety. For such requirements, the corresponding European safety standards have been considered. The safety requirements will be addressed in Section 2.16. Other requirements of the standard such as the operating condition line voltage are not applicable because the Gamma Cardio CG is powered by the USB plug connected to the PC. We consider first the **front-end module** requirements outlined in the first column of Table 2.9: Table 2.10 includes the Gamma Cardio CG performance and the standard performance from AAMI EC 11 or IEC/EN 60601-2-25 as indicated in the parenthesis of the second column. The first column of the table shows the system requirement identifier that is identical in the second digit in the corresponding AAMI EC 11 section. The last column shows the method of validation for the specific requirement, possible choices being: theoretical demonstration, inspection or specific tests.

Table 2.9 Performance of the measuring instrument from AAMI EC 11

Front-end module	Signal processing module	Presentation module
Observation method (R 3.2: lead definition)	Vertical scale (R 3.4: Gain)	Output display (R 3.6)
Input dynamic range (R 3.3)	Horizontal scale (R 3.5: Time base)	
Input impedance (R 3.9)	Accuracy of input signal reproduction (R 3.7)	
Common mode rejection (R 3.11)	Response to test signal (R 3.8: Standardizing voltage)	
	System noise (R 3.12)	
	Baseline stability (R 3.13)	

Table 2.10 Front-end requirements: Gamma Cardio CG vs. standards (AAMI, IEC)

System requirement	Requirement description (standard)	Standard performance	Gamma Cardio CG performance	Validation
R 3.2	**Lead definition (IEC)**	Standard 12 leads	Compliant (standard 12 leads recorded one channel at time, see Section 2.13.3)	Insp.
R 3.3	**Input dynamic range (AA)**:			
R 3.3.1	Maximum input dynamic range that ensures linear response from ECG	$\geq \pm 5$ mV	Compliant	Demo
R 3.3.2	Slew rate change	$\leq$320 mV/s	=400 mV	Demo
R 3.3.3	DC offset voltage range	$\geq \pm 300$ mV	Compliant	Demo
R 3.3.4	Allowed variation of amplitude in presence of a DC offset	$< \pm 5\%$	$< \pm 3\%$	Demo
R 3.3.9	**Input impedance** at 10Hz (for each lead) (AA)	>2.5 MΩ	≫600 MΩ	Demo
R 3.3.11	**Common mode rejection** (AA):			
R 3.3.11.1	Maximum interference from a common mode signal of 20 V_{rms}, 60 Hz with $\pm$ 300 mV DC	$\leq$1 mV p-p (RTI) (CMRR > 89 dB)	Compliant (CMRR $\geq$110 dB)	Test
R 3.3.11.2	The same test of R 3.3.11.1 adding 51 kΩ imbalance at input	$\leq$1 mV p-p	Compliant	Test

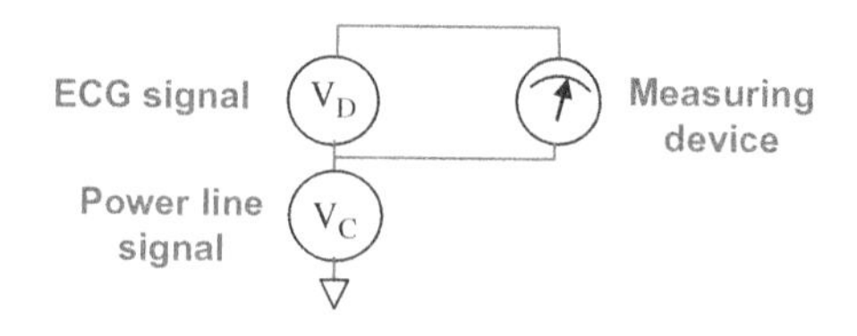

Figure 2.44 Differential and common mode signal.

The requirements of the table will be discussed in detail at the design level in Chapter 5. Here we give a general description. The observation **method requirement** (R 3.2) for the ECG is related to the definition of the 12 standard leads (I, II, III, aV_F, aV_R, aV_L, V_1–V_6) whose mathematical formulation is described in Section 2.13.3.

The **input dynamic range** refers to the maximum signals (wanted and unwanted) that can be handled by the device. The ECG device must be capable of displaying a maximum of ±5 mV wanted signal added to an unwanted DC offset voltage of maximum ±300 mV. This latter component is eliminated by the high-pass filter. The rationale of all the requirements with the associated clinical reference studies is reported in Appendix A of the standard.

The **input impedance** at the lead is an important parameter considering that the ECG signal generator can be modeled as a voltage generator with non-infinite internal resistance. In general, the lower the impedance, the higher the error on the measurements. With current FET based operational amplifiers this requirement can be easily obtained, as explained in Chapter 5.

Regarding **common mode rejection** it has to be considered that detecting the ECG signal is like 'finding a needle in a haystack' (Scott, 2000), since the ECG signal of ± 5 mV amplitude is embedded in a captured net power line signal of tens of volts, that is thousand of times higher then the ECG itself. As shown in Figure 2.44, the power line signal is a 'common mode voltage', that is, a signal captured by both the measuring electrodes. By contrast, the ECG signal is measured in a differential mode since it is present with opposite polarity at the two electrodes. An ideal front-end should eliminate completely common mode signals. Real front-ends are able to achieve a significant attenuation defined as 'common mode rejection'.

Table 2.11 shows the signal processing requirements with the Gamma Cardio CG performance and the standard performance, columns are organized as in Table 2.10.

In the **signal processing** functions, requirements related to vertical and horizontal **scaling** are included with associated errors and admissible values for scaling. Such requirements have been included, taking account of the much higher error of the old stylus based printers. The new generation ECG, like the Gamma Cardio CG, with all-digital printer processing can obtain much higher performance. The **accuracy** requirement contains performance related to the overall error and the cutoff passband frequencies (i.e., the frequency where the response is reduced by 3dB). Such values are set to 0.05–150 Hz to avoid significant errors, for example on the rapidly changing QRS complexes that may be unacceptably reduced by too reduced low frequency cutoff values. The high frequency cutoff allows ST segments shown in Figure 2.38 to remain almost undistorted, allowing diagnostic information to be retained. Even a cutoff low pass frequency of 0.67 Hz may reduce an ST elevation by 0.2 mV, thus hiding possible infarction information.

The square **test signal** also referred to as a 'standardizing voltage' is embedded in all the ECG devices to assess the proper calibration of the overall instrument. Distortions on this

Table 2.11 Signal processing requirements: Gamma Cardio CG vs. standards (AAMI)

System requirement	Requirement description (standard)	Target performance value	Gamma Cardio CG performance	Validation
R 3.4	**Gain (AA)**			
R 3.4.1	Gain selections	Selectable minimum at 20, 10, 5 mm/mV	Compliant	Insp.
R 3.4.2	Gain error	$\leq 5\%$	$\leq \pm 3\%$	Demo
R 3.4.3	Manual override of automatic gain control		Not applicable	
R 3.4.4	Gain change rate/minute	$\leq \pm 0.33\%$/min	Compliant (not detectable on ECG video)	Test
R 3.4.5	Total gain change/hour	$\leq \pm 3\%$	Compliant (not detectable on ECG video)	Test
R 3.5	**Time (AA)**			
R 3.5.1	Time base selections	Min. 25, 50 mm/s	5, 25, 50 mm/s	Insp.
R 3.5.2	Time base error	$\leq \pm 5\%$	Compliant	Test
R 3.7	**Accuracy** of input signal reproduction (AA):			
R 3.7.1	Overall error for signals up to $\pm$ 5 mV amplitude and 125 mV/s	$\leq 5\%$ and $\leq \pm 40\ \mu$V	Compliant	Test
R 3.7.2	Frequency Response	See Chapter 5	Compliant	Demo
R 3.7.3	Amplitude signal after a 3 mV 100 ms impulse	≤ 0.1 mV	Compliant	Demo
R 3.7.4	Displacement slope after a 3 mV, 100 ms impulse	≤ 0.3 mV/s	Compliant	Demo
R 3.7.5	Error in lead weighting factors	$\leq 5\%$	Compliant ($\leq 2\%$)	Demo
R 3.7.6	Hysteresis after 15 mm deflection from baseline	≤ 0.5 mm	Compliant	Demo
R 3.7.7	Response to 10 Hz, 20 mV p-p sinusoidal signal	Visible at 25 mm/s time base and 10 mm/mV gain	Compliant	Demo

(*continued*)

Table 2.11 (*Continued*)

System requirement	Requirement description (standard)	Target performance value	Gamma Cardio CG performance	Validation
R 3.8	**Standardizing voltage** (calibration square test signal) (AA)			
R 3.8.1	Nominal value	=1.0 mV	Compliant	Test
R 3.8.2	Rise time	≤1 ms	Compliant	Test
R 3.8.3	Decay time	≥100 s	Compliant	Test
R 3.8.4	Amplitude error	±5%	< ±3.5%	Test
R 3.12	**System noise (AA)**:			
R 3.12.1	Overall system noise RTI, p-p	≤30 μV	Compliant	Test
R 3.12.2	Multichannel crosstalk	<2%	Applicable only for multichannel ECG	N/A
R 3.13	**Baseline** control and stability (AA):			
R 3.13.1	Return time to baseline 10 s after reset	≤3 s	Compliant	Demo
R 3.13.2	Return time to baseline after lead switch	≤1 s	Compliant	Demo
R 3.13.3	Baseline drift rate RTI	≤10 μV/s	Compliant	Test
R 3.13.4	Total baseline drift RTI (2 min period)	≤500 μV	Compliant	Test

signal indicate problems in the signal processing chain. The decaying behavior of the top of the square signal in Figure 2.45 shows the overall high-pass (0.05 Hz cutoff) ECG response while the rapidly increasing/decreasing vertical signals show the behavior of the low-pass response (150 Hz cutoff). More advanced methods are implemented on the Gamma Cardio CG to test most of the ECG hardware software functions as explained in Chapter 6.

The **system noise** requirement takes into account all sources of noise and it is dimensioned to avoid interference that leads to false interpretations.

Baseline control describes the baseline reset functions required when overload of the high-pass filter occurs. After overload, which charges completely the capacitor first-order high-pass filters and the associated capacitors at $f_c = 0.05$ Hz need around 15 seconds for the capacitors to discharge so that signal returns to the zero value, as it should be in absence of an input signal. The 15 second value is obtained by the time constant τ:

$$\tau = 1/(2\pi f_c) \approx 3 \text{ sec} \tag{2.40}$$

Table 2.12 Output display requirements: Gamma Cardio CG vs. standards (AAMI, IEC)

System requirement	Requirement description (standard)	Target performance value	Gamma Cardio CG performance	Validation
R 3.6	**Output display (AA):**			
R 3.6.1	Visibility of a signals with minimum amplitude	$\geq \pm 5$ mV	Compliant	Insp.
R 3.6.2	Vertical size of the display per channel	>40 mm	Compliant	Insp.
R 3.6.3	Trace visibility of 25 Hz sinusoidal test signal 20 mm p-p amplitude	Sinusoid peaks distinguishable	Compliant	Test
R 3.6.4	Trace width after (R 3.6.3) test signal switch off	<1 mm	Compliant	Insp.
R 3.6.5	Pre-ruled paper division	10 div/cm	Compliant	Insp.
R 3.6.6	Error of rulings	±2%	Compliant	Insp.
R 3.6.7	Time marker error	±2%	Compliant	Insp.
R 3.6.8	Indication for degraded mode (IEC)	Indication of signal saturation of input stage	Compliant	Test

As explained in Chapter 5, the first-order filters have an exponential decay where the signal is almost completely zeroed after $5 \times \tau = 15$ seconds.

Since overload is usual in clinical practice, some mechanism is required to recover functionality of the instrument that basically resets the memory of the filter. In hardware resistance–capacitor filters, this may be done by fully discharging the capacitor to ground. In the software implementation, the filter memory has to be reset. The Gamma Cardio CG implements both automatic and manual systems to reset the baseline with performance compliant with the standard.

Table 2.12 shows the output display requirements with the Gamma Cardio CG performance and the standard performance, columns being organized as in Table 2.10.

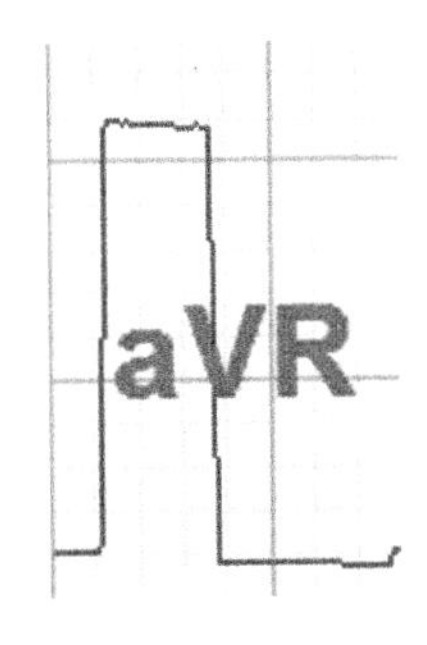

Figure 2.45 Square wave signal showing high frequency loss.

The **output display** requirements determine the performances on the visible ECG trace. In the Gamma Cardio CG, this refers to the ECG signal shown both on video and on the printer. Different resolutions are obtained on the PC screen and the printer and this has to be taken into account for requirement validation. The precision of the display is required since relevant diagnostic information is contained in the vertical and horizontal width of signal components (Section 2.14.2). Errors in ECG output may lead to diagnostic errors. Amplitude and interval measures are performed by physicians based on the pre-ruled paper divisions and therefore specific requirements must hold on errors within pre-ruled line distance. For paperless ECG interpretations, measures have to be taken on a video screen. For this reason, the Gamma Cardio CG offers a specific tool for performing measurements easily.

Note that, as indicated in the last column of Table 2.12, requirement R 3.6.8 is satisfied since the Gamma Cardio CG is compliant to another standard specifically devoted to safety of the medical electrical equipment. This standard (IEC/EN 60601) will be addressed in Chapter 8, where a comprehensive discussion on standards is included.

2.16 Safety Requirements

Device safety requirements are generated in response to the **essential requirements** contained in regulatory framework of the specific area. In the European Union the applicable framework is the European Medical Devices Directive (MDD) (EEC, 2007) whose Annex 1 contains the essential requirements. All medical devices sold in the European market 'must meet the applicable essential requirements set in Annex I, taking into account the intended purpose of the devices' (MDD, Article 3, EEC 2007).

'The relevant Essential Requirements of Annex I . . . shall apply as far as **safety** and **performance** related device features are concerned' (MDD, Article 1, EEC 2007).

'Member States shall presume compliance with the essential requirements . . . in respect of devices which are in conformity with the relevant national standards adopted pursuant to the **harmonized standards** the references of which have been publishes in the Official Journal of the European Communities' (MDD, Article 5.1, EEC 2007).

Updated information on harmonized standards may be found in (EEC, 2011). It is then the responsibility and of the manufacturers to select proper standards applicable for the manufactured device, but it is also for their benefit since standard compliance normally reduces costs and risks.

Harmonized standards can be divided into **horizontal and vertical standards**. Horizontal standards are applicable to different classes of medical devices such as the electrically operated medical devices (i.e., IEC/EN 60601-1). Vertical standards are applicable to a specific class of products such as ECGs (i.e., IEC/EN 60601-2-25). Vertical standards have to be preferred when applicable.

In this book we will mainly focus on European and US frameworks since these systems, as well as the Japan system, are comprehensive regulatory systems that follow the 'Essential principles of safety and performance of medical devices' recommended by the Global Harmonization Task Force (GHTF) (WHO, 2003). The GHTF (http://www.ghtf.org/) is a partnership between regulatory authorities (EU, USA, Canada, Australia and Japan) with the aim of achieving greater uniformity among national medical device regulatory systems. It is also worth noticing that global statistics reveal that the USA, Japan and the EU countries are responsible for most of the medical device production (WHO, 2003).

In the following, the European MDD essential requirements are outlined where, in parenthesis, there is the identification number of the requirement and, following, the associated device requirement.

The value of this section is in the explanation of the EU essential requirements that are conceptually valid worldwide and in the generation of marketed product requirements in response to essential requirements.

The first six essential requirements are **general requirements** concerning safe design. Ergonomic features and considerations on user profiles (1) have to be considered to reduce risks to compromise patient clinical condition or safety. For this, the Gamma Cardio CG has different user interfaces according to the user profile (see requirement R 15).

Risk management has to be introduced throughout the product life cycle to minimize unwanted events (2). The performances of the device must not degrade throughout the specified lifetime (4). Transport and storage must not negatively affect the device (5). Side effects must constitute acceptable risk (6) and finally 'Demonstration of conformity with the essential requirements must include a clinical evaluation in accordance with Annex X'. This last requirement is satisfied in the Gamma Cardio CG by compliance to AAMI EC 11.

The following requirements, ranged from (7) to (13), are related to **design and construction**.

Requirement (7) deals with the **chemical, physical and biological properties** of the device: toxicity, flammability (7.1), contaminants (7.2), compatibility for contacts with other substances (7.3), use of medicinal product (7.4), leakage (7.5) and penetration (7.6) of substances. The generated device requirements are:

R 8	**Used accessories**
	All the used accessories will be class I Medical devices, compliant with MDD (EEC 2007).
R 9	**IEC/EN 60601-1 and IEC/EN 60601-2-25 compliance:**
	The Gamma Cardio CG will be compliant with the IEC/EN 60601-1 and IEC/EN 60601-2-25.

Requirement (8) is related to **infection and microbial contamination**. The corresponding sub-requirements are mostly non-applicable and do not generate further device requirements.

Requirement (9) is related to **Construction and environmental properties**: combination with other devices (9.1), risk related to the interaction with the environment (physical, electromagnetic compatibility – EMC) (9.2), the risks of fire or explosion (9.3).

The generated device requirement is:

R 10	**EMC standard compliance:**
	The Gamma Cardio CG will be compliant to IEC/EN 60601-1-2 EMC standard.

The **EMC requirements** refer to the device capability to perform

1. in the presence of defined electromagnetic interference (EMI)
2. without introducing intolerable interference in the working environment.

This topic will be discussed in Section 2.13.1, in Chapter 5 and in Chapter 8 focused on certification. The other items of the essential requirement (9) are satisfied by requirement R 9.

Requirement (10) is related to **devices with a measuring function**, already discussed in Section 2.14.2, and it generates requirements R 3 (Compliance with AAMI EC 11) and R 4 (user interface and measuring units).

Requirement (11) related to **protection against radiation** is not applicable to an ECG.

Requirement (12) is related to **requirements for medical devices connected to or equipped with an energy source**: devices incorporating electronic programmable systems (12.1), alarms (12.3, 12.4), EMC minimization (12.5), protection against electrical risks (12.6), protection against mechanical and thermal risks (12.7), protection against the risks posed to the patient by energy supplies or substances (12.8) and controls/indicators clearly specified (12.9). The generated requirements are

R 11 **Software validation:**

The Gamma Cardio CG software will be validated.

R 12 **Class II CF type device ♥:**

The Gamma Cardio CG will be designed and manufactured as a class II type CF device according to IEC/EN 6060-1 classification.

Requirement (13) is related to **information supplied by the manufacturer** that is included in labels, user manual and packaging envelope. The generated device requirement is

R 13 **User manual:**

The user manual will contain all the information required by the essential requirements of the MDD.

2.16.1 EMC Performance

The electromagnetic compatibility (EMC) requirements refer to the maximum allowed device **emissions** that are conducted via cabling or emitted through radiation, and to the capability of the device to perform satisfactorily even in the presence of electromagnetic interference encountered in the environment where the device is then intended to be used. This latter performance is referred to as **immunity**. Again, a standard allows a faster approach to being compliant. Compliance to the IEC/EN 60601-1-2 standard is considered (as per Requirement R 10). This is a good choice for worldwide device marketing since the European standard is close to the corresponding IEC international standard recognized also in the USA. Further details are addressed in Chapter 8.

EMC testing requires compliance to both emissions and immunity. Table 2.13 shows tests performed on the Gamma Cardio CG. In the USA, the FDA normally accepts European EMC compliance as evidence that a product complies with EMC standards.

The following tests are also considered in EMC compliance:

- Surge immunity (switching and lightning-induced transients; applicable to AC input and I/O cabling which runs outside of the building)
- Voltage dips and interruptions (immunity to fluctuations on the AC power input)
- AC power line harmonics.

These tests are applicable only for devices connected to the AC power mains.

The tests in Table 2.13 require specific expensive instruments such as an anechoic chamber (i.e., specially designed room stopping reflection of electromagnetic waves). Therefore, EMC compliance is better evaluated by specialized labs that perform tests at the end of a product development when the final hardware is available.

This introduces a critical risk since, at the end of hardware development, the product may be discovered to be not compliant to requirements and hardware reworking would then be needed. To avoid this, some numerical evaluations may help in estimating performance before availability of the final product.

In general, most of the EMC problems are related to cables that conduct and radiate energy captured by the inside of the equipment and introduce transients and energy from the outside environment. Proper shielding and protection of each cable, connection through

Table 2.13 EMC tests on the Gamma Cardio CG device

Test	Description
Conducted emissions	Unintentional emissions conducted back to the AC power mains through ECG cabling.
Radiated E-field emissions, 80–2500 MHz	Unintentional radiation emissions from the product in normal operating mode.
Electrostatic discharge (ESD)	Product immunity to ESD: – contact discharge – air discharge – indirect discharge
Electrical fast transient/burst	Product immunity to switching and transient noise tested on AC input and I/O cabling
Conducted RF immunity	Product immunity to low frequency fields generated by intentional transmitters through cabling (mobiles phones, radios, TV etc.) tested on AC input and I/O cabling
Radiated RF immunity	Product immunity to fields generated by intentional transmitters (mobiles phones, radios, TV etc.)
Power frequency magnetic field immunity	Product immunity to low frequency magnetic fields on all three axes of the product

Table 2.14 Key evaluation criteria and Gamma Cardio CG target performances

ID	Key evaluation criterion	Target performance of the device
1	Measurements quality	Highest horizontal resolution (target 1200 samples per second) (R 14)
2	Reliability of measurements	Guaranteed performance through compliance to AAMI EC 11 standard (R 3)
3	Additional cost of ownership	No additional cost (R 23, R 24) (no battery maintenance, no pre-ruled paper to be bought, no periodic maintenance required, no disposable accessories, since product is supplied with reusable electrodes)
4	Required training for usage	Immediate in usage for different user profiles even with limited training **(R 15)**
5	Reliability of the signal management	Cardiologists can always show ECGs with or without filters applied, to evaluate the effects of filters on ECGs: filters may be inserted or removed, even after recording (R 5, R 6, R 7)
6	Robustness	Suitable for harsh environment, developing world medical environment, ambulances
7	Immunity to interference	Noise net filter able to eliminate power net artifacts that completely hide the ECG trace (R 5)
8	Safety	Certified at the highest and most demanding medical device class for similar products (class IIb) (R 23)
9	Patient safety against electric shock	Certified at the highest and most demanding medical device class for similar products (class CF symbol ♥) that guarantees the highest degree of protection against electric shock (R 12)
11	Usage in mobility	Light and easily transportable (acquisition unit with weight lower than 250 grams and dimensions $\leq (15 \times 9 \times 2\,\text{cm})$ **(R 18)**
12	Export data functions	– Embedded saving in Acrobat® PDF format – Microsoft Windows® copy and paste for export in Microsoft Word® or other Windows applications **(R 19)**
13	Assisted/automatic interpretation	Measurement of amplitudes axes and intervals with diagnostic statements
14	Medical record archive	The Gamma Cardio CG will store the patients' information, the associated case history and past ECGs with a user-friendly approach
15	Telemedicine	– Internal mailer application available for immediate tele-medicine functions: ECGs mailing and fax – Integration with existing MAPI compliant mailer applications (Microsoft Windows Outlook Express, Lotus Notes etc.) **(R 21)**

ferrite components will usually guarantee compliance to the standard. These issues will be discussed in Chapter 5 for the electronic implementation and in Chapter 8 for certification.

2.17 Usability and Marketing Requirements

In this section, requirements related to the improvement of user experience are introduced. Such requirements aim also to increase market sales performance and therefore a contribution from marketing resources is generally envisaged. In Chapter 1, some distinctive features have already been discussed. In general, customers have typical **key evaluation criteria** to assess a product and perform a choice when buying among similar products. For such criteria, valid for most of medical instruments, the Gamma Cardio CG must have improved perceived **target performances** that in turn will generate **additional requirements**.

The proposed product aims to be an ECG with performance at level of professional systems but with a limited production cost considering that most of the functions will be performed by the PC and the associated software application. A complete survey of ECG products and associated features is found in Healthcare Product Comparison System (HPCS) published by the ECRI Institute (ECRI, 2012b).

Table 2.14 shows the generic key evaluation criteria and the target product performances. The number in parenthesis of the associated requirements is printed in bold when the requirement is new.

The requirements generated by the target performance column of Table 2.14 are introduced in the following. Note the difference in the marketing-like language of the previous sentences and the translation into formal requirements.

R 14 **Horizontal resolution:**

The Gamma Cardio CG must have a horizontal resolution of 1200 samples per second.

R 15 **Required training for usage:**

The Gamma Cardio CG must be immediate in usage for different profiles, even without specific training. In particular, the product will have:

1. a user interface that recalls, for cardiologists, the familiar electrocardiographs
2. an elementary step-by-step procedure, devoted to non-expert users (generic physicians, nurses etc.), with associated checklist, to perform the test and send/print/store the ECG
3. some main errors (open lead, saturation indication) signaled with proper work around
4. automatic/manual isoelectric line reset function
5. main functions available through touchscreen interface
6. user interface requiring minimal MS Windows know-how.

R 16 **Harsh environment:**

The Gamma Cardio CG must be designed to be used in harsh environments.

R 17	**Class IIb certification:** The Gamma Cardio CG must be certified at class IIb (93/42, 2007/47/EEC).
R 18	**Weight and dimension of the acquisition unit:** The acquisition unit must have: weight $\leq$250 g and dimensions $\leq (15 \times 9 \times 2\,\text{cm})$.
R 19	**Export data functions:** The Gamma Cardio CG will also be able to save ECG and associated interpretation sentences in Acrobat PDF format, and ECGs can also be exported through Microsoft Windows copy and paste method 7 automatic/assisted interpretation tools.
R 20	**Patient archive:** The Gamma Cardio CG will store patients' information, associated case history and ECGs with an easy patient information retrieval interface.
R 21	**Mailer capabilities:** The Gamma Cardio CG must implement an internal mailer application; in addition, it must run seamless existing MAPI-compliant mailer applications.

2.18 Environment Issues

The environment imposes additional constraints on the product. The **operating conditions** of this medical device are defined in standards IEC/EN 60601-1 and IEC/EN 60601-2-25.

Some assumptions have to be set up in response to the essential requirements of the Medical Device Directive (requirement 13). These assumptions must be warnings included as warning in the user manual. It is worth evidencing some specific warnings related to

1. the working environment
2. the PC's minimum features
3. the PC's allowed connection.

Regarding **the working environment**, medical device designers have to define all the possible events that may produce an unacceptable risk with non-null probability in the risk analysis. In some cases, it may be sufficient to specify warning sentences in the user manual. A typical example is given in the following where some warnings related to the Gamma Cardio CG are shown:

⚠ *Warning: The Gamma Cardio CG:*

- *is not designed for use in the presence of flammable anesthetic mixtures, or in environments with explosion hazards,*
- *is not protected against the penetration of water,*

- *should not be installed in adverse environments (high humidity, direct sunlight and continuous, excessive presence of dust, sulfur),*
- *must not undergo sterilization processes.*

The other items are related to the increasing trend of using PC based platforms within medical instruments to reduce cost, risks and development time.

The first issue is the **operating system** that may be mainly Unix/Linux or MS Windows® derived. There are pros and cons in both choices and depend on many factors. For the Gamma Cardio CG the following assumption has been chosen:

The Gamma Cardio CG system will operate on a ***Personal Computer*** *with Windows XP or successive versions, a preferred video resolution (even virtual) of 1024 × 768 (1024×600 minimum) and a USB port. An Internet connection and software for mail management compliant to MAPI (Microsoft Windows Outlook Express, Lotus Notes etc.) is required to use of telemedicine functions.*

Another interesting topic in using computers with medical instruments is related to **patient shock hazard**.

For the Gamma Cardio CG the acquisition unit is the only module that is connected to the patient. The unit is connected then to the PC that itself is connected to the mains power. The acquisition unit electrically isolates the patient from the computer through a **double insulation** (indicated by the symbol ⧈). The double insulation is an 'insulation that provides two means of protection': a basic insulation (that is the 'basic protection against electric shock') and 'a supplementary insulation that is an independent insulation applied in addition to basic insulation in order to provide protection against electric shock in the event of a failure of basic insulation' (IEC/EN 60601-1).

The unit is also a low voltage module designed for maximum electrical insulation (patient's applied parts CF type with symbol ⊣♥⊢). This means that the patient is protected, even in the presence of one fault. The PC may be in any case a critical element when touched in specific fault conditions.

The problem is related to the PC that may be touched by the patient. In this case, a patient shock risk may hold.

Some computers, and in particular the majority of laptops, do not have metal parts that can be touched by the user and therefore they have a plug without a ground connection (a plug with two prongs). These computers do not require specific additional precautions when used in medical environments.

In the computers equipped with metal parts accessible to the user, the plug has a third terminal for grounding (three-prong plug). If computers are not connected according to regulatory requirements, in case of fault, the patient may receive an electric shock by touching the metal case of the PC. This event is independent of the use of a PC-ECG like Gamma Cardio CG. To avoid this problem, computers with grounding (three-prong plug) must always be connected to an outlet with an effective ground system designed according to applicable laws.

In general, for a PC that has to be used in medical environment three options hold to meet the IEC/EN 60601-1:

1. computers must be located outside the patient environment, that is the patient cannot touch computers directly or through other people, or

2. computers must be connected to ground with a permanent cable detachable only with a specific tool, or
3. computers must be connected to an outlet with additional isolating transformer.

In any case, the computer must meet the safety standards relating to information technology equipment (IEC/EN60950).

2.19 Economic Requirements

The economic requirements may be in conflict with other requirements since they may express constraints on the overall cost of the product to be developed.

The cost constraints may be related to

- production cost (materials, man labor, services)
- after-sales maintenance cost (also including warranty)
- cost of accessories.

In summary, considering what has already been discussed in Chapter 1:

R 22	**Production cost:**
	The overall production cost of the Gamma Cardio CG must be lower than 250 euros.

The overall production cost is decomposed into

1. materials to produce the ECG as specified in requirement R 2
   - acquisition unit
   - 10 reusable electrodes
   - patient cable
   - software application running on the PC
   - packaging
   - instruction manual
   - PC cable connection.
2. Man hours and services to produce directly the product:
   - to assemble and test the acquisition unit
   - to prepare the overall package.

Note that the previous requirement specifies only the overall cost and does not set constraints on the single components of cost (i.e., acquisition unit cost etc.). The requirements have to be limited only to those features whose non-compliance determines the failure of the project. Increasing the number of requirements/constraints will reduce the degrees of freedom in design that in turn allow introducing improved solutions to fulfill the requirement: the more the requirements, the more the risk of project failure.

Note also that the previous requirement implies that, from the earliest stages, product designers have to conceive product so that manufacturing costs are taken into account.

The **after-sales maintenance costs** are related to the economic effort that the owner will sustain to have product in operation throughout the product life.

R 23	**Periodic maintenance cost:** The Gamma Cardio CG must not technically require periodic maintenance operations and associated periodic costs.
R 24	**Compulsory specific consumable:** The Gamma Cardio CG must not require compulsory specific consumable in operative usage. The only components requiring substitution if no more efficient are the patient cable and the reusable electrodes.

These latter requirements are a competitive advantage of the Gamma Cardio CG with respect to usual stand-alone ECG devices. The Gamma Cardio CG owner will not need to spend extra money on maintenance (periodic maintenance, battery change) thus reducing the total cost of ownership.

In addition, the ECG paper is not compulsory for the Gamma Cardio CG as it can print pre-ruled grids on conventional, less expensive PC printer paper. Moreover, ECG printing is not required since all operations can be paperless: interpretation on video, storing in non-volatile memory and ECG delivery via mail through the telemedicine functions.

The final economic evaluation is related to the overall investment required to develop a product. Such evaluation is not part of product requirement but has to be estimated at startup and controlled continuously to avoid the business becoming unprofitable.

From Chapter 1 'Business context for design' defining P the average price of the product and C the production cost we have

$$N \geq \frac{\text{Total cost to design product}}{(P - C)} \tag{2.41}$$

where N is the minimum number of products to be sold so that the business is profitable (break-even point).

A design cost growth will increase the minimum number of products to be sold to achieve business profitability. Such an increase may turn to be unfeasible given the market/company conditions.

References

AHA (2007), Kligfield P. *et al.*, Recommendations for the Standardization and Interpretation of the Electrocardiogram., American Heart Association.

AAMI (1991), Association for the Advancement of Medical Instrumentation, American National Standard, Diagnostic Electrocardiographic Devices, ANSI/AAMI EC11-1991.

Barnes A. R. (1938), Pardee HEB, White PD. *et al.* Standardization of precordial leads. Am Heart J 1938.

Brugada P. (1992), Brugada J. Right Bundle Branch Block, Persistent ST Segment Elevation and Sudden Cardiac Death: A Distinct Clinical and Electrocardiographic Syndrome. J Am Coll Cardiol 1992;20.

Campbell, M. J., Machin D., Walters, S. J. (2007) Medical Statistics – A textbook for the health sciences – Fourth Edition, John Wiley & Sons Eds.

CEU (2011), Council of The European Union: 'Council conclusions on innovation in the medical device sector'.
CFR (2012), US Code of Federal Regulations, Title 21: (Food and Drugs) chapter I, subchapter H, part 870, subpart C, section 870.2340 – Electrocardiograph.
Dubin, D. *et al.* (2000), Rapid Interpretation of EKG's, Sixth Edition, Cover Inc.
ECRI (2012a), Healthcare Product Comparison System (HPCS) ECRI Institute, www.ecri.org.
ECRI (2012b), Electrocardiographs, Multichannel; Interpretive; Single-Channel, Healthcare Product Comparison System (HPCS) ECRI Institute.
EEC (2011), Medical devices European standards, harmonized standards available at link: http://ec.europa.eu/enterprise/policies/european-standards/documents/harmonised-standards-legislation/list-references/medical-devices/
Einthoven W. (1906), Le telecardiogramme. Arch. Int. de Physiol. 1906; 4: 132–164 (translated into English in Am. Heart J. 1957; 53).
Einthoven W. (1912), The different forms of the human electrocardiogram and their signification. Lancet.
Fairbanks R. J. (2004), Caplan S, 'Poor Interface Design and Lack of Usability Testing Facilitate Medical Error,' Joint Commission Journal on Quality and Safety.
Fye W. B. (1994), A history of the origin, evolution, and impact of electrocardiography. Am J Cardiology.
Geselowitz D. B. (1964), Dipole theory in electrocardiography. Am. J. Cardiol. 14: (9) 301–6.
Grimes (2004). Clinical engineers: stewards of healthcare technologies. IEEE Eng Med Biol.
Harrison (2011), Harrison's principles of internal medicine, McGraw Hill.
Kelleher B. (2003), The seven deadly sins of medical device development. Medical Device & Industry Magazine: http://www.devicelink.com/mddi/archive/01/09/002.html.
Kohn L. *et al.* (2000), 'To Err Is Human: Building a Safer Health System', National Academic Press.
Malmivuo J. and Plonsey R. (1995), Bioelectromagnetism Principles and Applications, Oxford University Press.
MIL-STD-498 (1994), see Wikipedia for updated links on document templates.
Nuwer M. (1997), Assessment of digital EEG, quantitative EEG, and EEG brain mapping: report of the American Academy of Neurology and the American Clinical Neurophysiology Society. Neurology.
Philips (2003), The Philips 12-Lead Algorithm Physician's Guide.
Scott Wayne (2000), Finding the Needle in a Haystack, Analog Dialogue, 34.
SSS (1994), DI-IPSC-81431 Data Item Description: System/Subsystem Specification (SSS), US Dept. of Defense.
Standish Group (1995), The Chaos Report.
Von Maltzahn W. W. (2006), Medical Instruments and Devices, Biomedical Engineering Handbook, J. D. Bronzino, CRC Press.
Webster John G. (2009), Medical instrumentation: application and design, John Wiley & Sons.
WHO (2003), World Health Organization. Medical Device Regulations Global Overview and Guiding Principles.
WHO (2012), World Health Organization. Family of International Classifications of Diseases. Accessed 2012: http://www.who.int/classifications/en/.

3

Biomedical Engineering Design

'the difference between poor conceptual designs and good ones may lie in the soundness of design method. . . . Great Design comes from great designers.'
F. P. Brooks

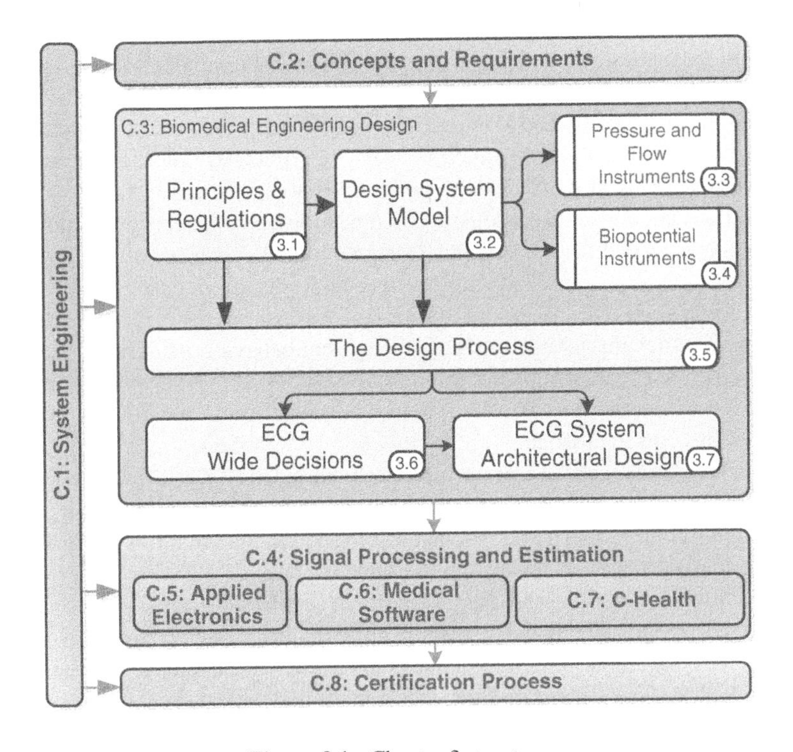

Figure 3.1 Chapter 3 structure.

Medical Instrument Design and Development: from Requirements to Market Placements, First Edition.
Claudio Becchetti and Alessandro Neri.

Chapter Organization

This chapter focuses on product design. Referring to the chapter organization in Figure 3.1, at this stage, the team specifies **how** the device will be made to work according to the requirements discussed in Chapter 2 that detail **what** the device must do. The design specifications must comply with constraints of the application field that, for medical devices, imply guidelines and **regulations** outlined in Section 3.1.

In the case of diagnostic devices, the design of the core components can be derived from a general device system **model** whose modules are (a) the sensors (or transducers), (b) the hardware and (c) the firmware to manage the hardware and to perform fast signal processing. The applications software on the PC platform is an additional component used more frequently for data management, network communication and the user interface (Section 3.2).

This model can then be particularized over the specific system to be designed. In this chapter, we will focus on **instruments** related to pressure and flow (Section 3.3) and biopotentials signals (Section 3.4).

The design is the **process** where the system architecture, the subsystems and the interfaces are defined from the requirements. The conceptual gap between requirements and design is relevant. There is generally a great effort to produce the design from the requirements: many solutions seem feasible with unclear trade offs and the problem formulation is complex and ambiguous.

For this reason, the '**divide and conquer**' approach is an effective strategy for reducing this complexity. The conceptual distance between requirements and design specifications can be divided into many simpler derived problems that can be easily faced one at a time. The set of solutions encompasses the overall design. A process is therefore required to define the sequential steps needed to generate the design solution. This is discussed in Section 3.5, where the design process is divided into 'system-wide design decisions' and the 'system architectural design'.

The **system-wide design decisions** details the system behavior, its external interfaces and the database structure regardless of the internal implementation. The **system architectural design** focuses on the internal system implementation defining the subsystems with their functional relations and the associated requirements. The implementation part (Sections 3.6 and 3.7) shows how such steps are performed for a marketed ECG device, the Gamma Cardio CG, that is fully disclosed by Gamma Cardio Soft (www.GammaCardioSoft.it).

Design activity encompasses some relevant topics such as human interface design (Section 3.6.2) and communication interface design (Section 3.7.2).

Since medical device market certification process requires specific documentation to be prepared, Section 3.8 introduces the structure of the technical file required for certification with the associated documents. Such topics will be detailed in Chapter 8.

By the end of this chapter, readers should be able to understand the design process and to derive a design from requirements, bearing in mind the instruments discussed in this chapter and the ECG implementation.

Part I: Theory

3.1 Design Principles and Regulations

In this section, we outline the main design principles and regulation concepts, while leaving to Chapter 8 a more formal discussion. When dealing with a diagnostic instrument that manages physiological parameters, a relevant part of the specifications is associated with patients' and operators' safety. The general principle is then: '***Prium non-nocere***' (i.e., most importantly, don't do any harm).

Notably, to collect data, a measurement system creates some perturbation on the measured element (i.e., the patient). In general terms, the presence of a measuring system will represent an increased risk (for patients and operators), that needs to be fully taken into account when designing an instrument intended for clinical applications. In line the with main regulatory frameworks, the first specification constraint is that *the perturbation due to the instrument shall not affect the patient's safety significantly with respect to the expected benefits of the measure.* The key concept is '**acceptable risk**', that is, the device must not imply unacceptable risks based on a risk–benefit analysis of its usage. This topic is discussed more formally in Chapter 8 in the framework of the certification process. We introduce here the risk that is defined by the Oxford dictionary as the 'possibility of loss, injury, or other adverse or unwelcome circumstance'. In clinical practice, risks can be generated under the typical usage of the diagnostic instrument (i.e., the normal condition), or with the possibility that a device problem arises (i.e., a fault condition). Both these situations may affect the patient's safety.

In general terms, the following key factors have to be addressed, when designing biomedical instrumentation:

1. patients' and operators' **safety**, in order to make direct risk associated acceptable when using instrument, in normal conditions, and in fault or abnormal conditions
2. instrument **quality and performance**, in order to minimize the effects of an increased indirect risk, associated with the use of the instrument. We refer here to the case of instruments that may have a degraded quality, and can thus lead to errors in diagnosis, thus exposing the patient to unwelcome circumstances.

The quality of the instrument is directly related to the parameters associated with the measuring part: the **intrinsic error** associated with a measurement may hinder the correct diagnosis of pathology, or may underestimate or overestimate its gravity.

Another source of risk is associated with the perception of the presence of a damage or of an unwanted circumstance. The **human–machine interface** has to be accurately designed to avoid risks in device usage and in data perception.

Referring to medical electrical equipment, the technical standard series IEC/EN 60601 is specifically devoted to safety and effectiveness requirements. The standard was first published in 1977, and now it represents a *de facto* constraint when a medical electrical instrument needs to be designed and placed on the market. The standard is composed of a general part, **IEC/EN 60601-1**, with ten different collateral sections, and numerous vertical standards that address specific medical instruments.

Table 3.1 EU medical device classes

Class	Risk	Level of control
Class I	low	sole responsibility of the manufacturers (apart from sterile and measuring devices)
Class IIa	medium risk	the intervention of a **notified body** should be compulsory at the **production stage**
Class IIb	high risk potential	inspection by a **notified body** is required with regard to the **design and manufacture** of the devices
Class III	high risk potential	inspection by a notified body is required with regard to the design and manufacture of the devices **explicit prior authorization** with regard to conformity is required for them to be placed on the market

In general terms, we will associate medical devices to **classes** based on the complexity of the design, the characteristics associated with their use and the potential hazards for patients and operators. There are differences in medical device classifications according to the applicable regulation framework (US, EU etc.) although a similar conceptual approach holds worldwide. The **EU classification** is a system 'based on the vulnerability of the human body taking account of the potential risks associated with the technical design and manufacture of the devices' (EEC, 2007). In the European market, there are four classes, ranging from low risk to high risk. Associated with these classes there are corresponding different **levels of control** that are needed to ensure safety and effectiveness, as shown in Table 3.1.

The European Regulation framework specified in the Medical Device Directive (EEC, 2007) defines a set of rules for classification with the corresponding level of control setting the responsibility of the manufacturers. Similarly, in the USA, there are three 'regulatory classes based on the level of control necessary to assure the safety and effectiveness of the device'. Additional details are discussed in Chapter 8.

As shown in Table 3.1, the level of control increases when a more harmful medical device is intended to be placed on the market. In the design phase, a risk analysis is needed to identify possible harms to patients or operators and to introduce **control measures** to minimize the effects. These measures will be implemented to minimize

- the *probability of having a harm* due to specific event (such as an accident)
- the *severity of that harm.*

Measures intended to reduce harm probability are defined as 'prevention measures', while 'protection measures' reduce harm severity.

In general, risk analysis is synthesized by an '**acceptability matrix**', where harm probability is categorized into probability levels (e.g., rare, improbable, remote, probable, frequent). Harm severity is associated to specific levels: negligible, marginal, serious, critical. The risk analysis process is outlined in Section 3.5.4, and the practical application is discussed in the implementation part for the ECG device. Chapter 8 details the discussion integrating the topics related to regulations and standards.

3.2 General Design System Model

The model shown in Figure 3.2, introduced in Section 1.4, is a possible template to represent the architecture of a medical instrument. This model is also applicable to the vast majority of medical devices and electronic equipment.

In this model, there are at least two **actors**:

- the patient representing the source, generating data for the instrument
- the healthcare operator who is able to manage and/or interpret the information presented by the instrument.

In the scheme, the second actor may be *also* the patient, in self-care or personalized health applications. In this case, the patient is able to interact directly with the instrument. This condition is becoming more frequent, with the increasing number of self-monitoring devices for personalized healthcare. This affects the design of the interfaces that are not only to be designed to interact with a specialized operator, but also they need to be suitable for untrained users. The **device model** of Figure 3.2 includes five modules.

The **sensor** module may be a simple sensing element such as an electrode in the case of biopotential recording, or an element that transduces some physical quantity into an electrical signal such as in the case of pressure or flow monitoring. The design of the sensor module will be focused on choosing the appropriate sensing technology suitable for the required performance: sensitivity, accuracy, noise level and physical properties. The sensing/transduction technology is detailed in Chapter 5.

The **hardware** includes the electronic circuits needed to process the analog signal that has been captured by the sensing module.

Sensing and hardware modules are distinct and kept separated from a functional point of view although separation may not hold in certain cases. The performance of the hardware module will involve the specifications of the internal processing (e.g. sampling rate, bandwidth, SNR . . .) and of the two interfaces (with sensors and with the communication network). A detailed description of is given in Chapter 5 which is focused on the **applied electronics**.

The **firmware** is basically the program code and the associated data stored in a non volatile memory that cannot be modified by the user. The firmware is designed to make hardware programmable devices (FPGA, Microprocessors, . . .) work according to the required logic and processing. Firmware is usually devoted to hardware management and signal processing. Firmware implementation is preferred rather than pure electronic implementation for dealing

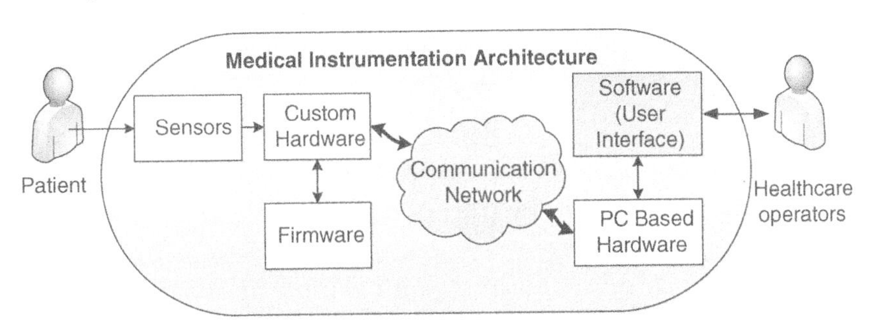

Figure 3.2 Medical instrument architecture.

with processing that cannot be performed on software applications running on PC platforms. Firmware is discussed in Chapter 6, which is devoted to software.

The hardware, firmware and sensor components are more often part of a **communication network** that links devices connected to the patient and the software control application used by the healthcare operators. This configuration decouples the place where the medical service is needed (i.e., the patient's location) and the place where the service is performed (i.e., the doctor's location). This telemedicine approach, enabled by information and communication technology allows to offer medical services where needed. Chapter 7, which is focused on **medical informatics** and cloud computing for healthcare, will discuss also these topics.

The **software application** module usually implements the interface with the users of the medical instrument. The module usually includes the man–machine interface, the low data rate or off-line data processing, and the functions for data sharing across communication networks. The software also exchanges messages with the firmware to manage the hardware board. Manufacturers are increasingly using PC platforms with applications software since the development cost of high-level functions is significantly lower with respect to a hardware implementation and PC based platforms have a superior performance to cost ratio thanks to the large volumes. Medical software is discussed in Chapter 6, while medical informatics, with a focus on e-Health distributed applications, is discussed on in Chapter 7.

3.3 Pressure and Flow Instruments

There are usually two distinct parts in medical devices designed to capture blood pressure and flow: the transduction (the sensor of the model in Figure 3.2) and the conditioning part, as shown in the left and right sides of Figure 3.3. The transduction part contains circuits devoted to the transduction of the physical quantity to be measured (e.g., the pressure) into an electrical quantity (typically, a voltage). In the conditioning part, the electrical quantity that has been picked up, is suitably amplified and filtered, with necessary reduction of interference and noise.

In some cases, since the transduction element may be sensitive to other unwanted signal sources, some circuit sections are devoted to measuring the interference so as to **compensate** for its effect on the primary sensing element.

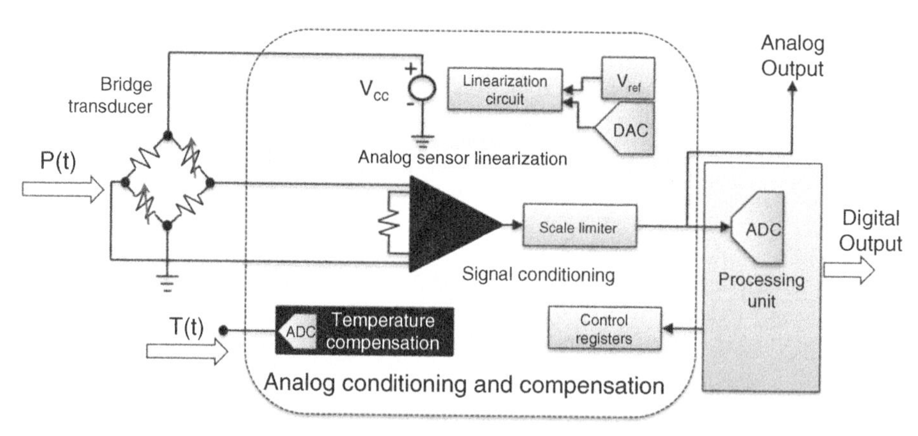

Figure 3.3 Acquisition module for pressure measurements.

For example, strain gauges used in pressure measurements are sensitive to temperature. Temperature monitoring is therefore necessary to compensate for input variations due to temperature changes as shown in the bottom part of Figure 3.3.

Transducers for sensing forces or pressure are usually implemented by strain gauges or piezoelectric elements. Such elements have a *non-linear response*, that is, when measuring a physical quantity $P(t)$, the relationship with their electrical response $R(t)$ is governed by a general sensitivity curve such as

$$R(t) = S(P(t)), \tag{3.1}$$

where the function $S(.)$ is not linear.

In this case, the analog circuit design may need to include some **linearization** circuitry in the conditioning or recording part. If linearization cannot be implemented in the analog part, the design will need to include input linearization, by simulating the inverse function $S^{-1}(.)$. As usual, the design will also take care of the **presentation and application level** where collected data are visualized to the users, stored further, offline processed and possibly shared across healthcare stakeholders.

There are *several possible configurations* for the sensing elements that may be based on resistive, capacitive or inductive transduction. The most common transduction configuration is the **Wheatstone bridge**, where four sensing elements are linked in such a way that, if appropriately powered, there is a voltage output which is a directly proportion to the physical quantity to be measured: in this case, there is an inherent compensation for temperature or other disturbing factors, provided that the four sensing element are equivalent, that is, their electrical characteristics are the same. Simpler configurations are in the form of a half Wheatstone bridge configuration. In other cases, the sensing element can be directly associated with the driving resistance that determines the gain of an instrumentation amplifier. In this case, there is a change in the amplifying factor proportional to the input electric signal that is related to the variation of the physical quantity. Figure 3.4 shows a simplified schema of this configuration.

Sensors for measuring physical quantities, such as in the case of blood and pressure measurements, are usually passive, that is, they change their electrical characteristics (e.g., capacity, resistance or inductance) as a consequence of the variation of the physical quantity.

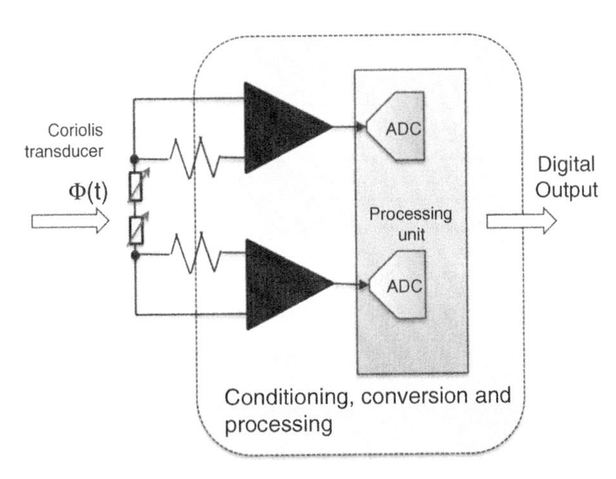

Figure 3.4 Mass flowmeter based on instrumentation amplifier.

A voltage or current signal proportional to the physical quantity is obtained through a stable voltage/current source that 'powers' the sensor.

By contrast, biopotentials are already electrochemical signals translated into electrical signals by chemical reaction in the sensors (i.e., the electrodes). The electrochemical signal is directly and autonomously generated by the source (i.e., the human body). This topic is discussed in Chapter 5. The following section introduces the design of blood pressure measurements that span from non-invasive self-care products to invasive devices used in intensive care units for critical cardiovascular conditions.

3.3.1 Blood Pressure Instruments

In the case of non-invasive long-term monitoring, arterial blood pressure is generally measured by quantifying the pressure needed to cause cessation of the blood pulse in a specific body region, according to what Riva-Rocci discovered in 1896 (Lewis, 1941). A specific actuator applies an external pressure to occlude completely the vessel, thus blocking blood pulse pressure. To do this, a limb is subject to a compression through the use of an active cuff. This compression is then transferred to the blood vessels. Assuming that there is no difference between the external and the limb-underlying tissue pressure, the vessel's internal pressure can be modified by varying the pressure exerted by the cuff itself. If the arterial blood pressure is higher than the cuff pressure, the lumen is open and a blood pulse can be detected. If the cuff pressure is higher than the arterial blood pressure, the lumen is closed and no pulse can be detected after the occlusion.

Blood pressure is then detected indirectly by somehow monitoring pulse activity. Possible approaches are direct vessel palpation, the **auscultatory method** with the use of pulse sound through a stethoscope, and the oscillometric method used in automated pressure monitoring devices. In the **oscillometric measurement** of blood pressure, the arm is placed within a cuff, and the measurement is performed using a standard arm cuff together with an inline pressure sensor. This method relies on the principle that when vessel pressure is the same of the cuff pressure, the pulse pressure oscillation can be detected. In essence, the cuff is inflated to a pressure greater than the vessel. Then, the cuff is slowly deflated. The conditions shown in Table 3.2 hold for determining systolic (maximum) and diastolic (minimum) pressure.

Table 3.2 Pressure measuring conditions

	Cuff pressure	Vessel circulation	Pressure oscillation
Pressure deflated ↓	Greater than vessel pressure	Vessel occlusion	Not present
	Equal to maximum vessel pressure	Start blood circulation in vessel	Start of oscillation
	Between maximum and minimum vessel pressure	Blood circulation in vessel	Maximum amplitude oscillation at **MAP**
	Equal to minimum vessel pressure	Blood circulation in vessel	Cessation of oscillation
	Less than vessel pressure	Blood circulation in vessel	Not present

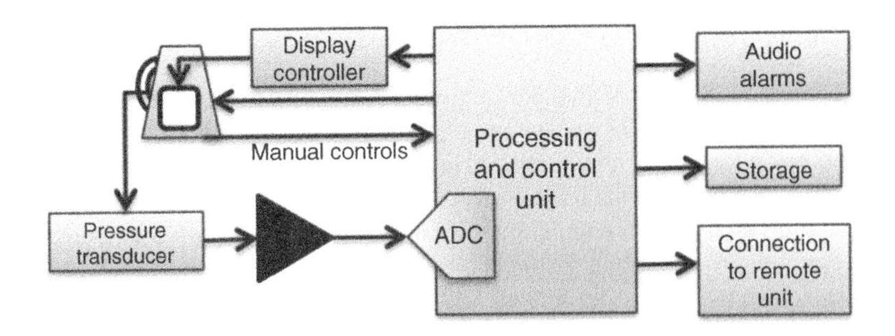

Figure 3.5 Simplified schema for non-invasive blood monitoring.

The cuff pressure recorded by the sensor is high pass filtered above 1 Hz to observe the pulsatile oscillations as the cuff slowly deflates. It has been determined that the maximum of the oscillations actually corresponds with a cuff pressure equal to the mean arterial pressure (MAP). Systolic pressure is located at the point where the oscillations are a fixed percentage of the maximum oscillations, with the systolic detection ratio at 0.55, and diastolic pressure at 0.85 of the maximum oscillations. In Figure 3.5, a representation of a non-invasive system for blood pressure monitoring, based on oscillometry, is represented.

In this scheme, the processing unit will measure the pressure that is sensed by the transducing part, and it will control the actuators for cuff inflation and deflation. The cuff is controlled to avoid inflating to a pressure that is much higher than the systolic pressure. A feedback sensor mechanism may be used to stop inflating once the sensor no longer detects any pressure variation.

The processing unit will then deflate the cuff with a constant rate of decrease. If the cuff pressure is higher than the arterial systolic pressure, the sensing elements will not sense any pulse oscillations. Once the cuff pressure is between systolic and diastolic pressure, the sensor will detect an oscillating pattern similar to the one described in the Chapter 2. Once the cuff pressure goes below the diastolic pressure, the detected signal will tend to vanish, since there is no significant pressure on the sensor.

Most non-invasive instruments for blood pressure measurement also estimate heart rate, which is a significant parameter from a clinical perspective. This parameter is easily evaluated by the sensor signal when the pressure lies between the systolic and diastolic values. The heart rate is usually computed by detecting the peaks and their time difference from the signal waveform recorded by the sensor. A successive processing is required to average the instantaneous heart rate estimations over time.

Invasive measurements yield more accurate measurements that may be useful in more safety-critical contexts such as critical care and operating rooms.

In the case of **invasive measurements**, the recording system shows differences in the sensor structure that may take the form of a miniaturized strain gauge embedded into a small chamber, which receives blood from a catheter inserted into an arterial line. The recording system is basically the same. In practical terms invasive measures imply additional risks such as thrombosis, infection and bleeding. Sensors have to be designed and manufactured for sterile management and blood contact. The recording system has to achieve a reliability suited to its intended environment and use (e.g., operational rooms). The higher safety class of the device implies a more complex life cycle (design, manufacturing etc.) due to the increased safety risks.

3.3.2 *Flow Measurements*

For **macrocirculatory** measurements, the variable which is usually monitored is the **cardiac output** (CO), which is defined as the volume of blood ejected by the left ventricle per unit of time (i.e., minutes). Left and right ventricle outputs are equal under normal circumstances. Cardiac output is then equal to the product of stroke volume (i.e., blood ejected by one ventricle at each heart beat) times the number of beats per minutes (i.e., heart rate). Dimensionally, CO is a measure of flow = ejected volume/time = cross sectional area × speed of fluid. The parameter is expressed in liters per minute (l/min), and it is divided by the body weight or body surface area to obtain a normalized value with respect to patient size. A typical resting value for a wide variety of mammals is 70 ml/min per kg (Bronzino, 2006). During sporting activities, CO increases to satisfy the body's additional demand of oxygen. In well-trained athletes, CO can increase to five times the normal value at maximum activity. In normal conditions, CO increases with heart rate, up to 180–200 beats/minute. In disease states, CO does not increase over a certain threshold that may be reached at 120 beats per minute. Arrhythmias such as atrial fibrillation or flutter may pathologically increase heart rate affecting negatively the CO. This condition is usually pharmacologically treated by reducing the heart rate.

CO is mainly measured by the **dilution invasive method**. This approach is based on the upstream injection of a detectable marker substance via an artherial catheter. The downstream dilution of the marker over time, referred to as the 'dilution curve', is used to evaluate mathematically the CO parameter. A suitable marker has to be selected so that the measure of its dilution in the blood is accurate and reliable.

By injecting a fixed quantity of marker into an idealized flowing stream having the same velocity across the diameter of the tube, the dilution curve has a rapid rise and an exponential fall. The flow is simply the amount of the marker divided by the area of the dilution curve (Bronzino, 2006). The area of the dilution curve can be sensed in different ways based on the marker type. If the indicator is different from the blood in terms of its temperature, the method is called thermodilution, and the transducing element can derive the dilution curve based on the profile of the temperature over time.

When the marker is simple water at room temperature, multiple measures can be performed since water is not toxic. This is generally accomplished through *thermistors* placed on the catheter inserted into the pulmonary arteria. Figure 3.6 shows the waveform

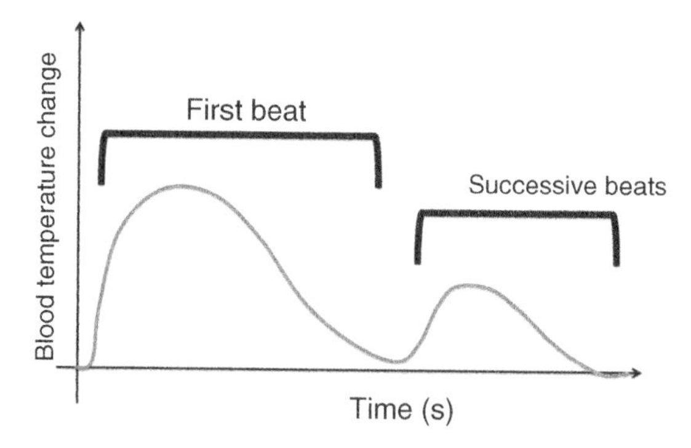

Figure 3.6 Blood temperature change.

associated with a thermodilution procedure, for two consecutive beats. Since the heat transfer is a process that spans more than just one beat, the waveform associated with the marker will repeat, albeit with a smaller amplitude, over a number of successive beats: the processing unit will assess only the first area for CO calculation.

Microcirculation is the delivery of blood to the smallest vessels present in the system. In clinical practice, this often refers to the study of blood flow in the vessels delivering blood to the heart, that is the *coronary system*. Due to the rather small environment, the estimation of flow in the coronary system is usually done indirectly, by using medical imaging systems, such as cineangiography with coronary catheterization, with transthoracic Doppler echocardiography or with positron emission tomography. It would be difficult to estimate flow directly with the catheter, as the catheter itself substantially reduces the lumen, thus disturbing the measured variable.

3.3.3 Measuring Oxygen Concentration

Pulse oximeters are common medical devices that indirectly measure the amount (e.g., the relative percentage) of oxygen saturation of hemoglobin in arterial blood. This non-invasive approach is a faster and safer alternative to measuring the parameter through a blood sample. In pulse oximeters, a section of the body (e.g., a finger or an earlobe) is chosen such that it can be penetrated by light emission through a short path. Then, two light emitting diodes of different working wavelength are activated on the selected tissue. On the other side of the tissue, a photo detector transduces the amount of received light into an electrical signal. Two different wavelengths are used, since oxyhemoglobin absorbs in the infrared range, whereas deoxyhemoglobin absorbs in the red range. The measured signal from the photo detector has a pulsing pattern, which replicates the cyclic cardiac pattern. The processing unit converts the electric signal into a sampled sequence that is processed to obtain an estimation of the partial saturation of oxygen. Figure 3.7 shows a conceptual design scheme for a medical instrument for SpO_2 measurement.

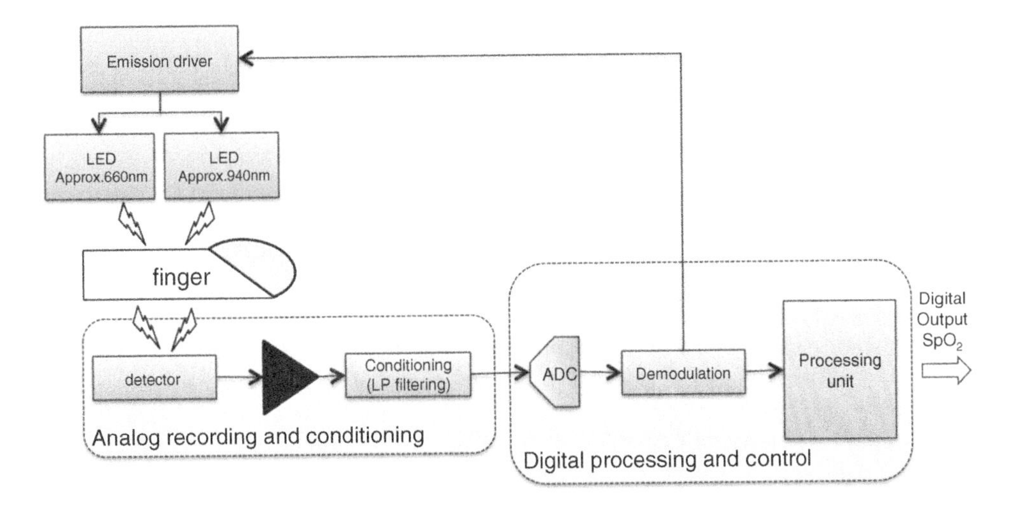

Figure 3.7 Pulse oximeter design scheme.

3.4 Biopotential Instruments

Biopotential instruments are a class of devices that are able to measure the electrical activity associated with the activity of body sections or organs. The principles behind biopotential instruments are similar: differences are based on the features of the signal to be measured (frequency band, input dynamics, requested accuracy, noise reduction strategies). For example, electrocardiography (ECG) and electromyography (EMG) share the same functional scheme including

1. a sensing section, with electrodes either to be placed on the body surface or to be inserted into the body tissue
2. a conditioning section, which contains an appropriate number of amplifying stages and analog and active filtering
3. a digital section containing an analog to digital converter and digital processing components.

However, such devices have differences, for example, in the cutoff frequency values of the low pass filter since the clinically significant signal has a different frequency bandwidth. Other differences are in the circuits, due to the different safety operational requirements. In the following sections, the electroencephalograph (EEG) will be discussed while the ECG will be addressed in detail in the implementation section.

3.4.1 Electroencephalographs

An EEG is a medical instrument that records the electrical activity of the brain in specific disorders such as the epilepsy, convulsions and brain death. The EEG makes use of various electrodes placed on the patient's scalp to measure and display the electrical signals that are generated by the nerve cells within the brain.

In continuous recording, EEG monitors show signal amplitude differences between electrodes over a period of time in order to detect possible pattern changes associated with the presence of disorders, such as hypotension, hyperventilation and hypoxia. In the operating room, EEGs are used to monitor anesthesia level and brain perfusion (i.e., the blood flow to the brain). When brain circulation is inadequate, there is a change in the EEG signal, notably a decrease of high-frequency activity.

In order to monitor the electrical activity associated with brain activity, scalp **electrodes** are usually placed on the patient through a conductive adhesive or paste, and then they may be attached to the body surface by using collodion. In clinical practice, the number of electrodes associated with the standard international 10–20 system is 19, with two earlobe electrodes. This results in the presentation of 16 different traces for EEG. Alternative electrode systems (that are less common in clinical practice) follow the 10-10 system (with up to 74 electrodes), and the 10-5 system (with up to 345 electrodes). The last two are also referred to as high density EEG (HD-EEG).

The method to derive EEG channels from electrode sources is referred to as a 'montage'. The **montage** depends on the number of electrodes that are placed on the body surface and the combination of picking up places for the representation. Clinically used approaches are

- the *bipolar* montage, where the voltage difference is gathered through two active electrodes placed on the scalp, also denoted as *scalp-to-scalp* montage
- the *referential* montage, where the voltage difference is gathered by using an active electrode placed on the scalp and an inactive one placed in a non-active area, usually the earlobe – this is also know as *scalp-to-reference.*

Both the montages are usually needed for accurate EEG interpretation. Alternative montages are also experimented with in research laboratories more than in clinical facilities. For example, in the *Laplacian* montage, a trace is calculated as a voltage difference between one active electrode and a weighted average of the surrounding electrodes.

A general schema of an EEG is shown in Figure 3.8, where the different stages of amplification and filtering are outlined. Electrodes from the scalp are connected to an **EEG head box**, a separate panel that is placed in the proximity of the individual's head. This panel contains lead connectors, and usually a first-stage amplifying unit to improve the signal-to-noise parameter; the amplifier is generally used to measure the low voltage values picked up from the scalp, reducing noise effects. The need for an amplifying stage that is as close as possible to the pick-up signal location is a direct consequence of the fact that unwanted signal attenuation at the amplifying input is generally caused by the ratio between impedances of the electrodes and the amplifying stage.

Figure 3.9 shows a simplified representation of the input stage of an EEG monitor. Z_{el} is the electrical impedance associated with the electrode and the cable losses, Z_{cab} is the impedance of the shunting capacitance of the cable and Z_{in} is the input impedance of the amplifying stage.

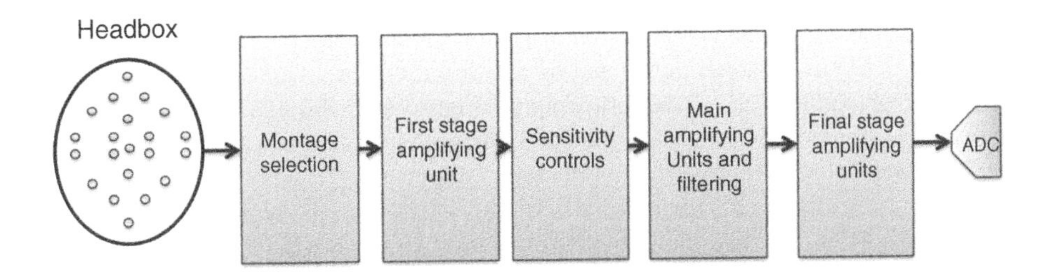

Figure 3.8 Conceptual representation of an EEG device.

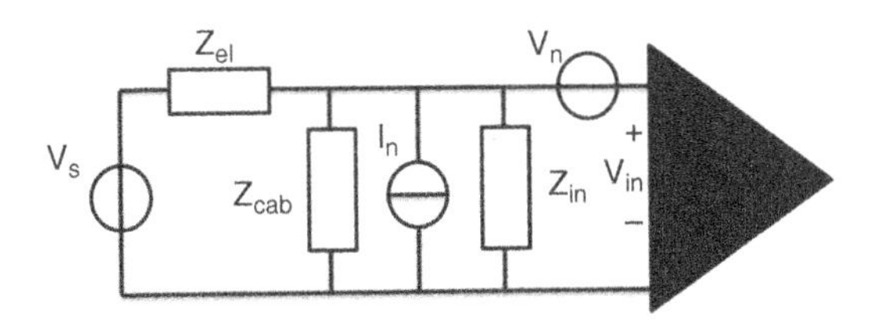

Figure 3.9 EEG input stage model.

With reference to Figure 3.9, the ratio between the source V_S, and the voltage that is present at the input of the amplifier V_{in} is given by

$$ {}^{V_{in}}\big/_{V_S} = \frac{1}{1 + {}^{Z_{el}}\big/_{Z_{in}} + {}^{Z_{el}}\big/_{Z_{cab}}} \tag{3.2} $$

This ratio tends to one if the total electrode impedance (which includes resistive losses due to the cable) is small compared to the amplifying input impedance and the impedance of the shunting capacitance of the cable. This suggests selecting high impedance input stages for design. Minimum recommended impedance is 10 MΩ; commercial products have much higher impedance.

I presence of noise sources V_n and I_n, the noise induced by the current source is directly related to the electrical impedance, and this implies that the **signal-to-noise** ratio decreases as a function of the electrode impedance.

Once the signal is collected through the head box, other amplifying stages are required, together with **filtering** stages needed to remove unwanted out-of-band noise. In general, considering the frequency ranges of the EEG signals, the useful bandwidth is between 0.16 Hz and 100 Hz. In special cases, such as the study of auditory stimuli, higher frequencies need to be detected, and filter response needs to be adapted. This may also happen when very slow changes of the EEG patterns have to be monitored. In this case, frequencies lower than fractions of a hertz have to be collected, thus making high pass filtering impractical.

The amplifying stages have to be properly designed to amplify EEG so that the analog to digital converter (**ADC**) can digitize analog signals without artifacts. This is discussed in Chapters 4 and 5.

EEG has a structure that is similar to any other biopotential instrument and, in particular, to the ECG, which is deeply analyzed in the implementation parts of this book.

EEGs can record many channels from sensors placed on the scalp. The minimum number of recorded channels is 16, and commercial products may include up to 128 channels (ECRI, 2012). Figure 3.10 shows some additional detail on the channel front-end EEG electronic scheme. Detailed explanations of the following used terms are available in Chapter 5. Focusing on one EEG channel, the head box and the montage selection connect the source electrodes to two inputs that are sent to circuit protection components. These components are suited to eliminating out-of-band interference and fast transients (e.g., electrostatic discharges) that may damage the EEG, and they may produce annoying artifacts on the EEG signal.

The source bipolar differential signal that usually ranges between 10 and 300 μV is amplified by an instrumentation amplifier that reduces the common mode interference.

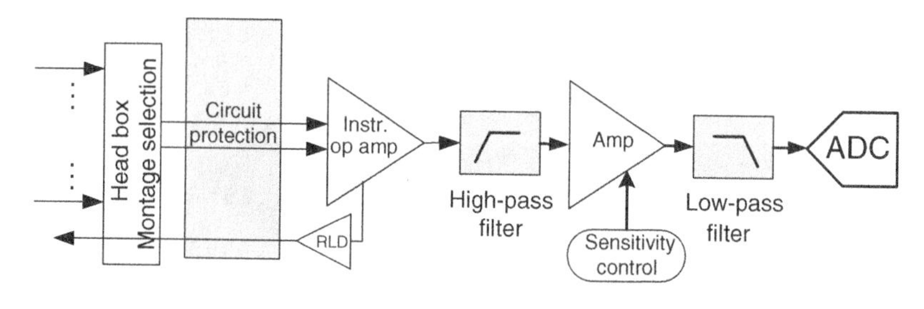

Figure 3.10 EEG conceptual single channel scheme.

This interference is due to unwanted signals that are equally present on both inputs of the amplifier. Power line interference is a typical unwanted signal that may be induced on both the amplifier inputs with an amplitude in the order of volts or tens of volts. The instrumentation amplifiers addressed in Chapter 5 are specific components that amplify the difference between the two signals in input while reducing significantly the signal components that are equal at both input (the so called Common Mode Signal). Such components is usually reduced by 10,000–100,000 times in commercial components. At this aim, the input circuit is designed so that the signal to be measured is present with opposite sign at the inputs while (i.e. the differential mode signal) while interference is added at both the inputs equally. The right leg drive (RLD) circuit additionally reduces the power line interference as described in Chapter 5. The control of the amplification and the frequency response is due to additional amplifiers and to low pass and high pass filters. Minimal specifications reccomend a frequency response of 0.3–70 Hz. Commercial products offer a wide selection of frequency range that may be changed from DC to 0.01 Hz low cutoff frequency to over 512 Hz high pass cutoff frequency and 20 kHz for research applications. Regarding digital conversion, the data resolution and the sampling rate have to be specified. Minimum performances are 200 samples/second with 12-bit resolution. Again commercial products, thanks to the advances in electronic technology, outperform such specifications with sampling rates as high as 8 kHz and 22-bit ADC (ECRI, 2012). Digital processing is then performed to obtain the signals to be shown on the monitor or printed on the paper with correct vertical and horizontal scale. Since power mains interference may be significant, an additional notch filter is considered to eliminate only the frequency of the net interference (50 or 60 Hz). The circuit of Figure 3.10 does not consider electrical safety protection that has to be included to avoid electric shocks. Such measures, common also to ECGs, will be addressed in Chapter 5.

3.4.2 Electromyographs

In the case of electromyography (EMG), the conceptual electrical scheme is similar to the EEG shown in Figure 3.10 and ECG examined in Chapter 5. The electrical activity to be detected is associated with the activity of skeletal muscle. As pointed out in the previous chapter, this activity can be recorded either invasively, by means of fine wire electrodes inserted through the skin in the neighborhood of the muscle, or by using surface **electrodes**, that are applied onto the body surface. In this latter case, which we will denote as *surface EMG*, the electrodes are generally small Ag/AgCl disks (with an area of tens of mm^2) with some conductive gel that is used to improve the skin–electrode interface. Capacitive-sensing ceramic dry electrodes with no paste are less used in clinical practice compared with Ag/AgCl. Once the biopotential associated with the activity of the muscle is picked up, different configurations can be used. In the **unipolar configuration**, the voltage is measured between an electrode placed in proximity to the muscle to be monitored, and a reference electrode, usually placed on the body surface, on a place where negligible muscle activity is supposed to be present (body prominences are generally used for this).

In the **differential configuration**, a differential amplifier is used to measure the voltage difference between two points on the 'muscle belly' (i.e., the fleshy central part of a skeletal muscle); the distance between the electrodes is generally assumed to be a few centimeters. The last and recommended in clinical practice is the so-called *double differential*

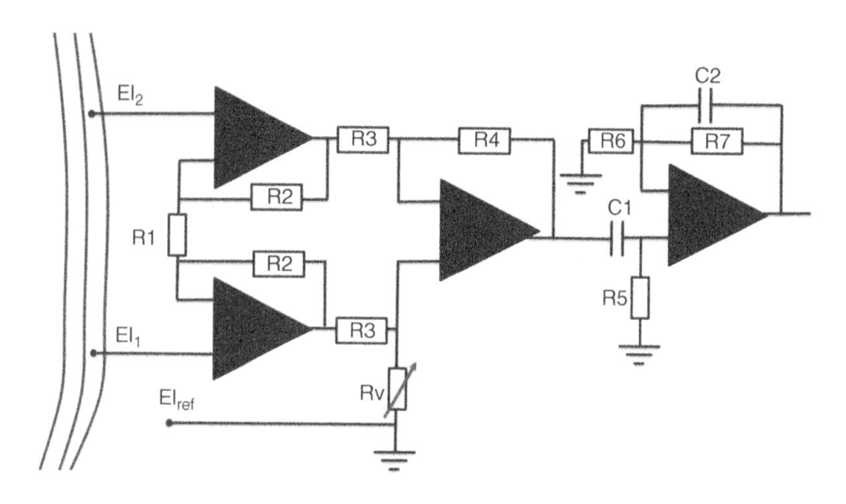

Figure 3.11 Instrumentation amplifier for EMG recordings.

configuration, where a differential amplifier is followed by a second stage amplifier in the so-called instrumentation amplifier configuration discussed in more detail Chapter 5.

Figure 3.11 shows an example of an instrumentation amplifier that can be conveniently used to record muscle electrical activity in a double differential configuration. This configuration has a first-stage gain equal to $(1+R_2/R_1)$, a second-stage amplifier with gain R_4/R_3, while the third stage is an active RC filter with gain $(1+R_7/R_6$ at the bandwidth center), and corner frequencies $1/(2\pi R_5C_1)$ and $1/(2\pi R_7C_2)$. R_v is used to maximize the common mode rejection through the reference electrode El_{ref}, while El_1 and El_2 are the electrodes to be placed on the muscle. This configuration is common to all biopotential measuring instruments. It is used to enhance selectivity and to reduce crosstalk (i.e., picking up the activity of neighboring muscles that are not being measured).

In general terms, the front-end amplifier that is to be used for surface EMG recordings must show an input impedance at least two orders of magnitude higher than the electrode–skin impedance: thus, input amplifiers with some hundreds of MΩ are generally used. In order to reject the presence of common mode signals in differential configurations, a high common mode rejection ratio (CMRR) is usually suggested with typical values of around 100 dB. If this is not possible, common mode reduction can be obtained by using a common mode feedback, which applies the principle of the *driven right leg* presented in Chapter 5.

In regards to the filter specifications, active filters include a high pass stage (with a cutoff frequency in the range 10–20 Hz), and a low pass stage (with a cutoff frequency in the range 400–450 Hz). High roll-off slopes are usually requested (>35 dB/decade), in order to cancel movement artifacts in the low frequencies, and appropriately exclude aliasing in the ADC. All these topics will be discussed in Ch. 5.

3.5 The Design Process

The use of a design process is **recommended in any product development** to improve quality and reduce risks and costs of reworking. In medical devices, a design process with the associated controls is often compulsory (FDA 2002, MDD 2007).

Many design techniques are available in the scientific literature. Since design cannot be considered a standard activity applicable to any class of products, design standards and guidelines mainly suggest frameworks that have to be tailored according to the product context and the applicable requirements.

A main first tailoring criterion to avoid cost and activity escalation is based on the '**least burdensome approach**' also supported by the US Food and Drug Administration, who say that design process tailoring must consider the least heavy procedures to comply applicable scientific and legal requirements (FDA, 2002).

In this chapter, the design process will be based mainly on the MIL 498 standard (MIL, 1994; SSDD, 1994), FDA 'Design Control Guidance For Medical Device Manufacturers' (FDA, 1997) and 'General Principles of Software Validation' (FDA, 2002).

As outlined in Figure 3.12, the first step is the conceptual design that has to define a coherent set of high-level requirements, functions, data structures, behaviors and layout to assess feasibility of the product development. This stage, introduced in Section 3.5.1, is performed

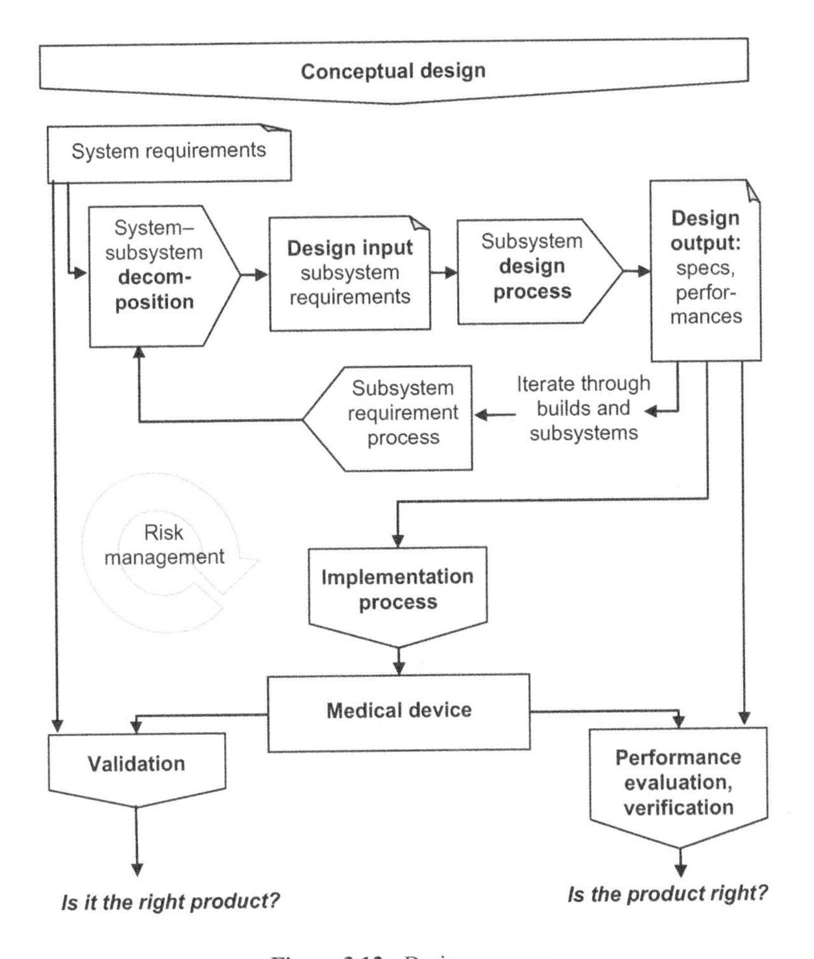

Figure 3.12 Design process.

at the beginning of the development project and should be updated when system requirements have been formalized.

Referring to Figure 3.12, the main design process stems from **system requirements** that are defined as the unique constraints that determine the non-acceptance of the system when non-fulfilled. The product acceptance is performed in the **validation** stage through qualification testing that confirms fulfillment of each requirement with respect to the implemented device.

System requirements are the **design input** of the first design process iteration. Design is performed iteratively focusing first on the system and then on the identified subsystems.

In the implementation part, we will address the Gamma Cardio CG as a system, with the requirements identified in Chapter 2. The subsystems to be designed are the acquisition hardware unit, the associated firmware and the software application hosted on the PC (Figure 3.14). The design is also performed incrementally by improving details at different iterations.

The iterations, called 'builds', have been defined in Chapter 1, Figure 1.27.

The **design process** includes the activities to obtain design outputs from inputs. The design process can be considered as a 'messy' type problem as discussed in Section 1.2, since either the problem formulation either the solutions are ambiguous. There are significant degrees of freedoms in defining the final solution, and the gap between design input and output is wide. At this aim, the '**divide and conquer**' paradigm (Section 1.5) can be employed to reduce the problem complexity. In essence, the main problem is broken down iteratively into sub-problems until each sub-problem complexity is affordable. In design, the complexity reduction is performed through system–subsystem decomposition and through the different builds that create the design solution incrementally.

Nevertheless, the complexity may remain high even with problems of smaller dimension. For this purpose, a proper design methodology may make the difference between poor conceptual designs and good designs. Often good design may be not sufficient to fulfill challenging requirements. Since design is a *creative* activity, 'great' designers may be a key success factor to achieve an effective design from challenging requirements. (Brooks, 1987).

The **design output** may include functional performance, interface specifications, drawing, processes and procedures for manufacturing and delivery, software documentation, source code etc. Most of this material has to be included in the documentation required to market the medical devices according to national regulation. In the USA, for example, the FDA requires a 'design history file' containing most of this information. In Table 3.6, the structure of the European equivalent design history file (the **Technical File for the Medical Devices**) is reported. This structure has been employed for the certification of the Gamma Cardio CG.

Referring to Figure 3.12, the design output may become the requirement specifications for another subsystem after a requirement analysis process. For example, the system design specifications will be processed in order to obtain requirements for the subsystems.

Defined design output specifications are also verified through the **performance evaluation** process at different builds/iterations.

In order to reduce the complexity of design, the subsystem design process described in Figure 3.12 can be further decomposed. As discussed in Chapter 1, standards may help either for processes or for the associated document templates. For example the output of the design process is specified in the corresponding MIL 498 template document: **System Subsystem Design Description** (SSDD, 1994).

The template includes two main chapters that correspond to two design sub-process steps: the *system-wide design decisions* dealt with in Section 3.5.2 and the *system architectural design* discussed in Section 3.5.3.

Risk management is a process that has to be performed in parallel with the design. Risks have to be assessed, evaluated and controlled continuously throughout the design. Section 3.5.4 outlines the risk analysis process that is further discussed in Chapter 6 in the context of the medical software. Chapter 8, relating to certification, has a more formal explanation based on regulations and standards.

3.5.1 The Conceptual Design

Before going into detailed design activity, the **conceptual design** process is performed to clarify device features and to assess product feasibility. The conceptual design is a first draft design, describing all the product features at a high level. According to Brook's 'law': 'all major mistakes are made on the first day of the project' (Brooks, 1995). This stage is addressed to avoid later errors that, as detailed in the first chapter, require an effort that is usually exponentially proportional to the time interval between the project start and the time when the error is discovered. This means that in product design: '*the earlier the errors are detected, the less the effort spent to fix them*'. Considering that medical devices are safety-critical products, errors may also affect patients' and operators' health. This is an additional critical reason to discover errors as early as possible. Conceptual design may be a powerful approach for discovering errors since it is a first level system design that can be easily evaluated by a wider audience, including health practitioners and operators, who may help to discover weak points.

The conceptual design should output a document also referred to as the operational concept description (OCD) that should 'describes a proposed system in terms of the user needs it will fulfill, its relationship to existing systems or procedures, and the ways it will be used.' (MIL 1994b). The document should include a *coherent* set of

1. requirements;
2. functions and data structures;
3. possible behaviors;
4. layout (form, structure); and
5. physical properties associated with the requirements.

In terms of **requirements**, in the case of diagnostic equipment, the conceptual design will take the information regarding what needs to be measured at high level following the concepts described in Chapter 2. For instance, in the case of an invasive blood flow measurement system, we may want the device to measure an amplitude range of 10 – 400 mm Hg at a minimum 50 Hz sampling rate.

The second step will focus on **functional design**, where high-level conceptual requirements are translated into modules able to execute specific functions. The output of this step is the draft functional block diagram generated by the requirements finalized in previous step. A minimal data structure is also formalized to support the defined functions when performing the device behaviors addressed in the next step.

Since most of the diagnostic instruments rely on an interaction with the user, and different behaviors are needed depending on this interaction, the following third step is the **definition of the behaviors** that each module needs to have as a response to this interaction.

The last two phases of the conceptual design refer to the **physical implementation**, specifically the embodiment, and the characteristics associated with the embodiment. At the end of the conceptual design stage, the biomedical engineer will output a detailed high-level product specification associated with the instrument that has to be produced.

The conceptual design involves technical aspects increasingly. At requirement level, considering a device for blood flow measurement, we need to be able to

1. gather flow data in the environment
2. capture flow data behavior over time and possibly reject other physiological and unwanted sources
3. present this information to the user in a concise and non-misleading approach.

The output of the functional design is a model, where module functions and data structures are outlined with their relationships. Functional design output should clarify for different behaviors of the device:

- which data are exchanged between modules
- what happens in functional terms when there is an interaction with the user, or with an information source
- how the different modules are integrated at functional level.

At the end of the functional level design, the conceptual design model will include the connections between modules both at the communication and at the functional level. In this phase, logical errors associated with the conceptual design are identified and corrected.

The functional behavior of a device is used to derive physical device specifications: size, form factor, weight, ease of use, etc. At the end of this process the design concept should be validated from various perspectives: medical effectiveness, safety, business sustainability, technical feasibility, compliance to applicable regulations. etc.

3.5.2 System-wide Design Decisions

The **System-wide design decisions** describe the design of the system and its behavior regardless of the internal implementation. Note that some decisions may already be specified in the system requirements and some others may be postponed to subsystem design activities. The main decisions to be addressed are (adapted from SSDD, 1994)

- system *behavior:* response to input, performance
- system *interfaces* with input and output specifications
- system *databases* and data presentation to users.

Other decisions may be related to safety and critical requirements or to the mechanical hardware systems items. The design decisions may become **internal requirements** eventually imposed on the subcontractors who implement the associated subsystems.

The designer is responsible for verifying that such decisions are fulfilled through proper internal testing. Alternatively, the designer may transform such decisions into formal requirements through a contractual process. The 'system-wide design decisions' step clarifies the system behavior and enables the identification of subsystems in the following step: 'system architectural design'. In the system–subsystem design description, the divide and conquer approach is performed by considering first system behavior as a whole ignoring internal implementation, and then focusing on the internal components in the architectural design.

3.5.3 System Architectural Design

In the **system architectural design**, the system components (i.e., subsystems) and the associated behaviors are first identified. Then the following steps are addressed:

- subsystem static functional relations
- identification of system requirements associated to subsystems (also called traceability)
- computer hardware resource specifications.

The control of the coherence of the design is performed through two additional steps: the concept of execution and the subsystems interface design. The **concept of execution** describes how the subsystems work together with their associated interfaces. This step must evidence possible inconsistency with the design since, for example, the design of specific interfaces may be in conflict with the dynamic relationships of the components (execution/data flow, transition state etc.) described in the concept of execution.

The example of the Gamma Cardio CG may further clarify the general design process outlined in Figure 3.12. The Gamma Cardio CG requirements detailed in Chapter 2 are the design input of the (sub-)system design process. The output of this process includes interfaces, functional and performance specifications over the system and the associated subsystems to be developed (hardware and firmware of the acquisition unit and the PC applications software). These design decisions are then transformed into requirements of the subsystems to be developed. Each subsystem, e.g., the software application has then a subsystem requirement specification that constitutes the input of the software application design. The design process of the software application will produce the design output: specifications and diagrams specifying functional behaviors, relations, timing and so on. This output is the internal specification set that has to be verified through testing or performance analysis. At the end of the process, the Gamma Cardio CG is validated through the qualification tests associated with the system requirements. This process is detailed in the implementation part.

3.5.4 Risk Management

Risk management is a dynamic process that is continuously iterated throughout the product life cycle, including the design step. Risk management implies

1. risk **analysis**, identifying hazards, considering all the combinations of events that may produce harm, defining the severity of harm and the probability of occurrence
2. risk **evaluation**, assessing the acceptability of risks and deciding whether protection measures are required

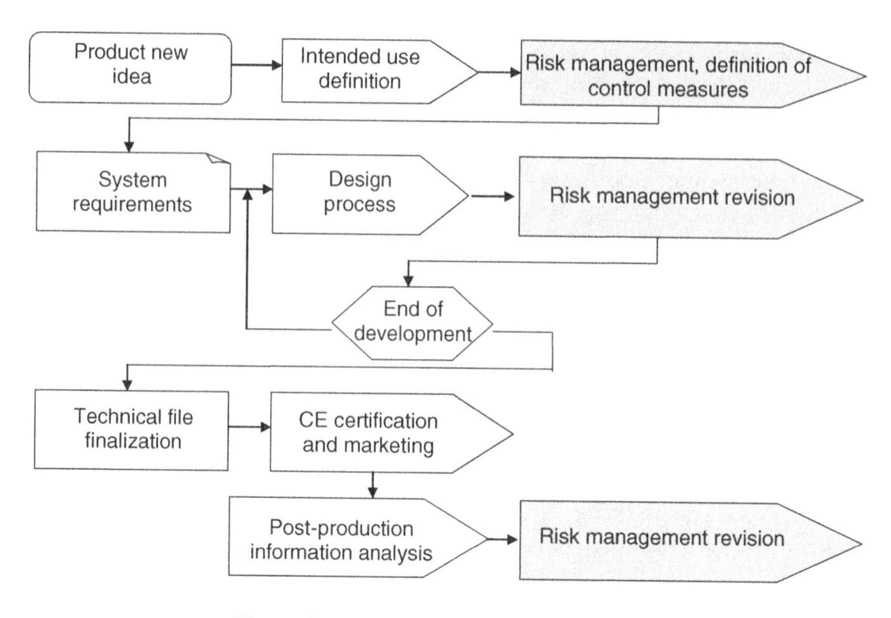

Figure 3.13 The risk management process.

3. risk **control** that focuses on the implementation of protection measures for unacceptable risks
4. risk **monitoring**, where risks are assessed periodically to evidence any changes (new risks, different estimation etc.)

The evidence of this process is included in the 'risk management files' which, for each product, contains all the records and documents generated by the process.

It is worth outlining here how this process is integrated into the product life cycle and in the design step. Figure 3.13 shows the relations between the risk management and the other processes for the Gamma Cardio Soft company that has developed the Gamma Cardio CG product.

The Company has to establish and maintain a systematic procedure to review information obtained by medical devices or similar devices in post-production. These data must be evaluated for safety significance. The post-production information must evidence

- if there are hazards not previously recognized
- if estimated risks, related to a specific hazard, are no longer acceptable
- if the original assessment is in any case to be invalidated.

If one of the above conditions is satisfied, the outcomes must be reported as input elements of the process of risk management revision. If there is a residual risk (i.e., 'risk remaining after risk control measures have been taken' – IEC 14971) or the risk acceptability has changed, the impact on the associated risk control measures has to be reassessed.

In particular, the risk management file will be updated in the following cases:

- change or introduction of new legislation
- evidence from scientific publications that additional methods exist for reducing risks

- introduction of new technologies
- changes in process, material or critical suppliers
- complaints
- reporting of accidents or 'near misses'.

In the design process, the risk management file therefore has to be mainly reviewed for changes in technologies or critical components. Chapter 8 will introduce risk management more formally, based on ISO 14971, (Medical devices, Application of risk management to medical Devices). Chapter 6 will also address risk management related to medical software, based on IEC 62304, (Medical device software – Software life cycle processes).

Part II: Implementation

3.6 ECG-wide Decisions

The ECG-wide decisions stage is the first element of the design process. The system is not decomposed yet and therefore decisions ignore the internal implementation of the ECG. In the first step, the system behavior clarifies the role of the system and its external interfaces that are then discussed in the following sections. This book will not go into the details of all the interfaces. The topics of this implementation part are presented as much as possible in terms of general guidelines suitable for any medical device. Then the example of the Gamma Cardio CG is described to show how theory is translated into a real implementation.

3.6.1 The Gamma Cardio CG Use Case

The Gamma Cardio CG use case is derived by a textual description of an approved usage of the ECG (BSC, 2010). The use case helps defining the successive design steps and it is also useful to understand how ECG will be used by ECG stakeholders. Referring to Figure 3.14, the ECG **acquisition** can be performed by a trained operator or by the cardiologist that prepares the patient for the ECG exam.

In this step, the operator connects patient cable to the 10 electrodes and to the acquisition unit, and then positions the 10 electrodes on the patient to obtain the standard 12-lead resting ECG. The skin areas where the electrodes are located have to be cleaned with alcohol to avoid artifacts on the ECG. When the patient preparation is concluded, the operator asks the patient to relax to avoid muscle artifacts and activates the acquisition on the PC screen where the Gamma Cardio CG software application is running. If the operator is not a qualified

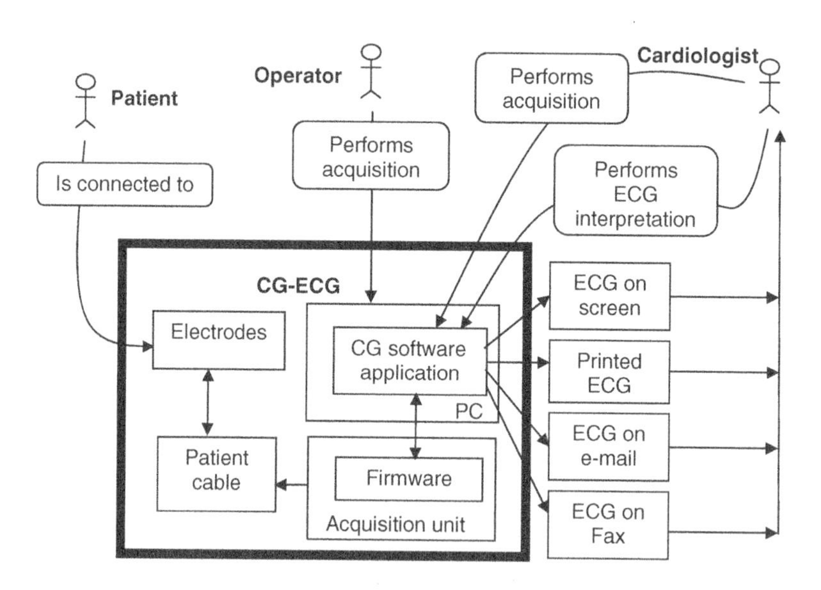

Figure 3.14 Gamma Cardio CG use case.

physician, he can use the simplified step-by-step user interface for an assisted ECG recording. After the acquisition of the 12 leads, the operator verifies if any of the lead recordings are not adequate for report and, if so, repeats the specific lead acquisition. After recording, the operator will send the ECG to the cardiologist for interpretation. The ECG may be sent by fax or e-mail, or shared on a communication network. Alternatively, the ECG report can printed out or shown on the screen. The cardiologist then performs an interpretation that can be added to the report within the ECG file: the cardiologist's procedure for determining the interpretation was described in the previous chapter.

3.6.2 Human Factors and the User Interface Design

Human factors relates to the interaction of users with product technologies. Human factors engineering (HFE) is the discipline that applies human factors to product design. It is also referred to as **usability** (as mentioned in the previous chapter) or **ergonomics** as cited in the European Medical Device Directive (EEC 2007). To improve security and enhance user experience, the HFE is applied to the man–machine interface (MMI) design. The MMI, also called the user interface, includes all the device components that interact with users (software interface, system logic, instructions etc.).

Human factor engineering is a key factor for users when buying products, either for aesthetic factors or because an improved MMI greatly reduces the training time and increases the willingness to use the product. In addition, the MMI is also an indicator of product quality for the customer since they perceive the external part quality as a proxy for the intrinsic product quality level.

In medical devices, HFE is also related to safety: a user interface can reduce risks of incorrect operation that in the past caused serious injury or death in medical devices (Fairbanks, 2004; Kohn *et al.*, 2000). Human factors for medical devices is a wide topic. Here we will consider a practical approach that will be explained through the example of the Gamma Cardio CG.

As discussed in Chapter 2, poor MMI design in medical devices has been revealed as one of the most common sources of medical errors (Fairbanks & Caplan, 2004; Kohn *et al.*, 2000). For this reason, the European Medical Device Directive has been updated to explicitly introduce ergonomics in the first essential requirement related to product design. Safety shall be improved by: (Appendix I, EEC 2007):

'– *reducing, as far as possible, the risk of use error due to the* ***ergonomic features*** *of the device and the environment in which the device is intended to be used (design for patient safety), and*
– ***consideration*** *of the technical* ***knowledge***, ***experience***, ***education*** *and* ***training*** *and where applicable the medical and physical conditions of intended users (design for lay, professional, disabled or other users).'*

Specific standards and guidelines have been introduced to improve usability and reduce risks, such as the usability standard for medical electrical devices (IEC/EN 60601-1-6) or the US FDA 'Medical Device Use-Safety: Incorporating Human Factors Engineering into Risk Management' (FDA, 2000 and 2011). The ISO/EN 9241 standard on Ergonomics of Human System should also be taken into account.

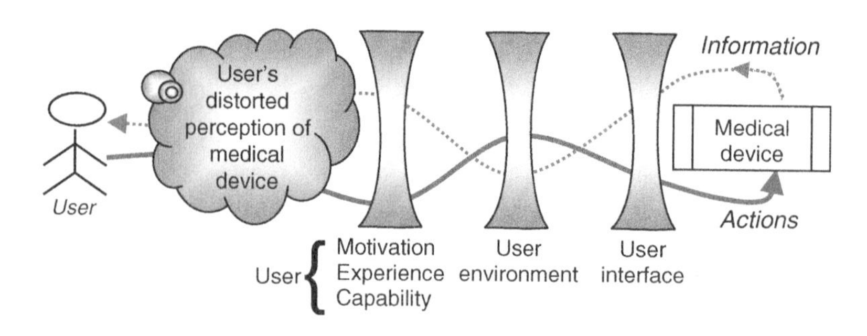

Figure 3.15 User interface distorting factors.

The model of human interaction considers that specific **causes** or **contributing factors** may lead to a **wrong usage** of medical devices with possible unsafe or ineffective outcomes (FDA 2011). Such elements are related to the users, the environment and the device–human interface.

Referring to Figure 3.15, the information generated by the medical device (e.g., alarms and displayed data) may first be distorted by the implemented presentation performed by the device user interface. This '**distorting lens**' effect may be due to a misleading display presentation. In recent years, the use of computer-based devices has led to the implementation of office-automation-like interfaces that increase the risk of errors. These interfaces are usually designed to be attractive and fast in use, but to the detriment of the precision of the interaction. The information and the corresponding actions performed by the users are modified by a second 'lens' representing the environmental conditioning. **Environmental factors**, such as noise, inadequate lighting, distractions and adverse temperature may negatively affect user interaction, perhaps reducing user attention, so that, for example, alarms may be ignored concerning specific environment conditions.

The third distorting effect is the **user**, who may fall into error through lack of experience, motivation or functional capability. Users may perceive distorted information because they associate a different meaning to specific device feedback, due to their different perceptual models. So the user interface must take into account the different profiles of users that are intended to use the device.

The overall effect is that users may perceive distorted input to and feedback from device. The user then responds with actions based on wrong input. Such actions may be in turn be implemented wrongly because the user may make mistakes due to the distorting effect of user environment and user interface factors.

This short analysis suggests that a proper design must be performed so that the interface is intuitive for all expected users in all expected environments. Designers must aim to reduce risks from dangerous intentional or unintentional product misuse.

3.6.2.1 The User Interface Design Process

A more predictable design result can be obtained using a specific process that merges both a risk management process (as in FDA, 2011) and the human factor engineering process (as in FDA, 1996) to achieve user interface effectiveness. Figure 3.16 outlines the required process, whose steps are quite straightforward.

1. conduct **exploratory studies** on similar devices
2. select all **requirements** associated to the interface
3. define the **design** from the requirements
4. identify the distorting factors (Figure 3.15), risks and associated **hazards**
5. prioritize associated critical **risks**
6. design and **implement** the interface with **risk mitigation** elements for major hazards
7. **verify** the implementation
8. **if** risk **residual** is not **acceptable,** go back to step 3 to consider different approaches
9. **validate** the product for the safe and effective usage of intended users in the expected environment

Figure 3.16 Process for user interface design.

It is worth noting that product validation mainly has to be performed through **usability tests**: a representative sample of users must execute specific scenarios on the device and the associated results have to be analyzed to individuate errors. For interested readers, the suggested approach and template report is also included in (FDA, 2011). The number of involved users and scenarios has to be determined, according to some assumptions over the error probability. The lower the probability, the higher the number of required users for the test. An interesting study reported in (FDA, 2011) shows that a sample population of 15 elements may be sufficient to find the 90% of computer errors. This number may suggest the order of magnitude of the sample population for medical device validation.

3.6.2.2 The Design of the Gamma Cardio CG User Interface

The process of Figure 3.14 will be used to derive the design of the interface. In the ECG case, the product is consolidated through 40 years of clinical experience of products placed on the market. Considering the first step of Figure 3.15, **exploratory studies** may provide various indications of the risks and the required design. Here, the FDA guide (FDA, 1996) may be helpful for practical recommendations based on more common mistakes. The following considerations and guidelines will be used in the user interface design:

1. User preferences do not often lead to safer and more effective approaches.
2. Attractive computer based interfaces may be faster, but can be error prone to an unacceptable level.
3. Windows-based interfaces allow designing 'more crowded' interfaces with higher numbers of functions. This may lead to an overwhelming and unhelpful cognitive overload for users, that in turn can become more error prone.
4. More risky and recurrent activities must have the simplest man-machine interaction (e.g., touchscreen based) even if perceived as less attractive or trivial.

After the analysis of the exploratory studies, the designer has to select all the requirements set in Chapter 2 that affect the user interface. Some of the requirements are associable to the user interface. Requirement R3 (AAMI EC 11 compliance 1991, Table 2.9 and Table 2.12) defines the dimensions of the output and the pre-ruled paper (R3.6) and the required values for vertical and horizontal scale (Table 2.11, R3.4 and R3.5). Their measuring unit is specified in R4. Filters to be implemented are in R5, R6, R7 (see section 2.17).

Regarding compliance to standards IEC/EN 60601-1 and IEC/EN 60601-1-25 (R9), the interface must respect recommendations on the colors of indicator lights included in IEC/EN 60601-1 ('green: ready to use, yellow: Caution – prompt response by the operator is required, red: Warning – immediate response by the operator is required'). Requirement R15 (required training for usage) specifies most of the elements of the user interface. The sub-requirements are addressed below. R19, R20, R21 (see section 2.17) introduce additional functions (export data, patient archive, mailer system) that have to be accessed from a proper user interface.

The **design decisions specifications** on the interface are then obtained, taking into account the previous guidelines, the explanatory study and the relevant requirements associated to the user interface.

From the use case of Figure 3.14, it is clear that there are two main functions: acquisition and interpretation. The **design specification** related to the acquisition function is:

there must be four areas (R15.5–15.6):

1. *The* ***ECG command bottom area*** *that mimics the usual ECG interfaces through a touchscreen approach. Commands are specified in requirements R5, R7, R4.*
2. ***the 'virtual ECG paper' area*** *where an ECG pre-ruled paper and the ECG trace are drawn as in a mechanical printer. The area must also display the main errors (R15.3) and output according to AAMI EC 11(R3)*
3. ***the right side tool/indications area*** *were cardiologists will find the usual ECG tools and display information such as*
   i. *caliper to measure distances,*
   ii. *scissors to cut unwanted ECG segments*
   iii. *baseline reset button (R15.4)*
   iv. *beat rate estimation*
4. ***the top side PC based advanced commands*** *to print, analyze and mail. (R19, 20, 21)*

Commands are implemented through a touchscreen approach and no menu is available.

The associated user interface is depicted in Figure 3.17. The interface clearly shows the four areas: the ECG commands (bottom), the paper area (middle), the tool area (right) and the advanced functions at the top. Note that the interface is drawn to look like the more classical ECG products on the market in the past two decades. The interface may therefore be less attractive, but it is immediate and therefore safe, since the probability of errors is greatly reduced. Windows menus and crowded functions have been avoided, and when using touchscreens, the more relevant functions can be activated with the fingers, thanks to the dimension of the buttons. Requirement R15.2 implies also 'an elementary step-by-step procedure with associated checklist to perform the test and the additional functions devoted to non-expert users (generic physicians, nurses etc.)'. This can be easily designed, considering the main steps discussed in the use case of Section 3.6.1. Figure 3.18 shows the implemented interface that has been designed to be used also for minimally trained personnel. Note that each step is numbered as in a checklist.

When the user has addressed a specific step, the associated checklist square becomes ticked.

The other ECG function included in the use case is the interpretation function. The design and implementation of this function in the Gamma Cardio CG follows the same process of the acquisition function design and it is therefore omitted for the sake of brevity.

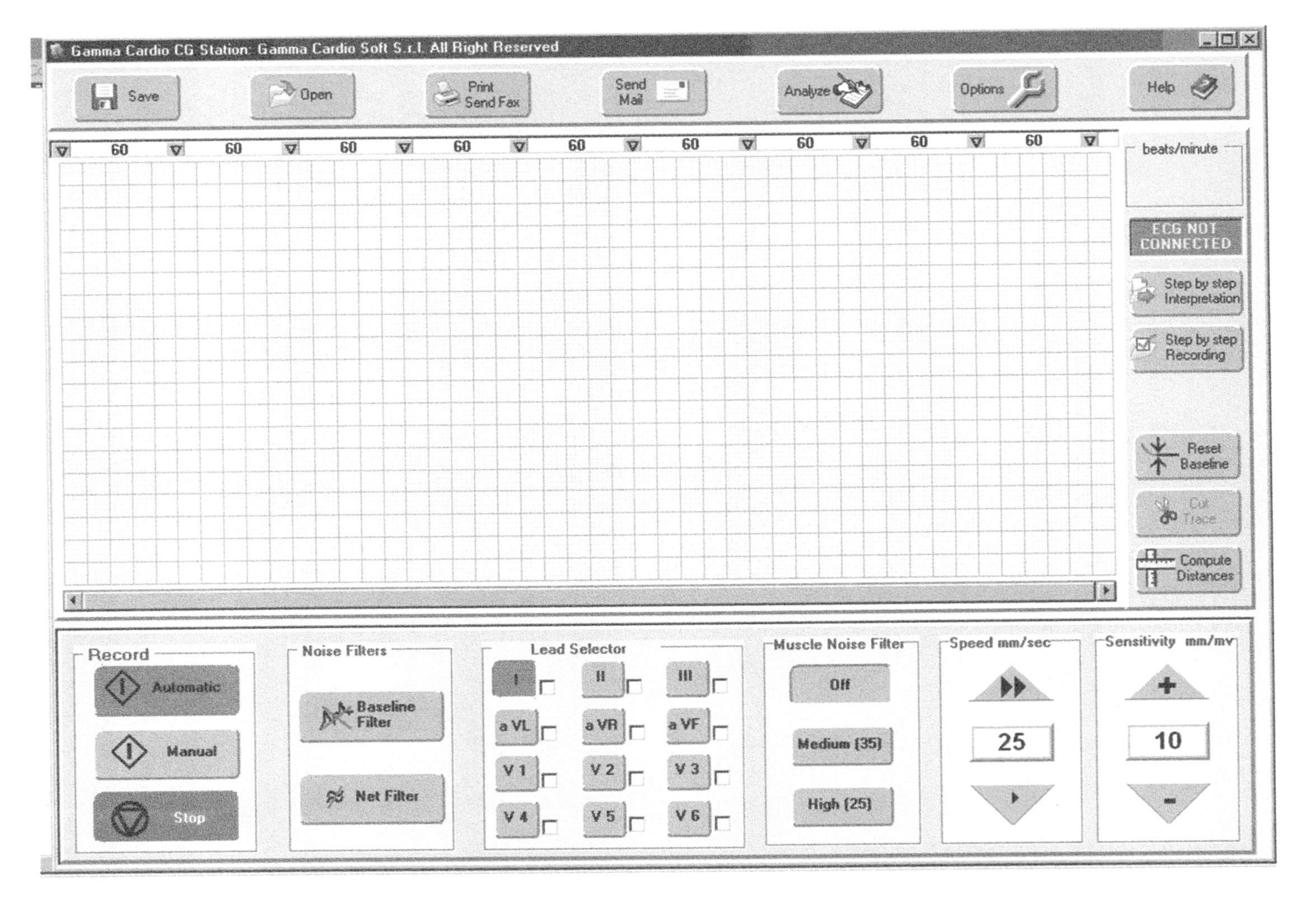

Figure 3.17 Acquisition user interface of the Gamma Cardio CG.

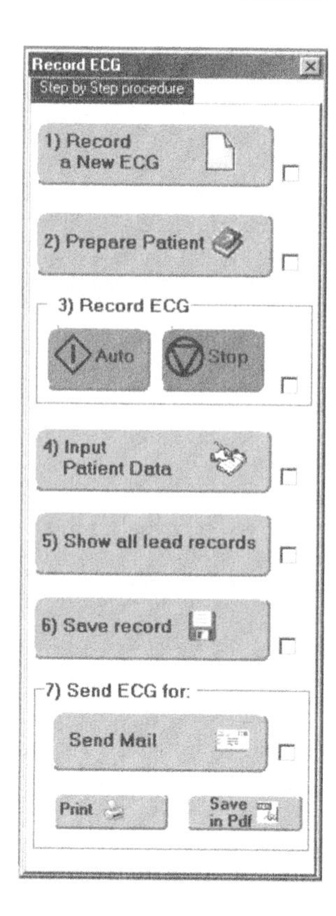

Figure 3.18 Step by step procedure.

The next step is the identification of the distorting factors (Figure 3.15) with the **risks** and the associated hazards when operating on the user interface. ISO/EN 14971 Annex C contains a list of questions to evidence such elements with specific reference to human factors (questions C.2.29–C.2.29.8 adapted from ISO/EN 14971 Annex C).

Some questions and the associated answers from the Gamma Cardio CG technical file are here outlined for explicative purposes.

C.2.29: User interface critical for effective device usage?
At worst, the filter or lead selection may be wrong but this does not induce measuring errors.

C.2.29.1 User interface contribution to use error?
The user interface is easy in use and it reproduces the commands and displays of an ordinary ECG.

C.2.29.2 Distractions from intended environment may generate use errors?
Yes, but commands have to be confirmed, and they are displayed in a different way once selected.

The overall risk analysis has not shown any critical issues on the user interface, taking into account the specific implemented design. The accuracy of the reproduction of the ECG signal is considered separately. The product has been validated through usability testing with different user profiles.

The user interface is more often a critical component of medical devices, since errors from wrong use are always more relevant in terms of safety. The application of a proper design process, including risk management, may mitigate such risks. In addition, proper design may reduce the overall cost of development and increase the user perception of product quality.

3.6.3 Patient Interface: the Biopotential Electrodes

The patient interface is the part of the device that interacts with patient. For the Gamma Cardio CG, the interaction comprises

1. the electrodes connected to the patient
2. possible contact with other part of the device
3. the simplified user interface depicted in Figure 3.18 when the user is trained.

Item 3 was discussed in the previous section. Regarding the possible contacts with the devices, there is the risk of electric shock that has to be addressed, following IEC/EN 60601-1. This risk is also reduced by introducing specific warning in the user manual (see for example Section 2.18 of this book). The core patient interface is related to the 10 electrodes connected to the patient cable that in turn is connected to the acquisition device (see Figure 3.14).

The definition of the design and the implementation of this interface or similar kinds of interfaces may be obtained following the process outlined in the previous section (see Figure 3.16). In the following, the first step (i.e., the exploratory study) is performed.

3.6.3.1 The ECG Electrodes

As discussed in Chapter 2, the heart generates an electric potential. This field can be recorded using electrodes at the skin surface connected to a voltmeter. The current flowing in the body is a ionic current, i.e. negative and positive charges flowing within the body. Electrodes transduce this ionic current into an electronic current made up of electron flow. This is obtained by a chemical reaction at the interface of the electrodes built of conductive metal connected to the device and the electrolyte that is located between the electrode and the skin. The electrode–skin interface may have a negative impact on heart potential measurement.

Referring to Figure 3.19, the heart potential is recorded by the ECG that can be simply modeled by a voltmeter measuring potential against time. The voltmeter is constituted by a current meter (galvanometer) in series with the instrument resistor R_I. As shown in Figure 3.19, the measured potential V_M is given by the measured current I_M multiplied by R_I according to Ohm's Law. We may first model the electrode–skin interface as a resistor R_S in

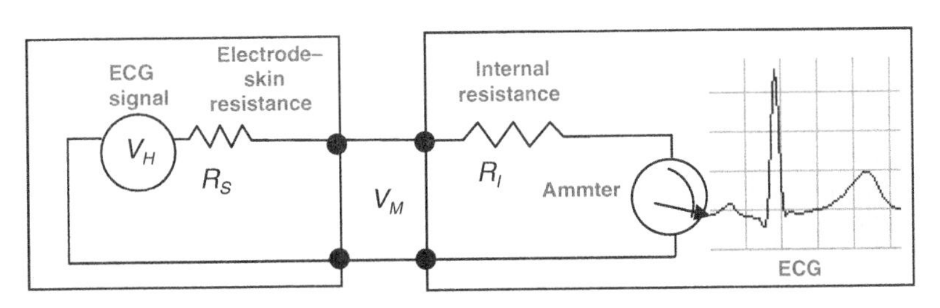

Figure 3.19 Conceptual measurement of heart potential.

series with the voltage generator V_H represented by the heart. From that, the current along the heart generator is equal to the current within the voltmeter and we have:

$$V_M = V_H \frac{R_I}{R_I + R_S} = V_H \cdot \left(\frac{1}{1 + R_S/R_I} \right) \tag{3.3}$$

From eq. 3.3 the measured potential V_M is similar to the heart potential only if the resistance of the instrument is much greater than the resistance of the skin interface ($R_I \gg R_S$). This suggests lowering the skin electrode resistance using good materials, and preparing the skin by cleaning with soap and water and drying the skin with a dry cloth or gauze. Studies on skin resistance report that a 200 kΩ impedance is expected and 700 kΩ is also possible (AAMI, 1991). Requirement 3.3.9 of Table 2.9 (AAMI, 1991) specifies that the signal must not be reduced more than 20% due to input impedance for a skin electrode resistance simulated by a 620 kΩ resistance. Taking into account equation (3.3) the minimum input resistance must be 2.48 MΩ.

The skin–electrode interface also has a capacitive component in parallel to the resistance and a DC offset potential raising to ±300 mV as shown in the model of Figure 3.20. The overall ECG instrument impedance is evaluated with a 10 Hz sinusoid to also take into account the capacitive effects.

Luckily, The Gamma Cardio CG has a much higher impedance (664 MΩ) than the minimum requested and therefore signal reductions are greatly reduced even in presence of elevated skin–electrode impedance, yielding only 1/1000 signal reduction in the AAMI EC 11 test (AAMI, 1991).

The capacitance of the skin electrode impedance has another side effect. The ECG must also be used during defibrillation. In this case, a 5 kV source charges an inductor that is then discharged to the patient (see Chapter 5 for a detailed discussion). When the inductor discharges, the capacitance of electrodes is charged and an offset is added to the ECG signal. Since the ECG input stages clamp the signals to within a specified voltage range, the offset may exceed this range, preventing ECG signal from displaying.

The Requirement R9 (compliance to IEC/EN 60601-1-25 section 51.101) specifies that, after 5 seconds, the ECG signal must be readable. Therefore, the electrodes must have a sufficiently low capacitance, so that signal recovery is achieved in the allowed time after defibrillation procedures. This performance requires specific design of the acquisition unit and the use of special electrodes based on silver/silver chloride (Ag/AgCl) chemical reaction introduced in Chapter 5.

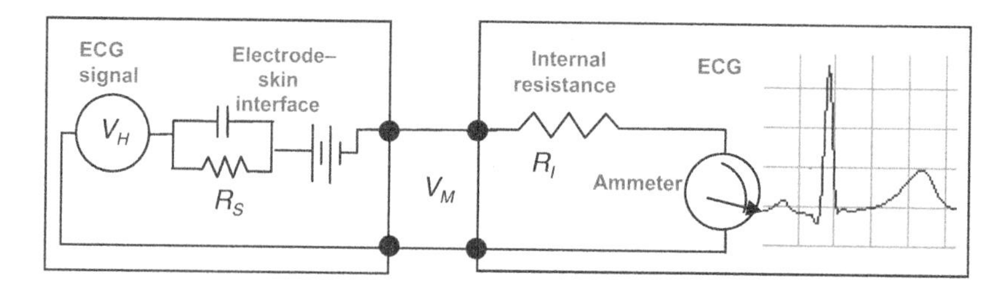

Figure 3.20 Electrode skin interface model.

Considering all the requirements of Chapter 2 that apply to the electrodes, the specifications are

1. electrodes must be *class I Medical device* according to Medical Device Directive (R8)
2. electrodes must be defibrillator proof (R9)
3. electrodes must not be a cytotoxic irritant or a sensitizer in compliance with the requirement of ISO 10993-1 (R8).

In addition, since ECG validation depends on the use of electrodes of a specific type, the user manual must report the type of electrodes that are compatible with the specific ECG model.

The same applies to the patient cable, which is a critical component of an ECG. The patient cable connects the electrodes to the acquisition unit. The first issue is that the ECG signal is of very low amplitude (max $\pm$ 5 mV) compared to the noise present in the environment such as the power line noise with an amplitude in the order of tens of volts (see Section 2.13.3). Proper design of the patient cable can greatly reduce such noise. The other main issues related to cable are the color recommendations and defibrillation protection.

Color and labelling recommendations on the plugs that connect the patient cable to the electrodes vary between European and American markets, as shown in Table 3.3. This has to be taken into account when considering products sold in the USA or Europe.

Table 3.3 Patient cable color recommendations

Europe (IEC)		**USA (American Heart Association)**		**Position on the body surface**
Label on the plug	**Color of the plug**	**Label on the plug**	**Color of the plug**	
R	red	RA	white	right arm
L	yellow	LA	black	left arm
F	green	LL	red	left leg
N	black	RL	green	right leg
C1	white/red	V1	brown/red	IV° right intercostal/ parasternal space
C2	white/yellow	V2	brown/yellow	IV° left intercostal/ parasternal space
C3	white/green	V3	brown/green	V° intercostal space between C2 and C4
C4	white/brown	V4	brown/blue	V° intercostal area/ hemiclavear line
C5	white/black	V5	brown/orange	left front axillary line/line on horizontal plane of C4
C6	white/purple	V6	brown/purple	left mid-axillary line/line on horizontal plane of C4

The final issue relates to defibrillation protection. The extra energy of the defibrillator may be absorbed by proper varistors and resistors embedded in the cable and in the acquisition unit. Defibrillation protection is then obtained through a coordinated design between cable and acquisition unit. Therefore, only the specified cable indicated for the given ECG model guarantees defibrillation protection. Electrodes and patient cable are critical ECG components. Associated risks are related to defibrillation protection, signal degradation due to noise, contamination and infection. As a risk mitigation action, the user manual must report the following:

- *Do not connect the metal parts of the electrodes to the earth plate.*
- *Do not use electrodes other than those supplied and indicated in the user manual. Electrodes of poor quality can degrade the recorded ECG.*
- *Do not use a patient cable type different from that supplied and indicated in the user manual. The system is protected against defibrillator discharge only if the specific cable model is used. The cable is also designed to minimize electrical interference that may affect the ECG trace.*

Verify that the patient cable is properly connected to the acquisition unit, and that the cable is undamaged in all its parts and is properly connected to the electrodes.

In particular, check for signs of wear near the connections with the electrodes: damaged cables greatly reduce the ECG signal quality.

The electrodes, cables and other accessories must be cleaned thoroughly after each use to prevent infection and/or contamination of the patient and/or the user. The cables must be cleaned as specified by the User Manual.

3.7 The ECG System Architectural Design

The system architectural design is the second design stage according to the MIL 498 standard (MIL, 1994a). This activity will identify and address subsystems, and it will generate the requirements and the specifications of the subsystems and their associated interfaces. The system architectural design includes the following steps (SSDD, 1994):

1. identification of the system (Gamma Cardio CG) components (i.e., **subsystems**)
2. subsystem static functional relations
3. identification of system requirements associated to subsystems (also called traceability)
4. definition of computer hardware resources
5. concept of execution describing how subsystems work together
6. subsystems interface design.

3.7.1 Subsystem Identification

The subsystem identification of the Gamma Cardio CG was described in detail in Section 1.15 and it is summarized in requirement R1, which identifies subsystems. Figure 3.21, adapted from Chapter 1, outlines the subsystem decomposition and the functional relations among subsystems.

From left to right, the first subsystems are the set of electrodes and the patient cable. These subsystems, bought from specialized companies, have been addressed in the patient interface section (Section 3.6.3). The patient cable is connected to the acquisition unit that comprises the hardware platform on which the firmware is hosted. Both the hardware and the firmware

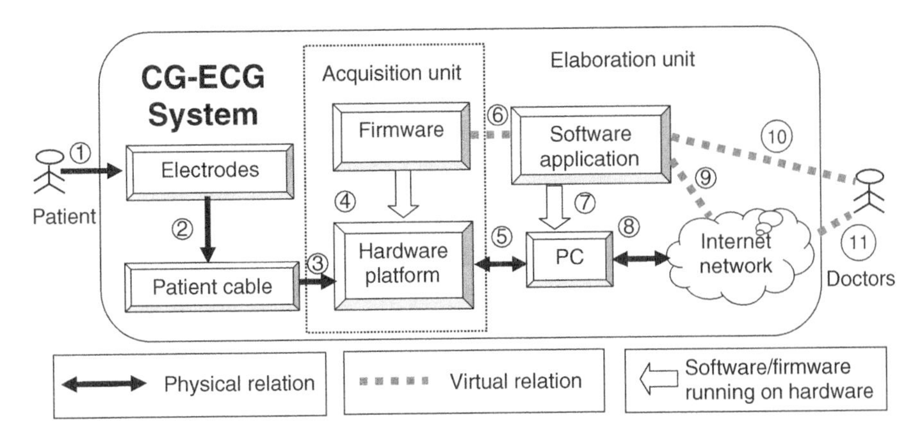

Figure 3.21 Gamma Cardio CG subsystem identification.

of the acquisition unit are subsystems to be developed. The acquisition unit is connected with a commercial USB cable to the PC that runs the Gamma Cardio CG software application. Three subsystems have to be developed:

- hardware of the acquisition unit
- firmware of the acquisition unit
- software application.

The associated documentation also has to be addressed for certification as outlined in Section 3.8.

3.7.2 The Communication Interfaces

The subsystem static functional relations are identified in Figure 3.21 with circled numbers. These relations are specified through the definition of interfaces that are worth some additional discussion. Medical devices are always very much 'interconnected devices' that, in turn, are composed of interconnected components. Such interconnections allow communication between components. Communication interface paradigms are used to allow components and medical systems to interact. Some basic concepts of communication interfaces are outlined in the following.

An **interface** between two entities defines the specifications that the two entities have to comply to achieve an effective communication. Such specifications may include mechanical, electronic and process features. The OSI (Open Systems Interconnection) model outlined in Table 3.4 may help in understanding and defining interfaces.

The **OSI model** standardized by the International Organization for Standardization (ISO) (ISO, 1994) defines the interface functions of a communication system in terms of stacked layers. Layers of the same type contain similar communication functions and provide services to the upper layer while using services of the lower layer.

The OSI model encompasses seven layers that address specific functions, as outlined in Table 3.4. Two layers of the same level of two interconnected systems virtually communicate with each other using their lower layers. For example, two computers may communicate

Table 3.4 ISO OSI layers

7. **Application layer**
Communication partner identification, resource availability, synchronization for the upper software applications. Associated with this layer are *HTTP* and *FTP* protocols.

6. **Presentation layer**
Adaptation of data semantic and syntax to provide independence of data for the upper layer. A *gateway* device may be used to connect different applications at this level or at a lower level. For example, two different mailing systems require a gateway to adapt and exchange mails.

5. **Session layer**
Session management between end user application processes. Used in remote procedure calls (RPC) contexts.

4. **Transport layer**
Reliable data transfer services between hosts. *TCP* and *UDP* Internet protocols are associated with this layer.

3. **Network layer**
Data transfer between different networks. The *IP* protocol is associated with this layer. A router device is used to interconnect different networks at this level.

2. **Data layer**
Functions and procedures to transfer data from point to point: data frame organization, dataflow, error detection-correction. Local area network (*LAN*) protocols are associated with this layer. A switch communication device is used to connect systems at this level.

1. **Physical layer**
Electrical and physical/mechanical specifications: cable, connectors, electrical signals, modulation. The *repeater* devices are used at this level to regenerate the communication signal between two systems.

through the transport layer (TCP protocol interface). Virtually, the two TCP protocol interfaces communicate with each other. Physically, the TCP protocol of one computer interacts with its lower layers to exchange data. The OSI model suggests how communication interfaces are organized; the use of a publicly available document template such as (IDD, 1994) is recommended for interface design description.

3.7.2.1 Gamma Cardio CG Subsystem interfaces

The Gamma Cardio CG subsystem interfaces are identified in Figure 3.21 with the circled numbers. Standards or consolidated choices have been used for interface implementation in order to lower cost and risk. For example, the patient cable to acquisition unit interface is implemented using the de facto standard approach. Standards are recommended in medical devices; debugging and integration required to set up complex non-standard interfaces is one of the most expensive and risky activities within development tasks. Additional effort is generally required in medical devices to certify non-standard interfaces.

At this development stage, it is important to specify all the interfaces to avoid reworking in the following stages. Here are the details of the Gamma Cardio CG interfaces (item numbers correspond to interface numbers shown in Figure 3.21):

1. Patient to electrode:
   Discussed in Section 3.6.3.
2. Electrodes to patient cable:
   This interface defines the specifications of the connection between electrodes and the patient cable. There are some de facto standards. Usually it is selected a 4 mm banana plug for reusable electrodes with recommended color and label identification as in Table 3.3. Disposable electrodes may have a snap connection and adaptors for banana plugs.
3. Patient cable to acquisition unit:
   This interface includes both mechanical and electronic specifications. The physical connector is usually the standard 15-pin D-Sub ECG type. Some electronic components are embedded in the cable so that the defibrillation-proof requirement (IEC/EN 60601-2-25) is fulfilled with associated components in the acquisition unit.
4. Firmware to hardware platform
   The firmware component is run on the hardware microcontroller. This interface specifies the type of processor and compiler and the software environment.
5. Hardware platform to PC
   To lower cost and risks, the standard USB interface is selected.
6. Firmware to software application
   This interface relies on the USB data layer. In this interface, taking into account the use case of Section 3.6.1, the messages containing the ECG signal must be received by the software application. In addition, the following control messages must be exchanged:
   a. Configure_Unit
   b. Get_Unit_State
   c. Start_Acquisition
   d. Stop_Acquisition
   e. Write_Memory_On_Unit
   f. Read_Memory_On_Unit

Further details will be given in Chapter 6.

7. Software application to PC
   The software application is run on a PC. The interface specifies the operating system (i.e. Microsoft Windows), the compiler and the software environment.
8. PC to Internet
   This interface connects the PC with Internet resources. Standard Internet mail protocols (i.e., SMTP, POP3, IMAP) have to be available on the PC. These protocols rely on an IP connection that can be obtained via any physical layer (LAN, Wi-Fi, mobile connection, ADSL, public switched telephone line PSTN, Bluetooth etc.).
9. Software application to Internet
   This interface allows the sending of ECGs through mail using the previous interface. The software application uses the Internet mail protocols to send mails with attached ECG records.
10. Software application to the doctor
    This is the user interface covered in Section 3.6.2.

11. Internet to the doctor
    This interface is represented by the usual Internet mailer system through which the doctor receives an e-mail with the attached ECG file

All the previous interfaces are sufficiently detailed, apart from the two interfaces: firmware to software application (interface 6) and patient cable to acquisition unit. These interfaces deserve more details on the associated subsystems to be fully specified and therefore they will be addressed in the following sections.

3.7.3 Acquisition Hardware Requirements

The following section defines the purpose of each of the three subsystems and identify the system requirements and the system-wide design decisions allocated to these subsystems.

The acquisition unit hardware has the purpose of capturing the ECG signal from the patient and transmitting it to the software application. The associated system requirements defined in Chapter 2 are easily seen to be (R3, 9, 10, 12, 14, 15, 16, 17, 22 and 23). In the following, the requirements of the **acquisition unit** and the associated **design** specifications are outlined. Details of design are discussed in Chapter 5

R 3 AAMI EC 11 standard compliance
Referring to Table 3.5, the standard includes many sub-requirements. Some requirements have to be implemented by hardware, firmware or software subsystems.

Table 3.5 AAMI EC 11 requirements on subsystems

<table>
<tr><th>Hardware</th><th>Firmware</th><th>Software</th></tr>
<tr><td>Observation method
(R 3.2: lead definition)</td><td rowspan="10">The firmware must guarantee that sampling is periodicly triggered and that ECG samples are correctly delivered to the software application</td><td rowspan="11">The software application must guarantee that ECG samples are not distorted and that they are correctly shown according to vertical and horizontal scale requirements:
(R3.4: Gain),
(R3.5: Time base)</td></tr>
<tr><td>Input dynamic range (R3.3)</td></tr>
<tr><td>Input impedance (R3.9)</td></tr>
<tr><td>Common mode rejection
(R3.11)</td></tr>
<tr><td>Response to test signal
(R3.8: Standardizing voltage)</td></tr>
<tr><td>Baseline stability (R3.13)</td></tr>
<tr><td>Accuracy of input signal
reproduction (R3.7)</td></tr>
<tr><td>Vertical scale (R3.4: Gain)</td></tr>
<tr><td>System noise (R3.12)</td></tr>
<tr><td>Overload protection (R3.14)</td></tr>
<tr><td colspan="2">Horizontal scale (R3.5: Time base)</td></tr>
<tr><td></td><td></td><td>(R3.6) Output display</td></tr>
</table>

For example, the ECG frequency response requirement may be fully satisfied by analog filters implemented in the hardware. Alternatively, some filter components may be digitally implemented in the firmware or in the software. In order to simplify design and qualification, it is decided that all the AAMI EC 11 requirements that can be fulfilled by the hardware acquisition unit are fulfilled at this level. This may increase the costs of hardware but, in turn, it guarantees an easier integration. In addition, this choice reduces safety risks, since software performance may depend on the specific PC hosting the software application. Table 3.5 shows how requirements are mapped between subsystems.

R 14 ***Horizontal resolution***

The horizontal resolution of 1200 samples per second has to be achieved by the A/D converter of the acquisition unit triggered by firmware and the corresponding data transfer has to be feasible through the interface between the firmware and the software application.

R 9 ***IEC/EN 60601-1 and IEC/EN 60601-2-25 compliance***

The compliance determines specific procedures and tests to be executed on the mechanical and electric performance of the acquisition unit. A detailed discussion of these standards is included in Chapter 8. In the following, examples are outlined: some specifications relate to the identification and marking of the device and to the associated documents. Other specifications are related to the device case, which must be resistant to liquid penetration to be suitable for the operating room or for the outpatient environment. Regarding electric shock, since the acquisition unit is not plugged directly into the power supply, a double fault (within the device and the connected PC) is needed to produce dangerous electric potentials. This reduces the constraints on the device. Risks of connector exchange are not present since connector shapes are different.

R 10 ***EN 60601-1-2 EMC standard compliance***

The EMC requirement is applicable only to hardware. See Table 2.12 for details.

R 12 ***Class II CF type device***

This requirement refers to the hardware input stage. The acquisition unit must have a limited patient leakage current. The patient leakage current is the current that flows in the patient through the patient device interface.

R 17 ***Class IIb certification***

This requirement determines the process of certifying the product according to the medical device European directive.

R 15 ***Required training for usage***

The applicable part of the requirement specifies that the hardware must implement the saturation indication and the automatic/manual isoelectric line reset function.

R 16 ***Harsh environment***

The requirement is associated completely with the acquisition unit that must be designed properly (no mobile part, proper manufacturing, proper case design).

R 18 ***Weight and dimension of the acquisition unit***

The requirement is associated completely with the acquisition unit.

R 22 ***Production cost***

The requirement is mainly associated with the acquisition unit, since considering the list of materials comprising the Gamma Cardio CG (requirement R1), the other components have either a negligible cost (manual, labels, software, CD) or the cost is easily determined since they are bought-out products (electrodes, patient cable, PC cable). To reduce production cost related to device qualification requiring human activity, an auto test function must be implemented that must test the modules included in the acquisition unit and the critical modules of the software application. The hardware must include a signal generator and test circuits to detect the electronic board problems. This auto test must reduce drastically the test procedures requiring human intervention.

R 23 ***Periodic maintenance cost***
Since software has no operative maintenance cost, the requirement is associated only with the acquisition unit.

3.7.4 Firmware Requirements

The acquisition unit firmware has the purpose of controlling the acquisition unit hardware functions and exchanging data and control signals with the software application subsystem. In particular, the following functions are to be embedded into the firmware:

1. ECG acquisition
2. ADC control
3. management of the control lines for lead selection and calibration signal
4. management of the communication with the software application
5. monitoring of the signal saturation
6. automatic management of the baseline reset
7. management of the internal autotest function
8. management of the internal signal generator.

The associated system requirements are easily seen to be (R3, 9, 11, 14 and 22). In the following, the requirements and the associated design specifications are outlined.

R 3 ***AAMI EC 11 compliance***
Applicable system requirements are R3.5 (Time) and R3.13 (Baseline Stability) of EC 11 requirements in Chapter 2.

R 11 ***Software validation***
This requirement is also associated with the firmware.

R 14 ***Horizontal resolution***
The associated requirements are mainly fulfilled by the hardware acquisition unit. As shown in Table 3.5, the firmware must guarantee a periodic sampling to obtain an undistorted digital version of the analog signal. Since the horizontal resolution is set at 1200 samples/sec, the sampling period is $1/1200 \cong 0.83\ \mu$sec. The firmware must then trigger the AD converter exactly every 0.83 μsec. Jitter on the sampling period will result in a distorted signal that in turn may produce errors in ECG interpretation. Proper firmware design has to be considered so that jitter is negligible. In addition, the firmware must support sufficient data transfer to allow 1200 samples/sec to be transferred to the application unit.

R 9,15 ***IEC/EN 60601-2-25, Required training for usage***
The firmware must monitor and communicate the saturation indication (degraded mode).

R 22 ***Production cost***
To reduce production cost related to device qualification, an auto test function must be implemented that must test the modules comprising the acquisition unit and the critical software application modules. The firmware must implement the logic to test the hardware circuits. Details are provided in Chapter 6.

3.7.5 Software Application Requirements

The software application offers all the ECG's basic functionalities to the users. In addition, the software application includes additional functions for ECG management. The main software application functionalities are

1. configuration and management of the acquisition unit
2. ECG acquisition and signal elaboration
3. visualization of ECG in on line (during acquisition) and off-line mode
4. creation of reports with associated patient data
5. report preview, print, fax and save
6. storage of ECGs reports in files and in the internal database
7. creation of reports in Acrobat® pdf format
8. e-mail management for ECGs exchanges
9. help system management
10. autotest system management.
11. step by step procedures for fast ECG acquisition and interpretation

The associated system requirements are easily seen to be related to the user interface (R16), the signal and output processing (R3, 4, 15), the filters (R5, 6, 7), the additional functions (R20, 21, 22) and the software validation (R11).

In the following, the requirements and the associate design specifications are outlined.

R 15 ***Required training for usage***
This requirement has an impact on the user interface of the software application (Section 3.6.2) and the associated internal functions.

R 3 ***AAMI EC 11 compliance***
Referring to Table 3.5, the software application must guarantee that ECG samples are not distorted within the internal signal processing. The signal has to be represented according to vertical and horizontal scale requirements: (R3.4: Gain; R3.5: Time base). In addition, sub-requirement R3.6 has to be satisfied for the output display.

R 14 ***Horizontal resolution***
The software application must handle a data flow rate up to 1200 samples/sec. This requirement must be satisfied by the software application components: the internal storage management, the signal processing and the protocol that exchanges data with the firmware application.

R 4 ***User interface and measuring units***
The measuring units will be standard (mm/mV and mm/ms).

R 5, R 6, R 7 Filters

The filters to be implemented in the user interface and in the signal processing of the software applications are: noise net filter, low pass filter and isoelectric line stability filter.

R 19, R 20, R 21 Additional functions

The software application will implement export data functions with Acrobat® pdf, Microsoft copy and paste method, patient interface database and mail capability.

R 11 Software validation

The Gamma Cardio CG software application must be validated.

R 22 Production cost

To reduce production cost related to device qualification, an auto test function must be implemented that will test the modules comprising the acquisition unit and the critical software application modules. The software must control the acquisition unit tests, show results and check if values are admissible.

3.7.6 Concept of Execution among Subsystems

The concept of execution describes how subsystems work together while performing the intended system functions. This step evidences possible inconsistencies between the design of the specific subsystems. For example, the interface design may be in contrast with the dynamic relationships of the subsystems (execution/data flow, transition state etc.) while executing specific functions. The following use case, derived from the system use case of Section 3.6.1, outlines the activities (shown in italics) between the subsystems and the interfaces identified in Figure 3.21.

As a **preliminary action**, a doctor or a trained operator runs the PC and the Gamma Cardio CG software application.

The software application verifies the features of the hosted PC, shows the user interface and starts monitoring for the existence of the acquisition unit.

The operator then connects the acquisition unit to the PC through the USB cable.

The PC activates the USB port and powers up the acquisition unit.
The software application detects and configures the acquisition unit. The acquisition unit returns its status to the software application which verifies the proper acquisition unit configuration.

The operator connects the patient cable to the acquisition unit. Finally, operator connects the electrodes to the patient cable and places the electrodes on the patient's body.

The operator pushes the automatic acquisition button to **start ECG recording**.

The software application sends the start acquisition message containing the required recording duration and the selected lead to the acquisition unit firmware. The firmware configures the acquisition unit hardware to record the test signal. It then reads the digital data from the ADC converter and transmits the ECG to the software application through the USB interface. The acquisition unit firmware configures the specific lead and the ADC and it starts reading data from the ADC registers. Data are then transmitted to the software application. The software application processes the signal so that it has the required vertical/horizontal scale with the required filter configuration. It then shows the signal on the screen. The firmware monitors the isoelectric ECG level and if it is out of the expected range, it resets isoelectric line. In addition, it monitors the amplitude of the signal and communicates possible out of range values.

The software application stops the ECG recording at the end of the required duration or after operator stop.

3.8 Gamma Cardio CG Technical File Structure

To explain a real example of the certification issues related to design, Table 3.6 shows the structure of the technical file for medical device certification for the EU market.

Table 3.6 Gamma Cardio CG technical file structure

Sec.	Topic	Gamma Cardio CG associated compulsory contents
1	General description and intended use	product description, intended use MDD classification (for Gamma Cardio CG class IIb) accessories expirations terms
2	Referenced documents	list of applied standards
3	Labeling plan	
4	Risk analysis	referenced regulations risk analysis plan 'questions that can be used to identify medical device characteristics that could impact on safety' according to ISO 14971:2009, appendix C objective items to support risk analysis revision of risk analysis surveillance
5	Instructions for use	user manual with: warnings, device intended use, installation and connection instructions, technical specifications, instructions for ECG recording electrodes palcement
6	Product technical description	device electronic structure; critical component specifications; datasheets device mechanic structure software validation and architecture
7	Design verifications	Conformity to standards (EMC etc.)
8	Manufacturing and final inspection	manufacturing steps intermediate inspections and tests
9	Clinical validation	validation process and data result according to AAMI EC 11 standard.
10	Conformity declaration	

Similar technical documentation has to be supplied in other non European markets. The third column includes the information supplied for the Gamma Cardio CG. The documents that are associated to the technical file for the Gamma Cardio CG are

1. risk analysis
2. user manual
3. hardware block diagram of the acquisition unit
4. electrical diagram of the acquisition unit
5. software validation document
6. clinical validation document test report
7. test report for compliance to EN 60601-1 EN 60601-1-25 (R9)
8. test report for compliance to EMC EN 60601-1-2 (R10)
9. instructions for manufacturing and testing; labeling plan
10. electrode certificates
11. patient cable certificate
12. datasheets of the critical components (optoisolators, DC-DC converters).

References

AAMI (1991), Association for the Advancement of Medical Instrumentation, American National Standard, Diagnostic electrocardiographic devices, ANSI/AAMI EC11-1991.

Brooks F. (1987), 'No Silver Bullet – Essence and Accidents of Software Engineering', IEEE Computer.

Brooks F. (1995), The Mythical Man-Month: Essays on Software Engineering, Addison-Wesley.

Bronzino J. D.editor, (2006), The Biomedical Engineering Handbook, Taylor & Francis.

BSC (2010), British Society of Cardiology, Recording a standard 12-lead electrocardiogram An Approved Method.

EEC (2007), Council Directive 93/42/EEC on medical devices, amended by 2007/47/EC.

ECRI (2012), Healthcare Product Comparison System (HPCS) ECRI Institute, www.ecri.org.

Fairbanks R. J. and Caplan S. (2004) 'Poor Interface Design and Lack of Usability Testing Facilitate Medical Error,' Joint Commission Journal on Quality and Safety.

FDA (1996), An introduction to Human Factors in Medical device: Do it by design, www.fda.gov, accessed 2012.

FDA (1997), Design Control Guidance For Medical Device Manufacturers, available at www.fda.gov accessed 2012.

FDA (2000), Medical Device Use-Safety: Incorporating Human Factors Engineering into Risk Management; Guidance for Industry and FDA Premarket and Design Control Reviewers available at www.fda.gov accessed 2012.

FDA (2002), General Principles of Software Validation; Final Guidance for Industry and FDA Staff available at FDA web site www.fda.gov accessed 2011.

FDA (2011), Draft Guidance for Industry and Food and Drug Administration Staff – Applying Human Factors and Usability Engineering to Optimize Medical Device Design.

IDD (1994), Interface Design Description (IDD), DI-IPSC-8143, U.S. Department of Defense.

ISO (1994), ISO/IEC 7498-1:1994, Information technology – Open Systems Interconnection – Basic Reference Model: The Basic Model.

Kohn L. et al. (2000), 'To Err Is Human: Building a Safer Health System', National Ac. Press.

Lewis W. H. (1941), Bulletin of the New York Academy of Medicine 17, 871.

MIL 498 Guide, (1994b), MIL-STD-498 Overview and Tailoring Guidebook.

MIL-STD-498 (1994a), Software Development and Documentation Standard, Amsc NO. N7069, U.S. Dpt. of Defense.

SSDD (1994), System/Subsystem Design Description, DI-IPSC-81432, US Dep. of Defense.

4

Signal Processing and Estimation

'Make everything as simple as possible, but not simpler'
Albert Einstein

Chapter Organization

This chapter deals with the theoretical framework and the methodologies supporting the digital processing of biomedical signals. Both the methodologies and techniques are applicable to a wider range of biomedical applications, including those related to multidimensional signals like those considered in medical imaging. Here we will include examples related to ECG digital signal processing.

As illustrated by Figure 4.1, nowadays analog processing is confined to (noisy) signal acquisition, conditioning and analog-to-digital conversion, while information processing is

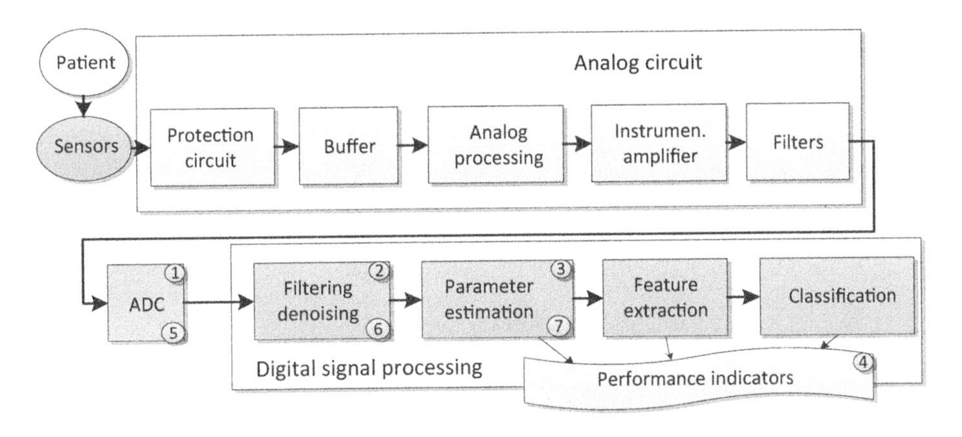

Figure 4.1 Medical device architecture.

Medical Instrument Design and Development: from Requirements to Market Placements, First Edition.
Claudio Becchetti and Alessandro Neri.

essentially realized in digital form. Signal conditioning is essentially focused on the task of providing adequate signal-to-noise ratio, which is determined by the noise figure and the gain of the first stage, and to remove the out-of-band noise and interference, thus preventing contamination of in-band components due to aliasing effects.

Digital processing may include in-band noise and artifact removal, parameter estimation, feature extraction and classification. Moreover, following an evolution similar to that experienced in other fields, such as communications, computer vision, radar and remote sensing, processing is performed by digital programmable devices at both signal and semantic levels. Adoption of this kind of architecture has been made possible, from the theoretical point of view, by the equivalence between analog and digital processing for signals arising in biomedical engineering, and, from the implementation point of view, by the availability of low-cost digital signal processors providing the required computational power. This in turn implies that innovation mainly resides in processing methodologies and techniques rather than in the implementation.

Signal processing and estimation is pervasive in all the conceptual blocks of Figure 4.1 related to a medical device. The framework discussed in this chapter can be considered as the mathematical background required to handle signals and to assess performance through a whole medical system comprised of hardware, software and telecommunication components, as shown in Figure 4.2. The material of this chapter is organized following the conceptual flow of the signal and information processing performed by a typical electronic (medical) device as shown in Figure 4.1 where the circled numbers refer to the sections in which the topics are discussed in the theory and in implementation parts.

In Section 4.1, after a short introduction on the **discrete representation of analog signals**, we deal with the conditions under which the full equivalence between analog and digital signal processing is valid. In Section 4.2, we then briefly discuss the basic properties of the discrete Fourier transform (**DFT**) and its fast implementation, the fast Fourier transform (**FFT**), which is the major factor in the success of digital signal processing.

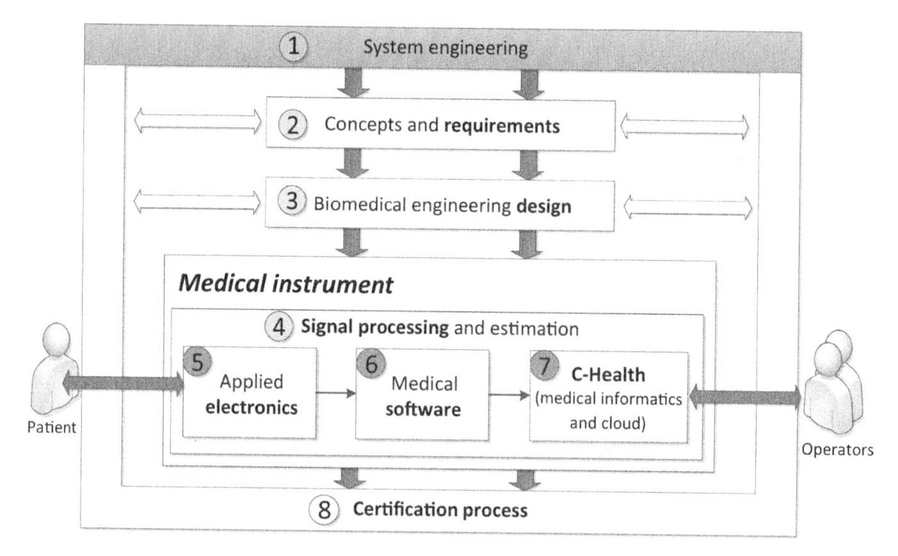

Figure 4.2 Book chapters model.

Since data provided by sensors are affected by noise and distortion whose behavior is usually known only in statistical form, in Section 4.3, we formulate the problem of the extraction of useful information from the data provided by sensors as an optimization problem in the **estimation theory** framework.

Then, in Section 4.4 we address **performance assessment** of real systems through mathematical models and discuss the role of the Fisher's information matrix and the related Cramér–Rao lower bound that limits the achievable accuracy of any possible processing, given the observed quantity, the measurement setup and the sensor characteristics. The section introduces criteria to assess the quality of the models that have to be *'as simple as possible'* according to Occam's razor, *'but not simpler'*, to be effective.

In the second part, the general theory is refined and detailed in view of the implementation of devices and/or software applications performing specific tasks of the signal processing chain. Thus, in Section 4.5, the implementation guidelines for **analog to digital conversion** are considered. In Section 4.6 the general purpose solutions for parameter estimations are applied to signal **denoising**.

Finally, in Section 4.7, to illustrate the process of the extraction of some parameter of interest from the observed waveform, the optimal solution for the **time of arrival** estimation problem is derived and its performance analyzed.

Part I: Theory

4.1 Discrete Representations of Analog Systems

Digital processing of biomedical signals mainly stems from the possibility of representing an analog signal by means of a countable set of coefficients as soon as we are in presence of some limitation either in the temporal duration or in the frequency domain occupancy. A first example of signals whose temporal duration limitation implies a countable number of degrees of freedom consists of the class of periodic signals, for which the Fourier's series expansion holds.

Theorem 4.1. *Given a periodic signal $x(t)$, almost continuous and absolutely integrable in $[-T/2, T/2)$, where T is the period, the following expansion holds*

$$x(t) = \sum_{n=-\infty}^{+\infty} X_n e^{j2\pi f_n t} = \sum_{n=-\infty}^{+\infty} X_n[\cos(2\pi f_n t) + j\sin(2\pi f_n t)], \tag{4.1}$$

where

$$f_n = n \cdot F = \frac{n}{T}, \tag{4.2}$$

and the Fourier expansion coefficients are given by

$$X_n = \frac{1}{T}\int_{-T/2}^{T/2} x(t)e^{-j2\pi f_n t}dt < +\infty. \tag{4.3}$$

To enlighten the limitations inherent in the periodic nature of a signal, we observe that any periodic signal $x(t)$ can be thought of as generated by the signal $s(t)$ with finite temporal support $[-T/2, T/2)$ defined as follows

$$s(t) = \begin{cases} x(t) & -\frac{T}{2} \le t < \frac{T}{2}. \\ 0 & otherwise \end{cases} \tag{4.4}$$

In fact, we obviously have

$$x(t) = \sum_{k=-\infty}^{+\infty} s(t - kT). \tag{4.5}$$

Let us denote in the following with $rect_T(t)$ a rectangular pulse of duration equal to T, centered on $t = 0$, that is

$$rect_T(t) = \begin{cases} 1 & -\frac{T}{2} \le t < \frac{T}{2}. \\ 0 & otherwise \end{cases} \tag{4.6}$$

Observing that a signal with finite support of size T can also be considered as a signal with finite support of size $\tilde{T} \geq T$, the following property can be easily demonstrated.

Corollary 4.1. *Given an almost continuous, absolutely integrable signal $s(t)$ with finite support $[-T/2, T/2)$, the following expansion holds*

$$s(t) = \sum_{n=-\infty}^{+\infty} S_n e^{j2\pi f_n t} rect_{\tilde{T}}(t), \tag{4.7}$$

where

$$f_n = n\tilde{F} = \frac{n}{\tilde{T}}, \tag{4.8}$$

with $\tilde{T} \geq T$, and the expansion coefficients are given by

$$S_n = \frac{1}{\tilde{T}} \int_{-\infty}^{+\infty} s(t) e^{-j2\pi f_n t} rect_{\tilde{T}}(t)\, dt < +\infty \tag{4.9}$$

Thus, considering that

$$\left\langle e^{j2\pi f_n t} rect_{\tilde{T}}(t), e^{j2\pi f_m t} rect_{\tilde{T}}(t) \right\rangle = \int_{-\infty}^{+\infty} e^{j2\pi \frac{n-m}{\tilde{T}} t} rect_{\tilde{T}}(t) dt = \tilde{T} \delta_{n,m}, \tag{4.10}$$

where

$$\delta_{n,m} = \begin{cases} 1 & n = m \\ 0 & otherwise \end{cases}, \tag{4.11}$$

we can state that the set of functions $\{e^{j2\pi f_n t} rect_{\tilde{T}}(t)\}$ constitutes an orthogonal complete basis for almost continuous, absolutely integrable signals with finite temporal support. On the other hand, given a time invariant linear filter, the relationship between its input $x(t)$ and its output $y(t)$ is given by the convolution between $x(t)$ and its impulse response $h(t)$, namely,

$$y(t) = x(t) * h(t) = \int_{-\infty}^{\infty} x(\tau) h(t-\tau) d\tau. \tag{4.12}$$

If both $x(t)$ and $h(t)$ have finite temporal supports $[-T_x/2, T_x/2)$ and $[-T_h/2, T_h/2)$, respectively, their convolution has as temporal support the interval $[-T_y, T_y)$ with $T_y = T_x + T_h$.

Starting from the Fourier series properties it can also be easily demonstrated that the expansion coefficients of the convolution of two signals with finite support can be obtained by multiplying the expansion coefficients of the two signals with respect to a finite support of duration $T' \geq (T_x + T_h)$.

Thus, the output of an analog filter with impulse response $h(t)$ with finite temporal support, corresponding to an input signal $x(t)$ also with finite temporal support can be equivalently generated by:

1. computing the expansion coefficients of $h(t)$
2. computing the expansion coefficients of $x(t)$

3. multiplying the corresponding expansion coefficients
4. computing expansion (4.7) corresponding to the step 3 coefficients.

Although steps 3 and 4 can be performed in the digital domain, to effectively apply digital signal processing to signals with finite duration the expansion coefficients have to be computed, which in turn implies the evaluation of the integrals (4.9). A more viable situation arises when the finite support constraint applies in the frequency domain. In this case, restating Corollary 4.1 in the frequency domain leads to the famous Whittaker–Nyquist–Kotelnikov–Shannon sampling theorem for band-limited signals. In fact, let $S(f)$ be the Fourier transform of the signal $s(t)$:

$$S(f) = \int_{-\infty}^{+\infty} s(t)e^{-j2\pi f t}dt. \tag{4.13}$$

If $S(f)$ has a finite support $[-B, B)$, then, based on Corollary 4.1, it admits the following representation:

$$S(f) = \sum_{n=-\infty}^{+\infty} S_{-n}e^{-j2\pi n\frac{f}{2B}} \, rect_{2B}(f). \tag{4.14}$$

Then, computing the inverse Fourier transform of both sides we obtain:

$$\begin{aligned} s(t) &= \sum_{n=-\infty}^{+\infty} S_{-n} \int_{-\infty}^{+\infty} e^{j2\pi\left(t-\frac{n}{2B}\right)f} rect_{2B}(f)df = \sum_{n=-\infty}^{+\infty} S_{-n} \frac{e^{j2\pi\left(t-\frac{n}{2B}\right)B} - e^{-j2\pi\left(t-\frac{n}{2B}\right)B}}{j2\pi\left(t-\frac{n}{2B}\right)} \\ &= \sum_{n=-\infty}^{+\infty} S_{-n}2B\frac{\sin\left[2\pi B\left(t-\frac{n}{2B}\right)\right]}{2\pi B\left(t-\frac{n}{2B}\right)} = \sum_{n=-\infty}^{+\infty} S_{-n}2B\,\mathrm{sinc}\left[2\pi B\left(t-\frac{n}{2B}\right)\right] \end{aligned} \tag{4.15}$$

where

$$sinc(x) = \frac{\sin(x)}{x}, \tag{4.16}$$

and the expansion coefficients can be computed as

$$S_{-n} = \frac{1}{2B}\int_{-\infty}^{+\infty} S(f)e^{-j2\pi(-n)\frac{f}{2B}}df = \frac{1}{2B}s\left(\frac{n}{2B}\right). \tag{4.17}$$

Thus the following theorem, known in the literature as the Whittaker–Nyquist–Kotelnikov–Shannon theorem holds.

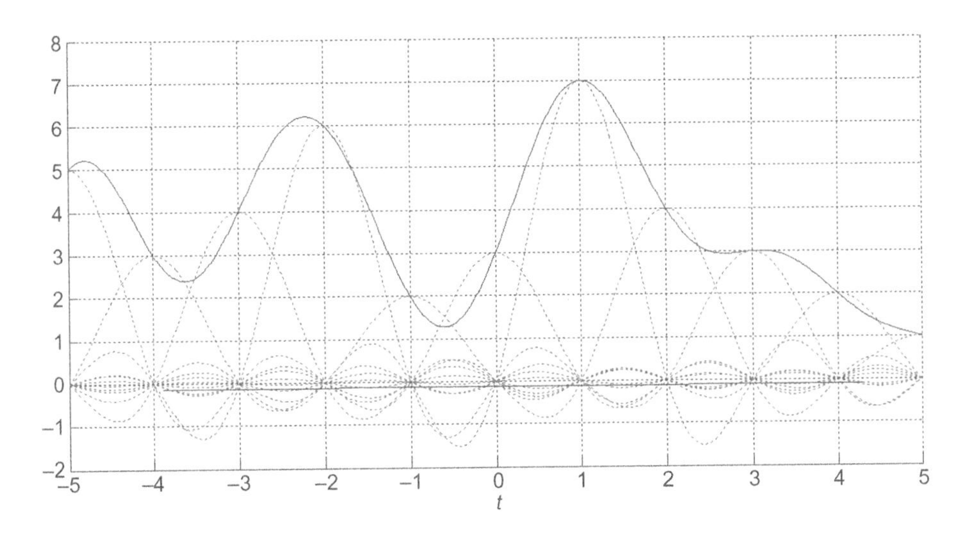

Figure 4.3 Band-limited signal expansion example ($B = 0.5$).

Theorem 4.2. *Let $s(t)$ be a band-limited signal in the frequency band $[-B,B]$. Then the following expansion holds*

$$s(t) = \sum_{n=-\infty}^{+\infty} s\left(\frac{n}{2B}\right)\operatorname{sinc}\left[2\pi B\left(t - \frac{n}{2B}\right)\right] \tag{4.18}$$

In other words, the original signal can be reconstructed by linearly combining a countable number of time shifted replicas of the *sinc* waveform. An example of the application of the Whittaker–Nyquist–Kotelnikov–Shannon theorem to a band-limited signal is shown in Figure 4.3. We note that for band-limited signals the expansion coefficients can be obtained by direct sampling of the signal waveform.

However, as detailed in the implementation section, direct sampling of the signal is prone to high frequency noise whose aliased version may affect the digital signal samples. To prevent this effect, the acquired signal should be band-limited by means of an analog low pass filter before the analog-to-digital conversion. However, in addition to the simplicity of extracting the countable expansion coefficients from the signal waveform, for band-limited signals the remarkable equivalence between analog and digital filtering holds. For the sake of simplicity, let us initially discuss the case that arises when the input signal $x(t)$ and the impulse response $h(t)$ are both band-limited signals in the frequency band $[-B,B]$. Then, as formalized by Theorem 4.3, the system output $y(t) = x(t)^* h(t)$ is a band-limited signal whose sample time series is proportional to the discrete convolution between the input sample time series and the impulse response sample time series.

Theorem 4.3. *Let $x(t)$ and $h(t)$ be two band-limited signals in the frequency band $[-B,B]$. Then, for the samples of their convolution $y(t)$ the following property holds*

$$y\left(\frac{n}{2B}\right) = \frac{1}{2B}\sum_{k=-\infty}^{+\infty} x\left(\frac{k}{2B}\right) h\left(\frac{n-k}{2B}\right). \tag{4.19}$$

Proof: By definition, the samples of $y(t)$ are given by

$$y\left(\frac{n}{2B}\right) = x(t) * h(t)|_{t=\frac{n}{2B}} = \int_{-\infty}^{+\infty} x(\tau)h\left(\frac{n}{2B} - \tau\right)d\tau. \tag{4.20}$$

Application of the sampling theorem to $x(t)$, $h(t)$, and $y(t)$ gives

$$x(\tau) = \sum_{k=-\infty}^{+\infty} x\left(\frac{k}{2B}\right)\text{sinc}\left[2\pi B\left(\tau - \frac{k}{2B}\right)\right], \tag{4.21}$$

$$h\left(\frac{n}{2B} - \tau\right) = \sum_{m=-\infty}^{+\infty} h\left(\frac{m}{2B}\right)\text{sinc}\left[2\pi B\left(\frac{n-m}{2B} - \tau\right)\right]. \tag{4.22}$$

Therefore we have

$$y\left(\frac{n}{2B}\right) = \sum_{k=-\infty}^{+\infty}\sum_{m=-\infty}^{+\infty} x\left(\frac{k}{2B}\right)h\left(\frac{m}{2B}\right)\int_{-\infty}^{+\infty} \text{sinc}\left[2\pi B\left(\tau - \frac{k}{2B}\right)\right]\text{sinc}\left[2\pi B\left(\tau - \frac{n-m}{2B}\right)\right]d\tau \tag{4.23}$$

Let us observe that

$$\begin{aligned}
&\int_{-\infty}^{+\infty} \text{sinc}\left[2\pi B\left(\tau - \frac{k}{2B}\right)\right] \quad \text{sinc}\left[2\pi B\left(\tau - \frac{n-m}{2B}\right)\right]d\tau \\
&= \left\langle \text{sinc}\left[2\pi B\left(t - \frac{k}{2B}\right)\right], \text{sinc}\left[2\pi B\left(t - \frac{n-m}{2B}\right)\right]\right\rangle \\
&= \left\langle \frac{1}{2B}e^{-j2\pi\frac{k}{2B}f} rect_{2B}(f), \frac{1}{2B}e^{-j2\pi\frac{n-m}{2B}f} rect_{2B}(f)\right\rangle \\
&= \frac{1}{2B}\delta_{k-(n-m)}.
\end{aligned} \tag{4.24}$$

Consequently,

$$y\left(\frac{n}{2B}\right) = \frac{1}{2B}\sum_{m=-\infty}^{+\infty} x\left(\frac{n-m}{2B}\right)h\left(\frac{m}{2B}\right) \tag{4.25}$$

□

Since, by definition, a band-limited signal $x(t)$ is invariant with respect to ideal low pass filtering, that is

$$x(t) = x(t) * 2B\,\text{sinc}(2\pi Bt), \tag{4.26}$$

for the output $y(t)$ of a linear filter with impulse response $h(t)$ we have

$$\begin{aligned} y(t) &= x(t) * h(t) = [x(t) * 2B\,\text{sinc}(2\pi Bt)] * h(t) \\ &= x(t) * [2B\text{sinc}(2\pi Bt) * h(t)] = x(t) * h^{(2B)}(t) \end{aligned} \tag{4.27}$$

where $h^{(2B)}(t)$ is the band-limited version of $h(t)$. Thus, when dealing with band-limited input signals, any analog filter can always be replaced by a digital filter whose discrete impulse response is obtained by sampling the band-limited version of $h(t)$ at a rate equal to $2B$, namely,

$$\frac{1}{2B} h^{(2B)}\left(\frac{n}{2B}\right). \tag{4.28}$$

4.2 Discrete Fourier Transform

Although digital filtering of band-limited signals is completely equivalent to its analog counterpart, the computational complexity can be still too high for practical implementation. In fact, let N_x be the number of non-null samples of $x(t)$ and N_h the number of non-null samples of $h^{(2B)}(t)$, then the computation of the discrete convolution requires $min(N_x, N_h)$ multiplications and $[min(N_x, N_h) - 1]$ sums for each sample. Thus if the analog signal is sampled at the Nyquist frequency the number of multiplications and additions in one second is $min(N_x, N_h) \times 2B$.

However, to reduce the number of computations we can operate in the frequency domain. In fact, similarly to the analog case, discrete (circular) convolution in one representation domain corresponds to simple multiplication in the transformed domain. To demonstrate this property we resort to the discrete Fourier transform. In fact, given the discrete time, finite sequence

$$\mathbf{s} = \begin{bmatrix} s_0 & s_1 & \dots & s_{N-1} \end{bmatrix}^T \tag{4.29}$$

where

$$s_k = s\left(\frac{k}{2B}\right), \qquad k = 0, 1, \dots, N-1 \tag{4.30}$$

are samples extracted from a band-limited signal, the signal $s^0(t)$, defined as

$$s^0(t) = \sum_{k=0}^{N-1} s_k \delta\left(t - \frac{k}{2B}\right) \tag{4.31}$$

where $\delta(t)$ is the Dirac delta function, has a finite temporal support and its Fourier transform coefficients compute as follows:

$$S_n = \frac{2B}{N} \sum_{k=0}^{N-1} \int_{-\infty}^{+\infty} s_k \delta\left(t - \frac{k}{2B}\right) e^{-j2\pi \frac{2Bn}{N} t}\, dt = \frac{2B}{N} \sum_{k=0}^{N-1} s_k e^{-j2\pi \frac{nk}{N}}. \tag{4.32}$$

In the previous formulas, the Dirac delta function $\delta(t)$ is the defined as:

$$\delta(t) = \begin{cases} 0 \quad if \quad t \neq 0 \\ \int_{-\infty}^{\infty} \delta(t)dt = 1 \end{cases} \tag{4.33}$$

The previous equation suggests the following definition of Discrete Fourier Transform.

Definition 4.1. *Given a finite discrete time sequence* $\mathbf{x} = [x[0] \quad x[1] \quad \dots \quad x[N-1]]^T$ *of length N, its discrete Fourier transform is defined as:*

$$X[n] = \sum_{k=0}^{N-1} x[k] \cdot W_N^{-nk}, \qquad n = 0, \dots, N-1. \tag{4.34}$$

where

$$W_N^{nk} = e^{j\frac{2\pi}{N}hk}. \tag{4.35}$$

Then it can be demonstrated that the following representation theorem holds.

Theorem 4.4. *Given a finite discrete time sequence* $\mathbf{x} = [x[0] \quad x[1] \quad \dots \quad x[N-1]]^T$ *of length N, the following expansion holds*

$$x[k] = \frac{1}{N} \sum_{n=0}^{N-1} X[n] W_N^{nk}. \tag{4.36}$$

Denoting with $\mathbf{W}_N$ the $N \times N$ matrix with elements

$$[\mathbf{W}_N]_{n,k} = W_N^{nk} = e^{-j\frac{2\pi}{N}hk}, \tag{4.37}$$

the direct and inverse DFT can be written in matrix form as

$$\mathbf{X} = \mathbf{W}_N \mathbf{x}, \quad \mathbf{x} = \frac{1}{N} \mathbf{W}_N^{\dagger} \mathbf{X}. \tag{4.38}$$

Where † denotes the transpose complex conjugate operator (also called the Hermitian operator).

Theorem 4.5. *Given two finite discrete time sequences*

$$\mathbf{x} = [x[0] \quad x[1] \quad \dots \quad x[N-1]]^T, \quad \mathbf{y} = [y[0] \quad y[1] \quad \dots \quad y[N-1]]^T, \tag{4.39}$$

let

$$\mathbf{X} = [X[0] \quad X[1] \quad \dots \quad X[N-1]]^T, \quad \mathbf{Y} = [Y[0] \quad Y[1] \quad \dots \quad Y[N-1]]^T. \tag{4.40}$$

respectively be their DFT. Then, the coefficients Z[k] of the DFT of their discrete circular convolution

$$z[k] = \sum_{n=0}^{N-1} x[n]y\left[(k-n)_{\mathrm{mod}N}\right] \tag{4.41}$$

are given by

$$Z[k] = X[k]Y[k] \tag{4.42}$$

Proof: By definition

$$\begin{aligned}
Z[n] = \sum_{k=0}^{N-1} z[k] \cdot W_N^{-nk} &= \sum_{k=0}^{N-1}\sum_{m=0}^{N-1} x[m]y\left[(k-m)_{\mathrm{mod}N}\right] \cdot W_N^{-nk} \\
&= \sum_{k=0}^{N-1}\sum_{m=0}^{N-1} x[m]y\left[(k-m)_{\mathrm{mod}N}\right] \cdot W_N^{-nk} W_N^{+nm} W_N^{-nm} \\
&= \sum_{k=0}^{N-1}\sum_{m=0}^{N-1} x[m] W_N^{-nm} y\left[(k-m)_{\mathrm{mod}N}\right] \cdot W_N^{-n(k-m)}
\end{aligned} \tag{4.43}$$

On the other hand, since $W_N^{hN} = 1$, we have $W_N^{-n(k-m)} = W_N^{-n\left[(k-m)_{\mathrm{mod}N}\right]}$. Therefore, for $Z[n]$ we obtain

$$\begin{aligned}
Z[n] &= \sum_{k=0}^{N-1}\sum_{m=0}^{N-1} x[m] W_N^{-nm} y\left[(k-m)_{\mathrm{mod}N}\right] \cdot W_N^{-n[(k-m)\mathrm{mod}N]} \\
&= \sum_{m=0}^{N-1}\sum_{h=0}^{N-1} x[m] W_N^{-nm} y[h] \cdot W_N^{-nh} = X[n]Y[n]
\end{aligned} \tag{4.44}$$

□

The fundamental properties of the discrete Fourier transform are summarized in Table 4.1 where * denotes complex conjugation.

We observe that straightforward DFT computation requires the evaluation of N^2 complex multiplications and $N(N-1)$ sums. However, the DFT computational complexity is reduced to $O(N \log_2 N)$ by adopting a **fast Fourier transform** (FFT) algorithm.

Although today several versions of the FFT algorithm exist, each optimized on the basis of properties implied by specific conditions satisfied, for instance, by the DFT size (e.g., N equal to a power of 2), for the sake of compactness, here we illustrate the algorithm proposed by Cooley and Tukey in 1965. Publication of the Cooley–Tukey FFT algorithm represented a significant innovation in signal processing, dramatically accelerating the replacement of analog signal processing by its digital counterpart. As discussed in (Heideman, 1984), an algorithm for the efficient computation of the coefficients of a finite Fourier series, similar to the Cooley–Tukey FFT, can be found in an unpublished manuscript of the German mathematician Carl Friedric Gauss, probably dated 1805, appearing only in his collected works after his lifetime. The rationale of the fast Fourier transform algorithm is to partition the time samples into two subsets respectively comprising the $N/2$ even samples $x_e[n]$ and the $N/2$

Table 4.1 DFT properties

Property	Time domain	Frequency domain
Linearity	$ax[n] + by[n]$	$aX[k] + bY[k]$
Time shifting	$x[n - n_0]$	$e^{-j\frac{2\pi}{N}kn_0}X[k]$
Frequency shifting	$e^{j2\pi/Nk_0 n}x[n]$	$X[k - k_0]$
Real signals	$Im\{x[n]\} = 0$	$X[N - k] = X^*[k]$
Temporal first-order difference	$x[n] - x[n - 1]$	$\left(1 - e^{-j\frac{2\pi}{N}k}\right)X[k]$
Circular convolution	$\sum_{n=0}^{N-1} x[n]y\left[(k - n)_{\bmod N}\right]$	$X[k]\,Y[k]$
Parseval's relation	$\sum_{n=0}^{N-1} \lvert x[n]\rvert^2 = \sum_{k=0}^{N-1} \lvert X[k]\rvert^2$	

odd samples $x_o[n]$ of $x[n]$, so that we can write:

$$
\begin{aligned}
X[n] &= \sum_{k=0}^{N-1} x[k] \cdot W_N^{-kn} = \sum_{h=0}^{\frac{N}{2}-1} x[2h] \cdot W_N^{-2hn} + \sum_{h=0}^{\frac{N}{2}-1} x[2h+1] \cdot W_N^{-(2h+1)n} \\
&= \sum_{h=0}^{\frac{N}{2}-1} x[2h] \cdot W_{N/2}^{-hn} + W_{N/2}^{-n} \sum_{h=0}^{\frac{N}{2}-1} x[2h+1] \cdot W_{N/2}^{-hn} \\
&= \sum_{h=0}^{\frac{N}{2}-1} x_e[h] \cdot W_{N/2}^{-hn} + W_{N/2}^{-n} \sum_{h=0}^{\frac{N}{2}-1} x_o[h] \cdot W_{N/2}^{-hn} \\
&= X_e[n] + W_{N/2}^{-n} \cdot X_o[n].
\end{aligned}
\tag{4.45}
$$

where
$$x_e[h] = x[2h] \text{ and } x_o[h] = x[2h+1]$$

This procedure can be recursively applied $\log_2 N$ times. Since at each step $N/2$ products have to be computed, the FFT algorithm requires a computation of $\frac{N}{2} \cdot \log_2 N$ complex products. This complexity is further reduced by a factor of 2, when dealing with real signals. In fact, in these cases the DFT obeys the symmetry constraint, that is $X[N-k] = X^*[k]$ (see Table 4.1) and only $N/2$ DFT coefficients have to be effectively computed. Similarly, the symmetry constraint can be imposed at the first stage of the inverse transform, thus obtaining the same computational complexity saving.

Let us now compare the computational complexity of filters respectively implemented in the time and frequency domains. For this, we consider an input signal of length N, and an M-taps filter in the discrete time domain, (i.e., a filter whose impulse response is constituted by M non-null samples) with $M \leq N$. Then, computation of their circular convolution of length N, requires the computation of $M \times N$ real multiplications.

Since we are dealing with real signals (in the time domain), when circular convolution is computed by means of its equivalent operator in the frequency domain we have to perform $N/2$ complex multiplications of the FTT coefficients plus the computation of the FFT of the real signal to be filtered and the inverse FFT of the output, considering that the filter impulse response FFT can be precomputed off-line. Thus, due to symmetry, a total amount of about $N + N\log_2 N$ real multiplications have to be performed.

To normalize the complex evaluation with respect to the length of the signal to be filtered, let us rewrite the length of the filter impulse response as $M = \alpha N$, with $0 < \alpha \leq 1$. Thus, filtering in the frequency domain has a lower complexity than direct convolution computation as soon as $\alpha > (1 + \log_2 N)/N$ while the computational saving is about $\alpha N^2 - N(1 + \log_2 N)$. Incidentally, we observe that saving in computation increases with the square of N. For instance, in the ECG case where we may have $N = 512$, use of the FFT is effective for $M > 10$.

Theorem 4.3 states the **equivalence** of the analog continuous time convolution and the discrete time convolution, while Theorem 4.5 states the equivalence between circular digital convolution and the DFT product. However, the discrete convolutions appearing in (4.25) and (4.41) are different, and extra care has to be taken to obtain the equivalence between the output of an analog filter and the product of the DFTs.

So, let us consider the case in which the sequences of the samples of the input signal $x(t)$ and of the filter impulse response $h^{(2B)}(t)$, have a finite extension. More specifically, let N_x be the length of the sequence of non-null samples of $x(t)$ and N_h the length of non-null samples of $h^{(2B)}(t)$. Then the discrete time convolution (4.25) is a sequence of length $(N_x + N_h - 1)$. Thus, let $\tilde{x}[n]$ be the sequence of length N given by

$$N = 2\mathrm{Max}(N_x, N_h) \tag{4.46}$$

obtained by zero padding $x[n]$, that is

$$\tilde{x}[n] = \begin{cases} x[n] & 0 \leq n \leq N_x - 1 \\ 0 & Otherwise \end{cases} \tag{4.47}$$

and $\tilde{h}^{(2B)}[n]$ the sequence of length N obtained by zero padding$h^{(2B)}[n]$, that is

$$\tilde{h}^{(2B)}[n] = \begin{cases} h^{(2B)}[n] & 0 \leq n \leq N_h - 1 \\ 0 & Otherwise \end{cases} \tag{4.48}$$

Then, it can be easily verified that, due to the zero padding, the discrete time convolution between $x[n]$ and $h^{(2B)}[n]$ equals the circular convolution of $\tilde{x}[n]$ and $\tilde{h}^{(2B)}[n]$, that is

$$\sum_{m=-\infty}^{+\infty} x\left(\frac{m-n}{2B}\right) h^{(2B)}\left(\frac{m}{2B}\right) = \sum_{k=0}^{N-1} \tilde{x}\left[(k-n)_{\mathrm{mod}N}\right] \tilde{h}^{(2B)}[k] \tag{4.49}$$

The whole procedure is summarized in Table 4.2.

Table 4.2 Equivalent digital signal processing filter in the DFT domain

Input	$x[n] = x\left(\frac{n}{2B}\right), \quad n = 0, 1, \ldots, N_x - 1$
	$h^{(2B)}[n] = h^{(2B)}\left(\frac{n}{2B}\right) = [2B\,\mathrm{sinc}(2\pi Bt) * h(t)]_{t=\frac{n}{2B}}, \quad n = 0, 1, \ldots, N_h - 1$
Compute	1. $\tilde{x}[n] = \begin{cases} x[n] & 0 \leq n \leq N_x - 1 \\ 0 & N_x \leq n \leq N - 1 \end{cases}$
	2. $\tilde{h}^{(2B)}[n] = \begin{cases} h^{(2B)}[n] & 0 \leq n \leq N_h - 1 \\ 0 & N_h \leq n \leq N - 1 \end{cases}$
	3. $N = 2\mathrm{Max}(N_x, N_h)$
	4. $\tilde{X}[k] = FFT\{\tilde{x}[n]\}$
	5. $\tilde{H}^{(2B)}[k] = FFT\left\{\tilde{h}^{(2B)}[n]\right\}$
	6. $\tilde{Y}[k] = \tilde{X}[k]\tilde{H}^{(2B)}[k], \quad k = 0, 1, \ldots, N$
	7. $\tilde{y}[n] = FFT^{-1}\{\tilde{Y}[k]\}$
Output	$y[n] = \tilde{y}[n], \quad n = 0, 1, \ldots, N_x + N_h - 2$

4.2.1 Discrete Fourier Transform Statistics

Let us first recall that a discrete-time random process, also referred to here as a random time series, is a sequence of random variables $\{x[n], n \in N\}$, indexed by integers, defined in a common probability space. The random time series is said to be stationary if its statistical properties are invariant to time shift.

In addition, a stationary time series is said to be strict-sense ergodic if the strong law of large numbers applies, so that its statistical properties can be deduced from a single, sufficiently long realization, thanks to the equivalence between the time average of any function and its expectation.

Let $x[n]$ be a random zero mean stationary ergodic periodic time series with period N. Let $r_x[n]$ denote its autocorrelation function. Since $x[n]$ is assumed to be periodic, the covariance function is periodic with period equal to N. On the other hand, since $x[n]$ is ergodic and its expectation is null, the autocorrelation function is equal to the covariance function of the process. Therefore, in this case the covariance function is also periodic. This kind of signal is a special case of a wider class of random processes named **wide-sense cyclostationary** processes for which a periodicity property is required for second-order statistics only.

Definition 4.2. *A random process is said to be wide-sense cyclostationary with period N if its expectation $m_x[k] = E\{x[k]\}$ and its covariance function:*

$$r_x[k] = E\{(x[n] - Ex[n])(x^*[n+k] - E\{x^*[n+k]\})\} \tag{4.50}$$

are periodic with period N, that is, $m_x[k] = m_x[k_{\mathrm{mod}N}]$ and $r_x[k] = r_x[k_{\mathrm{mod}N}]$.

We note that wide-sense cyclostationarity does not require that the signal itself is periodic but only its second-order statistics. This is a relevant property in biomedical applications where many physiological signals like ECGs are repetitive by nature, but, due to continuous variations of internal and external conditions, they cannot be considered periodic from a formal point of view, and periodicity can be considered an emerging behavior, valid in the average.

Let us now denote with $\mathbf{x} = [x[0] \quad x[1] \quad \ldots \quad x[N-1]]^T$ the array formed by N temporal samples of x[n] and with $\mathbf{X} = \mathbf{W}_N\mathbf{x}$ its DFT. Thus, due to the linearity of the transformation the expectation of $\mathbf{x}$ is $E\{\mathbf{X}\} = \mathbf{W}_N E\{\mathbf{x}\}$. On the other hand, the covariance matrix of X is given by

$$\tilde{\mathbf{R}} \stackrel{def}{=} E\left\{[\mathbf{X} - E\{\mathbf{X}\}][\mathbf{X} - E\{\mathbf{X}\}]^{\dagger}\right\} = \mathbf{W}_N \mathbf{R} \mathbf{W}_N^{\dagger} \tag{4.51}$$

where $\mathbf{R}$ is the Toepliz matrix with elements

$$R_{n,m} = r_x\left[(n-m)_{\text{mod}N}\right]. \tag{4.52}$$

We recall that a matrix is said to be Toepliz if and only if $R_{n,m} = R_{n-m}$.

Incidentally we observe that equation (4.51) can be equivalently written in scalar form as

$$\tilde{R}_{h,k} = \sum_{m=0}^{N-1}\sum_{n=0}^{N-1} e^{-j\frac{2\pi}{N}hn} R_{n,m} e^{j\frac{2\pi}{N}km} = \sum_{m=0}^{N-1}\sum_{n=0}^{N-1} e^{-j\frac{2\pi}{N}(hn-km)} R_{n,m} \tag{4.53}$$

If $x(t)$ is wide-sense cyclostationary the following theorem holds.

Theorem 4.6. *The DFT* $\mathbf{X}$ *of an array* $\mathbf{x}$ *of* N *samples extracted from a zero mean wide-sense cyclostationary random process* $x[n]$ *with covariance function* $r_x[n]$, *is a random variable (r.v.) with uncorrelated components, whose variance is N times the Power Density Spectrum* $P_X[k]$ *of* x, *defined as the DFT of* $r_x[n]$:

$$P_X[k] = \sum_{s=0}^{N-1} e^{-j\frac{2\pi}{N}ks} r_x[s] \tag{4.54}$$

that is

$$Cov\{X[h], X[k]\} = N\, PX[h]\delta_{h,k} \tag{4.55}$$

Proof: Let us pose $\begin{cases} t = n \\ s = (n-m)_{\text{mod}N} \end{cases}$ with inverse function $\begin{cases} n = t \\ m = (t-s)_{\text{mod}N} \end{cases}$

Then, equation (4.53) can be rewritten as

$$
\begin{aligned}
\tilde{R}_{h,k} &= \sum_{t=0}^{N-1}\sum_{s=0}^{N-1} e^{-j\frac{2\pi}{N}\left[ht-k(t-s)_{\text{mod}N}\right]} R_{t,(t-s)_{\text{mod}N}} \\
&= \sum_{t=0}^{N-1}\sum_{s=0}^{N-1} e^{-j\frac{2\pi}{N}\left[(h-k)t+kt-k(t-s)_{\text{mod}N}\right]} R_{t,(t-s)_{\text{mod}N}} \\
&= \sum_{t=0}^{N-1} e^{-j\frac{2\pi}{N}(h-k)t} \sum_{s=0}^{N-1} e^{-j\frac{2\pi}{N}k\left[t-(t-s)_{\text{mod}N}\right]} R_{t,(t-s)_{\text{mod}N}} \\
&= \sum_{t=0}^{N-1} e^{-j\frac{2\pi}{N}(h-k)t} \sum_{s=0}^{N-1} e^{-j\frac{2\pi}{N}ks} r_x[s] = P_x[k] \sum_{t=0}^{N-1} e^{-j\frac{2\pi}{N}(h-k)t}
\end{aligned}
\tag{4.56}
$$

Since

$$
\sum_{t=0}^{N-1} e^{-j\frac{2\pi}{N}nt} = \begin{cases} N & n=0 \\ 0 & otherwise \end{cases} \tag{4.57}
$$

we have

$$
\tilde{R}_{h,k} = \begin{cases} N\,P_X[k] & h=k \\ 0 & h \neq k \end{cases} \tag{4.58}
$$

□

We observe that linearity of the Fourier transform also implies that if $\mathbf{x}$ is Gaussian, its DFT is also Gaussian and therefore the following property holds.

Theorem 4.7. *The DFT* $\mathbf{X}$ *of an array* $\mathbf{x}$ *of N samples extracted from a Gaussian wide-sense cyclostationary random process* $x(t)$ *with covariance function* $r_x(t)$*, is a Gaussian r.v. with mutually independent components, whose variance is N times the power density spectrum of* $\mathbf{x}$.

The direct consequence of the above property is that if the observed signal can be modeled as a Gaussian cyclostationary random process, the implied mutual independence allows us to process each frequency component separately.

However, this property not only drastically simplifies signal denoising, but it also implies that due to the mutual independence, extrapolation of the spectrum of a Gaussian signal outside the observed frequency band is not possible. Thus super-resolution techniques are ineffective when applied to Gaussian cyclostationary signals. Moreover, it has been demonstrated that the same limitation applies to Gaussian stationary random processes, (see Wong, 1985).

Also, in many practical situations in which super-resolution is successfully applied, Gaussianity is supposed to hold for the observation noise only, and deviation from Gaussianity of the signal of interest is the turnkey for super-resolution.

4.3 Estimation Theory Framework

In general, observation of physical and/or physiological data implies some signal distortions and alterations caused by the medium supporting the measure and also partially by the equipment employed for signal acquisition and analog to digital conversion. Consequently, the observed data has to be processed in order to remove possible observation noise and to restore the original waveform.

Since disturbances and alterations are often known just in a statistical form, extraction of the useful information from the data provided by sensors may be performed in the estimation theory framework.

Let $\mathbf{x}$ and $\mathbf{z}$ respectively be the quantity of interest and the observable. Then we model $\mathbf{x}$ as a sample of an N-dimensional r.v. $\mathbf{X}$. In addition, we assume that the relationship between the quantity of interest and the observable is described by means of a probabilistic model, as the conditional probability density function (p.d.f.) $p_{\mathbf{Z}/\mathbf{X}}(\mathbf{z}/\mathbf{x})$ of $\mathbf{Z}$ given $\mathbf{X}$. Then, estimation of the quantity of interest from the observable can be specified as the determination of a function:

$$\hat{\mathbf{X}} : \mathbf{Z} \rightarrow \mathbf{X} \tag{4.59}$$

that maps any observed value $\mathbf{z}$ into a specific value $\hat{\mathbf{x}}(\mathbf{z})$ on the basis of some the optimality criterion. So, a cost function $C[\mathbf{x}, \hat{\mathbf{x}}(\mathbf{z})]$ modeling in a quantitative way the effects of an estimation error has to be specified. Then, in the Bayesian approach, the objective functional to be minimized is constituted by the *absolute* risk R_B defined as the expectation of the cost with respect to both the quantity of interest and the observable, that is

$$R_B[\hat{\mathbf{x}}(\cdot)] = E_{\mathbf{X},\mathbf{Z}}\{C[\mathbf{x}, \hat{\mathbf{x}}(\mathbf{z})]\} = \int_{\mathbf{X}}\int_{\mathbf{Z}} C[\mathbf{x}, \hat{\mathbf{x}}(\mathbf{z})]\, p_{\mathbf{X},\mathbf{Z}}(\mathbf{x}, \mathbf{z}) d\mathbf{z} d\mathbf{x} \tag{4.60}$$

where $p_{\mathbf{X},\mathbf{Z}}(\mathbf{x}, \mathbf{z})$ is the joint p.d.f. of $\mathbf{X}$ and $\mathbf{Z}$, and $E_{\mathbf{X},\mathbf{Z}}\{\}$ denotes the expectation. Therefore, the Bayesian estimate of $\mathbf{x}$ given $\mathbf{z}$ is

$$\hat{\mathbf{x}}_B = Arg\left\{\min_{\hat{\mathbf{x}}(\cdot)} R_B[\hat{\mathbf{x}}(\cdot)]\right\}. \tag{4.61}$$

In general it may become extremely difficult to specify a cost functional that effectively models the effects of an incorrect estimate. More often, a simplified version that can be easily managed from a mathematical viewpoint is preferred. In particular, in many situations of practical interest, it is convenient to assume that the cost function only depends on the estimation error $\mathbf{e}(\mathbf{x})$ defined as:

$$\mathbf{e}(\mathbf{x}) = \hat{\mathbf{x}}(\mathbf{z}) - \mathbf{x}. \tag{4.62}$$

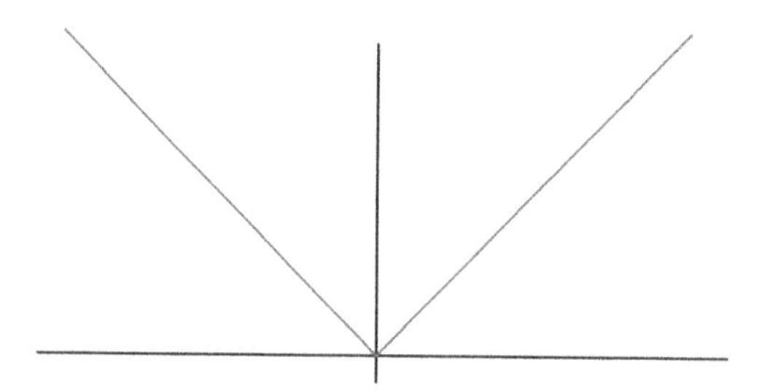

Figure 4.4 Absolute error cost function.

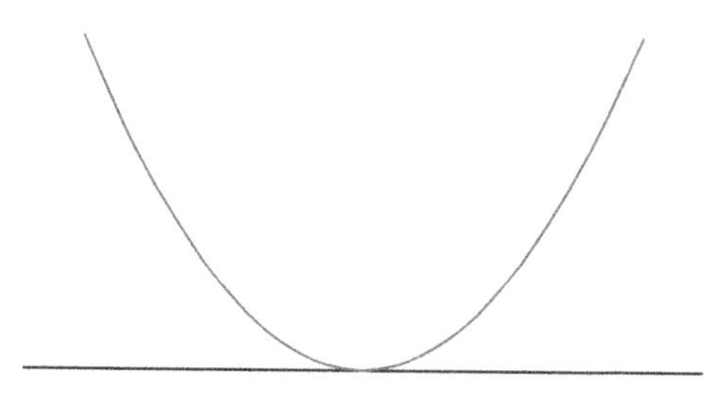

Figure 4.5 Quadratic cost function.

In this case, it is quite reasonable to assume that the cost function is a non-negative convex, even, function of $\mathbf{e}(\mathbf{x})$. Cost functions that are most frequently used are: quadratic, absolute error and uniform, as shown in Figures 4.4 – 4.6.

For the sake of compactness of the computation of the optimal estimators related to those cost functions, we observe that, since the joint p.d.f. $p_{\mathbf{X},\mathbf{Z}}(\mathbf{x}, \mathbf{z})$ can always be factored in terms of the conditional p.d.f. of $\mathbf{X}$ given $\mathbf{Z}$ and the p.d.f. of $\mathbf{Z}$ (i.e., $p_{\mathbf{X},\mathbf{Z}}(\mathbf{x}, \mathbf{z}) = p_{\mathbf{X}/\mathbf{Z}}(\mathbf{x}/\mathbf{z})p_{\mathbf{Z}}(\mathbf{z})$), the Bayesian absolute risk can be rewritten as follows:

$$\begin{aligned} R_B[\hat{\mathbf{x}}(\cdot)] &= E_{\mathbf{X}}\{E_{\mathbf{Z}/\mathbf{X}}\{C[\mathbf{x}, \hat{\mathbf{x}}(\mathbf{z})]/\mathbf{x}\}\} = \int_{\mathbf{X}}\int_{\mathbf{Z}} C[\mathbf{x}, \hat{\mathbf{x}}(\mathbf{z})]\, p_{\mathbf{X},\mathbf{Z}}(\mathbf{x}, \mathbf{z})d\mathbf{z}d\mathbf{x} \\ &= \int_{\mathbf{Z}}\left\{\int_{\mathbf{X}} C[\mathbf{x}, \hat{\mathbf{x}}(\mathbf{z})]\, p_{\mathbf{X}/\mathbf{Z}}(\mathbf{x}/\mathbf{z})d\mathbf{x}\right\}p_{\mathbf{Z}}(\mathbf{z})d\mathbf{z} \\ &= E_{\mathbf{Z}}\{E_{\mathbf{X}/\mathbf{Z}}\{C[\mathbf{x}, \hat{\mathbf{x}}(\mathbf{z})]/\mathbf{z}\}\} = E_{\mathbf{Z}}\{r'[\mathbf{z}, \hat{\mathbf{x}}(\cdot)]\}, \end{aligned} \tag{4.63}$$

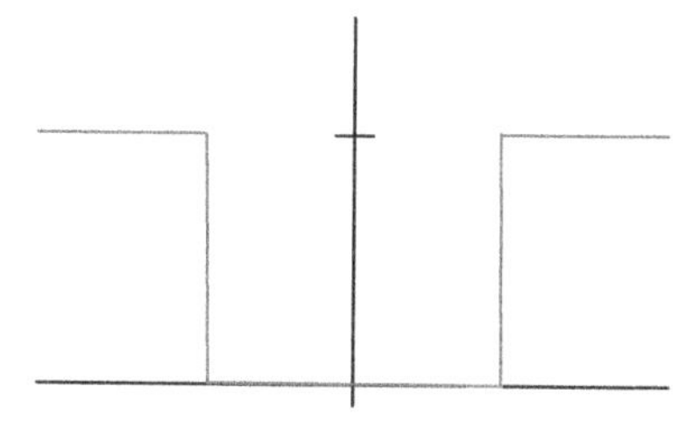

Figure 4.6 Uniform cost function.

in terms of the *posterior conditional risk*

$$r'[\mathbf{z},\ \hat{\mathbf{x}}(\cdot)] = E_{\mathbf{X}/\mathbf{Z}}\{C[\mathbf{x},\ \hat{\mathbf{x}}(\mathbf{z})]/\mathbf{z}\}. \tag{4.64}$$

For non-negative cost functions, the estimator $\hat{\mathbf{x}}_B(\mathbf{z})$ that minimizes the absolute risk, also minimizes, for each $\mathbf{z}$, the posterior conditional risk. Thus, $\hat{\mathbf{x}}_B(\mathbf{z})$ can be derived by minimizing $r'[\mathbf{z},\ \hat{\mathbf{x}}(\cdot)]$. In the following sections we derive the optimal estimators associated to the most relevant cost functions.

4.3.1 Minimum Mean Square Error Estimate

In general, the minimum mean square error (MMSE) estimator is usually considered as the first choice by engineers. The popularity of this estimator is partly due to its effectiveness, as well as to the availability of a well-established theoretical background, covering both implementation and performance, with particular reference to linear systems, and of a wide range of algorithms that have translated the abstract general solution into computationally efficient implementations for many areas of application.

We present here the general solution, starting from the case of one-dimensional random variables. In this case, the quadratic cost function is defined as

$$C[x, \hat{x}] = \gamma(\hat{x} - x)^2, \tag{4.65}$$

where γ is an arbitrary positive constant.

The substitution of the above expression into the definition of the *posterior conditional risk* gives

$$\begin{aligned} r'[z,\ \hat{x}(\cdot)] &= \gamma E_{X/Z}\left\{[\hat{x}(z) - x]^2/z\right\} = \\ &= \gamma E_{X/Z}\{\, x^2/z\} - 2\,\gamma E_{X/Z}\{\, x\,\hat{x}(z)/z\} + \gamma E_{X/Z}\left\{[\hat{x}(z)]^2/z\right\} \\ &= \gamma E_{X/Z}\{\, x^2/z\} - 2\gamma\,\hat{x}(z)\, E_{X/Z}\{\, x\,/z\} + \gamma[\hat{x}(z)]^2. \end{aligned} \tag{4.66}$$

Since the posterior conditional risk is a quadratic function, its maximum corresponds to the value $\hat{x}(z)$ of x, for which the first order derivative is null, that is

$$\frac{\partial}{\partial \hat{x}}\gamma\left[E_{X/Z}\{x^2/z\} - 2\hat{x}(z)E_{X/Z}\{x/z\} + [\hat{x}(z)]^2\right] = 0 \tag{4.67}$$

which in turn can be rewritten as

$$2\hat{x}(z) - 2E_{X/Z}\{x/z\} = 0. \tag{4.68}$$

Thus the Bayesian estimator for the quadratic cost function $\hat{x}^{MMSE}(z)$, also known in literature as minimum mean square error (MMSE) estimator, is the posterior conditional expectation of x given z:

$$\hat{x}^{MMSE}(z) = E_{X/Z}\{x/z\} = \int_X x\, p_{X/Z}(x/z)\, dx \tag{4.69}$$

The above result can be easily extended to the N-dimensional. In fact, in this case the quadratic cost functions is, by definition, a quadratic form

$$C[\mathbf{x}, \hat{\mathbf{x}}] = (\hat{\mathbf{x}} - \mathbf{x})^T \mathbf{Q}\,(\hat{\mathbf{x}} - \mathbf{x}), \tag{4.70}$$

where $\mathbf{Q}$ is a symmetric positive-definite matrix. Incidentally, we observe that, in a linear algebra framework, this cost function represents the weighted L^2 norm of the estimation error vector. Then, for the posterior conditional risk we have:

$$\begin{aligned} r'[z, \hat{x}(\cdot)] &= E_{\mathbf{X}/\mathbf{Z}}\{[\hat{\mathbf{x}}(\mathbf{z}) - \mathbf{x}]^T \mathbf{Q}\,[\hat{\mathbf{x}}(\mathbf{z}) - \mathbf{x}]/\mathbf{z}\} \\ &= E_{\mathbf{X}/\mathbf{Z}}\{\mathbf{x}^T\mathbf{Q}\mathbf{x}/\mathbf{z}\} - 2\,E_{\mathbf{X}/\mathbf{Z}}\{\mathbf{x}^T/\mathbf{z}\}\mathbf{Q}\hat{\mathbf{x}}(\mathbf{z}) + \hat{\mathbf{x}}^T(\mathbf{z})\mathbf{Q}\hat{\mathbf{x}}(\mathbf{z}). \end{aligned} \tag{4.71}$$

Thus observing that, since $\mathbf{Q}$ is symmetric, we have

$$\frac{\partial \mathbf{x}^T\mathbf{Q}\mathbf{x}}{\partial \hat{x}_k} = 0, \tag{4.72}$$

$$\frac{\partial \mathbf{x}^T\mathbf{Q}\hat{\mathbf{x}}}{\partial \hat{x}_k} = \frac{\partial}{\partial \hat{x}_k}\left(\sum_{i=1}^{N}\sum_{j=1}^{N} q_{ij}x_i\hat{x}_j\right) = \sum_{i=1}^{N} q_{ij}x_i = \sum_{i=1}^{N} q_{ji}x_i, \tag{4.73}$$

$$\frac{\partial \hat{\mathbf{x}}^T\mathbf{Q}\hat{\mathbf{x}}}{\partial \hat{x}_k} = \frac{\partial}{\partial \hat{x}_k}\left(\sum_{i=1}^{N}\sum_{j=1}^{N} q_{ij}\hat{x}_i\hat{x}_j\right) = \sum_{i=1}^{N} q_{ij}\hat{x}_i + \sum_{i=1}^{N} q_{ij}\hat{x}_j = 2\sum_{i=1}^{N} q_{ij}\hat{x}_j. \tag{4.74}$$

These scalar properties can be written in a compact equivalent vector form as

$$\frac{\partial \mathbf{x}^T\mathbf{Q}\mathbf{x}}{\partial \hat{\mathbf{x}}} = \mathbf{0}, \qquad \frac{\partial \mathbf{x}^T\mathbf{Q}\hat{\mathbf{x}}}{\partial \hat{\mathbf{x}}} = \mathbf{Q}\mathbf{x}, \frac{\partial \hat{\mathbf{x}}^T\mathbf{Q}\hat{\mathbf{x}}}{\partial \hat{x}_k} = 2\mathbf{Q}\hat{\mathbf{x}}. \tag{4.75}$$

Substituting of relations above into the expression of the vector of the partial derivatives of the risk with respect to $\hat{\mathbf{x}}(z)$gives:

$$\begin{aligned} &\frac{\partial}{\partial \hat{\mathbf{x}}}\left[E_{\mathbf{X}/\mathbf{Z}}\{\mathbf{x}^T\mathbf{Q}\mathbf{x}/\mathbf{z}\} - 2E_{\mathbf{X}/\mathbf{Z}}\{\mathbf{x}^T/\mathbf{z}\}\mathbf{Q}\hat{\mathbf{x}}(\mathbf{z}) + \hat{\mathbf{x}}^T(\mathbf{z})\mathbf{Q}\hat{\mathbf{x}}(\mathbf{z})\right] \\ &\qquad = -2\mathbf{Q}E_{\mathbf{X}/\mathbf{Z}}\{\mathbf{x}/\mathbf{z}\} + 2\mathbf{Q}\hat{\mathbf{x}}(\mathbf{z}). \end{aligned} \tag{4.76}$$

Then, making this vector null, we finally obtain the condition

$$\mathbf{Q}\left[\hat{\mathbf{x}}(\mathbf{z}) - E_{\mathbf{X}/\mathbf{Z}}\{\mathbf{x}/\mathbf{z}\}\right] = \mathbf{0}. \tag{4.77}$$

Thus, it is demonstrated that, even in the N-dimensional case, the MMSE estimator is the posterior conditional expectation of $\mathbf{x}$ given $\mathbf{z}$:

$$\hat{\mathbf{x}}^{MMSE}(\mathbf{z}) = E_{\mathbf{X}/\mathbf{Z}}\{\mathbf{x}/\mathbf{z}\} = \int_{\mathbf{X}} \mathbf{x}\, p_{\mathbf{X}/\mathbf{Z}}(\mathbf{x}/\mathbf{z}) d\mathbf{x}. \tag{4.78}$$

Note that, although the solution is rather simple from a conceptual point of view, implementation of the MMSE implies the computation of the posterior conditional expectation. The design of digital, computationally efficient MMSE estimators applicable to specific contexts has therefore received specific attention by many researchers. The most popular of these are the Wiener filter representing the MMSE estimator for stationary Gaussian random processes observed in additive, stationary, Gaussian random noise, and the Kalman filter representing the MMSE estimator of the status of a linear, time-variant dynamical system driven by a non-stationary Gaussian random processes, observed in additive Gaussian random noise. Details of those specific estimators are discussed in the implementation part of this chapter.

4.3.2 Minimum Mean Absolute Error Estimate (MMAE)

Let us now examine the case of the absolute error cost that in the one-dimensional case is specified as

$$C[x, \hat{x}] = \gamma|\hat{x} - x|, \tag{4.79}$$

where γ is any positive arbitrary constant. Thus, in a linear algebra framework, this cost function represents the L^1 norm of the estimation error vector. Substitution of (4.79) in (4.64) gives the following expression for the posterior conditional risk:

$$r'[\mathbf{z}, \hat{x}(\cdot)] = \gamma \int_{-\infty}^{+\infty} |x - \hat{x}(\mathbf{z})| p_{X/\mathbf{Z}}(x/\mathbf{z}) dx. \tag{4.80}$$

Observing that

$$\frac{\partial C[x, \hat{x}]}{\partial \hat{x}} = \gamma \frac{\partial|\hat{x} - x|}{\partial \hat{x}} = \gamma \operatorname{sign}[\hat{x} - x], \tag{4.81}$$

where *sign* [.] is defined as

$$\operatorname{sign}(x) = \begin{cases} -1 & x < 0 \\ 0 & x = 0, \\ 1 & x < 0 \end{cases} \tag{4.82}$$

for the first-order derivative of the posterior conditional risk w.r.t. $\hat{x}(z)$, we obtain

$$\begin{aligned} \frac{\partial r'[\mathbf{z}, \hat{x}(\cdot)]}{\partial \hat{x}} &= \gamma \int_{-\infty}^{+\infty} \operatorname{sign}[x - \hat{x}(\mathbf{z})] p_{X/\mathbf{Z}}(x/\mathbf{z}) dx \\ &= -\gamma \int_{-\infty}^{\hat{x}(\mathbf{z})} p_{X/\mathbf{Z}}(x/\mathbf{z}) dx + \gamma \int_{\hat{x}(\mathbf{z})}^{+\infty} p_{X/\mathbf{Z}}(x/\mathbf{z}) dx. \end{aligned} \tag{4.83}$$

Therefore, by making the derivative null, the following condition for the minimum mean absolute error estimate $\hat{x}^{MMAE}(z)$ is obtained:

$$\int_{-\infty}^{\hat{x}^{MMAE}} p_{X/\mathbf{Z}}(x/\mathbf{z})\,dx = \int_{\hat{x}^{MMAE}}^{+\infty} p_{X/\mathbf{Z}}(x/\mathbf{z})\,dx. \tag{4.84}$$

Thus the estimate minimizing the average L^1 norm of the error is given by the posterior conditional median value.The estimate minimizing the average L^1 norm coincides with the one minimizing the L^2 norm if and only if the expectation and the median of the posterior p.d.f. of the parameter to be estimated given the observations coincide.

4.3.3 *Maximum A Posteriori (MAP) Probability Estimate*

The maximum a posteriori probability is a quite diffuse procedure that cannot be considered Bayesian in the strict sense. Nevertheless, it represents the limit of the Bayesian solution associated with a uniform cost function

$$C[x,\hat{x}] = \begin{cases} 0 & \text{if } |\hat{x}-x| \leq \Delta \\ 1 & \text{if } |\hat{x}-x| > \Delta \end{cases} \tag{4.85}$$

when the interval amplitude Δ goes to 0, as shown in Figure 4.7.

Let us observe that, for the uniform cost, the conditional posterior risk is

$$\begin{aligned} r'[z,\hat{x}(\cdot)] &= E_{X/Z}\{C[\hat{x}(z)-x]/z\} = \Pr\{|\hat{x}(z)-x| > \Delta/Z = z\} \\ &= 1 - \Pr\{|\hat{x}(z)-x| \leq \Delta/Z = z\} \end{aligned} \tag{4.86}$$

Thus, it can be easily verified that estimate minimizing the risk is the value that maximizes the posterior probability that x falls inside the interval

$$[\hat{x}(z)-\Delta,\ \hat{x}(z)+\Delta]. \tag{4.87}$$

Therefore, when Δ goes to 0, the Bayesian estimate tends to the value for which the a posteriori conditional p.d.f. takes its maximum, that is

$$\hat{x}_{MAP} = arg\left\{ \underset{x\in X}{Max}\ p_{X/Z}(x/z) \right\}. \tag{4.88}$$

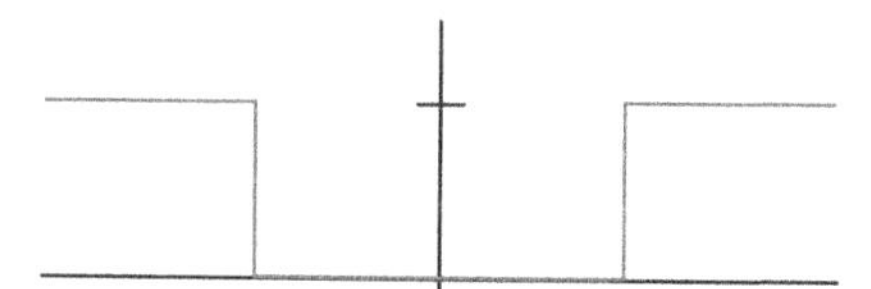

Figure 4.7 Uniform cost function.

We observe that if the posterior conditional p.d.f. is symmetric and unimodal, the MAP MMSE and MMAE estimates coincide.

4.3.4 Maximum Likelihood Estimation (MLE)

In many practical cases Bayesian techniques cannot be applied, either because a reliable statistical model of the process to be estimated is not available or because the analytical solution of the related optimization problem is too complex.

A viable alternative in these cases is the maximum likelihood procedure. The concept of likelihood and of its maximization to perform estimates was introduced by Sir Ronald Aylmer Fisher in his classic paper 'On the Mathematical Foundations of Theoretical Statistics', presented in 1921 and published in 1922, (see Fisher, 1921). Following Fisher, the estimate is obtained by maximizing the likelihood functional $\Lambda(\mathbf{z}/\mathbf{x})$ defined as the ratio between the conditional p.d.f. $p_{\mathbf{Z}/\mathbf{X}}(\mathbf{z}/\mathbf{x})$ of the observation $\mathbf{z}$ given $\mathbf{x}$, and any arbitrary function $V(\mathbf{z})$ that does not depend on x, namely,

$$\Lambda(\mathbf{z}/\mathbf{x}) = \frac{p_{\mathbf{Z}/\mathbf{X}}(\mathbf{z}/\mathbf{x})}{V(\mathbf{z})}. \tag{4.89}$$

Therefore the maximum likelihood estimator is

$$\hat{\mathbf{x}}_{ML} = arg\left\{\underset{\mathbf{x}\in\mathbf{X}}{Max}\ \Lambda(\mathbf{z}/\mathbf{x})\right\}. \tag{4.90}$$

Since the logarithm is a monotonically increasing function, the ML estimate can also be written in the following form:

$$\hat{\mathbf{x}}_{ML} = arg\left\{\underset{\mathbf{x}\in\mathbf{X}}{Max}\ \ln \Lambda(\mathbf{z}/\mathbf{x}) + \beta\right\}. \tag{4.91}$$

with β an arbitrary (finite) positive constant.

Let us now consider any transformation $\mathbf{y} = T(\mathbf{z})$ of the original observation $\mathbf{z}$. The question now is under which conditions we can use $\mathbf{y}$ to estimate $\mathbf{x}$, instead of $\mathbf{z}$, still obtaining the same result as in case of direct use of the maximum likelihood estimation based on the observation $\mathbf{z}$. The above constraint can be formulated in mathematical form as

$$\Lambda(\mathbf{z}/\mathbf{x}) = \Lambda(T(\mathbf{z})/\mathbf{x}), \quad \forall \mathbf{x}, \mathbf{z} \tag{4.92}$$

which in turn implies that

$$\frac{p_{\mathbf{Z}/\mathbf{X}}(\mathbf{z}/\mathbf{x})}{V(\mathbf{z})} = \frac{p_{\mathbf{Y}/\mathbf{X}}(\mathbf{y}/\mathbf{x})}{Q(\mathbf{y})} \tag{4.93}$$

or equivalently

$$p_{\mathbf{Z}/\mathbf{X}}(\mathbf{z}/\mathbf{x}) = g(\mathbf{z}) p_{\mathbf{Y}/\mathbf{X}}(T(\mathbf{z})/\mathbf{x}). \tag{4.94}$$

where

$$g(\mathbf{z}) = \frac{V(\mathbf{z})}{Q(T(\mathbf{z}))} \tag{4.95}$$

is a function of $\mathbf{z}$ only.

The condition given by equation (4.94) represents Fisher's factorization criterion, and any transformation T, for which condition (4.94) holds, is said to be a ***sufficient statistic***.

Since, by definition, sufficient statistics can replace observations without any performance loss as far as maximum likelihood estimation is concerned, the task is to find sufficient statistics for which the computational complexity is low.

4.4 Performance Indicators

Performance assessment can be considered a cornerstone in the design of all equipment and systems. In the Bayesian approach to signal restoration and denoising, and more generally to signal estimation, performance optimization is the driver of the whole process and the absolute and conditional risks are the key performance indicators to be used when comparing optimal and suboptimal solutions. In practice, besides them, first- and second-order moments of the estimation error are still extensively employed to verify that a given solution complies with user needs and requirements. In fact, the average norms (e.g., L^1, L^2, L^∞) of the estimation error aim to quantify the general concept of accuracy, which, by definition, defines the closeness of the true quantity value and its estimate in a non-quantitative manner (JCGM, 2008). This clarifies why the average *non-central second-order moment* of the estimation error, also named *mean squared error* (MSE),

$$MSE = E_{\mathbf{Z},\mathbf{X}}\{[\hat{\mathbf{x}}(\mathbf{z}) - \mathbf{x}][\hat{\mathbf{x}}(\mathbf{z}) - \mathbf{x}]^T\}, \tag{4.96}$$

has received special attention with respect to its evaluation, approximation and bounds.

However, the mean squared error can always be written in terms of the first-order moment of the estimation error (referred to in the literature as *bias*):

$$\mathbf{B} = E_{\mathbf{Z},\mathbf{X}}[\hat{\mathbf{x}}(\mathbf{z}) - \mathbf{x}] \tag{4.97}$$

and the second-order central moment, also called the *estimation error covariance matrix*:

$$Cov\{\hat{\mathbf{x}}(\mathbf{z}) - \mathbf{x}\} = E_{\mathbf{Z},\mathbf{X}}\{[\hat{\mathbf{x}}(\mathbf{z}) - \mathbf{x} - \mathbf{B}][\hat{\mathbf{x}}(\mathbf{z}) - \mathbf{x} - \mathbf{B}]^T\}. \tag{4.98}$$

The mean square error is then

$$E_{\mathbf{Z},\mathbf{X}}\{[\hat{\mathbf{x}}(\mathbf{z}) - \mathbf{x}][\hat{\mathbf{x}}(\mathbf{z}) - \mathbf{x}]^T\} = Cov\{\hat{\mathbf{x}}(\mathbf{z}) - \mathbf{x}\} + \mathbf{B}\mathbf{B}^T. \tag{4.99}$$

We observe that the *estimation error covariance* can be considered a quantitative indicator of the estimation *precision* that defines the closeness of agreement between estimates performed on replicate observations of the same quantity under specified conditions, while the *bias* describes the systematic component of the error.

Since analytical computation of the mean squared error is often difficult, use of computer simulations represents a viable solution for assessing the performance of a given piece of equipment or a part of it, hardware or software. Nevertheless, both analytical evaluation and computer simulation require a deep description of the solution. However, during the preliminary design phases, details are often unavailable and it is a common practice to resort to approximations and bounds that can provide a forecast of the achievable performance versus the key design parameters.

In this framework, the Cramér–Rao inequality, that specifies the lower bound for the mean squared error, can be considered the cornerstone of performance assessment of medical diagnostic equipment. In fact, this inequality allows us to evaluate a bound for the second-order statistics of the estimation error on the basis of *Fisher's information* whose computation only requires the knowledge of the joint p.d.f. of the quantity to be estimated and the observation.

Thus it represents an invaluable support during the feasibility study because it allows us to predict the achievable performance on the basis of the waveforms and sensor characteristics, without any knowledge of the actual optimal estimator.

It also supports the tradeoff between performance and complexity of the solution because it allows us to determine the efficiency losses of possible suboptimal solutions.

An example of applications of the Cramér–Rao bound can be found in medical ultrasound where cross correlations and similar operations are used for displacement and Doppler estimation. The Cramér–Rao bound determines the lowest magnitude of errors that can be achieved in the estimated values.

Let us first formulate the bound in the one-dimensional case.

Theorem 4.8. *Given an estimator* $\hat{x}(\mathbf{z})$ *with finite bias* B *of a one-dimensional r.v. X, if*

$$\frac{\partial p_{\mathbf{Z},X}(\mathbf{z},x)}{\partial x} \quad \text{and} \quad \frac{\partial^2 p_{\mathbf{Z},X}(\mathbf{z},x)}{\partial x^2} \tag{4.100}$$

exist and are absolutely integrable w.r.t. both x *and* $\mathbf{z}$, *then for the mean squared error of the estimate the following inequality holds*

$$E\left\{[\hat{x}(\mathbf{z}) - x]^2\right\} \geq J_{\mathbf{Z},X}^{-1}, \tag{4.101}$$

where

$$J_{\mathbf{Z},X} = E_{\mathbf{Z},X}\left\{\left[\frac{\partial \ln p_{\mathbf{Z},X}(\mathbf{z},x)}{\partial x}\right]^2\right\} \tag{4.102}$$

is the Fisher's information which, in this case, can also be written in the equivalent form

$$J_{\mathbf{Z},X} = -E_{\mathbf{Z},X}\left\{\frac{\partial^2 \ln p_{\mathbf{Z},X}(\mathbf{z},x)}{\partial x^2}\right\}. \tag{4.103}$$

Proof: Let us denote by $b(x)$ the conditional expectation of the estimation error given x, that is

$$b(x) = E_{\mathbf{Z}/X}[\hat{x}(\mathbf{z}) - x] = \int_{\mathbf{Z}} [\hat{x}(\mathbf{z}) - x] p_{\mathbf{Z}/X}(\mathbf{z}, x) d\mathbf{z}. \quad (4.104)$$

Then, by definition, the bias B is given by

$$B = E_{\mathbf{Z},X}[\hat{x}(\mathbf{z}) - x] = \int_X \int_{\mathbf{Z}} [\hat{x}(\mathbf{z}) - x] p_{\mathbf{Z},X}(\mathbf{z}, x) d\mathbf{z} dx. \quad (4.105)$$

Thus, it immediately follows that B can be expressed in terms of $b(x)$ as

$$B = E_X[b(x)] = \int_{-\infty}^{+\infty} b(x) p_X(x) dx. \quad (4.106)$$

Since, by hypothesis $|B| < \infty$, convergence of the above integral implies that

$$\lim_{x \to \pm\infty} b(x) p_X(x) = 0. \quad (4.107)$$

Since for a given estimator, B is a finite constant, taking its first-order derivative w.r.t. x we obtain:

$$\frac{\partial}{\partial x} \int_X \int_{\mathbf{Z}} [\hat{x}(\mathbf{z}) - x] p_{\mathbf{Z},X}(\mathbf{z}, x) d\mathbf{z} dx = \frac{\partial B}{\partial x} = 0. \quad (4.108)$$

On the other hand, by hypothesis, the derivative of the joint p.d.f. is absolutely integrable and we can exchange the integral and differentiation order, then obtaining

$$\int_X \int_{\mathbf{Z}} [\hat{x}(\mathbf{z}) - x] \frac{\partial p_{\mathbf{Z},X}(\mathbf{z}, x)}{\partial x} d\mathbf{z} dx - \int_X \int_{\mathbf{Z}} p_{\mathbf{Z},X}(\mathbf{z}, x) d\mathbf{z} dx = 0. \quad (4.109)$$

Since the integral of the joint p.d.f. over the whole space (X,**Z**) is equal 1, we have

$$\int_X \int_{\mathbf{Z}} [\hat{x}(\mathbf{z}) - x] \frac{\partial \ln p_{\mathbf{Z},X}(\mathbf{z}, x)}{\partial x} p_{\mathbf{Z},X}(\mathbf{z}, x) d\mathbf{z} dx = 1. \quad (4.110)$$

Recalling that

$$\frac{\partial p_{\mathbf{Z},X}(\mathbf{z}, x)}{\partial x} = \frac{\partial \ln p_{\mathbf{Z},X}(\mathbf{z}, x)}{\partial x} p_{\mathbf{Z},X}(\mathbf{z}, x), \quad (4.111)$$

we can also write

$$\int_X \int_{\mathbf{Z}} [\hat{x}(\mathbf{z}) - x] \sqrt{p_{\mathbf{Z},X}(\mathbf{z}, x)} \frac{\partial \ln p_{\mathbf{Z},X}(\mathbf{z}, x)}{\partial x} \sqrt{p_{\mathbf{Z},X}(\mathbf{z}, x)} d\mathbf{z} dx = 1. \quad (4.112)$$

The above relation can be written in the simple compact form

$$\left| \int_X \int_{\mathbf{Z}} f(\mathbf{z}, x) g(\mathbf{z}, x) d\mathbf{z} dx \right|^2 = 1, \tag{4.113}$$

by posing

$$f(\mathbf{z}, x) = [\hat{x}(\mathbf{z}) - x] \sqrt{p_{\mathbf{Z},X}(\mathbf{z}, x)}, \tag{4.114}$$

$$g(\mathbf{z}, x) = \frac{\partial \ln p_{\mathbf{Z},X}(\mathbf{z}, x)}{\partial x} \sqrt{p_{\mathbf{Z},X}(\mathbf{z}, x)}. \tag{4.115}$$

Then, the Cramér–Rao lower bound immediately follows from the Schwarz inequality

$$\int_X \int_{\mathbf{Z}} |f(\mathbf{z}, x)|^2 d\mathbf{z} dx \int_X \int_{\mathbf{Z}} |g(\mathbf{z}, x)|^2 d\mathbf{z} dx \geq \left| \int_X \int_{\mathbf{Z}} f(\mathbf{z}, x) g(\mathbf{z}, x) d\mathbf{z} dx \right|^2. \tag{4.116}$$

In fact, from equations (4.113) and (4.116) we have

$$\int_X \int_{\mathbf{Z}} |f(\mathbf{z}, x)|^2 d\mathbf{z} dx \int_X \int_{\mathbf{Z}} |g(\mathbf{z}, x)|^2 d\mathbf{z} dx \geq 1, \tag{4.117}$$

which can also be rewritten as

$$\int_X \int_{\mathbf{Z}} [\hat{x}(\mathbf{z}) - x]^2 p_{\mathbf{Z},X}(\mathbf{z}, x) d\mathbf{z} dx \geq \left[\int_X \int_{\mathbf{Z}} \left| \frac{\partial \ln p_{\mathbf{Z},X}(\mathbf{z}, x)}{\partial x} \right|^2 p_{\mathbf{Z},X}(\mathbf{z}, x) d\mathbf{z} dx \right]^{-1}. \tag{4.118}$$

Finally, since

$$\int_X \int_{\mathbf{Z}} p_{\mathbf{Z},X}(\mathbf{z}, x) d\mathbf{z} dx = 1, \tag{4.119}$$

differentiating w.r.t. x we have

$$\int_X \int_{\mathbf{Z}} \frac{\partial p_{\mathbf{Z},X}(\mathbf{z}, x)}{\partial x} d\mathbf{z} dx = 0, \tag{4.120}$$

or equivalently

$$\int_X \int_{\mathbf{Z}} \frac{\partial \ln p_{\mathbf{Z},X}(\mathbf{z}, x)}{\partial x} p_{\mathbf{Z},X}(\mathbf{z}, x) d\mathbf{z} dx = 0. \tag{4.121}$$

Differentiating again w.r.t. x we have

$$\int_X \int_{\mathbf{Z}} \frac{\partial^2 \ln p_{\mathbf{Z},X}(\mathbf{z},x)}{\partial x^2} p_{\mathbf{Z},X}(\mathbf{z},x) d\mathbf{z} dx + \int_X \int_{\mathbf{Z}} \frac{\partial \ln p_{\mathbf{Z},X}(\mathbf{z},x)}{\partial x} \frac{\partial p_{\mathbf{Z},X}(\mathbf{z},x)}{\partial x} d\mathbf{z} dx = 0 \quad (4.122)$$

which can be rewritten as

$$\int_X \int_{\mathbf{Z}} \frac{\partial^2 \ln p_{\mathbf{Z},X}(\mathbf{z},x)}{\partial x^2} p_{\mathbf{Z},X}(\mathbf{z},x) d\mathbf{z} = - \int_X \int_{\mathbf{Z}} \left| \frac{\partial \ln p_{\mathbf{Z},X}(\mathbf{z},x)}{\partial x} \right|^2 p_{\mathbf{Z},X}(\mathbf{z},x) d\mathbf{z} dx, \quad (4.123)$$

□

which states the equivalence between the two inequalities.
For a deeper insight of the meaning of Fisher's information, let us recall that

$$\ln p_{\mathbf{Z},X}(\mathbf{z},x) = \ln \left[p_{\mathbf{Z}/X}(\mathbf{z}/x) p_X(x) \right] = \ln p_{\mathbf{Z}/X}(\mathbf{z}/x) + \ln p_X(x). \quad (4.124)$$

Thus defining as *a priori Fisher Information* J_X, on X the quantity

$$J_X \stackrel{def}{=} E_X \left\{ \left[\frac{\partial \ln p_X(x)}{\partial x} \right]^2 \right\} = -E_X \left\{ \frac{\partial^2 \ln p_X(x)}{\partial x^2} \right\}, \quad (4.125)$$

and as posterior Fisher Information $J_{z/z}$, on X the quantity

$$J_{\mathbf{Z}/X} \stackrel{def}{=} E_{\mathbf{Z},X} \left\{ \left[\frac{\partial \ln p_{\mathbf{Z},X}(\mathbf{z}/x)}{\partial x} \right]^2 \right\} = -E_{\mathbf{Z},X} \left\{ \frac{\partial^2 \ln p_{\mathbf{Z},X}(\mathbf{z}/x)}{\partial x^2} \right\}, \quad (4.126)$$

From (4.124) we have that the overall Fisher information $J_{\mathbf{Z},X}$ can be split in the two additive contributions: the a priori and the posterior information:

$$J_{\mathbf{Z},X} = J_X + J_{\mathbf{Z}/X}. \quad (4.127)$$

4.4.1 Efficient Estimators

By definition, an estimator is said to be efficient if and only if the mean squared error equals the Cramér–Rao lower bound.

Let us observe that, based on the Schwarz inequality properties, the equality is achieved if and only if

$$\frac{\partial \ln p_{\mathbf{Z},X}(\mathbf{z},x)}{\partial x} = k[\hat{x}(\mathbf{z}) - x], \quad (4.128)$$

or equivalently

$$\frac{\partial \ln p_{X/\mathbf{Z}}(x/\mathbf{z})}{\partial x} + \frac{\partial \ln p_{\mathbf{Z}}(\mathbf{z})}{\partial x} = k[\hat{x}(\mathbf{z}) - x]. \tag{4.129}$$

Since the p.d.f. of **Z** does not depend on X, we have

$$\frac{\partial \ln p_{X/\mathbf{Z}}(x/\mathbf{z})}{\partial x} = k[\hat{x}(\mathbf{z}) - x]. \tag{4.130}$$

Integration w.r.t. x then gives:

$$\ln p_{X/\mathbf{Z}}(x/\mathbf{z}) = c_1 + k\hat{x}(\mathbf{z})x - \frac{k}{2}x^2. \tag{4.131}$$

Therefore if an efficient estimator exists, the p.d.f. of X given **Z** is necessarily Gaussian.

4.4.2 *Fisher's Information Matrix*

To deal with the performance assessment of the estimation of N-dimensional parameters or random variables, let us first generalize the concept of Fisher's information.

Let us denote by $\nabla_{\mathbf{x}}$ the partial derivatives array operator: $\nabla_{\mathbf{x}} = \left[\frac{\partial}{\partial x_1}, \frac{\partial}{\partial x_2} \cdots \frac{\partial}{\partial x_N}\right]^T$.

Then, the Fisher's information matrix on **X**, given the observation **Z**, is defined as

$$\begin{aligned}\mathbf{J}_{\mathbf{Z},\mathbf{X}} &\stackrel{def}{=} E_{\mathbf{Z},\mathbf{X}}\left\{\left[\nabla_{\mathbf{x}} \ln p_{\mathbf{Z},\mathbf{X}}(\mathbf{z},\mathbf{x})\right]\left[\nabla_{\mathbf{x}} \ln p_{\mathbf{Z},\mathbf{X}}(\mathbf{z},\mathbf{x})\right]^T\right\} \\ &= -E_{\mathbf{Z},\mathbf{X}}\left\{\nabla_{\mathbf{x}}\nabla_{\mathbf{x}}^T \ln p_{\mathbf{Z},\mathbf{X}}(\mathbf{z},\mathbf{x})\right\}.\end{aligned} \tag{4.132}$$

The previous definition and its equivalent form involving second-order derivatives can also be written in scalar form as

$$\left[\mathbf{J}_{\mathbf{Z},\mathbf{X}}\right]_{ij} = E_{\mathbf{Z},\mathbf{X}}\left\{\frac{\partial \ln p_{\mathbf{Z},\mathbf{X}}(\mathbf{z},\mathbf{x})}{\partial x_i}\frac{\partial \ln p_{\mathbf{Z},\mathbf{X}}(\mathbf{z},\mathbf{x})}{\partial x_j}\right\} = -E_{\mathbf{Z},\mathbf{X}}\left\{\frac{\partial^2 \ln p_{\mathbf{Z},\mathbf{X}}(\mathbf{z},\mathbf{x})}{\partial x_i \partial x_j}\right\}. \tag{4.133}$$

On the other hand, defining as an a priori Fisher's information matrix the quantity

$$\begin{aligned}\mathbf{J}_{\mathbf{X}} &= E_{\mathbf{X}}\{[\nabla_{\mathbf{x}} \ln p_{\mathbf{X}}(\mathbf{x})][\nabla_{\mathbf{x}} \ln p_{\mathbf{X}}(\mathbf{x})]^T\} \\ &= -E_{\mathbf{X}}\{\nabla_X \nabla_{\mathbf{x}}^T \ln p_{\mathbf{X}}(\mathbf{x})\},\end{aligned} \tag{4.134}$$

and as posterior Fisher's information matrix the quantity

$$\begin{aligned}\mathbf{J}_{\mathbf{Z}/\mathbf{X}} &= E_{\mathbf{Z},\mathbf{X}}\left\{\left[\nabla_{\mathbf{x}} \ln p_{\mathbf{Z}/\mathbf{X}}(\mathbf{z}/\mathbf{x})\right]\left[\nabla_{\mathbf{x}} \ln p_{\mathbf{Z}/\mathbf{X}}(\mathbf{z}/\mathbf{x})\right]^T\right\} \\ &= -E_{\mathbf{Z},\mathbf{X}}\left\{\nabla_{\mathbf{x}}\nabla_{\mathbf{x}}^T \ln p_{\mathbf{Z}/\mathbf{X}}(\mathbf{z}/\mathbf{x})\right\},\end{aligned} \tag{4.135}$$

the overall Fisher's information is the sum of the a priori and posterior information:

$$\mathbf{J}_{\mathbf{Z},\mathbf{X}} = \mathbf{J}_{\mathbf{X}} + \mathbf{J}_{\mathbf{Z}/\mathbf{X}}. \tag{4.136}$$

In fact,

$$-E_{\mathbf{Z},\mathbf{X}}\left\{\frac{\partial^2 \ln p_{\mathbf{Z},\mathbf{X}}(\mathbf{z},\mathbf{x})}{\partial x_i \partial x_j}\right\} = -E_{\mathbf{Z},\mathbf{X}}\left\{\frac{\partial^2 \ln p_{\mathbf{Z}/\mathbf{X}}(\mathbf{z}/\mathbf{x})}{\partial x_i \partial x_j}\right\} - E_{\mathbf{X}}\left\{\frac{\partial^2 \ln p_{\mathbf{X}}(\mathbf{x})}{\partial x_i \partial x_j}\right\}. \tag{4.137}$$

Let $\mathbf{y} = \mathbf{A}\mathbf{x}$ be an invertible linear transformation of $\mathbf{x}$, so that the inverse of $\mathbf{A}$ exists and $\mathbf{x} = \mathbf{B}\mathbf{y}$ with $\mathbf{B} = \mathbf{A}^{-1}$; then the simple rule of N-dimensional p.d.f. transformation applies, thus obtaining

$$p_{Z,Y}(\mathbf{z},\mathbf{y}) = p_{Z,X}(\mathbf{z},\mathbf{B}\mathbf{y})|\det(\mathbf{B})|. \tag{4.138}$$

This in turn implies that the gradient of the logarithm of the joint probability of $\mathbf{Z}$ and $\mathbf{Y}$ can be written in terms of the gradient of the logarithm of the joint probability of $\mathbf{Z}$ and $\mathbf{X}$ as

$$\begin{aligned}\frac{\partial \ln p_{Z,Y}(\mathbf{z},\mathbf{y})}{\partial y_j} &= \frac{\partial \ln p_{Z,X}(\mathbf{z},\mathbf{B}\mathbf{y})}{\partial y_j} + \frac{\partial \ln |\det(\mathbf{B})|}{\partial y_j}\\ &= \sum_i \frac{\partial \ln p_{Z,X}(\mathbf{z},\mathbf{x})}{\partial x_i}\frac{\partial x_i}{\partial y_j} + \frac{\partial \ln |\det(\mathbf{B})|}{\partial y_j} = \sum_i B_{ij}\frac{\partial \ln p_{Z,X}(\mathbf{z},\mathbf{x})}{\partial x_i},\end{aligned} \tag{4.139}$$

where we accounted for the fact that det($\mathbf{B}$) is constant w.r.t. y and therefore its derivative is null, and that for a linear transformation we have

$$\frac{\partial x_i}{\partial y_j} = B_{ij}. \tag{4.140}$$

The scalar relation (4.139) can be rewritten in matrix form as

$$\nabla_{\mathbf{y}} \ln p_{Z,Y}(\mathbf{z},\mathbf{y}) = \mathbf{B}, \nabla_{\mathbf{x}} \ln p_{Z,X}(\mathbf{z},\mathbf{x}). \tag{4.141}$$

Therefore, for the Fisher's information matrix $\mathbf{J}_{\mathbf{Z},\mathbf{Y}}$, we can write

$$\begin{aligned}\mathbf{J}_{\mathbf{Z},\mathbf{Y}} &\stackrel{def}{=} E_{\mathbf{Z},\mathbf{Y}}\left\{\left[\nabla_{\mathbf{Y}} \ln p_{\mathbf{Z},\mathbf{Y}}(\mathbf{z},\mathbf{y})\right]\left[\nabla_{\mathbf{Y}} \ln p_{\mathbf{Z},\mathbf{Y}}(\mathbf{z},\mathbf{y})\right]^T\right\}\\ &= E_{\mathbf{Z},\mathbf{X}}\left\{\mathbf{B},\left[\nabla_{\mathbf{x}} \ln p_{\mathbf{Z},\mathbf{X}}(\mathbf{z},\mathbf{x})\right]\left[\nabla_{\mathbf{x}} \ln p_{\mathbf{Z},\mathbf{X}}(\mathbf{z},\mathbf{x})\right]^T\mathbf{B}\right\}.\end{aligned} \tag{4.142}$$

Thus, Fisher's information under linear invertible transformations satisfies the tensor rule:

$$\mathbf{J}_{\mathbf{Z},\mathbf{Y}} = \mathbf{B}, \mathbf{J}_{\mathbf{Z},\mathbf{X}}\mathbf{B}. \tag{4.143}$$

Concerning bounds on estimator performance, here we just state the Cramér–Rao lower bound for unbiased estimator.

Theorem 4.9. *Given an unbiased estimator* $\hat{\mathbf{x}}(\mathbf{z})$*of an N-dimensional r.v.* $\mathbf{X}$*, if*

$$\frac{\partial p_{\mathbf{Z},\mathbf{X}}(\mathbf{z},\mathbf{x})}{\partial x_i} \quad and \quad \frac{\partial^2 p_{\mathbf{Z},\mathbf{X}}(\mathbf{z},\mathbf{x})}{\partial x_i \partial x_j} \tag{4.144}$$

exist and are absolutely integrable w.r.t. both $\mathbf{x}$ *and* $\mathbf{z}$*, then the difference between the estimation Error Covariance matrix* $\mathbf{R}_\varepsilon$ *and the inverse of the Fisher's information matrix* $\mathbf{J}_{\mathbf{Z},\mathbf{X}}$ *is a positive-semidefinite matrix, that is*

$$\mathbf{s}^T \left[\mathbf{R}_\varepsilon - \mathbf{J}_{\mathbf{Z},\mathbf{X}}^{-1} \right] \mathbf{s} \geq 0, \qquad \forall \mathbf{s} \neq \mathbf{0}. \tag{4.145}$$

Note that since the matrix is positive-semidefinite if and only if all its principal minors have non-negative determinants, the bound also implies that for the variance of the estimation error of each component: $\sigma_{\varepsilon_i}^2 = [\mathbf{R}_\varepsilon]_{ii}$, the following bound holds:

$$\sigma_{\varepsilon_i}^2 \geq \left[\mathbf{J}^{-1}\right]_{ii} \tag{4.146}$$

Thus the diagonal elements of the inverse of the Fisher's information matrix constitute lower bounds for the mean square errors of the single components.

On the other hand, since similar relations can be written for principal minors of higher order, the bound given by Theorem 4.9 imposes tighter constraints to the error covariances than those expressed by (4.146) alone.

To enlighten this property let us consider the particular case in which the estimation error is a zero mean Gaussian random variable. Then its p.d.f. reduces to

$$p_E(\boldsymbol{\varepsilon}) = \frac{1}{\sqrt{(2\pi)^n \det[\mathbf{R}_\varepsilon]}} \exp\left\{ -\frac{1}{2} \boldsymbol{\varepsilon}^T \mathbf{R}_\varepsilon^{-1} \boldsymbol{\varepsilon} \right\}. \tag{4.147}$$

Thus the set of points of equal probability density are defined by the relation

$$\boldsymbol{\varepsilon}^T \mathbf{R}_\varepsilon^{-1} \boldsymbol{\varepsilon} = C. \tag{4.148}$$

The above equation defines an hyperellipsoid, also known in the literature as a concentration hyperellipsoid. This term stems from the property that the probability that the error vector lies inside the hyperellipsoid is

$$P = 1 - \exp\left\{ -\frac{C^2}{2} \right\}. \tag{4.149}$$

Demonstration of this property for the 2D case can be found in (Van Trees, 2001).

Thus, the Cramér–Rao lower bound (4.145) states that the concentration hyperellipsoid (4.146) lies either outside or on the bound hyperellipsoid defined by the relation

$$\boldsymbol{\varepsilon}^T \mathbf{J}_{\mathbf{Z},\mathbf{X}} \boldsymbol{\varepsilon} = C. \tag{4.150}$$

4.4.3 Akaike Information Criterion

So far we have assumed full knowledge of the deterministic and/or statistical rules that relate the quantity of interest and the observable. Nevertheless, in practice the first step of the analysis is the estimation of the model itself.

To distinguish the process of estimating the quantity of interest from the process of estimating the underlying model, we address this last task as *model identification*.

In principle, if there were no noise, and the relationships were deterministic, we could apply Ockham's razor, stating that between competing hypotheses one should select the simpler one, which in our case turns out to be the model with fewer parameters or, equivalently, fewer degrees of freedom.

In fact, we are always dealing with disturbances and noise, and a criterion to compare two approximate models has to be adopted. For this we may resort to criteria that combine the average cost and model complexity as the information criterion, first published by Hirotugu Akaike in 1974. The AIC (an information criterion) was originally formulated as an extension of the maximum likelihood criterion, and states that in the presence of several models we should select the one minimizing the quantity

$$AIC(\boldsymbol{\theta}^{(i)}) = -2 \ln \Lambda(\mathbf{z}/\mathbf{x}; \boldsymbol{\theta}^{(i)}) + 2k^{(i)}, \tag{4.151}$$

where

$$\ln \Lambda(\mathbf{z}/\mathbf{x}; \boldsymbol{\theta}^{(i)}) = \ln \frac{p_{\mathbf{Z}/\mathbf{X},\Theta^{(i)}}(\mathbf{z}/\mathbf{x}; \boldsymbol{\theta}^{(i)})}{V(\mathbf{z})} \tag{4.152}$$

is the loglikelihood functional, based on the i-th model, and $k^{(i)}$ is the number of independently adjusted parameters $\boldsymbol{\theta}^{(i)}$ of said model to get $\mathbf{x}$.

The rationale for Akaike's choice stems from the fact that, when the true model is unknown, the quantity $\ln \Lambda(\mathbf{z}/\mathbf{x}; \boldsymbol{\theta}^{(i)})$replaces the correct likelihood functional in the maximum likelihood procedure.

This leads to an error in the log likelihood functional evaluation equal to:

$$\Lambda(\mathbf{z}/\mathbf{x}) - \Lambda(\mathbf{z}/\mathbf{x}; \boldsymbol{\theta}^{(i)}) = \ln \frac{p_{\mathbf{Z}/\mathbf{X}}(\mathbf{z}/\mathbf{x})}{p_{\mathbf{Z}/\mathbf{X},\Theta^{(i)}}(\mathbf{z}/\mathbf{x}; \boldsymbol{\theta}^{(i)})}. \tag{4.153}$$

We note that the bias of the log likelihood functional approximation is therefore

$$\begin{aligned} E\left\{\Lambda(\mathbf{z}/\mathbf{x}) - \Lambda(\mathbf{z}/\mathbf{x}; \boldsymbol{\theta}^{(i)})\right\} &= \int\int p_{\mathbf{Z},\mathbf{X}}(\mathbf{z},\mathbf{x}) \ln \frac{p_{\mathbf{Z}/\mathbf{X}}(\mathbf{z}/\mathbf{x})}{p_{\mathbf{Z}/\mathbf{X},\Theta^{(i)}}(\mathbf{z}/\mathbf{x}; \boldsymbol{\theta}^{(i)})} d\mathbf{x} d\mathbf{z} \\ &= \int\int p_{\mathbf{Z},\mathbf{X}}(\mathbf{z},\mathbf{x}) \ln \frac{p_{\mathbf{Z},\mathbf{X}}(\mathbf{z},\mathbf{x})}{p_{\mathbf{Z},\mathbf{X};\Theta^{(i)}}(\mathbf{z},\mathbf{x}; \boldsymbol{\theta}^{(i)})} d\mathbf{x} d\mathbf{z}. \end{aligned} \tag{4.154}$$

This quantity is known in the information theory literature as the Kullback–Leibler divergence and accounts for the average information loss when the parametric statistical model replaces the true one.

As demonstrated by Akaike, for models with small Kullback–Leibler divergence, the bias asymptotically tends, apart from the sign, to the dimension of the parameter space k.

Thus, in the Akaike approach, the tradeoff between model accuracy and complexity is a direct consequence of the maximum likelihood approach. As suggested by Akaike himself in the original paper, the above criterion, purposely named *An* information criterion (AIC), can be adapted to different cost functions.

Among the other commonly used information criteria we cite the Bayesian information criterion (BIC) and Akaike's final error prediction (FEP) criterion.

In BIC, the functional to be minimized is

$$BIC(\boldsymbol{\theta}^{(i)}) = -2 \ln \Lambda(\mathbf{z}/\mathbf{x}; \boldsymbol{\theta}^{(i)}) + k^{(i)} \log N \tag{4.155}$$

where N is the size of the data set employed for the model identification (Schwarz, 1978). BIC penalizes complex models more heavily than AIC. In FEP, the log likelihood functional is replaced by the average squared L^2 norm of the prediction error, that is the fitting error when predicting the test data set. Let V be the loss functional computed on the observed data set

$$V(\boldsymbol{\theta}^{(i)}) = \det \left[\frac{1}{N} \sum_{m=1}^{N} \left\| \mathbf{z}_m - \hat{\mathbf{z}}\left(\hat{\mathbf{x}}; \hat{\boldsymbol{\theta}}^{(i)} \right) \right\|^2 \right]. \tag{4.156}$$

Then the FEP is defined as

$$FEP(\boldsymbol{\theta}^{(i)}) = V\left(\boldsymbol{\theta}^{(i)} \right) \frac{N+k}{N-k} \tag{4.157}$$

While BIC is a consistent estimator, that is it converges to the true values with probability 1, AIC may outperform BIC in the true model order selection for small data samples. A detailed comparison of AIC and BIC can be found in (Burnham and Anderson, 2004). We finally observe that the cited criteria can be extended in order to account for additional constraints on the computational complexity and required accuracy. For instance, we could select the model with minimal complexity subject to the constraint that the average weighted L^2 norm of the estimation error is below a given threshold specifying the required accuracy. Thus, in this case, instead of minimizing the norm of the estimation error, we minimize the system complexity.

Part II: Implementation

4.5 Analog to Digital Conversion

The results reported in the previous sections indicate that digital signal processing represents an effective means of manipulating analog signals in a fast, efficient and effective way with full equivalence, in the linear case of the results achievable by means of analog devices. For more details, visit the PhysioNet site (Goldberger *et al.*, 2000), that provides free web access to open-source algorithms with the associated collection of recorded physiologic signals (PhysioBank) and software (PhysioToolkit).

The results mentioned apply to the discrete time signals extracted from their continuous time counterparts by ideal sampling. However, in practice, the samples are represented by a finite number of bits. Therefore, one action to be performed in the design stage of the digital system is the specification of the analog to digital conversion.

Thus the next two sections are devoted first to the devices to be inserted to prevent contamination of the original signal with noise originated by high frequency components and second to the quantizer design.

4.5.1 Indirect Sampling versus Direct Sampling

In principle, if the signal to be converted into digital form contained only the signal of interest, sampling at time $k \times T$ could be performed just reading the value of the waveform at that time. However, in general, the useful signal is also contaminated by a wideband additive noise. Therefore sampling design has to take into account the effects produced on the noise components. For this we observe that equation (4.18) can also be written as

$$s(t) = \sum_{n=-\infty}^{+\infty} s\left(\frac{n}{2B}\right)\delta\left(t - \frac{n}{2B}\right) * \text{sinc}[2\pi Bt] \tag{4.158}$$

which states that the original waveform s(t) can be obtained by filtering the sampled signal

$$s^s_{\frac{1}{2B}}(t) = \sum_{n=-\infty}^{+\infty} s\left(\frac{n}{2B}\right)\delta\left(t - \frac{n}{2B}\right) = s(t)\sum_{n=-\infty}^{+\infty} \delta\left(t - \frac{n}{2B}\right) \tag{4.159}$$

with the low pass filter with impulse response $h(t) = \text{sinc}[2\pi Bt]$. On the other hand, denoting by $\pi_T(t)$ the periodic signal

$$\pi_T(t) = \sum_{n=-\infty}^{+\infty} \delta(t - nT), \tag{4.160}$$

its Fourier transform is

$$F\{\pi_T(t)\} = \frac{1}{T}\sum_{n=-\infty}^{+\infty} \delta\left(t - \frac{n}{T}\right). \tag{4.161}$$

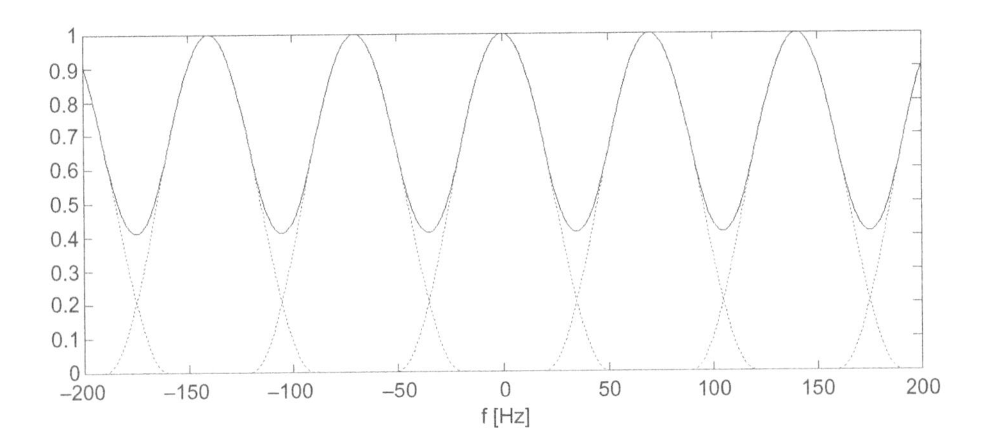

Figure 4.8 Sampling effect in the frequency domain (B = 50 Hz, 1/T = 70 Hz).

Consequently, for the Fourier transform of the sampled signal we have

$$F\{s_T^s(t)\} = F\{s(t)\pi_T(t)\} = \frac{1}{T}S(f) * \sum_{k=-\infty}^{+\infty} \delta\left(f - \frac{k}{T}\right) = \frac{1}{T}\sum_{k=-\infty}^{+\infty} S\left(f - \frac{k}{T}\right). \tag{4.162}$$

The above equation states that, after sampling the spectrum of the signal becomes periodic with period equal to 1/T. The aim of the ideal low pass reconstruction filter is therefore the removal of the high frequency replicas of the original spectrum.

However, as illustrated in Figure 4.8, when direct sampling is performed, if the original signal is affected by noise with higher bandwidth, its high frequency components are subject to aliasing effects, and ghost components in the low frequency band appear. As a consequence if a band-limited signal contaminated by high frequency noise is directly sampled, the output is affected by the high frequency noise folded into the low frequency band. On the other hand, the expansion coefficients of equation (4.18) can be interpreted as the generalized Fourier coefficients for which the following expression holds:

$$s\left(\frac{n}{2B}\right) = 2B\left\langle s(t), \mathrm{sinc}\left[2\pi B\left(t - \frac{n}{2B}\right)\right]\right\rangle = [s(t) * 2B\,\mathrm{sinc}(2\pi Bt)]_{t=\frac{n}{2B}} \tag{4.163}$$

Thus, to comply with the general rule for the computation of the expansion coefficients of a given signal with respect to an orthogonal basis, the samples should be taken at the output of an ideal low pass filter. This scheme is usually referred to in the literature as indirect sampling.

Since the signal is band-limited, it is not altered when passing through the ideal low pass filter, and direct and indirect sampling schemes can be considered fully equivalent with respect to the signal of interest. Different considerations apply to additive high frequency noise. In fact, while this kind of noise is filtered out by indirect sampling, its aliased replica is present at the output of the direct sampling. For this reason, indirect sampling, that is the insertion of a low pass filter before sampling, is always strongly recommended.

4.5.2 Quantizer Design

By their nature, analog signals may assume a continuum of values. On the other hand, digital processing implies that in the analog to digital conversion each sample is represented by a finite number of bits.

As illustrated in Figure 4.9, the real axis is partitioned into $N=2^R$ intervals $\{V_i = (\xi_{i-1}, \xi_i], i = 1, \ldots, N\}$with $\xi_0 = -\infty$ and $\xi_N = \infty$. The scalar quantizer then associates a value $\hat{x}_i$ to each point of the interval V_i.

Proper design of the partition $\{V_i\}$ and of the quantized values $\{\hat{x}_i\}$ requires, as first conceptual step, the definition of a ***distortion measure***$d(\mathbf{x}, \hat{\mathbf{x}})$

$$d : \mathbf{X} \times \hat{\mathbf{X}} \rightarrow R^+ \tag{4.164}$$

that models the cost associated with the replacement of the original value $\mathbf{x}$ with its quantized version $\hat{\mathbf{x}}$. Once the distortion measure has been specified, the design of the quantizer can be stated in one of the two following forms:

- Given a class of signals whose statistics are known and a set of possible quantizers $\{\hat{\mathbf{x}}(\mathbf{x}; R)\}$, each characterized by a predefined number R of bits and the associated *average distortion* $D(R)$:

$$D(R) = E_{\mathbf{X}}\{d[\mathbf{x}, \hat{\mathbf{x}}(\mathbf{x}; R)]\} \tag{4.165}$$

  determine the quantizer associated to the minimum *average distortion.*
- Given a class of signals, whose statistics are known, and a set of possible quantizers, determine the minimum number R of bits that can satisfy the constraint that the average distortion does not exceed a predefined value.

Although we have to determine both the partition $\{V_i\}$ and the quantized values $\{\hat{x}_i\}$ let us first determine the best quantization values associated with a given partition$\{V_i\}$.

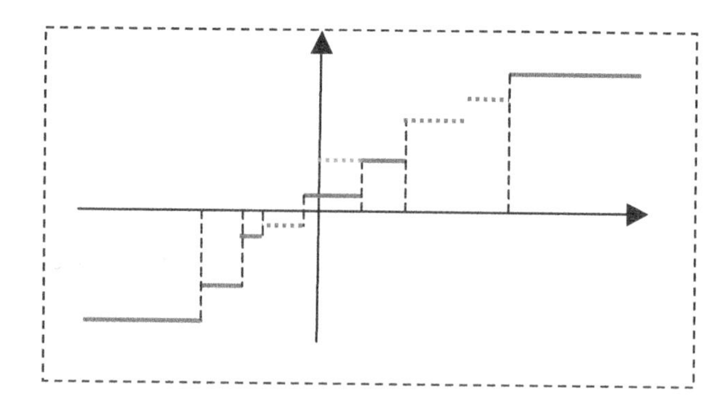

Figure 4.9 Scalar quantizer.

Due to its wide adoption, we evaluate the best quantization rule for the case of quadratic distortion. In this case, given the quantizer, the average distortion is

$$D(R) = \sum_{i=1}^{N} \int_{V_i} d(x, \hat{x}_i) p_X(x) dx = \sum_{i=1}^{N} \int_{V_i} (x - \hat{x}_i)^2 p_X(x) dx \tag{4.166}$$

Therefore the optimal quantized values $\{\hat{x}_i\}$ can be evaluated by imposing

$$\frac{\partial D(R)}{\partial \hat{x}_i} = -2 \int_{V_i} (x - \hat{x}_i) p_X(x) dx = 0 \tag{4.167}$$

which in turn implies that

$$\hat{x}_i = \frac{\int_{V_i} x p_X(x) dx}{\int_{V_i} p_X(x) dx} = \int_{V_i} x \frac{p_X(x)}{\int_{V_i} p_X(x') dx'} dx. \tag{4.168}$$

Thus, for any interval V_i the optimal quantization level is the conditional expectation of x, given the condition that the observed value lies in V_i. It reduces to the midpoint of the interval if the p.d.f. of x is constant in the given interval. The above result can be employed to devise an iterative optimization procedure known in the literature as the Max–Lloyd algorithm that allows determining both the partition $\{V_i\}$ and the quantized values$\{\hat{x}_i\}$.

Let $\left\{V_i^{(0)} = \left(\xi_{i-1}^{(0)}, \xi_i^{(0)}\right], i = 1, \ldots, N\right\}$ be the initial approximation of the optimal partition. We observe that, when no further a priori information is available, a uniform partition can be selected (i.e., partitioning with intervals of the same length the dynamic range of x). Then, for each interval V_i the optimal quantized values $\hat{x}_i$ are computed by applying equation (4.168). Then the endpoints of the interval are updated, setting them as midpoints of the intervals whose endpoints are the quantized values, as detailed below.

Max–Lloyd quantizer algorithm

1. *Set* $k = 0$
2. *Repeat*

   a. $\hat{x}_i^{(k+1)} = \dfrac{\int_{\xi_{i-1}^{(k)}}^{\xi_i^{(k)}} x p_X(x) dx}{\int_{\xi_{i-1}^{(k)}}^{\xi_i^{(k)}} p_X(x) dx}$

   b. $\xi_i^{(k+1)} = \dfrac{\hat{x}_i^{(k+1)} + \hat{x}_{i+1}^{(k+1)}}{2}$

   c. $k = k + 1$

   until $Max_i \left\{\left|\xi_i^{(k+1)} - \xi_i^{(k)}\right|\right\} > \Delta\xi$

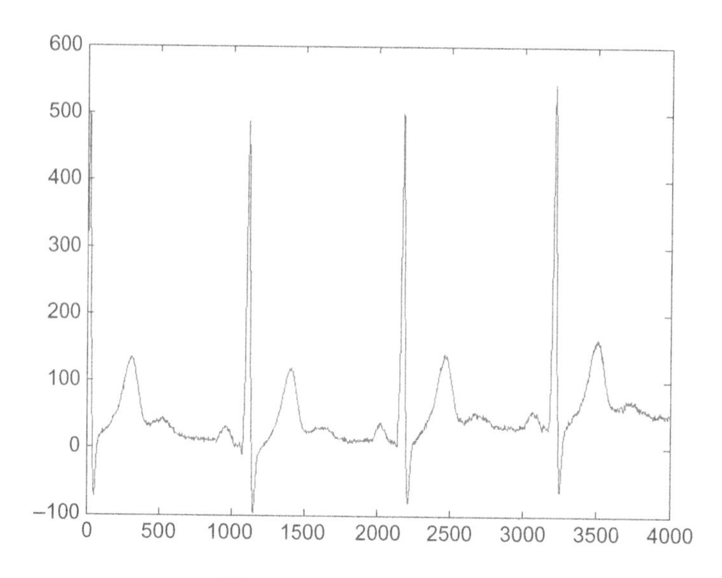

Figure 4.10 Original signal.

An example of the quantization effects on ECG signals versus the number of levels of the Max–Lloyd quantizer is shown in Figures 4.10–4.13.

Let us now evaluate the distortion in the reference case of signals uniformly distributed in the interval $[-A, A]$, that is

$$p_X(x) = \begin{cases} \frac{1}{2A} & -A \leq x \leq A \\ 0 & otherwise \end{cases} \tag{4.169}$$

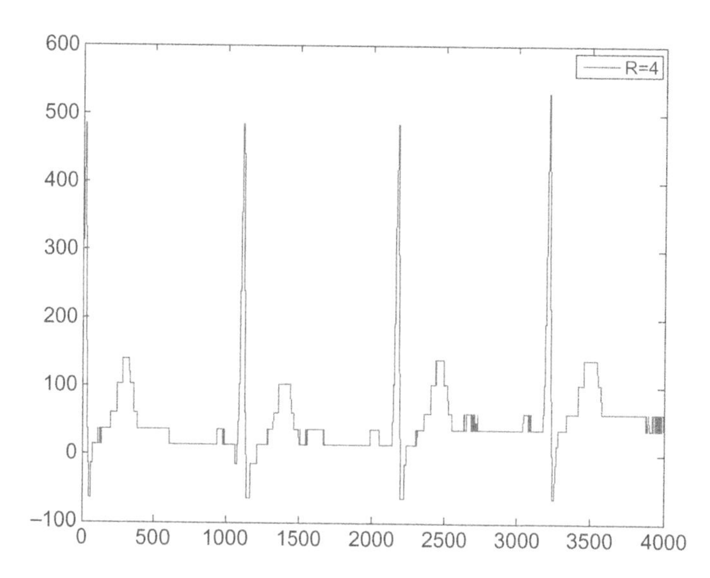

Figure 4.11 Max Lloyd quantizer R = 4.

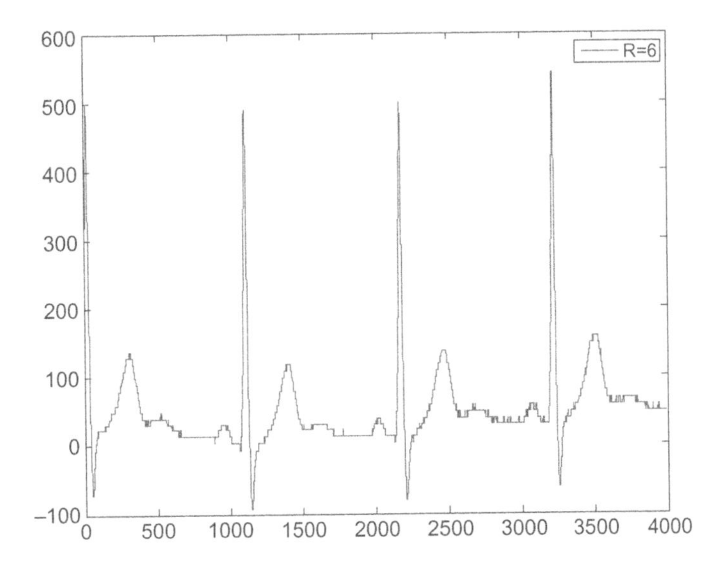

Figure 4.12 Max Lloyd quantizer R = 6.

Let Q_i be the size of the i-th interval. Then, since the p.d.f. is constant, the optimal quantized value is the midpoint of the interval,

$$\hat{x}_i = \xi_{-1} + \frac{Q_i}{2}. \tag{4.170}$$

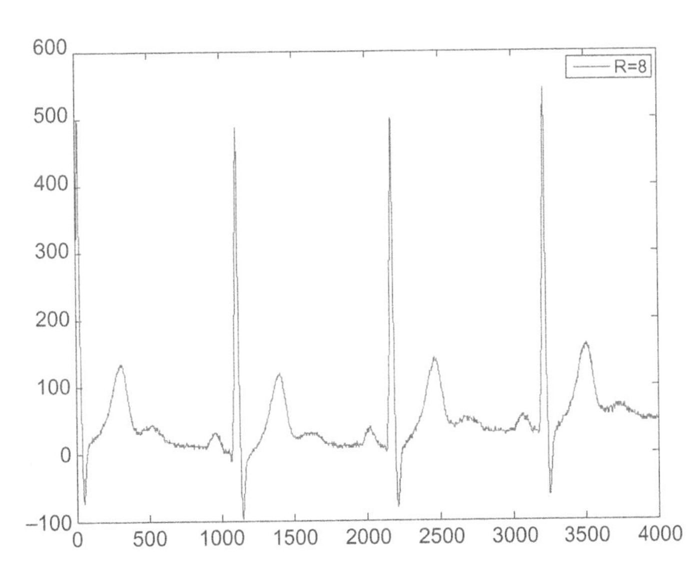

Figure 4.13 Max Lloyd quantizer R = 8.

Thus the average distortion is

$$D(R) = \frac{1}{2A}\sum_{i=1}^{N}\int_{\xi_{i-1}}^{\xi_{i-1}+Q_i}\left(x - \xi_{i-1} + \frac{Q_i}{2}\right)^2 dx = \frac{1}{2A}\sum_{i=1}^{N}\int_{-\frac{Q_i}{2}}^{\frac{Q_i}{2}} x^2 dx = \frac{1}{2A}\sum_{i=1}^{N}\frac{Q_i^3}{12}. \quad (4.171)$$

The optimal partition minimizing the average distortion with the additional constraint

$$\sum_{i=1}^{N} Q_i = 2A \quad (4.172)$$

can be computed by applying the Lagrange multipliers method, thus minimizing the functional

$$J = \frac{1}{2A}\sum_{i=1}^{N}\frac{Q_i^3}{12} + \lambda\left(2A - \sum_{i=1}^{N} Q_i\right), \quad (4.173)$$

where λ is the Lagrange multiplier. Thus differentiating J w.r.t. Qi we have

$$\frac{\partial J}{\partial Q_i} = \frac{Q_i^2}{8A} - \lambda = 0, \quad (4.174)$$

which in turn implies that

$$Q_i = \sqrt{8\lambda A}. \quad (4.175)$$

The value of the Lagrange multiplier can be computed by imposing the constraint (4.172), thus obtaining

$$\sum_{i=1}^{N}\sqrt{8\lambda A} = 2A. \quad (4.176)$$

Therefore,

$$\lambda = \frac{A}{2N^2}. \quad (4.177)$$

Thus the optimal partition is the uniform partition with intervals of size

$$Q_i = \frac{2A}{N}. \quad (4.178)$$

Consequently the mean squared error associated with the uniform quantizer is

$$D(R) = \frac{1}{2A}\sum_{i=1}^{N}\frac{Q_i^3}{12} = \frac{1}{2A}N\frac{8A^3}{12N^3} = \frac{A^2}{3N^2} = \frac{A^2}{3}2^{-2R}. \quad (4.179)$$

Expressing the MSE in dB we have

$$[D(R)]_{dB} = 10\log_{10}\frac{A^2}{3} - (20\log_{10}2) \times R \cong 10\log_{10}\frac{A^2}{3} - 6R. \tag{4.180}$$

Thus the MSE associated to the uniform quantization of a uniformly distributed signal decreases of about 6 dB for each additional bit employed in the analog to digital converter.

4.6 Signal Denoising

In the following, we will apply the estimators presented in the previous sections to the restoration of signals in additive noise. We begin with the simple one-dimensional case and then we extend the solution to the denoising of Gaussian signals in Gaussian, colored additive noise.

4.6.1 *White Gaussian Signals in Additive White Gaussian Noise*

Let us consider first the case of a single sample x whose observation z is affected by an additive Gaussian noise y.

If x is a sample from a Gaussian random variable, then for *MMSE* estimator, the following theorem holds.

Theorem 4.10. *Let* X *and* Y *be two statistically independent Gaussian random variables with expectation* m_x *and* m_y*, and variance* σ_x^2 *and* σ_y^2 *respectively. Let* z *be the observation consisting of their sum:* $z = x + y$.

Then the MMSE estimator $\hat{x}^{MMSE}(z)$*of* x *is*

$$\hat{x}^{MMSE}(z) = E_{X/Z}\{x/z\} = m_x + \frac{\sigma_x^2}{\sigma_x^2 + \sigma_y^2}(z - m_x - m_y) \tag{4.181}$$

and the variance of the estimator is given by

$$\sigma_{x/z}^2 = E_{X/Z}\left\{\left(x - m_{x/z}\right)^2/z\right\} = \frac{\sigma_x^2\sigma_y^2}{\sigma_z^2} = \frac{\sigma_x^2}{1 + \frac{\sigma_x^2}{\sigma_y^2}} \tag{4.182}$$

For the proof see (Van Trees, 2001). Let us remark that the MMSE estimator is a linear function that consists of two terms:

- the first term equals the expectation of X and therefore accounts for the a priori information on X.
- The second is proportional to the deviation of the observation z from its expected value given by $(m_x + m_y)$ with a gain proportional to the variance of the signal to be estimated, and reflects the information gained with the observation.
- The variance of the estimator goes to zero when the variance of the observation noise goes to zero.

4.6.2 *Denoising of Gaussian Cyclostationary Signals*

When both the useful signal and the observation noise are samples from mutually independent Gaussian wide-sense cyclostationary random processes, that is $z[m] = x[m] + y[m]$, for the DFTs of $x[m]$, $y[m]$, and $z[m]$ the results of Theorem 4.7 apply, and denoising can be performed separately for each frequency component. Since $Z[k] = X[k] + Y[k]$, the expected value of $X[k]$ given the DFT $\mathbf{Z}$ of the observation, can be computed as

$$E_{X/Z}\{X[k]/\mathbf{Z}\} = \frac{P_x[k]}{P_x[k] + P_y[k]} Z[k]. \tag{4.183}$$

Thus the MMSE solution for cyclostationary processes is obtained by filtering the observed signal by a filter with transfer function $H_W[k]$, in the discrete frequency domain equal to

$$H_W[k] = \frac{P_x[k]}{P_x[k] + P_y[k]}. \tag{4.184}$$

This filter can be considered the extension of the MMSE filter for analog stationary processes, introduced by Norbert Wiener in the 1940s (Wiener, 1949).

For both effectiveness and simplicity, the Wiener filter has been widely applied in the past decades for ECG signal denoising (Kestler, 1998; Lander *et al.*, 1997; Nikolaev and Gotchev, 2000; Smital *et al.* 2011).

4.6.3 *MMSE Digital Filter*

Although a general denoising theory for both analog and digital, stationary and not stationary, Gaussian signals in additive Gaussian noise can be formulated, since we are dealing with digital signals with finite temporal support, we specify the general results in the digital framework, assuming that only a finite number of observations is available.

Let $\mathbf{x}$ be a sample of an N-dimensional r.v. and $\mathbf{z}$ the observation consisting of a linear transformation of $\mathbf{x}$ plus a noise $\mathbf{v}$:

$$\mathbf{z} = \mathbf{Hx} + \mathbf{v}. \tag{4.185}$$

In addition, we assume that both $\mathbf{X}$ and $\mathbf{V}$ are statistically independent Gaussian random variables with expectation $\mathbf{m_X}$ and $\mathbf{m_V}$, and covariance matrices $\mathbf{P}$ and $\mathbf{R}_V$ respectively. In this case, as demonstrated in the following, the MMSE estimator $\hat{\mathbf{x}}^{MMSE}(\mathbf{z})$ of $\mathbf{x}$ reduces to

$$\hat{\mathbf{x}}^{MMSE}(\mathbf{z}) = \mathbf{m_X} + \mathbf{K}(\mathbf{z} - \mathbf{m_Z}), \tag{4.186}$$

where

$$\mathbf{K} = \mathbf{PH}^T\mathbf{R}_Z^{-1} = \mathbf{PH}^T\left(\mathbf{R}_V + \mathbf{HPH}^T\right)^{-1} \tag{4.187}$$

and

$$E\{\mathbf{z}\} = \mathbf{m_Z} = \mathbf{Hm_X} + \mathbf{m_V}. \tag{4.188}$$

In addition, the variance of the estimator is

$$\mathbf{P}_{\mathbf{X}/\mathbf{Z}} = (\mathbf{I} - \mathbf{KH})\mathbf{P}. \tag{4.189}$$

Equation (4.188) states that the MMSE estimate of $\mathbf{x}$ is the sum of two contributions, the first of which is the a priori expectation of $\mathbf{X}$, that is $\mathbf{m_X}$, and the second is proportional to the *innovation* defined as the difference between the current observation $\mathbf{z}$ and its a priori expectation.

It could be demonstrated that estimate (4.186) constitutes the linear least square estimate, even when both noise and signal of interest are non-Gaussian. In this case, (4.188) is optimal with respect to all linear transformations of the observation. Nevertheless a non-linear transformation exhibiting better performance may exist, while, when the signal of interest and the noise are Gaussian, the minimum of the L^2 norm of the estimation error is achieved by linear transformation (4.186).

Before demonstrating the validity of (4.186), let us examine some particular cases frequently appearing in applications. The simplest situation arises when the observation noise is null and the prior estimate has uncorrelated components of similar magnitude. This case can be modeled by posing

$$\mathbf{P} = \sigma_X^2\mathbf{I}, \quad \mathbf{m_V} = \mathbf{0}, \quad \mathbf{R}_V = \mathbf{0}. \tag{4.190}$$

Under the above hypotheses the gain matrix reduces to

$$\mathbf{K} = \mathbf{H}^T\left(\mathbf{H}\mathbf{H}^T\right)^{-1}, \tag{4.191}$$

which is, by definition, the Moore–Penrose pseudo-inverse matrix of $\mathbf{H}$. In particular, when the expectation of $\mathbf{X}$ is also null, the estimate becomes:

$$\hat{\mathbf{x}}^{MMSE}(\mathbf{z}) = \mathbf{H}^T\left(\mathbf{H}\mathbf{H}^T\right)^{-1}\mathbf{z}. \tag{4.192}$$

On the other hand, since the inversion of $\mathbf{HH}^T$can be prone to numerical errors and numerical inversion procedures stability, it is general practice to introduce, in any case, a zero mean Gaussian white noise, modeling at least errors introduced by analog to digital conversion and digital signal processing errors. This case corresponds to the following model:

$$\mathbf{P} = \sigma_X^2\mathbf{I}, \quad \mathbf{m_V} = \mathbf{0}, \quad \mathbf{R}_V = \sigma_V^2\mathbf{I}, \tag{4.193}$$

for which the gain matrix becomes:

$$\mathbf{K} = \mathbf{H}^T\left(\mathbf{H}\mathbf{H}^T + \gamma\mathbf{I}\right)^{-1} \tag{4.194}$$

with γ

$$\gamma = \frac{\sigma_V^2}{\sigma_X^2}. \tag{4.195}$$

We note that equation (4.194) coincides with the formula introduced by Andrey Tikhonov (Tikhonov, 1963, Tikhonov *et al.* 1998) for regularizing the inversion of ill-posed problems.

However, Tikhonov regularization does not necessarily imply a Bayesian approach, and several criteria have been proposed to set the regularization parameter γ. In the Bayesian context, γ should be set in accordance to the signal-to-noise ratio as stated by (4.195).

Theorem 4.11. *Let* $\mathbf{X}$ *and* $\mathbf{V}$ *be two statistically independent Gaussian random variables with expectation* $\mathbf{m_X}$ *and* $\mathbf{m_V}$*, and covariance matrices* $\mathbf{P}$ *and* $\mathbf{R}_V$ *respectively. Let* $\mathbf{z}$ *be the observation consisting of a linear transformation of* $\mathbf{x}$ *plus* $\mathbf{v}$: $\mathbf{z} = \mathbf{Hx} + \mathbf{v}$.

Then the MMSE estimator $\hat{\mathbf{x}}^{MMSE}(\mathbf{z})$ *of* $\mathbf{x}$ *is*

$$\hat{\mathbf{x}}^{MMSE}(\mathbf{z}) = \mathbf{m_X} + \mathbf{K}(\mathbf{z} - \mathbf{Hm_X} - \mathbf{m_V}), \tag{4.196}$$

where

$$\mathbf{K} = \mathbf{PH}^T\mathbf{R}_Z^{-1} = \mathbf{PH}^T\left(\mathbf{R}_V + \mathbf{HPH}^T\right)^{-1}, \tag{4.197}$$

and the variance of the estimator is given by

$$\mathbf{P}_{X/Z} = (\mathbf{I} - \mathbf{KH})\mathbf{P}. \tag{4.198}$$

For the proof see (Van Trees, 2001).

4.7 Time of Arrival Estimation

The term *time of arrival* originated in the sonar and radar contexts where acoustic and electromagnetic pulses are respectively transmitted in order to probe the surrounding medium. The presence of an object of interest is then revealed by the reception of a delayed replica of the transmitted pulse backscattered by the said object. Estimation of the time of arrival of the backscattered pulse is then the key factor for locating the object of interest. In fact, the object range can simply be computed by multiplying the difference between the estimated time of arrival and the time of transmission by the pulse propagation velocity in the given medium (supposed to be known). The same principle is employed in ultrasound medical imaging when an evaluation of the size of some internal organ is automatically extracted from the sonogram. Moreover, accurate localization, either in time or in space, of any signal is a recurrent problem in medical signal processing.

Let us now examine the case in which we want to estimate the time at which an event characterized by a signal $s(t)$ occurs, for example the case of the extraction of features describing the ECG waves, or to estimate the time interval between two consecutive heart beats through the QRS complex detection (Köhler *et al.*, 2002). The term 'time of arrival' is borrowed from the sonar and radar contexts where the related estimation problem has been formalized and optimal solutions investigated. Since ultrasound medical imaging is based on the same physical principles as sonar, the estimators investigated here are then applicable to the automatic extraction of features related to object size, such as the tissue thickness.

We assume that the measured signal $r(t)$ is affected by a stationary ergodiç, zero mean, additive white Gaussian noise $n(t)$. For the sake of compactness we examine the case in which both $s(t)$ and $n(t)$ are band-limited signals in the band $[-W, W]$.

Let θ be the time delay to be estimated. Then the observed signal can be written as

$$r(t) = s(t-\theta) + n(t). \tag{4.199}$$

where $n(t)$ is a sample of a zero mean, Gaussian white noise with power density spectrum $P_n(f)$ given by:

$$P_n(f) = \begin{cases} \mathcal{N}_0 & -W \leq f \leq W \\ 0 & otherwise \end{cases}. \tag{4.200}$$

We further assume that the estimate has to be performed based on the observation of the signal received in the interval $[-T_1, T_2] = [-M_1/2W, M_2/2W]$. Thus, let us denote

$$\mathbf{s}^\theta = \left[s^\theta_{M_1} \cdots s^\theta_0 \cdots s^\theta_k \cdots s^\theta_{M_2}\right]^T, \mathbf{r}^\theta = \left[r^\theta_{M_1} \cdots r^\theta_0 \cdots r^\theta_k \cdots r^\theta_{M_2}\right]^T, \mathbf{n}^\theta = \left[n^\theta_{M_1} \cdots n^\theta_0 \cdots n^\theta_k \cdots n^\theta_{M_2}\right]^T$$

whose $M = M_1 + M_2 + 1$ components are

$$s^\theta_k = s\left(\frac{k}{2W} - \theta\right), \quad n_k = n\left(\frac{k}{2W}\right) \quad z_k = z\left(\frac{k}{2W}\right). \tag{4.201}$$

In order to evaluate the maximum likelihood estimator (introduced in Section 4.3.4),

$$\hat{\theta} = Arg\left\{\underset{\theta}{Max} \ln \Lambda(\mathbf{r}/\theta)\right\}, \tag{4.202}$$

we observe that since: $\mathbf{r} = \mathbf{s}^\theta + \mathbf{n}$, from the statistical independence between $n(t)$ and θ, it follows that: $p_{\mathbf{R}/\Theta}(\mathbf{r}/\theta) = p_{\mathbf{N}}(\mathbf{r} - \mathbf{s}^\theta)$.

Furthermore since $\mathbf{n}$ is a sample of a zero mean Gaussian M-dimensional random variable with uncorrelated components, its p.d.f. is:

$$p_{\mathbf{N}}(\mathbf{n}) = \frac{1}{(2\pi)^{M/2}\sigma_N^M} e^{-\frac{1}{2\sigma_N^2}\mathbf{n}^T\mathbf{n}}, \quad \text{with } \sigma_N^2 = 2W\mathcal{N}_0.$$

Therefore

$$p_{\mathbf{R}/\Theta}(\mathbf{r}/\theta) = \frac{1}{(2\pi\sigma_N^2)^{M/2}} e^{-\frac{\left[\mathbf{r}-\mathbf{s}^\theta\right]^T\left[\mathbf{r}-\mathbf{s}^\theta\right]}{2\sigma_N^2}}. \tag{4.203}$$

Thus, setting $V(\mathbf{r})$ in (4.89) as

$$V(\mathbf{r}) = \frac{1}{(2\pi\sigma_N^2)^{M/2}}, \tag{4.204}$$

the loglikelihood functional reduces to

$$\begin{aligned} \ln \Lambda(\mathbf{r}/\theta) &= \ln \frac{p_{\mathbf{R}/\Theta}(\mathbf{r}/\theta)}{V(\mathbf{r})} = -\frac{[\mathbf{r}-\mathbf{s}^\theta]^T[\mathbf{r}-\mathbf{s}^\theta]}{2\sigma_N^2} \\ &= -\frac{1}{2\sigma_N^2}\sum_{i=-M_1}^{M_2}(r_i - s_i^\theta)^2. \end{aligned} \tag{4.205}$$

Considering that for a band-limited signal $x(t)$ the following relationship holds

$$\frac{1}{2W}\sum_{i=-\infty}^{\infty}|x_i|^2 = \int_{-\infty}^{+\infty}|x(t)|^2 dt, \tag{4.206}$$

when $M_1 \to \infty$ and $M_2 \to \infty$
we have

$$\begin{aligned} \ln \Lambda[r(t)/\theta] &= \ln \frac{p_{\mathbf{R}/\Theta}(r(t)/\theta)}{V[r(t)]} = -\frac{1}{2\frac{\sigma_N^2}{2W}}\int_{-\infty}^{+\infty}|r(t)-s(t-\theta)|^2 dt \\ &= -\frac{1}{2\mathcal{N}_0}\int_{-\infty}^{+\infty}|r(t)-s(t-\theta)|^2 dt, \end{aligned} \tag{4.207}$$

or equivalently

$$\ln\ \Lambda[r(t)/\theta] = -\frac{1}{2\mathcal{N}_0}\left\{\int_{-\infty}^{+\infty}|r(t)|^2 dt + 2\int_{-\infty}^{+\infty} r(t)s(t-\theta)dt - \int_{-\infty}^{+\infty}|s(t-\theta)|^2 dt\right\} \tag{4.208}$$

On the other hand, since the signal energy is shift invariant, that is

$$\int_{-\infty}^{+\infty}|s(t-\theta)|^2 dt = \int_{-\infty}^{+\infty}|s(t)|^2 dt, \tag{4.209}$$

the maximum likelihood estimate of the time of arrival corresponds to the time instant for which the correlation between the received and the transmitted signal is maximum:

$$\begin{aligned} \hat{\theta} &= Arg\left\{\underset{\theta}{Max}\ln \Lambda(\mathbf{r}/\theta)\right\} = Arg\left\{\underset{\theta}{Min}\int_{-\infty}^{+\infty}|r(t)-s(t-\theta)|^2 dt\right\} \\ &= Arg\left\{\underset{\theta}{Max}\int_{-\infty}^{+\infty} r(t)s(t-\theta)dt\right\}. \end{aligned} \tag{4.210}$$

Let us now compute the Cramér–Rao lower bound. Considering that

$$\begin{aligned}\frac{\partial \ln p_{\mathbf{R}/\Theta}(r(t)/\theta)}{\partial \theta} &= \frac{1}{\mathcal{N}_0}\frac{\partial}{\partial \theta}\int_{-\infty}^{+\infty} |r(t)-s(t-\theta)|^2 dt \\ &= -\frac{1}{\mathcal{N}_0}\int_{-\infty}^{+\infty} [r(t)-s(t-\theta)]\dot{s}(t-\theta)dt = -\frac{1}{\mathcal{N}_0}\int_{-\infty}^{+\infty} n(t)\dot{s}(t-\theta)dt,\end{aligned} \tag{4.211}$$

we obtain

$$\begin{aligned}E_{R,\Theta}\left\{\left[\frac{\partial \ln p_{\mathbf{R}/\Theta}(r(t)/\theta)}{\partial \theta}\right]^2\right\} &= \frac{1}{\mathcal{N}_0^2}E_{R,\Theta}\left\{\left[\int_{-\infty}^{+\infty} n(t)\dot{s}(t-\theta)dt\right]^2\right\} \\ &= \frac{1}{\mathcal{N}_0^2}E_{R,\Theta}\left\{\int_{-\infty}^{+\infty} n(t_1)\dot{s}(t_1-\theta)dt_1 \int_{-\infty}^{+\infty} n(t_2)\dot{s}(t_2-\theta)dt_2\right\} \\ &= \frac{1}{\mathcal{N}_0^2}\int_{-\infty}^{+\infty}\int_{-\infty}^{+\infty} E_{R,\Theta}\{n(t_1)n(t_2)\}\dot{s}(t_1-\theta)\dot{s}(t_2-\theta)dt_1dt_2,\end{aligned} \tag{4.212}$$

On the other hand, since $n(t)$ is supposed to be stationary and ergodic, from the Wiener–Kintchin theorem we have

$$E_{R,\Theta}\{n(t_1)n(t_2)\} = F^{-1}\{P_n(f)\} = 2W\mathcal{N}_0\text{sinc}[2\pi W(t_2-t_1)]. \tag{4.213}$$

Furthermore, the time derivative band-limited signal is a band-limited signal with the same bandwidth so that it is invariant with respect to ideal low pass filtering, that is

$$\dot{s}(t) = \dot{s}(t) * 2W\text{sinc}[2\pi Wt], \tag{4.214}$$

which, in turn implies that

$$\int_{-\infty}^{+\infty} \dot{s}(\tau-\theta)\text{sinc}[2\pi W(t-\tau)]d\tau = \frac{1}{2W}\dot{s}(t-\theta). \tag{4.215}$$

Substitution of (4.213) and (4.215) into (4.212) finally yields

$$\begin{aligned}&E_{R,\Theta}\left\{\left[\frac{\partial \ln p_{\mathbf{R}/\Theta}(r(t)/\theta)}{\partial \theta}\right]^2\right\} \\ &= \frac{2W}{\mathcal{N}_0}\int_{-\infty}^{+\infty}\int_{-\infty}^{+\infty} \text{sinc}[2\pi W(t_1-t_2)]\dot{s}(t_2-\theta)dt_2\dot{s}(t_1-\theta)dt_1 \\ &= \frac{1}{\mathcal{N}_0}\int_{-\infty}^{+\infty} \dot{s}(t_1-\theta)\dot{s}(t_1-\theta)dt_1 = \frac{1}{\mathcal{N}_0}\int_{-\infty}^{+\infty} |\dot{s}(t)|^2 dt = \frac{1}{\mathcal{N}_0}E_{\dot{s}},\end{aligned} \tag{4.216}$$

where

$$\mathcal{E}_{\dot{s}} = \int_{-\infty}^{+\infty} |\dot{s}(t)|^2 dt, \tag{4.217}$$

is the energy of the time derivative of the signal.

Thus for the mean squared error of the time of arrival estimate the following lower bound holds:

$$E\left\{ \left[\hat{\theta}(\mathbf{z}) - \theta\right]^2 \right\} \geq \frac{\mathcal{N}_0}{\mathcal{E}_{\dot{s}}}. \tag{4.218}$$

We finally observe that, due to the Parseval theorem, denoting with $S(f)$ the Fourier transform of $s(t)$ the energy of the signal derivative can be written as

$$\mathcal{E}_{\dot{s}} = \int_{-\infty}^{+\infty} |\dot{s}(t)|^2 dt = \int_{-\infty}^{+\infty} |j2\pi f|^2 |S(f)|^2 df = 4\pi^2 \bar{f^2}\, \mathcal{E}_s \tag{4.219}$$

where

$$\bar{f^2} \stackrel{def}{=} \frac{\int_{-\infty}^{+\infty} f^2 |S(f)|^2 df}{\int_{-\infty}^{+\infty} |S(f)|^2 df} = \frac{\int_{-\infty}^{+\infty} f^2 |S(f)|^2 df}{\mathcal{E}_s} \tag{4.220}$$

is the mean squared effective signal bandwidth, also denoted in literature as the second-order moment of the signal spectrum.

Therefore the Cramér–Rao lower bound can also be written in the equivalent form

$$E\left\{ \left[\hat{\theta}(\mathbf{z}) - \theta\right]^2 \right\} \geq \frac{\mathcal{N}_0}{4\pi^2 \bar{f^2}\, \mathcal{E}_s}, \tag{4.221}$$

which states that the lowest inaccuracy is strictly related to the inverse of the mean squared effective bandwidth of the signal. In other words the larger the effective signal bandwidth, the smaller is the mean squared error of the estimate, for a given signal-to-noise ratio $\mathcal{E}_s/\mathcal{N}_0$.

Thus, given a requirement specified in terms of maximum acceptable variance (or standard deviation) of the time localization error, the above equation states that, when designing some new instrumentation only three elements impact on the fulfillment of that requirement:

- **Sensor noise characteristics:** This factor can be partially mitigated by inserting a low noise amplifier as the first stage of the instrumentation front-end. Since noise power introduced by cables is proportional to their length, when expressed in dBs, ideally one should integrate a low noise amplifier in the probe, whenever possible.
- **Energy of the transmitted signal:** Since the energy of the received signal appearing in the Cramér–Rao lower bound is proportional to the energy of the transmitted pulse,

receiver noise negative effects can be reduced by increasing the transmitter power, since performance depends on the ratio of the received signal power (dependent on the transmitted power) and the power of the noise introduced by the receiver itself. Nevertheless, constraints on the maximum radiated power imposed by national regulations exist, taking into account potential effects on patients' health.

- **Effective bandwidth:** Limitations on the effective bandwidth that can be employed may depend on the physical nature of the signals employed. For instance, acoustic waves at different frequencies are subject to different attenuations by human tissues (selective medium). Moreover, in the case of radio signals, use of the frequency spectrum is regulated by national and international authorities that assign frequency bands to different services and systems. In addition electromagnetic compatibility with instrumentation sharing the same frequency band and/or bound on interference caused to other systems operating in the adjacent frequency band has to be accounted for when setting the band.

References

Akaike, Hirotugu (1974), 'A new look at the statistical model identification'. IEEE Transactions on Automatic Control 19 (6): 716–723.

Burnham, K. P. and Anderson, D.R. (2004), 'Multimodel inference: understanding AIC and BIC in Model Selection', Sociological Methods and Research, 33: 261–304.

Fisher R. A. (1921), 'On the Mathematical Foundations of Theoretical Statistics', Philosophical Transactions of the Royal Society of London. Series A, Containing Papers of a Mathematical or Physical Character, vol. 222, (1922), pp. 309–368 Published by: The Royal Society, Stable URL: http://www.jstor.org/stable/91208

Goldberger A. L., Glass L, Hausdorff J. M., Ivanov P. Ch., Mark R. G., Mietus J. E., Moody G. B., Peng C.-K., Stanley H. E. (2000), PhysioBank, PhysioToolkit, and PhysioNet: Components of a New Research Resource for Complex Physiologic Signals. Circulation [Circulation Electronic Pages: http://circ.ahajournals.org/cgi/content/full/101/23/e215] Accessed June 2000. (https://www.physionet.org)

Heideman, M., Johnson, D. and Burrus, C. (1984), 'Gauss and the history of the fast Fourier transform,' ASSP Magazine, IEEE, vol. 1, no. 4, pp. 14–21.

JCGM 200:2008 (2008). International vocabulary of metrology – Basic and general concepts and associated terms (VIM). URL: http://www.bipm.org/utils/common/documents/jcgm/JCGM_200_2008.pdf

Kestler H.A. *et al.*, (1998), Denoising of high-resolution ECG-signals by combining the discrete wavelet transform with the Wiener filter, Proceedings, IEEE Conf. Comput. Cardiology,

Köhler B. U., Hennig, C., Orglmeister, R. (2002), The Principles of Software QRS Detection, IEEE Engineering in Medicine and Biology Magazine.

Lander P. *et al.,* (1997). Time-frequency plane Wiener filtering of the high-resolution ECG: Background and time-frequency representations, IEEE Trans. Biomed. Eng.

Nikolaev N. and Gotchev A. (2000), ECG signal denoising using wavelet domain Wiener filtering,' Proc. Eur. Signal Process. Conf. EUSIPCO- 2000, Tampere, Finland,

Schwarz, Gideon E. (1978), Estimating the dimension of a model. Annals of Statistics 6 (2): 461–464. DOI: 10.1214/aos/1176344136

Smital L. Vítek M., and Kozumplík J. (2011), Optimization of the wavelet Wiener filtering for ECG signals. In Proceedings of the 4th International Symposium on Applied Sciences in Biomedical and Communication Technologies (ISABEL '11).

Tikhonov A.N. (1963), The solution of ill-posed problems, Doklady Akad. Nauk SSSR, 151.

Tikhonov A., Leonov A., Yagola A., (1998), Non-linear ill-posed problems, Chapman and Hall, London.

Van Trees, H. L. (2001). Detection, Estimation, and Modulation Theory, Part I, John Wiley & Sons.

Wiener N. (1949), Extrapolation, Interpolation, and Smoothing of Stationary Time Series. New York: Wiley.

Wong E. Hajek B. (1985), Stochastic processes in engineering systems. Springer. 1985.

5

Applied Electronics

The devil is in the details

Chapter Organization

The material of this chapter is organized following the **signal** flow of a typical electronic (medical) device from the input sensors to the output platform that digitally processes and presents the sensor information. The input signal is elaborated through processing stages (functions) common to most of electronic devices. Referring to Figure 5.1, each stage is

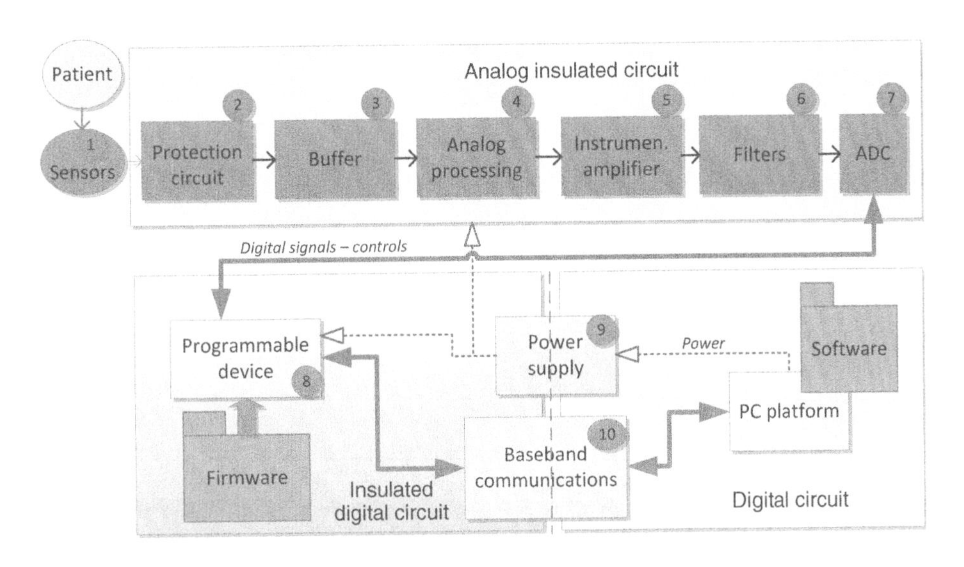

Figure 5.1 Chapter 5 structure (Medical device architecture).

Medical Instrument Design and Development: from Requirements to Market Placements, First Edition.
Claudio Becchetti and Alessandro Neri.

addressed in a specific section of the theory part identified by the circled number in the theory section. The implementation part has a corresponding section with the same unit number relating to the Gamma Cardio CG hardware stage. The Gamma Cardio CG from Gamma Cardio Soft (*www.gammacardiosoft.it*) is a 12-lead diagnostic product placed on the market whose details are fully disclosed in this book.

The electronic, telecommunication and signal theory concepts are introduced in this chapter when required by the processing stages. Through this approach, the reader may correlate more easily how core subjects are applied to typical electronic stages. For example, interferences affect more deeply the first input stages where the signal is not amplified and may be comparable with noise. Interference is then discussed in the circuit protection section (5.2) that is usually the first stage of a device. In the same section, there is also a reference to the basic concepts of resistors, capacitors, inductances, voltage clamping devices and Printed Circuit Boards (PCB) that are the building blocks of circuit protection circuits. The discussed topics are introduced according to the following logical order:

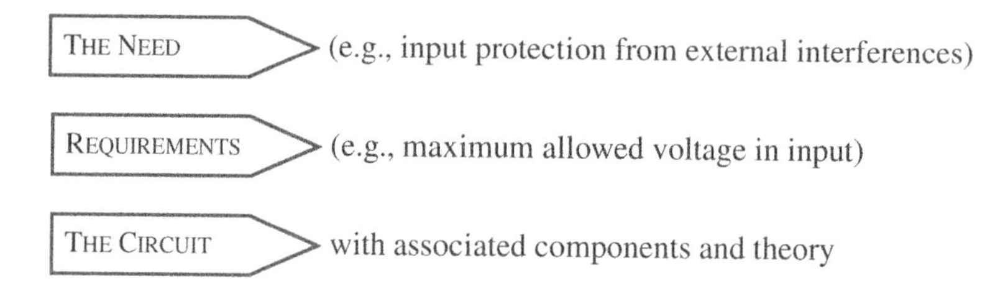

Part I: Theory

This chapter deals with electronic design in medical devices. Medical systems usually consist of specialized electronic boards controlled by the firmware embedded in programmable chips (microcontrollers etc.) and by software applications running on PC platforms.

Figure 5.1 shows an example where the specialized electronic board is represented by the upper analog circuit. Analog signals are translated into their digital counterpart by analog to digital converters that may be embedded in the programmable devices that may also perform digital signal processing. The digital signal is then transferred to the PC platform where the software performs low complexity processing and graphic presentation.

For cost and design reasons, medical device functions are preferably performed by software and firmware, while electronic components are always more limited. Software usually manages the user interface and the electronic microcontrollers logic.

This **system architecture** based on **mixed hardware–firmware–software** with a major element of software are driven by cost and time reduction in the design and manufacturing. Circuits with dedicated electronics have higher development cost in and add extra cost to the manufactured product. In turn, electronics offers higher computational performance and the capability of implementing some specific complex functions. Over the past few years, firmware and software have increased their performance in implementing functions with reduced development effort. The trend in complex electronic devices is then to shift functionalities to software when possible. Functions that require higher computational power or real-time response may be implemented in firmware given that more recent microcontrollers embed some analog functions (ADC, DAC, comparators etc.). Functions that cannot be implemented in firmware are now implemented using programmable chips (FPGAs, PSoC, Microcontrollers). Finally, functions that cannot be digitalized or cannot be handled by programmable logic are implemented through in ASICs or dedicated electronic circuits.

The ideal approach that is now envisaged in complex electronic systems consists of extending the **PC-based platform** to perform as many of the functionalities as possible, limiting the electronic dedicated circuits to functions that cannot be implemented in any other way. The PC platform is now computationally powerful and inexpensive for most of the applications, with reduced cost in software development thanks to tools such as rapid application development (RAD). This means that the role of dedicated electronics in recent years has been always more limited and focused on specific functions (analog front-end, fast processing, circuit protection etc.). **Network-centric architecture** is also growing. Devices such as medical instruments are expected to be conceived in two parts. The sensor part observing the patient is connected to the intelligence of the device through the network. The intelligence is going to be limited to the software applications running on the network while the sensor component is going to be reduced. This approach not only reduces costs but means that medical services can be offered where it is needed (the patient's own location), decoupling that from the places where the service is supplied (the physician's facility). This telemedicine approach, strongly supported by national authorities, has clear benefits, but also has extra concerns related to safety and compliance compared to usual clinical practice. These topics will be discussed in Chapter 7.

In this chapter, we will focus on the electronic functions that are to be implemented by the dedicated electronics (circuit protection, analog front-end, amplifiers, analog processing, filtering, power etc.). These functions are usually included somehow in most (medical) devices/instruments. In their application guides, vendors suggest specific implementations for electronic functions of medical devices and the associated components to be used (Texas, 2010; Analog Device, 2011a). The electronic functions will be addressed in a logical order that reflects the flow of the signal within a typical complex electronic device: from sensors and input protection via analog processing, toward ADC and microprocessor.

In real electronic design, the solution is often a **tradeoff** between the various performance to be achieved. For example, improving the protection immunity may increase noise and possibly decrease the input impedance. Unfortunately, all these three parameters (impedance, immunity and noise) have to be over a specific threshold as for example in the ECG (AAMI, 1991).

Note also that the real behavior of a component may have a negative impact on the circuit performance if disregarded. A simple example is the component tolerance. The real component values (capacitance, resistance etc.) is given by: nominal value $\times$ $(1 \pm \text{tolerance})$. Tolerance has a big impact on filters where small changes in value may greatly change frequency and phase response. In the case of the ECG, a wrong phase response may in turn change the shape of a wave, leading to a wrong interpretation. This simple example shows that not considering real behavior of components may generate unacceptable risks in terms of diagnosis with an associated incorrect treatment.

Real circuits often fail because **real component behavior** is not taken into account – that is, *'the devil is in the details'*. This chapter will provide recap boxes that highlight variations between ideal and real component behavior. Table 5.1 outlines suggestions for electronic design taking into account what already discussed.

By the end of this chapter, the reader should be familiar with the design of some of the main electronic functions and the associated electronic components. The reader should understand how these functions are used in a real product (the Gamma Cardio CG) in the implementation part. For sake of clearness, Table 5.2 summarizes the main symbols used in this chapter.

Table 5.1 The main steps in electronic design

1. Apply System-subsystem decomposition as explained in Chapter 1 to design simple subsystems
2. Use **proven** electronic **solutions** often included in application reports, datasheets, demo boards and application boards
3. Assess circuit performance considering **ideal** device **behavior**
4. Evaluate **real** circuit **performance** taking into account the real behavior of all devices
5. Compute real performance through **simulations**
6. Modify circuit and **iterate** from step 3 until performance requirements are met
7. Test performance with prototypes and demo boards before finale design

Table 5.2 Symbols and their meaning

Symbol	Meaning (symbol of measuring unit)	Symbol	Meaning (symbol of measuring unit)
$\|a\|$	absolute value of a	μ	micro scale unit $= \times 10^{-6}$
AC	alternating Current	m	milli scale unit $= \times 10^{-3}$
C	capacitance in farads (F)	M	magnitude of sine wave signal
dB	decibels usually in terms of power: $dB = 10\log_{10}\left((A_1)^2/(A_2)^2\right) = = 20\log_{10}(A_1/A_2)$ where A_1 and A_2 are the amplitude of the two signals	n	nano scale unit $= \times 10^{-9}$
DC	direct current	P	power in watt (W)
$e^{j\varphi}$	complex exponential function (Euler's formula): $e^{j\varphi} = \cos(\varphi) + j\sin(\varphi)$, $j = \sqrt{-1}$	p	pico scale unit $= \times 10^{-12}$
E	energy in joule (J)	Q	charge in coulombs
f	frequency in hertz (Hz)	$\Re(H(f))$	real part of the complex $H(f)$ function
f	femto scale unit $= \times 10^{-15}$	R	resistance in ohms (Ω)
φ	Phase of sine wave signal in radians	*rms*	root mean square
G	giga scale unit $= \times 10^9$	sec.	seconds
I, i	current in ampere (A)	T_k	temperature in kelvin (K)
$\Im(H(f))$	imaginary part of the complex $H(f)$ function	T	tera scale unit $= \times 10^{12}$
j	$j = \sqrt{-1}$	V	voltage in volts (V)
k	kilo scale unit $= \times 10^6$	Z	impedance in ohms (Ω)
L	inductance in henries (H)		

5.0 Architectural Design

We have already looked at Figure 5.1, which shows a typical simplified block diagram of an electronic (medical) device. Each block has a circled number which relates to a specific section in the theory section of this chapter. A corresponding section related to the Gamma Cardio CG hardware is described in the implementation part.

Referring to Figure 5.1, the overall electrical circuit can be divided into three parts: the digital circuit connected to the PC (bottom right side), and the two analog and digital insulated circuits that are separated from the non insulated circuit by a galvanic insulation (dotted line). This avoids a direct connection between the patient and the power distribution system even in presence of faults.

In the following, we outline a simple description of the architecture of Figure 5.1 indicating the sections where the topic are discussed.

The patient's vital parameters are transformed by **sensors** into electronic signals. Sensors are connected through a **protection circuit** that protects hardware and patient from overvoltage and interferences. A **buffer** stage is devoted to creating high impedance at the instrument input to avoid measurement errors as explained in Section 5.3. Input signals have to be combined together to obtain the target measurement signals. The **analog processing** network determines the proper combination of signals as described in Chapter 2 and in section 5.4. **Instrumentation amplifiers** increase target signals amplitude eliminating common mode interference (Section 5.5). **Filters** (Section 5.6) reduce other out of band interferences. The Analog to Digital device (Section 5.7) converts the continuous signal to a discrete time digital sequence that is sent to the **programmable device** (Section 5.8) through digital lines indicated in bold in Figure 5.1. The programmable device controls the analog circuit part through the control lines and sends the input signals to the PC platform through proper **communication** components (Section 5.10). The firmware that is loaded into the programmable component performs all the required logic processing. The parts that are connected to the patient are galvanically isolated through suitable devices (optocouplers, transformers). The **power** module (Section 5.9) has to provide power to the board and to isolate patient connected circuits. Power lines are indicated as dotted lines in Figure 5.1. Through the isolation barrier, the digital data are sent to the PC platform from which the application software controls the hardware and shows the sensors' inputs through the user interface. The next sections introduce the theory behind these modules.

5.1 Sensors

THE NEED A physiological system may be assessed through the measurement of specific biomedical quantities. Such quantities are usually estimated through the measurement of the six types of energy:

- mechanical
- magnetic
- thermal (temperature)
- electrical
- chemical
- radiation

For example, the electrical activity of the heart is assessed through measuring the electrical energy of the biopotentials generated by the heart itself. Medical electronic instruments are usually designed to estimate electrical variations. So a sensor is used, which is a device that converts a medical quantity evaluable by a specific energy emission into an electronic signal suitable for measurement. Sensors may be active or passive. Passive sensors measure directly one of the six types of energy. Active sensors supply energy to the system under evaluation.

Sensors can be considered as the interface between the biological environment and the medical device. This suggests that *the overall device performance is limited by the characteristic of the sensor* itself. Sensors are therefore critical to the medical instrument performance and, in addition, they may also impact on patient safety. Table 5.3 outlines the conceptual scheme to understand the role of the medical sensors in the case of biopotentials evaluation. The physiological subsystem is assessed by measuring biopotentials. A sensor, in

Table 5.3 Medical sensor model

Analyze ↓ Physiological subsystem	Through a ↓ Measured quantity (biopotentials)	Translated by ↓ Sensors	Using a ↓ Conversion process	That yields a ↓ Signal variation	Measured by a ↓ Medical instrument
Heart	0.5–4 mV	Electrodes	Chemical reaction that transforms body ionic current into electronic current	Electrical potential	ECG Electro-cardiography
Brain	5–0.3 mV				EEG Electro-encephalography
Stomach	10–1000 μV				EGG Electro-gastrography
Skeletal muscles	0.1–5.0 mV				EMG Electr-omyography
Eyes (retina resting potential)	50–3500 μV				EOG Electro-oculography
Eyes (retina)	50–900 μV				ERG Electro-retinography

this case an electrode, translates biopotentials into an electric signal that is measured by a specific medical instrument. The electrodes rely on a chemical reaction that transforms the body's ionic current into electronic current. The same conceptual scheme applies when the measured quantity is a mechanical, chemical or biochemical property. For example, the non-invasive blood pressure is measured through the pressure variation that can be translated by a sensor into an electric signal variation; this signal is then displayed by a sphygmomanometer. Sensors can use a range of different conversion techniques. For example, temperature may be estimated by relying on the properties of specific materials (i.e., metals or semiconductors) which change resistance with temperature. Other physical properties may be used, such as the Seebeck effect, thermocouples, capacitive variations and infrared radiations as discussed below.

Electrical signals are the most convenient input for medical instruments, since electronic variations can be easily processed by analog circuits and then digitalized for further hardware or software processing. More details on sensors for medical device can be found in (Bronzino, 2006).

REQUIREMENTS Figure 5.2 shows a simple model of a sensor. The medical input quantity is transformed into an electronic signal using a conversion technique that relies on some physical property. The ideal conversion will yield an output electrical signal proportional to the input quantity variations. Real measures include errors that affect the output. Such errors are also influenced by external factors b_1, b_2 . . . , as for example, the temperature or the external electromagnetic noise.

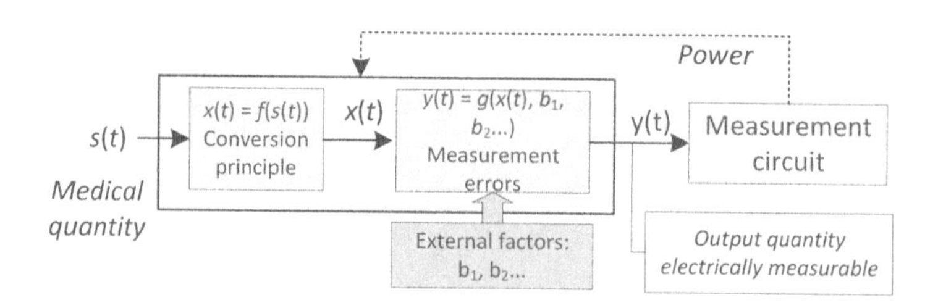

Figure 5.2 Sensor conceptual scheme.

Table 5.4 shows typical physical properties used to convert input quantities into electric signals. The input measurand (i.e., the quantity to be measured) is converted into an electronic measurable physical property, and a specific formula models the input–output relationship.

In the following, we detail some examples related to **temperature measurement** to clarify the conversion principles.

From Table 5.4, **temperature** may be measured by metal *thermistors* whose resistance changes with temperature. Temperature variation $T - T_0$ is converted into resistance R according to the formula 5.1, where R_0 is the value at temperature $T_0 = 27°\text{C}$ and TCR is the temperature coefficient of resistance dependent on the specific temperature (see equation (5.86) for an application):

$$R_T = R_0 \times (1 + \text{TCR} \times (T - T_0)) \tag{5.1}$$

Semiconductors, by contrast, have a non-linear formula linking resistance to temperature.

Thermocouples rely on the Seebeck effect. When two different conductors/semiconductors are in contact and the two points of contacts are at different temperature T_A and T_B, an electrical potential ΔV is generated between of the two points of contacts. The value of the potential may be modeled by the formula

$$\Delta V = K \times (T_A - T_B), \tag{5.2}$$

where K depends on the nature of the two materials in contact. Temperature variations are also experienced by *capacitors*. The capacitance value can be roughly considered as linearly proportional to temperature. The capacitance variation may be measured as a frequency variation when the specific capacitor is included in a resonant circuit. The same approach may be used with a *quartz crystal* that may be produced with a resonant frequency that is linearly dependent on the temperature. Oscillators using frequency-dependent quartz/capacitors have the benefit that frequency variations are easily and accurately measured by all-digital circuits (i.e., counters). This helps in producing simple digital instruments such as digital thermometers.

Digital thermometry may also rely on the property of the physical bodies to emit electromagnetic *radiation*. The amount of energy is dependent on the physical properties of the body itself and on the temperature. The temperature may be inferred by measuring the thermal radiation emitted by the body surface. 2D thermal maps of the body can also be

Table 5.4 Examples of conversion principles

Input quantity to be measured	Conversion principle
Temperature	Thermistors: Resistance variations in metals Resistance variations in semiconductors Thermocouples: Seebeck effect Capacitance variations Quartz resonant frequency variations Infrared radiations
Displacement	Resistance variations Inductance variation Capacitance variations Hall effect Ultrasonic time of arrival Optical properties
Velocity and acceleration	Doppler effect Piezoelectric effect
Force	Displacement on elastic materials
Pressure	Hydraulic pressure converted into displacement that is converted into a measurable electrical quantity Fiber optic deflection induced by pressure that induces a variation on the intensity of the light
Flow	Voltage variation proportional to the flow obtained in a uniform magnetic field Doppler effect
Chemical properties (composition, PH, concentration etc.)	Resistance (conductance) dependent on the concentration and the nature of the given solute Voltage/current (conductance) dependent on concentration and the cell potential of the given solute Optical variations

derived using this conversion principle, and they can help to show anomalous temperature distribution suggesting cancer and circulatory system diseases.

The diverse conversion properties have different performance and different usages. For example, thermistors quickly reach the temperature of the body to be measured due to the low mass of the sensor itself. This reduces the time of the measurements and so improves the efficiency of the medical staff, who can then assess more patients in a given time. On the other hand, measurements based on thermal radiation allows for the development of contactless sensors with great benefits in terms of safety.

Sensors introduce **errors** according to the specific conversion principle. In general, errors in medical instruments are roughly due to the measurement method, artifacts in the specific measurement process, and errors generated within the sensor and the associated measurement circuit/instrument.

Let us consider the last type of errors that can be more easily analyzed. Since sensors are the interface of the medical instruments, the sensors performances is directly related to the performance of the medical instrument itself. Performance is generally ascribed to the behavior in slowly varying (i.e., quasi-static) and dynamic input conditions. Regarding **quasi-static performances**, some instruments/sensors have an output value that is dependent only to the current input value regardless of the dynamics of the previous input values. In this case, the input-output relation $y(s, t)$, with s the input state of the systems and t the time, simplifies to

$$y(s,t) = f(s(t)) \Leftrightarrow y = f(s) \tag{5.3}$$

Table 5.5 outlines the main performance factors and the consequent errors of sensors related to quasi-static behavior.

Table 5.5 Quasi-static performance characteristics

Characteristics	Definition
Measuring input range	The interval of the input quantity that the sensor can measure; in general, a specific sensor performance is guaranteed within this range (linearity, accuracy etc.).
Accuracy	Degree to which a measurement or estimate is close to the true value. For statistical estimators, less bias $\Rightarrow$ more accuracy. Accuracy accounts for all the error components of the system (linearity, resolution, noise etc.) and may be measured as overall error performance (total error/true value)%.
Precision	Degree of reproducibility of the measurement. For statistical estimators, less variance $\Rightarrow$ more precision. Precision is also used to indicate the number of digits nominally available in the measure (in the context of the measurement)
Resolution	Minimum incremental quantity that can nominally be detected
Linearity	A linear system has the property $f(a_1 \cdot s_1(t) + a_2 \cdot s_2(t)) = a_1 \cdot f(s_1(t)) + a_2 \cdot f(s_2(t))$.
Hysteresis	Errors due to the dependence of the sensor output on the past input values
Sensitivity	Sensitivity in general defines the capacity to sense input variations; various definitions hold The sensitivity may be defined as the minimum quantity variation that determines the minimum output variation. The sensitivity may also be defined as the ratio of the output quantity variations to the associated input variations; in this sense, the sensitivity can be assumed to be the gain of the system
Reproducibility	Degree of agreement between measurements of the same input quantity over different time periods Calibration procedures may be required to maintain reproducibility over time

Table 5.6 Example of sensor performance (adapted from Freescale, 2010)

Characteristic	Symbol	Min	Typ	Max	Unit
Pressure (input) range	P_{OP}	0	—	10	kPa
Full scale span (output range)	V_{FSS}	24	25	26	mV
Sensitivity	$\Delta V/\Delta P$	—	2.5	—	mV/kPa
Linearity	—	−1.0	—	1.0	$\%V_{FSS}$
Offset	V_{off}	−1.0	—	1.0	mV
Offset stability	—	—	±0.5	—	$\%V_{FSS}$
Temperature effect on offset	TCV_{off}	−1.0	—	1.0	mV
Pressure hysteresis	H_P	—	±0.1	—	$\%V_{FSS}$
Temperature hysteresis	H_T	—	±0.1	—	$\%V_{FSS}$
Temperature effect on full scale span	TCV_{FSS}	−1.0	—	1.0	$\%V_{FSS}$
Response time (10–90%)	t_R	—	1.0	—	ms
Supply voltage	V_S	—	3	10	V_{dc}
Supply current	Io	—	6.0	—	mA_{DC}
Output impedance	Z_{out}	1400	—	3000	Ω
Warm-up	—	—	20		ms

Performance specifications change greatly between different devices and vendors, but datasheets usually show what is needed for a specific performance, to allow a faster design and, in general, a comparison between different vendors' products.

Table 5.6, adapted from (Freescale, 2010), shows an example of the performance specifications of a disposable sensor for invasive blood pressure measurement. The sensor requires a voltage supply to translate pressure into a proportional voltage value.

The first parameters are the measuring **input range** (pressure range) P_{OP} and the corresponding **voltage output span** V_{FSS} that specifies the difference between the output values when the input spans from the minimum to the full rated pressure.

The sensor **sensitivity** ($\Delta V/\Delta P$) specifies the static gain which, in this case, coincides with the full range gain: $sensitivity = (output\ range)/(input\ range) = 25/10\ \text{mV/kPa} = 2.5\ \text{mV/kPa}$. For all the input values included in the operating range, the output of an ideal linear sensor should be expressed as

$$Output = input \times sensitivity + offset\_constant. \tag{5.4}$$

This is a simple linear equation in the *input* variable. A real sensor is generally non-linear with a variable offset.

Non-linearity error is usually indicated as the maximum deviation with respect to the ideal linear equation (5.4). This error is expressed as a percentage with respect to the full scale span (%VFSS). Real sensors are also characterized by variable offset; Table 5.6 details the maximum variation (V_{off}), the stability over long periods (e.g., 1000 hours in percentage over full scale range) and the dependence on temperature (TCV_{off}).

Hysteresis performance is given here with respect to pressure and temperature cycles. When temperature is repeatedly changed between the minimum and the maximum value of the operating range, a non-zero output deviation (H_T) is measured even if the applied pressure does not change. **Temperature** also affects full-scale range (TCV_{FSS}). An incorrect output deviation is experienced when input pressure is at the same level but with previous pressure cycle variations between the allowed range (H_P).

The last few parameters of Table 5.6 are useful for designing the measuring circuit connected to the sensor that has to provide a supply voltage V_S with minimum current value. The input impedance of the measuring circuit has to be much higher than the output impedance of the sensor (Z_{out}) as explained in Section 5.3 equation 5.19, in addition, the medical device interfacing the sensor has to consider that the sensor has typically 20 ms warm-up before it can provide a correct measurement.

Accuracy accounts for the overall errors. In this case, accuracy may include the sum of errors related to linearity, hysteresis from both temperature and pressure, and the deviations of the offset and the full scale span due to temperature. Accuracy also indicates the minimum input variation that can be detected in practice. Although a sensor may nominally detect a minimum value determined by resolution and precision, this value cannot be detected accurately when it is lower than the overall error of the sensor as indicated by the accuracy.

Table 5.6 only includes measurement performance. Other types of performance are required for medical applications. For example, the sensor must have specific performance in terms of the materials being suitable for invasive medical applications.

From a legal point of view, it should be noted that component vendors do not guarantee performance of their products for medical device applications. The designer of the medical device system has the responsibility of guaranteeing that overall safety performances is obtained without relying on datasheet performance given by vendors.

The **dynamic characteristics** account for performance when the output is dependent on the current and past input values. Performance analysis is in general complex unless some assumptions can be made. In particular, if the system can be modeled as linear, then a simple mathematical analysis of the input-output transfer function can be obtained. In this case, the input–output quantities may be modeled by linear differential equations with constant coefficients. To simplify the analysis, the differential equations are transformed into simple algebraic equations using the Laplace transform. This avoids the computation and the solution of differential and integral equations. The Laplace transform is also useful since it generalizes the Fourier transform that is used for the frequency analysis of a system as shown in Section 5.2.4.

Other performances characterize the sensors. Dimensions may be relevant especially for invasive measurements. The sensor has to be suitable for the intended environment that may be particularly hostile not only in terms of mechanical stress but also in terms of electromagnetic interference. So, sensors must be sufficiently immune to ElectroMagnetic Interference (EMI) and should in general reduce the effects of noise. Some specific circuit configurations may help as explained in Section 5.5. Cost may be a determinant parameter especially for disposable sensors.

In general, since it is hard to find sensors with all the required performances, tradeoffs have to be considered in choosing sensors and designing the associated circuits.

5.2 Circuit Protection Function

THE NEED ➤ The I/O and power pins that connect an electronic device to the external world are a major source of failure.

This is due to electronic magnetic interference (EMI) such as electrical fast transient/burst or ElectroStatic Discharge (ESD). For any external interface connector port, an ESD is a constant threat to system reliability. EMI enters the electronic circuit through the device interfaces creating over-voltage or over-current that may lead to circuit failure. Interference is conducted via cabling or emitted/absorbed through radiation. This interference may more often provoke unwanted behavior of the device such as lock-up, data loss, reset or simply distortion over internal signals that may impact on performance. In the worse case, EMI may threaten the patient or the operator's life, for example through electric shock.

This suggests that **all the circuit board connections must somehow be protected** from the external world. In general, the protection circuit must guarantee immunity to the EMI in the environment where the device is intended to be used. The protection circuits must also guarantee that emissions of the device will not interfere with other electronic devices.

REQUIREMENTS ➤ Standard IEC/EN 60601-1-2 defines the type of interference that medical devices have to withstand.

Table 5.7 shows the main requirements of the standard which includes performance at allowed values and associated tests related to emissions, immunity, electrostatic discharge, radiated radio frequency electromagnetic fields, bursts and surges. If devices are not directly powered by AC lines, the protection is only on I/O connections and the requirements included in the last two rows of Table 5.7 are not applicable.

When interference is applied within the limits of the standard specified in Table 5.7, the device must continue to work without degradation (EEC, 2007); that is, the interference must not provoke

1. failure in diagnosis or treatment,
2. errors (artifacts, distortion) inducing wrong actions on diagnosis or treatment (such as misleading interpretation),
3. false alarms,
4. termination of the intended operations or activation of unintended operations,
5. changes in parameters or in the internal memory followed by false alarms,
6. changes in operating modes, and
7. failures on components or in the PCB, such as short-circuit or opening of PCB tracks.

Note that, IEC/EN 60601-1-2 requires that the immunity and emission performance of the device be disclosed. This performance must be obtained with all the intended accessories in the intended electromagnetic environment. The user has therefore the responsibility to use the device within the intended electromagnetic environment and with the prescribed accessories that the manufacturer has to explicitly indicate in the user manual.

Table 5.7 IEC/EN 60601-1-2 immunity requirements for conducted/radiated EMI

Test	Reference standard	Description	Compliance level
Electrostatic discharge (ESD)	EN 61000-4-2	Product immunity to ESD: contact, coupling plane, air discharge	±6 kV contact, coupling plane ±8 kV air
RF radiated immunity	EN 61000-4-3	Product immunity to electric fields generated by intentional transmitters (mobiles etc.)	3 V/m from 80 MHz to 2.5 GHz.
Electrical fast transient/burst	EN 61000-4-4	Product immunity to switching and transient noise tested on AC input and I/O cabling located outside the building	±2 kV
Surge immunity	EN 61000-4-5	Product immunity to lightning strike tested on AC input and I/O cabling	±1 kV differential mode ±2 kV common mode
RF conducted immunity	EN 61000-4-6	Product immunity to low frequency fields generated by intentional transmitters through cabling (mobiles phones, radios, TV etc.) tested on AC input and I/O cabling	3 V_{eff}/m from 150 kHz to 80 MHz
Power frequency H-field immunity	EN 61000-4-8	Product immunity to low frequency magnetic fields	3 A/m at 50/60 Hz

The circuit protection module must protect electronic components connected directly to the I/O lines from radio frequency (RF) interference and from over-voltage due to transients and ESD. THE CIRCUIT Over-voltage may create input voltage or current that exceeds the maximum allowed value of the components that are connected to the I/O lines. Countermeasures imply a mix of voltage limiters and resistors that reduce voltage and current (Bryant *et al.* 2000; Kester, 2005).

Referring to Figure 5.3, the components connected directly to the I/O must withstand the maximum allowed input voltage and current. A voltage limiter and the resistors R_{lim} reduce the current and voltage so that the other components are not affected by interference.

The component values are selected as follows. The **clamping voltage value** V_{lim} of the voltage limiter must be lower than the maximum voltage allowed by front-end. The **maximum current** through the front-end component is given by $I_{FE} = V_{lim} \times R_{FE}$. R_{FE} must be selected so that I_{FE} is lower than the maximum allowed current for the front-end component. The current I_{lim} flowing through the resistors R_{lim} is given by

$$I_{lim} = \frac{V_{Emi} - V_{lim}}{2 \cdot R_{lim}}. \quad (5.5)$$

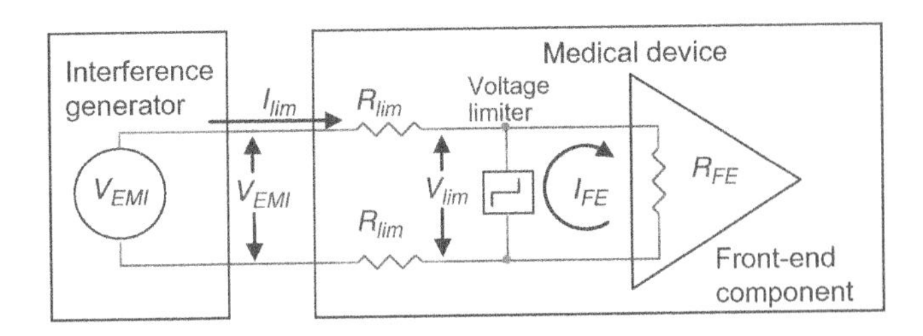

Figure 5.3 Measurement of heart potential.

The resistors must be selected so that the maximum **dissipated power** P_{Max} for each resistor is

$$P_{Max} > R_{lim} \cdot I_{lim}^2 = \frac{(V_{Emi} - V_{lim})^2}{4 \cdot R_{lim}} \tag{5.6}$$

Table 5.8 outlines the differences between ideal resistors and commercially available components. Electronic circuits often fail because the performance of real components is not

Table 5.8 Resistor ideal versus real behavior

Ideal performance behavior	**Resistor**
Voltage, Current Power	$V = R_{Nominal} \times I,\ P_{absorbed} = V \times I,$
Dissipated Power	$P_{absorbed},\ I,\ V =$ any value,
Impedance, noise	$L = 0\,H,\ C = 0\,F,\ P_{Noise} = 0\text{W}$
Real performance behavior	**Resistor**
Tolerance	$R_{real} = R_{Nominal} \times (1 \pm \text{tolerance})$
R variation:	R_{real} depends on humidity, Voltage, age and temperature – see equation (5.86)
Inductance and Capacitance	L > 0, may capture environmental high frequency interference
Max power	The maximum dissipated power is limited and depends on temperature, air cooling and so on. Maximum instantaneous power dissipation is higher than the continuous dissipation: this may help in protecting against transients. Exceeding maximum power may result in flames
Electronic noise	Resistors produce noise proportional to their values (see Johnson noise in Section 5.2.1)
Temperature range	Operating conditions are guaranteed only in the specified temperature range
Application specific features	Moisture resistant, anti surge, flame proof, non-inductive, pulse withstanding, high power resistances are available for specific applications

Table 5.9 THT and SMT/SMD

THT	SMT/SMD
Through-Hole Technology (**THT**) components have leads that are inserted through holes drilled in the PCB. THT devices are soldered on pads on the opposite side. These devices are currently being replaced by SMD which allows automatic and fast mounting. THT is used for specific prototype board since they allow a simpler manual assembly.	Surface Mount Technology Devices (SMT/SMD) are mounted directly on the surface of PCBs by special assembly machines.

taken into account. The table outlines some main critical issues; datasheets have to be carefully analyzed to take into account components' behavior and validate the specific circuit design.

Regarding the packaging of the components there are two technologies: surface mount technology (SMT) and through-hole technology (THT) devices as outlined in Table 5.9.

Note that bursts and ESD are transient with limited duration in time. Therefore, maximum power is applied for a very limited time.

The market offers special resistors developed to withstand transients that in general have a time length in the order of milliseconds. Resistors may have a limited long-term power dissipation (e.g., 1.5 W for surface mount resistors of size 2512) but they can dissipate short term higher power peaks (e.g., 250 W for 1 ms. for a total dissipated energy of 0.25 J or even 5 kW for a 1 μs: energy (joules) = power (watts) × time (seconds)).

Similar considerations apply to voltage limiting components. Such components conduct current only above a certain voltage threshold. Two constraints apply for these components:

1. they have to withstand the maximum current when in conduction
2. they have to be fast enough to switch to the conducting state when thresholds is passed to avoid failure in the device they are protecting.

Finally, increasing the input resistance R_{lim} in Figure 5.3 will reduce current in the front-end component (equation (5.5)) at the cost of an increase in the generated **noise** power. This may be unacceptable for some applications as shown in the following.

5.2.1 *Johnson Noise*

All conductors produce *thermal Johnson noise* due to Brownian motion of carriers in the conductor. The root mean square (rms) value of the noise voltage is proportional to temperature and the resistance as follows:

$$V_{rms} = \sqrt{4 \cdot K \cdot R \cdot T_k \cdot \Delta f} \cong 10^{-9} \cdot 0.13\sqrt{R \cdot \Delta f}, \qquad (5.7)$$

where K is the Boltzmann's constant (1.38×10^{-23}), R is the resistance in ohms, T_k is the temperature in kelvin (room temperature ~27 °C ~300 K) and Δf is the bandwidth in hertz over which the noise is measured.

Regardig circuit noise evaluation, the following four rules hold:

- Given statistically independent noise sources, the i-th power spectrum density $P_i = (V_{rms})_i^2$ of the i-th source is summed to the other sources to obtain the overall power spectrum. Therefore, the overall effect $(V_{rms})_{tot}$of independent sources is equal to the square root of the sum of squares of the single $(V_{rms})_i$:

$$(V_{rms})_{tot} = \sqrt{\sum_i (V_{rms})^2} \tag{5.8}$$

This is because since sources are statistically independent, the cross products of the noise sources are null

- For Gaussian noise, the standard noise deviation σ is equal to the square root of the variance σ^2 and to the *rms* power of the noise: $V_{rms} = \sigma$
- The 99.7% of the possible values of samples of a the noise time series are included in the $\pm 3\sigma$ range
- We may then assume that peak to peak (PP) values of the noise signals are equal to six times of the *rms* value:

$$V_{PP} \approx 6 \cdot V_{rms} = 6 \cdot \sigma \tag{5.9}$$

For example, a resistor of 100 kΩ at room temperature (~300 K) generates a root mean square noise voltage $V_{rms} = 0.5\ \mu V$ considering a 150 Hz bandwidth. This is roughly equivalent to $V_{PP} = 3\ \mu V$.

Resistors may be located at the input of a measuring device such as an ECG. In this case, noise (and therefore resistor value) must be limited because it may generate noise that is comparable to the low-level signals to be measured. For example, the standard AAMI EC 11 (AAMI, 1991) allows for a maximum of $V_{PP} = 30\ \mu V$. Two 100 kΩ resistors at input will yield $V_{PP} = 3\ \mu V\sqrt{2} = 4.2\ \mu V$, a value still acceptable. On the contrary, with a 10 MΩ resistor, the noise value will rise to $V_{PP} = 42\ \mu V$. This value would not satisfy AAMI requirements and would hide significant signal information.

5.2.2 *Transient Voltage Suppressors*

Over-voltage may be handled by various components also cited as transient voltage suppressors (TVS). Their role consists of clamping the transient over-voltage to an acceptable voltage level. When voltage is under the clamping value, an ideal TVS component should be in an open-circuit state. When the voltage is higher than the clamping value, an ideal component will immediately switch to short-circuit state, with unlimited current flow. Real components offer some limitations that have to be taken into account in selection. In open-state, some **leakage current** flows within the component and the component may behave as a **capacitance**. The TVS transition to closed circuit may not be immediate, thus producing a spike voltage transient on the circuits to be protected. In closed state, the components have a **maximum current flow** and **power dissipation**. Finally, some components may have **degradation** when facing continuous over-voltage.

The voltage clamping function may be performed through the components indicated in the Table 5.10, where main features with indicative values of clamping voltage are outlined.

Table 5.10 Voltage clamping technology components

Component technologies	Switching time	Dissipated power	Components	Voltage clamping
Semiconductors	fast	low to medium	simple diode	0.7 V silicon 0.3 V germanium
			zener diode (transorb)	1.8–200 V
			Schottky diode	5–600 V
			transistor	0.8 V
			thyristor (SCR, triac etc.)	25–700 V
MOV	medium	medium to high	Metal Oxide Varistor	5 V–1.8 kV
GDT	slow	High	Gas Discharge Tube	60 V–5 kV

Digikey part search (www.digikey.com) is a simple method to investigate the wide range of available components and their performance. Note that such components are grouped into the TVS class (transient voltage suppressors) of components. Table 5.11 summarizes the differences between the ideal component and real-world devices.

The choice of component depends on the specific application and the EMI environment. Regarding maximum current and dissipation, in the on state MOVs and GDTs (Gas Discharge Tube) offer the better performance over-current >10 kA. GDTs are now available in SMD packages and therefore they are suitable for automatic assembly.

Diodes and MOVs have faster switching time. On the other hand, GDTs have the better leakage and capacitance values. Diodes and transistors may be suitable for limited over-voltage protection. Again, datasheets have to be analyzed carefully to validate the circuit design performance.

Referring to Figure 5.3, during over-voltage all the conductors that connect to the components have to withstand high current, even if for a limited time. Electronic components are placed on the PCB. At minimum, in the PCB, conductive pathways must be designed so that the expected maximum current may flow without risks, the board shall be suited to withstand component thermal dissipation, the design shall take into consideration EMI performance in terms of immunity and emissions. For example, too thin PCB traces may not be able to conduct the extra current. In addition, if tracks are too close, arcs may be produced at over-voltage. This may generate flames and high voltages on parts applied to patients. In (Barnes, 2001), many suggestions are included to avoid such problems. In general, PCB design is a source of many problems at design. See for example Chapter 12 of (Zumbahlen, 2008), and Table 5.12 for an outline.

5.2.3 RF Filter Circuit Protection

An electronic circuit may be negatively affected when connected to other circuits that transmit RF interferences.

Table 5.11 Voltage clamping ideal versus real behavior

Ideal performance behavior	**Voltage clamping**
Voltage clamping–resistance relation	$V \geq V_{Clamp} : R = 0$ $V < V_{Clamp} : R = \infty$
Dissipated power	$V \geq V_{Clamp}$ P_{absorbed}, $I =$ any value ($P_{\text{absorbed}} = 0$ since to $R = 0$)
Real performance behavior	**Voltage Clamping**
Maximum values in conducting state	The maximum power/current energy that can be dissipated is limited. Maximum instantaneous dissipation is higher than the continuous dissipation. This is useful for protecting against transient over-voltage signals. Some component technologies degrade after several over-voltage occurrences
Switching time	Non-null time required to switch from off to on state. If the switching time is too high, dangerous transients may be generated
Capacitance in off state	Voltage clampers may behave as capacitances in the order of 10–100 pF. This may impact input circuits
Leakage current in off state	Leakage current is the amount of current flowing when voltage is below the clamping value
Application specific features	Surge protection, ESD, transient protection for digital and analog circuits

According to requirements of Table 5.7, the interference may be up to 3 V_{rms} in the frequency range 150 kHz–80 MHz.

Note that for a sinusoidal signal of amplitude A, the peak-to-peak value is $V_{pp} = 2 \times A$ and the root mean square value is $V_{rms} = A/\sqrt{2}$. $V_{rms}$is a preferred value for AC sources rather than V_{pp} since, on a resistor, the V_{rms}value dissipates the same power as a DC of equal voltage.

Let us assume, for example, an input circuit that has to amplify a signal at low voltage and low frequency. This circuit may accept a maximum amplitude signal as input. The presence of a 3 V_{rms} interference may saturate the input stage thus blanking off the signal to be amplified.

The RF protection circuit must reduce the conducted RF signal generated by external transmitters.

The basic protection circuit is simply a low pass filter that reduces interference at frequencies that are above the signal frequency band. Simple low pass filters may be made using capacitors (Table 5.13) or inductors (Table 5.14). Usually capacitors are preferred in filters because of their limited space and cost and their performance which is closer to ideal components (Horowitz, 1999). Inductors are better used in specific filter applications such as noise suppression or line filtering for high frequency noise.

Table 5.12 Printed circuit board ideal versus real behavior

Ideal performance behavior	Printed circuit board
Resistance, Impedance and Dissipated power on tracks	$R = 0$, $P_{dissipated} = 0$, I = any within conducting pathways $R = \infty$, $C = 0$, $L = 0$, between adjacent pathways
Maximum differential voltage	Any voltage between adjacent pathways
Interference	No RF interference on PCB (no RF emission by PCB, PCB immune to external interference)

Real performance behavior	Printed circuit board
Resistance of conducting paths	$R = \rho \times length/section$, ρ = resistivity of copper $= 1.724 \times 10^{-6}$ Ω/cm R may reduce analog-to-digital conversion performance
Resistance of adjacent paths	Leakage current may be conducted up to nano amperes
Capacitance C	Adjacent surfaces (tracks or planes) behave as a parasitic capacitor (often referred to 'stray capacitance'). $C = \varepsilon \frac{\text{area}}{\text{distance}}$, where ε is the dielectric constant of the insulating material. For distance = 1.5 mm we have C = 3 pF per cm2
Impedance	Impedance is a complex formula depending on the physical size of the paths. 1 cm of 0.25 mm PCB track 0.038 mm thick has an inductance of 9.59 nH. At 10 MHz, an inductance of 9.59 nH has an impedance of 0.6 Ω, which is over 1% error on a 50 Ω system.
Maximum voltage	Two adjacent tracks behave like capacitors; an excess voltage may produce arcing because of breakdown of the dielectric Arcs may happen between the shortest path between two conductive parts measured along the surface of the insulation (also referred as 'creepage' distance). Arcs may also happen at the shortest distance (i.e. 'clearance') between two conductors measured through air. To avoid arcs, a minimum creepage and clearance must be considered, based on the working voltage. IEC/EN 60601-1 defines such values for medical devices. For example, a minimum of 4 mm clearance has to be considered to cope with defibrillator over-voltage.
Maximum current	PCB paths are very thin conductors and excessive current increases temperature on the tracks. To avoid track overheating, the width has to be properly dimensioned. Tools for PCB dimensioning are easily found on the Internet.
Interference	The PCB tracks may act as antennas with specific layouts. Proper design has to be performed to avoid antenna effects on more sensitive parts of the circuits (as for example at the input stage of a measuring instrument, to avoid excessive noise being added to the signal)

Table 5.13 Capacitor ideal versus real behavior

Ideal performance behavior	**Capacitor**
Voltage–capacitance	$V = C_{\text{Nominal}}/Q$, Q = charge in coulomb
Current	$i(t) = C\frac{dV(t)}{dt}$
Impedance Z	$Z = 1/j\,2\pi fC$
Stored power	$P_{\text{stored}} = 0.5\ C \times V^2$
Limiting values	P_{stored}, I, V = any value, $L = 0$ H, $R = 0\ \Omega$, $P_{\text{Noise}} = 0$ W

Real performance behavior	**Capacitor**
Tolerance	$C_{\text{real}} = C_{\text{Nominal}} \times (1 \pm \text{tolerance})$
C variation	C_{real} changes with parameters such as temperature and frequency
Voltage rated	Capacitor insulation is guaranteed only under the specified working voltage
Equivalent series Resistance/ inductance ESR, ESL	A resistance/inductance is to be considered in series to the capacitor. Inductance affects high frequency performance
Leakage	Current leakage may be expected between capacitor terminal
Temperature range	Operating conditions are guaranteed only in the specified temperature range

5.2.4 *Circuit Frequency Response*

Let us now look at **linear time invariant** (LTI) circuits, that is, circuits containing only linear elements: ideal resistors, capacitors, inductors. For any input frequency, a linear circuit will change only the amplitude and phase of the input sine wave signal, leaving unchanged the sine wave's shape and the input frequency. Assuming that the input signal is $V_{in} = \sin(2\pi ft)$, the output signal is given by

$$V_{out} = M(f)\sin(2\pi ft + \varphi(f)), \tag{5.10}$$

with $M(f)$ and $\varphi(f)$, the magnitude response and phase response with respect to the frequency f, and t is the time. Note that instead of the frequency f, the angular frequency ω is also used, with $\omega = 2\pi f$. The magnitude response $M(f)$ indicates how a sine wave of frequency f is amplified or attenuated on passing through an LTI circuit. The phase response $\varphi(f)$ shows how the sine wave of frequency f is shifted in time. The **frequency response** is the response of a circuit to a sine wave signal in terms of modified amplitude $M(f)$ and phase $\varphi(f)$ for different frequencies f.

The quantities $M(f)$ and $\varphi(f)$ can be considered as the magnitude and the phase of the complex circuit response $H(f)$ where

Table 5.14 Inductor ideal versus real behavior

Ideal behavior	Inductor
Current	$V(t) = L\frac{dI(t)}{dt}$
Impedance Z	$Z = j\,2\pi fL$
Limiting values	P_{stored}, I, V = any value, $C = 0$ F, $R = 0\,\Omega$, $P_{\text{Noise}} = 0$ W
Dissipated/radiated Power	$P_{\text{dissipated-radiated}} = 0$

Real behavior	Inductor
Tolerance	$L_{\text{real}} = L_{\text{Nominal}} \times (1 \pm \text{tolerance})$
Resistance and capacitance effects	Real inductors have series resistance effects (usually specified at direct current (DC) and parallel capacitance
Quality factor Q (i.e. relative loss of inductors)	$Q = 2\pi fL/R_{effective}$, Q = ratio of the inductive reactance to effective resistance
Resonant filter	Inductors and associated parallel capacitance behave as resonant filter at a certain frequency. Datasheets specify self-resonant frequency
Energy dissipation	Energy dissipation due to wire resistance and hysteresis
Non-linear effect	Hysteresis and magnetic saturation may cause non-linear effects
Antenna effect	Inductors capture and radiated electromagnetic interference
Rated current	Maximum continuous DC current allowed through the inductors
Temperature range	Operating conditions are guaranteed only in the specified temperature range

$$\begin{aligned} M(f) &= \|H(f)\| = \sqrt{H(f) \times H(f)^*} = \sqrt{\Re(H(f))^2 + \Im(H(f))^2} \\ \varphi(f) &= \text{atan2}(\Re(H(f)), \Im(H(f))), \end{aligned} \tag{5.11}$$

with $H(f)^*$ the complex conjugate of $H(f)$, and $\text{atan2}(\Re(H(f)), \Im(H(f)))$ the function based on arctangent to obtain the phase of a complex number from its real and imaginary parts. It can be demonstrated (Horowitz, 1999) that

$$H(f) = \frac{V_{out}}{V_{in}}, \quad H(f) = M(f)\, e^{j\varphi(f)}, \tag{5.12}$$

Equation (5.12) is obtained taking into account Euler's formula (Table 5.2). In particular, V_{out} in equation (5.10) is obtained from V_{in} , considering that $\text{Sin}(x) = (e^{-jx} - e^{jx})/2j$, and $H(f)$ can be computed using the usual DC circuit laws (Ohm, Thevenin and Kirchhoff) on V_{out} starting from V_{in} and considering the impedance Z of capacitors and inductors. (DC is the Direct Current, that is the current that has constant value over time, or, in other world,

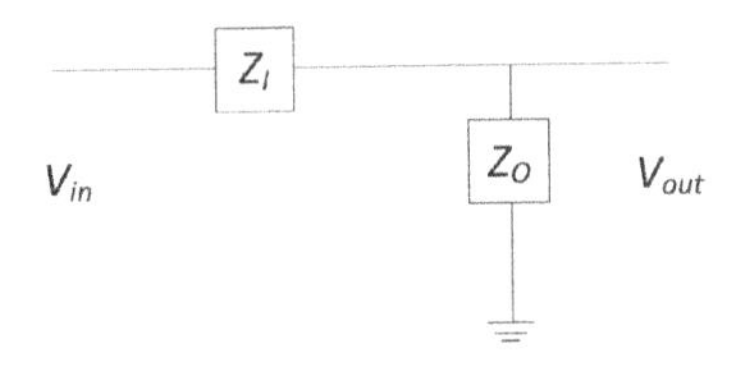

Figure 5.4 Generalized one stage filter.

when the frequency of the input signal is zero). Capacitors and inductors impedance are given by

$$Z_C = 1/(j2\pi fC), \quad Z_L = j2\pi fL, \tag{5.13}$$

where C and L are the capacitance and impedance values respectively. The generic **first-order filter** is obtained by the scheme shown in Figure 5.4 when only one reactive element is present (either an inductance or a capacitor). Simply speaking, the order of a filter is the number of independent reactive elements in the filter. More formally, By definition, the order of the filter is the order of the denominator polynomial (i.e. the number of zeroes) of the transfer function $H(f)$.

Assuming that the current in Z_I is equal to the current in Z_O (i.e., no current flows into V_{out}) and V_{in} is an ideal voltage generator (i.e., there is no series impedance before Z_I), the circuit response (i.e. the circuit transfer function) is given by

$$H(f) = \frac{V_{out}}{V_{in}} = \frac{Z_O}{Z_I + Z_O} = \frac{1}{Z_I/Z_O + 1}. \tag{5.14}$$

Low and high frequency filters are obtained by replacing Z_I or Z_O with capacitors or inductors, considering that, from (5.13), capacitors offer lower impedance at higher frequencies, while inductors have the reverse behavior.

A capacitor based low frequency filter is obtained by placing a capacitor of appropriate value $Z_O(= 1/(j2\pi fC))$ with Z_I being a simple resistance ($Z_I = R$). In this case, the capacitor will short-circuit higher frequencies to ground. From 5.14, 5.13, 5.11, the frequency response $H(f)$ and the magnitude response $M(f)$ of the circuit are

$$H(f) = \frac{1}{1 + j2\pi f RC}, \quad M(f) = \frac{1}{\sqrt{\left((2\pi f RC)^2 + 1\right)}}. \tag{5.15}$$

It is clear from the previous formula that when $f = 0$, $M(f) = 1$, that is, there is no attenuation at DC. As the frequency tends to infinity, the magnitude tends to zero and therefore the highest frequencies are increasingly attenuated.

It is also useful to have an indication of the frequency where the circuit starts its attenuation. This value, usually called '**cutoff**' frequency, is defined as the frequency where the voltage is attenuated by $\sqrt{1/2}$ with respect to the frequency where the signal is not attenuated (passband voltage). For the high pass circuit described by equation (5.15), we have

$$f_{\text{cutoff}} = 1/(2\pi RC). \tag{5.16}$$

The magnitude of the frequency response is better expressed in **decibels** (symbol dB), where the ratio of two signals A_1 and A_2 is expressed as dB $= 20\log_{10}(A_1/A_2)$. Since the magnitude of the frequency response is the ratio of the absolute value of the output and the input voltages we have

$$\left\|\frac{V_{out}}{V_{in}}\right\| = \|H(f)\| = M(f), \quad M(f)_{db} = 20\log_{10}(M). \tag{5.17}$$

Note that since the ratios are usually computed on the power of the voltage, the power ratio in decibels is given by

$$\text{Ratio in dB of power} = 20\log_{10}\left(\left\|\frac{V_{out}}{V_{in}}\right\|\right) = 10\log_{10}\left(\left\|\frac{V_{out}}{V_{in}}\right\|^2\right). \tag{5.18}$$

From 5.16 and 5.15 the cutoff frequency is the frequency where the output power is halved. Equation (5.18) and that $20\log_{10}\left(\sqrt{1/2}\right) \cong 3$ dB, the cutoff frequency is also the frequency where the power is reduced by 3 dB: $f_{\text{cutoff}} = f_{3\text{ dB}}$.

For example, assuming that the information is mainly included in the frequency range 0–150 Hz, as in the case of the ECG, we may consider a low pass filter that improves 'immunity to low frequency fields generated by intentional transmitters through I/O cabling (mobiles phones, radios, TV etc.)' in the range 150 kHz–80 MHz (IEC/EN 61000-4-6 standard). Let us consider $f_{\text{cutoff}} = f_{3\text{ dB}} = 150$ Hz, then, interference at 1.5 kHz will be attenuated by 10 times, at 15 kHz by 100 times, at 150 kHz 1000 times as per equations (5.15) and (5.17).

Referring to Figure 5.4, this means that an interference of $V_{\text{in}} = 3\ \text{V}_{\text{rms}}$ at 150 kHz will be attenuated at $V_{\text{out}} = 3\ \text{mV}_{\text{rms}}$.

Thanks to the concept of impedance, usual circuit laws can be applied (Ohm, Thevenin and Kirchhoff) using complex numbers. Watch out for a common error, since the magnitude of the sum of complex numbers is not the sum of the magnitudes $\|Z_1 + Z_2\| \neq \|Z_1\| + \|Z_2\|$, impedances have to be summed as complex numbers and then the magnitude has to be calculated from the result. For multiplication and division a simpler computation may be performed considering that $\|Z_1 \times Z_2\| = \|Z_1\| \times \|Z_2\|$, $\|Z_1/Z_2\| = \|Z_1\|/\|Z_2\|$.

5.3 Buffer Stage

THE NEED The source to be measured by an instrument may be modeled as an ideal voltage source with an impedance in series that is connected to an instrument as shown in Figure 5.5. An ideal voltage source may supply any current and the ideal voltage instrument has an infinite resistance between its pins. We have then

$$V_M = V_S \frac{Z_I}{Z_I + Z_S} = V_S \cdot \left(\frac{1}{1 + Z_S/Z_I}\right) \tag{5.19}$$

Note that the attenuation factor $\frac{Z_I}{Z_I+Z_S}$ in (5.19) changes magnitude and phase of the input signal V_S. The measured signal V_M may therefore even have a different shape with respect to V_S. From equation (5.19) it is clear that the internal instrument impedance Z_I has to be much greater than the impedance of source Z_S to avoid measurement errors.

In Figure 5.5, the voltage source may be a sensor generating a voltage signal. Similar considerations hold when the measurand is proportional to a current signal generated by a

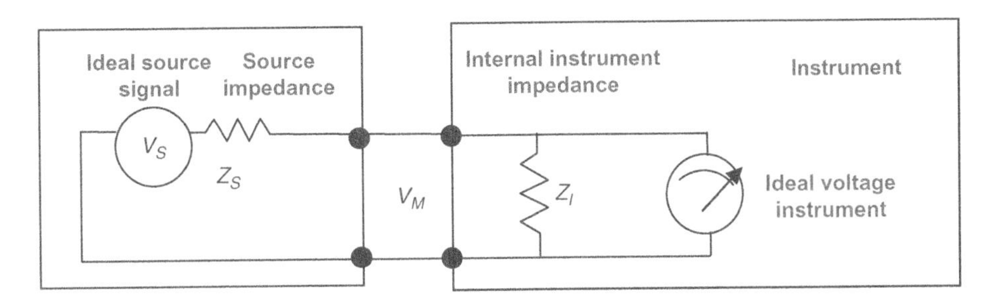

Figure 5.5 Measurement of heart potential.

specific sensor. In this case, the instrument's internal resistance is in series with an ideal current sensing instrument with zero impedance at its pins. It can be easily shown that the internal impedance of the instrument has to approach to zero to avoid errors.

For voltage sensors, a buffer stage may be required to offer high input impedance when the source sensor impedance is significant. The stage is called a 'buffer' because it *isolates* source impedance.

The buffer stage is also the first stage of the instrumentation and therefore noise and interference have to be considered with care. Noise introduced by this stage must be much lower than the input measured signal. In addition, the buffer stage must withstand over-voltage and interference that are not eliminated by the protection circuit.

The main **requirements** that have to be defined **for the buffer stage** are

1. input impedance
2. maximum over-voltage allowed at input
3. input dynamic range
4. contribution to system noise
5. frequency response and linearity.

The **input impedance** must be much higher than the source impedance. For an ECG, the source impedance may exceed 220 kΩ for 5% of the population and the input impedance must be greater to 2.48 MΩ to avoid unacceptable errors according to (AAMI, 1991).

The maximum allowed **over-voltage** depends on the features of the protection circuit and the clamping device that limits over-voltage to the specific value V_{Clamp} (see Tables 5.10 and 5.11). The buffer stage must withstand continuously a V_{Clamp} value, and higher transient values for a shorter time, considering that the clamping device at input may take some time to shortcut the over-voltage.

The buffer stage must withstand the **input dynamic range** plus all the allowed interference modifying the source signal within the prescribed limits. The stage shall also add limited noise w.r.t. the measurand.

The final main requirement is that the input signal must not be modified by **linear** or **non-linear performance** of the buffer stage. The buffer stage must not introduce attenuation at specific frequencies where the source signal is significant. If the signal shape is significant, as in the ECG, the buffer stage must have a **linear phase property**, that is, all frequency components in the passband area must have the same delay in time. In this case, there is no wave distortion due to different delays associated with the different frequency components. Linear phase filters are discussed in Section 5.6.1.

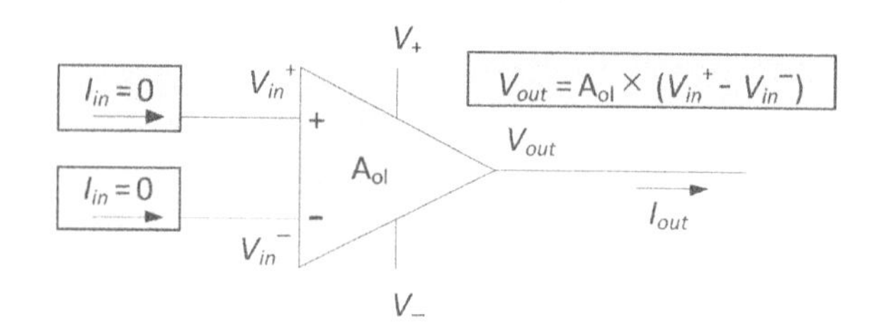

Figure 5.6 Ideal operational amplifier.

5.3.1 Operational Amplifiers

Operational amplifiers (op-amps) can be an effective building block for buffer stages. As shown in Figure 5.6, op-amps are electronic components with two differential inputs (V_{in}^+, V_{in}^-), ideally with infinite input impedance, and one output V_{out} with infinite **open loop** gain A_{ol} expressed as:

$$V_{out} = A_{ol} \times (V_{in}^+ + -V_{in}^-). \tag{5.20}$$

(open loop means that no load is connected to the output pins of the amplifier). These devices are powered by bipolar supply V_+, V_-, as for example ±5 V, or, in specific models, by a single (unipolar) supply.

The supply sources are referred to as '**rails**'. The output signal V_{out} will have a value included within the positive and negative rails: $V_- < V_{out} < V_+$. Some specific op-amps are 'rail to rail', that is, they can reach supply rails at low output currents: $V_- \leq V_{out} \leq V_+$.

Op-amps are widely used in medical devices especially in the analog interface between the device and the external sensors (electrodes, photodiodes, chemical arrays etc.). The market offers thousand of different models, with a wide range of specifications suitable for many different applications. For example, in battery-powered applications, there are many low power and small footprint models. Here are some op-amp applications in medical devices (Analog, 2011):

➢ ECG, EEG and heart rate monitors:
- instrumentation op-amp amplifiers (see Figure 5.15) with high common-mode rejection ratio (CMRR Table 5.15) to reduce common mode interference
- operational amplifiers for the buffer stage, amplification and filter stages

➢ Photodiode applications including glucose meters (amperometric and photometric):
- traditional operational amplifiers as buffers: low noise, low input bias current and low offset voltage amplifiers.

➢ Blood pressure applications:
- low power, high precision instrumentation amplifiers for the bridge sensor application.

As for any design, we must consider first the ideal op-amp behavior, and then analyze how real behavior affects the implemented circuit.

The ideal behavior is summarized by the two rules (Horowitz, 1999):

1. $I_{in} = 0$: input pins do not draw any current (5.21)

2. $V_{out} = A_{ol} \times (V_{in}^+ - V_{in}^-), A_{ol=\infty}$: output amplifies infinitely the difference of the inputs (5.22)

Table 5.15 Operational amplifier ideal versus real behavior

Ideal performance behavior	**The operational amplifier**
Voltage	$V_{out} = A_{ol} \times (V_{in}^{+} - V_{in}^{-})$
Voltage gain	$A_{ol} = \rightarrow \infty$
Current	$I_{in} = 0,\ I_{out} = \text{any}$
Impedance	Input impedance $= \infty$ Output impedance $= 0$
Frequency response	Flat response in magnitude, with zero phase shift

Real performance behavior	**The operational amplifier**
Input	Input impedance $< \infty$ Input bias current > 0 Maximum input voltage $< \infty$ Input offset voltage > 0 (required to have zero output voltage)
Output	$V_{out} = A_d \times (V_{in}^{+} - V_{in}^{-}) + A_{cm} \times (V_{in}^{+} + V_{in}^{-})$; $A_{cm} > 0$ Common mode rejection ratio (CMRR) $= A_d/A_{cm}$ $CMRR_{dB} = 10 \log_{10} (A_d/A_{cm})^2$, see Figure 5.13 and following PSRR = (variation in power supply)/(variation in input that generates in the output the same effect as the power supply variation) Negative rail $\leq V_{out} \leq$ positive rail Slew rate (max. rate at which output can change) $< \infty$ Maximum output current $< \infty$ Frequency/phase response dependent on frequency Temperature dependence Noise power > 0

Other ideal behaviors are outlined in Table 5.15. When the output is connected as a negative feedback to the input pins then the second rule implies that the output pin attempts to reduce the difference between inputs. In linear operative conditions this adds an extra rule:

3. $$V_{in}^{+} = V_{in}^{-} \tag{5.23}$$

This is clarified by the circuit in Figure 5.7 also known as 'voltage follower' configuration. From equation (5.22), when $A_{ol} \rightarrow \infty$, we have $V_{out} = V_{in}^{+}$.

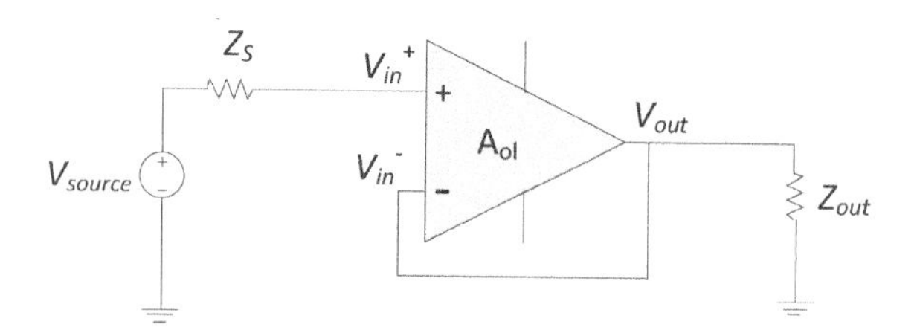

Figure 5.7 Voltage follower configuration.

The circuit also satisfies the ideal requirements of a buffer stage since the ideal input impedance is infinite while the output impedance is zero. This means that $V_{out} = V_{S,}$ for any values of Z_S and Z_{out} in Figure 5.7.

Again ***non-ideal*** *component* ***behavior*** *may even stop the circuit working in practice.* Table 5.15 outlines some of the main ways that the behavior departs from ideality. Designers must consider all these specific behaviors and the suggestions included in the datasheets for avoiding potential problems at the implementation stage. Solutions are often tradeoffs between contrasting effects.

For example, considering the simple circuit in Figure 5.7, there are factors that may stop the circuit working. Since the input current I_{in} is greater than zero, if the source impedance Z_S is high, a non-negligible voltage $V_{in} = Z_s \times I_{in}$ may appear at the input that is then amplified at the output.

This may saturate the op-amp's output. A resistor in parallel with the input may be a choice in some applications so that current I_{in} does not generate a significant voltage at input. This has the drawback that it reduces the input impedance and increases thermal noise, as discussed in Section 5.2.1. From Table 5.15, the voltage between the two inputs and between each input pin and ground must be within a specific range to avoid saturation or device destruction, but a protection circuit may help, as outlined in Section 5.1.

Real op-amps are also sensitive to variations in power supply. Real power supplies usually have ripples that will be propagated to the output of the op-amp. The power supply rejection ratio (PSRR) measures the change of the output signal due to power supply changes. To reduce such effects, suitable capacitors are connected between the op-amp power supply pins and ground. This capacitors are also called 'by-pass' because they short-cut the power interference to ground. Capacitors values of 100 nF connected between power supplies and ground address this problem and also prevent the op-amps from oscillating.

5.4 Analog Signal Processing

THE NEED Digital signals are more easily processed compared to their analog counterparts since operations are performed on discrete values sampled in time and quantized in amplitude. The usual operations, such as logical addition or multiplexing (selecting one output from several inputs) are more simply implemented on digitalized versions of the analogue signals. In addition, digital signals do not suffer from noise degradation as long as the noise does not exceed specific limits. In contrast, analog signals have continuous amplitude-time values, and their processing requires that signal waveforms be not distorted significantly with respect to their intended use. Since noise cannot be eliminated in analog processing, its interference has to be limited by careful design choices. In general, analog circuit design is much more critical. Unfortunately, analog circuits cannot be always replaced by digital processing. In this section, the main simple analog circuits – the summing amplifier, the analog switch and the multiplexer – will be addressed.

REQUIREMENTS Ideal analog processing should perform the required operations, introducing signal modifications that are acceptable for the intended use.

The main unwanted modifications in analog signal processing are

1. signal saturation
2. frequency–phase signal modifications
3. non-linear distortion
4. noise.

In the following sections, three circuits will be considered: the summing amplifier, the analog switch and the multiplexer. This circuits are good examples for analyzing basic design in analog processing.

5.4.1 *Summing Amplifier Circuit*

Figure 5.8 shows a circuit where voltage sources V_1, V_2, V_3 are summed together. Considering the op-amp's main rules (5.21), the point A draws no current to the op-amp input. Since A is at zero voltage level from eq. 5.23, this point is also called 'virtual ground'. Considering that all the currents introduced at node A from R_1, R_2, R_3 flow into R_4, we have

$$V_{out} = -R_4\left(\sum_i \frac{V_i}{R_i}\right), \text{ when } R_i = R \rightarrow V_{out} = -\frac{R_4}{R} \times \sum_i V_i. \tag{5.24}$$

This equation shows that the output is proportional to the sum of the input voltages, as required.

Some **problems** may arise from the real behavior of the circuit components. Equation (5.24) shows that the circuit is sensitive to the value of the input resistances R_1, R_2, R_3. A real circuit approaches ideal conditions when real resistance values are close to their nominal value. Resistances should therefore be of the precision type, that is, with low tolerance and

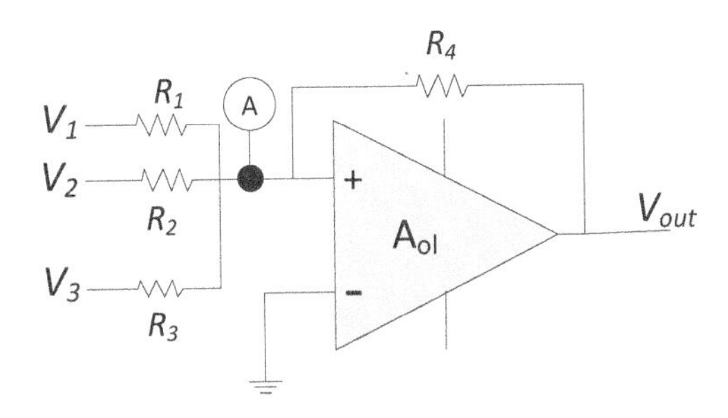

Figure 5.8 The summing circuit.

low dependence by temperature (see Table 5.8). Care must be taken also because these resistances introduce thermal noise, proportional to their resistance value (see Section 5.2.1). The limitations of real op-amps also suggest a specific value range for the feedback resistance R_4. A high value would generate a voltage offset due to the input bias current, additional noise and increased pickup of environmental interference. Low R_4 values may load the op-amp output stage since an higher output current $i_{out} = V_{out}/R_4$ is required to set point A to zero, and for real op-amps i_{out} is limited.

In general, the real behavior of op-amps outlined Table 5.15 has to be considered, to avoid malfunctions. For example, capacitors must always be used between the power supply lines and ground, and rails have to be checked to avoid output saturation.

5.4.2 Analog Signal Switching

Most medical instruments and devices include analog sensors to measure biometric functions such as blood pressure, body temperature or heart rate. Biometric sensors measure physical quantities, such as temperature, pressure, light intensity or fluid flow. The sensors convert the physical quantities into proportional voltage or current values. The signal is then processed and digitized for additional analysis. Signals often have to be combined or simply selected. For example, in pulse oximetry, the output signals of two photodiodes are used to assess blood absorption at different wavelengths. In ECG, the nine analog input signals have to be recombined to obtain the twelve output signals, that is the twelve standard ECG leads as explained in Chapter 2. Analog switches are employed to perform such processing. Analog signal switching is useful when many sources have to be selected or recombined dynamically. The circuit of Figure 5.9 shows a simple circuit using analog switches. Such components perform the functions of mechanical relays but without including moving parts.

As shown in Figure 5.9, the ideal switch should always yield output = input when it is in the *on* state. In the *off* state, the resistance between input and output should be infinite. Analog electronic switches are usually based on a suitable configuration of FETs that are controlled by digital lines, usually at CMOS or TTL logic levels. Ideal and real behaviors are summarized in Table 5.16; (Texas, 2011) adds extra details. According to the value of the digital control line, the FET configuration will assume a low or high resistance. The circuit of Figure 5.9 would be simple with ideal devices while it may have subtle problems

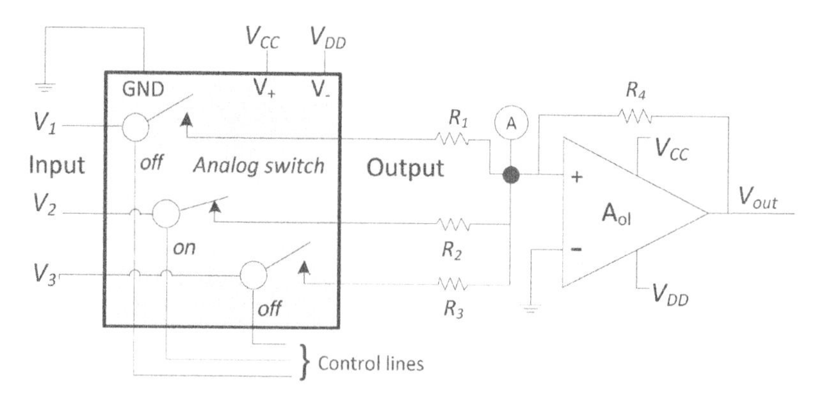

Figure 5.9 Circuit for dynamic recombination of analog input.

Table 5.16 Analog switch ideal versus real behavior

The Analog Switch

Performance	Ideal behavior	Main real limits
Signal voltage range	any	Generally rail to rail (V_- to V_+)
Transfer function	$V_{COM} = V_{NO}$	Saturation, distortion effects, non-constant frequency response
On-resistance	$R_{on} = 0\,\Omega$	On-resistance can start from milliohms to $\approx 1000\,\Omega$. It is dependent on voltage input and temperature
On-capacitance	$C_{on} = 0\,\text{pF}$	Order of tens of pF; the value limits the bandwidth of the input signal
Off-state leakage current	$I_{\text{on}} = 0\,\text{mA}$	Order of 10^{-9} A leakage
Off-capacitance	$C_{off} = 0\,\text{pF}$	Order of tens pF; high frequency signals pass through such non-null capacitance in the off state
Switching time (t_{ON}, t_{OFF})	$t_{ON} = 0\,s$ $t_{OFF} = 0\,s$	Order of nanoseconds
Charge injection	$Q = 0\,\text{pC}$	Order of pC; the switch injects an electric charge into the signal when switching, causing signal noise

in implementation. First consider that the *on* resistance is non-null for the analog switch and, worse, it changes with input signals.

Since the value of the output voltage V_{out} depends on the resistances, significant distortion may arise. The sum of the input voltage must not obviously saturate the maximum allowed voltage of the components. Exceeding this voltage may have negative effects on the FET, such as damage due to excess current. To avoid destruction, such components may also have specific constraints on power-supply sequencing startup (e.g., first activate the positive power supply and then activate the negative power supply). Switching may introduce significant glitches and noise, which may be unacceptable for the specific application. Again, the design approach must follow the main steps:

1. design with ideal devices in mind
2. consider real limits shown in the datasheet
3. compute the real performance.

At the end, test with simulation programs and prototypes.

Analog switches are also the base components for multiplexers, that allow the selection of one input source from many. Multiplexers are used in medical device multichannel systems, such as 12-lead ECGs, where a single ADC converts many multiplexed signals.

The key requirements for the multiplexer (mux) are typically low on-resistance and low transients while switching sources, also referred to as low charge injection.

5.5 Interference and Instrumentation Amplifiers

5.5.1 *Eliminating In-band Interference*

Sensor measurements are often affected by significant electromagnetic **interference** (EMI) contained in the environment. In recent years, there has been an increase in the numbers of both the sources of EMI and the electronically controlled medical devices that may be affected. The signals measured by medical sensors may contain interference, whose amplitude may be in the same order of magnitude as the measurand. For example, as shown in Figure 5.10, the ECG signal amplitude is in the order of millivolts, while interference from power lines can reach volts or tens of volts.

When the interference is located **outside the frequency** band of the desired signal, a filter can eliminate the interference. For example, conducted RF interference, generated by intentional transmitters through cabling (mobile phones, radios, TV etc.), is located between 150 kHz and 80 MHz (see Table 5.7). The medical and physiological parameters have frequency contents much lower than 150 kHz: electromyography which has the widest frequency range has components up to 10 kHz. **Low pass filters** may reduce interference from transmitters, while leaving the desired signal unchanged.

When the interference lies in the same frequency band of the **measurands**, other approaches have to be considered.

High amplitude transients, such as electrostatic discharge or fast transients included in Table 5.7, may be eliminated by using their features of high voltage and short duration. Voltage clamping devices may shortcut the input to ground at a given amplitude signal level, as

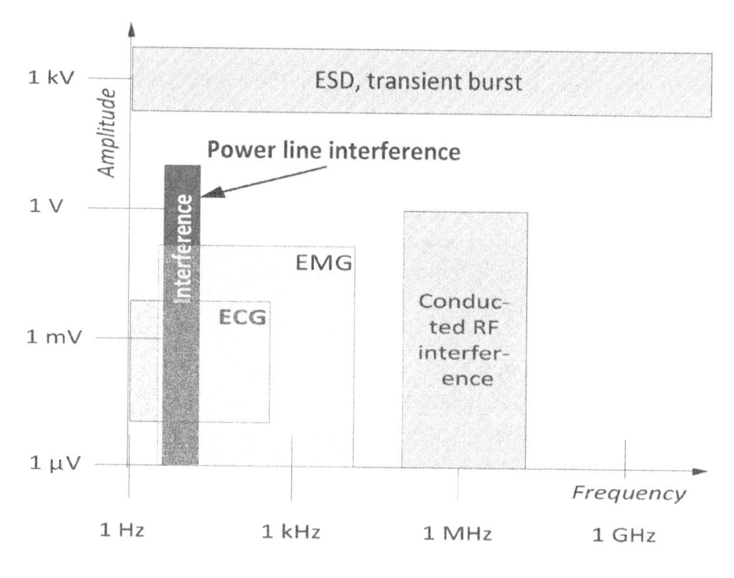

Figure 5.10 Medical parameters and interference.

discussed in Section 5.2. The side effect is that the measurand is not recorded when the transient occurs.

Other interferences may be continuous signals with frequency contents overlapping the band of the measurand. This is the case of electrical power lines that generate electric and magnetic interference. The medical devices must be immune to magnetic and electric fields generated by the power lines of a typical commercial or hospital environment. **Magnetic fields** are generated by current flowing in power lines or by transformers. Medical devices must work properly in the presence of a magnetic field of 3 A/m at 50/60 Hz frequency, radiated over all the three axes of the product (see Table 5.7). Magnetic fields are captured by any coil formed by the spiral geometry of wires or PCB tracks within the medical device. A voltage proportional to the magnetic field and to the coil surface area is then induced at the coil's terminals. This problem may be reduced by **shielding** the device by using a high magnetic permeability metal alloy envelope that surrounds all the sensitive device parts. This shielding envelope draws the magnetic field around the shielded volume, avoiding the magnetic field being captured by the internal coils.

Twisted wires may also reduce induced interference. Since the voltage induced in coils is proportional to the surface area of the coil itself, when wires are twisted, the area of the possible coils is reduced and so is the induced magnetic field.

Electric fields from power lines strongly influence medical devices. Their intensity is proportional to the line voltage and inversely proportional to the distance between the power line and the medical device. Unlike magnetic fields, electric fields are not proportional to the current and therefore they are always present, even when the electric devices are switched off in the environment.

Electric fields from power lines may generate current through patients, sensors, wires and within the medical device itself. These currents can induce voltage interference which can be an order of magnitude higher than the signal to be measured, thus providing an interference that may impede measurements in medical instruments. In medical devices, such currents can also generate errors in the device control circuits.

The generation of current is due to the capacitances between power lines and patient from one side and from wires and the medical device from the other side as shown in Figure 5.11, for the case of the ECG (adapted from Webster, 2009). Two adjacent conductors form a capacitor whose value is proportional to the surface area and inversely proportional to the distance.

We recall that the permittivity ε is the degree of the resistance encountered by an electric field when in a specific dielectric medium type such as the vacuum, paper) Taking into account the capacitor formulas in Table 5.13 and assuming the simple *model of a capacitor*

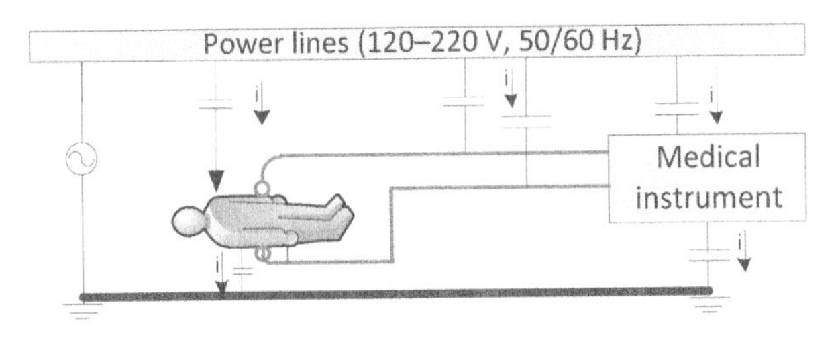

Figure 5.11 Current flows from power line.

with two parallel conductive plates separated by a dielectric with permittivity ε, the current i depicted in Figure 5.11 is given by

$$i = |V/Z| = V\frac{1}{1/2\pi fC} = \frac{V \cdot 2\pi f\varepsilon A}{d}, \tag{5.25}$$

which shows that the current is inversely proportional to frequency f, the voltage V and the area A of the capacitor surfaces that are involved. The current is proportionally reduced with the distance d from the power lines.

Let us consider the medical instrument case in the European power supply environment: $V = 220\,\mathrm{V}_{rms}$ at 50 Hz where *rms* stands for *root mean square* value. An equivalent plate surface of $A = 0.005\,\mathrm{m}^2$ at a distance $d = 1.5\,\mathrm{m}$ generates a current of $2 \times 10^{-9}\,\mathrm{A}$. If such current is discharged into a 50 kΩ, resistance, we have a voltage of $100\,\mu\mathrm{V}_{rms} \approx 282\,\mu\mathrm{V}_{\mathrm{pp}}$, since for a sinusoidal signal

$$V_{\mathrm{pp}} = V_{\text{peak to peak}} = V_{rms} \times 2\sqrt{2}. \tag{5.26}$$

This value ($282\,\mu\mathrm{V}_{\mathrm{pp}}$) is greater than the maximum allowed system noise: $30\,\mu\mathrm{V}_{\mathrm{pp}}$ for an ECG, according to the AAMI EC 11 standard. The interference may be reduced by shielding. **Shielding** consists of enclosing the device by a conductive surface so as to create a so-called 'Faraday cage' that blocks the propagation of external electric fields inside the device. The **Faraday cage** also reduces the RF electromagnetic fields propagation.

Shielding the device in Figure 5.11 allows us to eliminate the interference of electric fields generated by power lines and by RF interference, provided that the thickness of the shield is sufficiently high and the holes in the shield are smaller than the wavelength of the radiation.

A Faraday cage does not isolate the devices from slowly varying or static magnetic fields such as those generated by power supplies. Magnetic field interference may be reduced by high magnetic permeability metal alloys, as discussed above.

The **patient cables** of Figure 5.11 suffer from power line interferences as well. From equation (5.25) considering a 3 m cable, 2 mm width at a distance $d = 0.5\,\mathrm{m}$ from the power line at $V = 220\,\mathrm{V}_{rms}$ and 50 Hz frequency, we have around $i_{rms} = 7 \times 10^{-9}\,\mathrm{A}$ current flow. Figure 5.12 shows possible effects of this current. Interference is generated from the fact that electrodes attached to the patient show different impedances Z_1 and Z_2.

To test ECGs, the AAMI standard (AAMI, 1991) uses a value of $Z_1 - Z_2 = 51\,\mathrm{k\Omega}$ resistor in parallel with a 47 nF capacitor that, at 50 Hz, are equivalent to an impedance of 37.8 kΩ. (*Remember that impedances must always be summed as complex numbers.*) In the model of

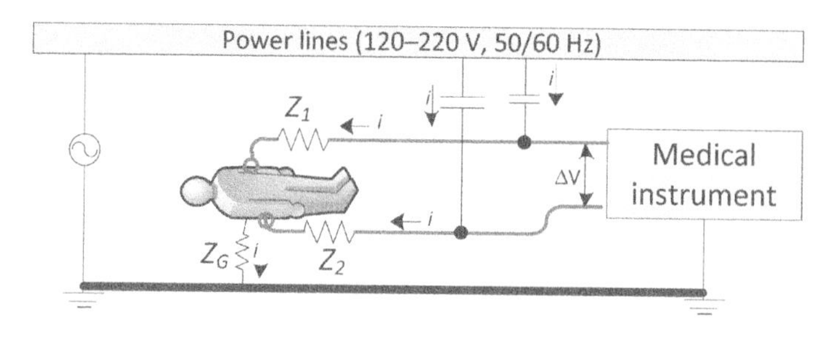

Figure 5.12 Voltage interference from unbalanced impedance.

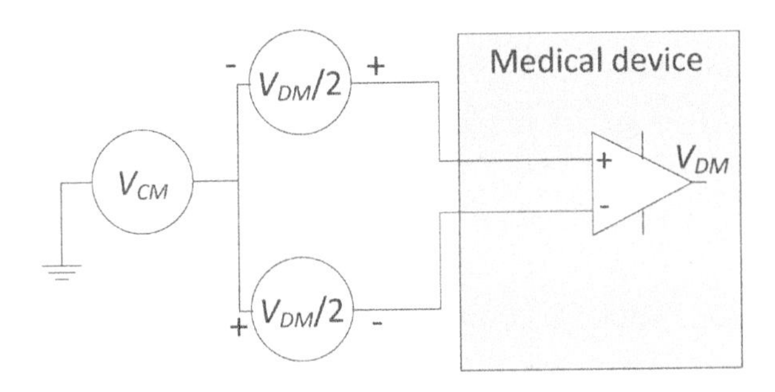

Figure 5.13 Differential and common mode signal.

Figure 5.12, the current i and the imbalance of the skin–electrode resistance will generate a voltage differential:

$$\begin{aligned}\Delta V &= i \times (Z_1 + Z_G) - i \times (Z_2 + Z_G) = i \times (Z_1 - Z_2) \\ \Delta V &= 7 \times 10^{-9}\,A \times 37.8\,k\Omega = 265\,\mu V_{rms} = 749\,\mu V_{pp}.\end{aligned} \tag{5.27}$$

Such value is much higher than the maximum allowed system noise: 30 μV_{pp}.

To avoid this problem, wires have to be shielded. In this case, the current cannot flow inside the cabling.

Taking into account Figure 5.12, another source of interference is the object to be measured and the associated sensors (patient and electrodes) where the current is induced from power lines. This interference coexists with the signal to be measured. A possible countermeasure consists of designing the device so that interference is maximally transformed into a **common mode signal** while the measurand is a differential signal.

As shown in Figure 5.13, in the ideal case, all the interference is a common mode signal V_{CM} that is therefore added to both lines connected to the measuring device. The measurand V_{DM} is carried by the two lines at equal amplitude but opposite voltage. Since the device measures only the difference between the two lines, the common mode interference is rejected. From a practical point of view, the design must at minimum imply twisted wires and a differential stage within the medical device so that interference can become a common mode signal that can be significantly reduced.

When wires are twisted, both the lines are exposed to the same amount of interference and therefore interference generates mainly a common mode signal. The differential input stage may be based on an op-amp in a differential configuration such as shown in Figure 5.14.

V_{out} is calculated considering the ideal rules of the op-amp detailed in equations (5.21) and (5.23). To simplify the computation, since this is a linear circuit we can consider first the effect of V_{in-} on V_{out}, assuming $V_{in+} = 0$, then the effect of V_{in+} on V_{out} assuming $V_{in-} = 0$ and, at the end, sum the partial output contributions to obtain the overall V_{out}. Assuming $V_{in+} = 0$, then $V_B = V_A = 0$ and since no current flows ideally inside the input of the op-amp, the current flowing through R_2 is equal to the current flowing through R_1 and therefore

$$\text{with } V_{in+} = 0 \rightarrow V_B = V_A = 0 \quad \text{and} \quad V_{out}^{-} = -\frac{R_2}{R_1} V_{in-} \tag{5.28}$$

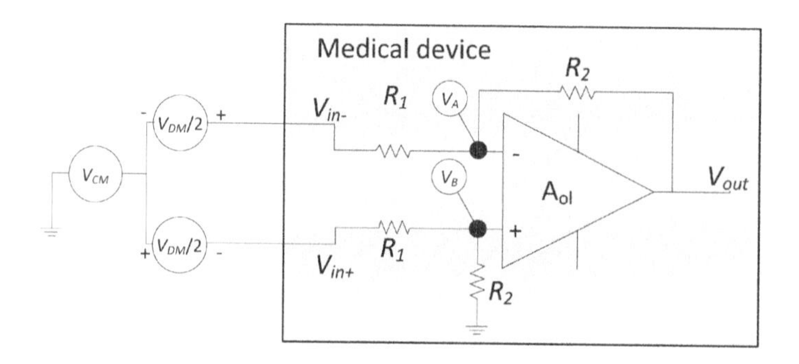

Figure 5.14 Subtraction circuit based on op-amp differential amplifier.

$$\text{with } V_{in-} = 0 \rightarrow V_B = V_A = \frac{R_2}{R_1 + R_2} V_{in+},$$
$$V^+_{out} = \frac{R_1 + R_2}{R_1} V_A \rightarrow V^+_{out} = \frac{R_2}{R_1} V_{in+} \tag{5.29}$$

Summing both the contributions at V_{out} we have

$$V_{out} = V^+_{out} + V^-_{out} = \frac{R_2}{R_1}(V_{in+} - V_{in-}) = \frac{R_2}{R_1} V_{DM}. \tag{5.30}$$

As expected, the scheme of Figure 5.14 completely rejects the common mode interference V_{CM} and amplifies the differential voltage V_{DM} by a factor R_2/R_1. Thus, the ideal CMRR (i.e., the ratio between the gain of V_{DM} and the gain of V_{CM}) is infinite. The previous scheme can be customized to perform other functions. **Inverting amplifiers** are obtained by setting V_{in-} to ground. Conversely, **non-inverting amplifiers** are obtained setting V_{in-} to ground and eliminating unnecessary resistances. The previous circuit is also helpful for raising the input signal to a defined voltage range. For example, single supply analog to digital converters accept positive inputs (e.g., 0–5 V) while input signals may be in the positive–negative range. The previous circuit can also perform '**level shifting**' by setting V_{in-} or V_{in+} to a predefined value.

Unfortunately, the input impedance of the previous scheme cannot be too high. This restricts the use of this scheme when the source impedance is high, as explained in Figure 5.5 because of the introduction of significant errors. So the previous scheme can be improved by adding a suitable configuration of buffer stages (op-amp followers) that offers very high input impedance as shown in Figure 5.15.

This configuration is known as an '**instrumentation amplifier**' and it is easily shown that the output voltage is given by

$$V^I_{out} = \frac{R_2}{R_1}\left(\frac{2R_3}{R_G} + 1\right) \times (V^I_{in+} - V^I_{in-}). \tag{5.31}$$

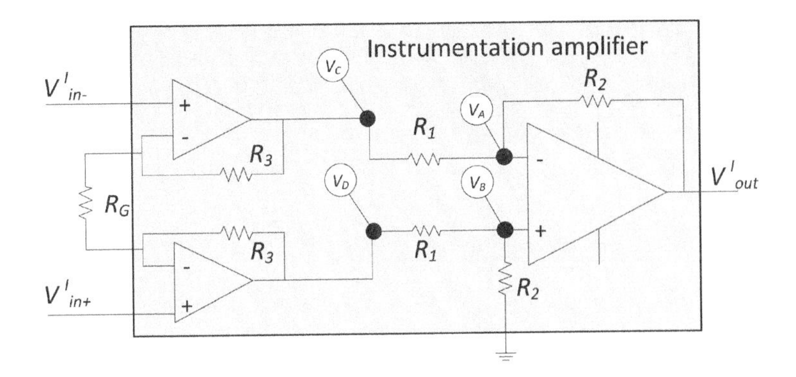

Figure 5.15 Instrumentation amplifier.

The ideal instrumentation amplifier has infinite rejection of the common mode signals, that is, common mode signals are eliminated at the output. In the real case, the circuits of Figures 5.14 and 5.15 have finite rejection of common mode voltage, mainly due to the imbalance of resistors. It can be shown that a limited imbalance even of 0.01% on resistor values that may be due to tolerance or temperature dependence can significantly degrade the CMRR (Scott, 2000). For this purpose, instrumentation amplifier circuits are provided on single chips containing resistors that are all laser trimmed to obtain the required precision. R_G is not integrated on the chip so that the gain of the amplifier can be set externally without affecting the CMRR.

With this configuration, a CMRR of as much as 120 dB may be achieved. In comparison, single op-amp chips have around 80 dB CMRR, but it may degrade due to resistance unbalance. To obtain high CMRR in real circuits, it is important that the parts of the two circuits that manage the two signals feeding the instrumentation amplifier are as symmetrical as possible. This prevents common mode interference becoming a differential mode signal. An unbalance between these circuits reduces the CMRR significantly. However, instrumentation amplifiers may still be not sufficient to overcome common mode interference, so other techniques may be used, as explained for the case of the ECG in the next section.

5.5.2 *Patient Model*

In the previous section, specific countermeasures have been set to reduce power supply interference. Unfortunately, these measures, such as the use of shielding and instrumentation amplifiers may not be sufficient. A residual interference induced by the patient itself may generate a significant source of errors, as in the ECG case.

In general, **a patient and device model** can help designers to understand the effects of design choices to reduce problems. The following sections will outline an example where the patient and the ECG model are outlined to understand the effects of design options and to determine the best choice to reduce such residual interference.

The interference model derived from the AMMI EC 11 standard and the ECG model may help to analyze effects and counter measures. On the left side of Figure 5.16, the patient interference model is depicted. Interference is produced by a 60 Hz signal generator and a series source capacitor C_s (usually of 200 pF equivalent to around 13 MΩ at 60 Hz) that

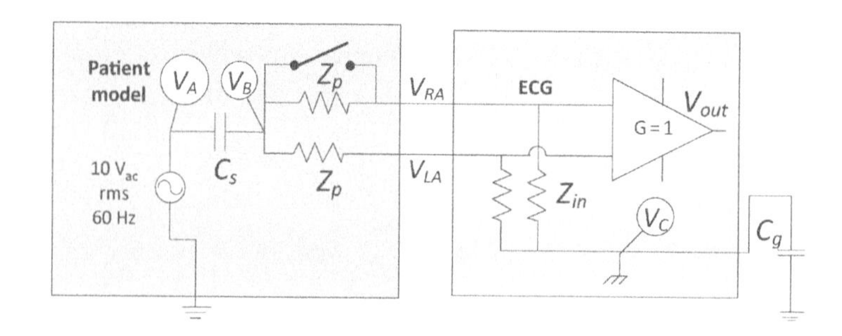

Figure 5.16 Patient and ECG model for power line interference.

simulates the voltage level of the patient when the current *i* flows from patient's capacitances to power line and ground as in Figure 5.11. Z_p is the impedance that simulates the skin–electrode effects (a 51 kΩ resistor in parallel with a 47 nF capacitor equivalent to an impedance of around 56 kΩ).

One Z_p impedance can be short-circuited to simulate a severe unbalance that may happen between the skin–electrode interfaces. On the right side, an ECG model is outlined.

5.5.3 The ECG Model

The ECG of Figure 5.16 is modeled as an ideal op-amp with impedances connected between input pins and ground. The device can be battery powered; in this case, we can consider that a stray capacitance connects the device to ground. If the device is connected to a power supply, the non-ideal behavior of the isolation components (power supply transformer, optocoupler etc.) may be modelled as an unwanted impedance that connects the isolated device part with the device ground. For example in the ECG device of Figure 5.17, this impedance is made up of C_2 and C_3 in parallel plus C_1 in series. The overall impedance is: $((C_2 + C_3) \times C_1)/(C_2 + C_3 + C_1)$. In both the battery and the line supply case, we can consider that a capacitor connects the device isolated part to ground. The value depends on the specific device configuration. For line-powered devices, as shown in Figure 5.17, there is usually a power supply transformer and a further isolation barrier that transfers signals and power without allowing a direct DC connection. This avoids dangerous leakage currents flowing to

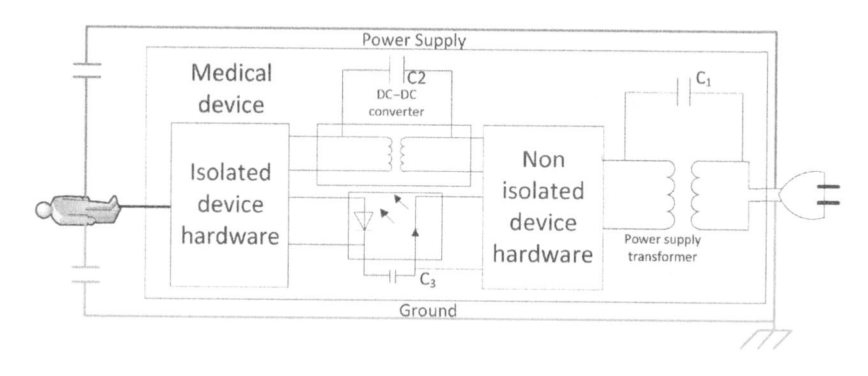

Figure 5.17 Medical device isolation model.

the patient due to improper grounding or hardware failure. This so-called 'galvanic' isolation discussed in Section 5.9 may be implemented with photodiodes and transformers.

The overall capacitance C_g can be therefore modeled by the capacitance of the power transformer C_1 (around 50 pF) in series with the parallel of the capacitance of the power supply isolator (40–80 pF) and the photodiodes (around 1 pF). The resulting capacitance is therefore in the order of 25 pF.

We can now consider the effects of the power supply interference and possible countermeasures. The power supply interference is present at output for two reasons: the input voltages generated by the unbalance of skin electrodes, and the common mode signal that is transferred at the output of the op-amp in Figure 5.16.

We recall that the the **common mode voltage** V_{CM} is defined as the half sum of the input voltages with respect to ground. When input impedances are equal V_{CM} is simply given by the current flowing on the input device impedance multiplied by the impedance itself (Ohm Law). The current flowing from source generator at V_A is given by the voltage generator divided by the overall resistance Z_{tot} with $Z_{tot} = Z_s + \frac{Z_p+Z_{in}}{2}$. This current flows halved on the V_{RA} and V_{LA} inputs. Considering that the impedance Z_s of the two capacitors C_s and C_g is given by $Z_s = 1/2\pi f C_s + 1/2\pi f C_g$, the common mode voltage at device input V_{CM} with respect to device ground point V_C is

$$V_{CM} = \frac{V_{RA} + V_{LA}}{2} = V_s \frac{Z_{in}}{Z_{tot}} \frac{1}{2} = V_s \frac{1}{\frac{2\times Z_s + Z_p}{Z_{in}} + 1}. \quad (5.32)$$

Assuming an op-amp gain $A_d = 1$ for simplicity, the output voltage of the op-amp is given by

$$V_{out} = V_{CM}/\text{CMRR} \quad (5.33)$$

where CMRR is the common mode rejection ratio defined in Table 5.15.

The switch in Figure 5.16 in the closed position simulates a **skin unbalance** that determines a voltage differential at device input ΔV_{SI}:

$$\Delta V_{SI} = V_{LA} - V_{RA} \approx \frac{1}{2} V_s \frac{Z_p}{Z_{tot}}. \quad (5.34)$$

This formula is simplified assuming that $Z_{tot} \gg Z_p$ and $Z_{in} \gg Z_p$.

To reduce the two interferences (common mode interference V_{CM} and skin unbalance ΔV_{SI}), the designer can control the values of Z_{in} and C_g, increasing Z_s. Z_{in} is the parallel combination of the cable impedance and the input stage capacitance. In general, the ECG input stage has a higher impedance than the cable shield capacitance (Sommerville, 1994). A resulting reactance of 100 pF may be expected by cabling. This translates into a 26 MΩ resistance at 60 Hz. A lower capacitance for the patient cable may be induced by the signal guarding technique (Horowitz, 1999) as discussed in Section 5.25.1.

The designer can control the values of the device input impedance Z_{in} and the impedance between device ground and power line ground C_g to increase Z_s. The previous formulas indicate that increasing the input ECG impedance Z_{in} will reduce the interference due to unbalance ΔV_{SI} (eq. 5.34) but it in turn it will increase the common mode voltage V_{CM} (5.32). As $Z_{in} \to \infty$, $Z_{tot} \to \infty$ and $\Delta V_{SI} \to 0$ in equation (5.34), but from equation (5.32) $V_{CM} = V_s$. The reduction

Table 5.17 Interference reduction for different techniques

	Common mode interference: values Referred To Input – (i.e. RTI)			Differential mode interference with input unbalance RTI	
	RTI mV_{rms}	RTI mV_{pp}	RTI values reduced by op-amp common mode attenuation mV_{pp}	mV_{rms}	mV_{pp}
ECG without right leg connection	3391	9591	*0.96*	4.93	*14*
ECG with right leg connected to ground	54	153	0.015	0.079	*0.223*
ECG with active RLD	0.448	1.266	0.00013	0.001	0.002

of the Z_{in} will have the reverse effect: $Z_{in} \rightarrow 0$, $\Delta V \rightarrow \infty$, but $V_{CM} \rightarrow 0$. The increase of C_g (i.e., Z_s) always has a positive effect on interference. This condition is obtained by increasing the isolation between patient ground and device ground using galvanic isolation or by using a battery supply.

The first row of Table 5.17, labeled 'ECG without right leg connection', shows some typical values obtained by the simplified equations of this section and confirmed by an analog electronic circuit simulators such as Spice. The values in the table show that some additional counter measures have to be implemented considering that the system errors must be lower than 30 μV to distinguish ECG components. With this configuration, we have 0.96 mV_{pp} of common mode signal referred to input considering the reduction of the common mode rejection given by op-amp. The differential voltage at input due to the imbalance between the skin–electrode interfaces is at 14 mV_{pp}. The next section will suggest further techniques to reduce interference.

5.5.4 *Right Leg Connection*

In the scheme of Figure 5.16, a high value of common mode interference V_{CM} was due to the current flowing from the voltage source having to discharge into the high value resistance Z_{in}, thus creating a considerable voltage drop even with limited current flow.

To improve performance, the common mode signal can be injected into the patient through the right leg connection as in the ECG. This reduces the common mode interference at both the inputs. At this aim, the patient right leg can be connected to the device ground through Z_{rl}. The skin–electrode impedance is always modeled with the same value Z_p. The benefit is that the current from the signal generator mainly flows through the right leg to ground and therefore does not create a high voltage differential V_{RA} and V_{LA} at the device input due to the presence of Z_{in}.

Referring to Figure 5.18, since $V_{RA} = V_{LA}$, the common mode voltage is $V_{CM}^{rl} \stackrel{def}{=} V_{RA} - V_C = V_{LA} - V_C$. The current i flowing at the V_A point is $i = V_s/Z_{tot}$, where Z_{tot} is the overall

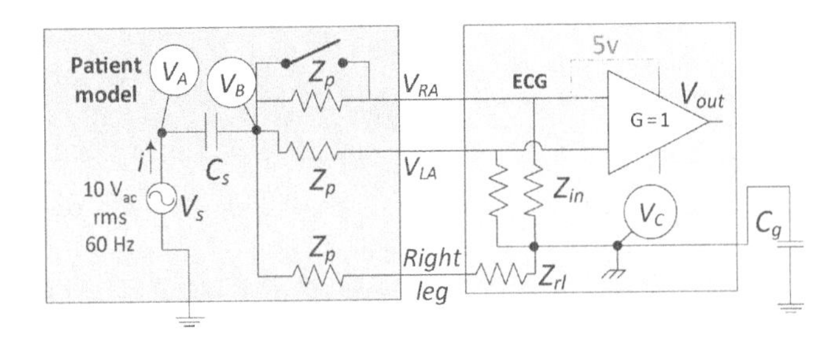

Figure 5.18 ECG model with right leg connection.

impedance seen by the voltage generator. Considering that $Z_{in} \gg Z_{rl}$, Z_{tot} is equal to $Z_{tot} \approx Z_s + Z_{rl} + Z_p$. Assuming $Z_{in} \gg Z_p$ we have

$$V_{CM}^{rl} = V_{RA} - V_C = \frac{V_B - V_C}{Z_p + Z_{in}} Z_{in} \approx V_B - V_C = V_s - i \times Z_s = V_s \frac{Z_{tot} - Z_s}{Z_{tot}}$$
$$V_{CM}^{rl} \approx V_s \frac{Z_{rl} + Z_p}{Z_{rl} + Z_p + Z_s} = V_s \frac{1}{\frac{Z_s}{Z_{rl} + Z_p} + 1}. \tag{5.35}$$

The closed switch in Figure 5.18 simulates an unbalance. Assuming that $Z_{in} \gg Z_p$, the unbalance determines a voltage differential at device input ΔV_{SI}^{rl}:

$$\Delta V_{SI}^{rl} = (V_B - V_C)\left(1 - \frac{Z_{in}}{Z_{in} + Z_p}\right) \approx V_s \frac{Z_{rl} \times Z_p}{(Z_s + Z_{rl}) \times Z_{in}} = V_{CM}^{rl} \frac{Z_p}{Z_{in} + Z_p}. \tag{5.36}$$

Comparing (5.32) and (5.35) we see that Z_{in} is roughly replaced by Z_{rl}, and a reduction of Z_{rl} will limit the common voltage at the input device while a change in Z_{in} has no effect on ΔV_{CM}^{rl} in the previous case of eq. (5.34). In turn, with right leg connection, the reduction of Z_{in} will reduce ΔV_{SI}^{rl}in (5.36).

Unfortunately, Z_{rl} cannot be zeroed because, if a fault condition occurs that short-circuits the device ground to the power line, the patient may be in contact with the device power. For example, the input may be short-circuited to the power supply as shown with the dotted line in top right part of Figure 5.18. Assuming a power supply of 5V, a current of $5V/Z_{rl}$ cannot be greater than the maximum safety limit that is 50 μA according to the IEC standard. This limits the minimum value of Z_{rl} to 100 kΩ.

Table 5.17 shows the results from this technique in the row 'ECG with right leg connected to ground'. The interference without imbalance is now at an acceptable level (15 μV) but with imbalanced interference, it reaches an unsustainable value (223 μV) with respect to the measurand. The previous formulas have to be used to understand the qualitative effects of the design choice since in the real circuit parasitic effects reduce performance.

Real measures on ΔV^{rl} show that this circuit is still not sufficient to achieve adequate performance. Better performance is obtained by reducing Z_{rl} without having the side effect of possible extra current to the patient.

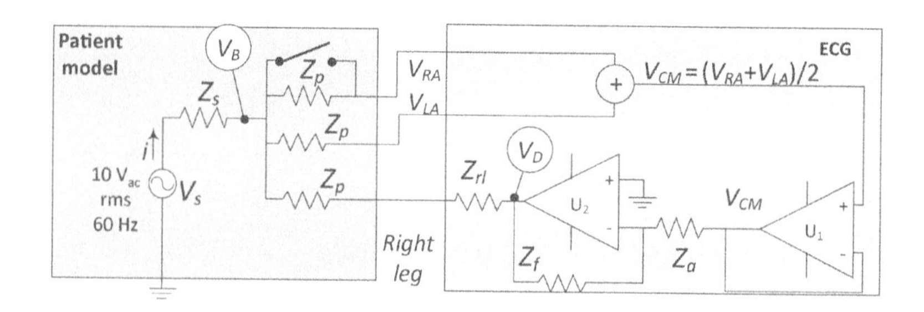

Figure 5.19 Right leg driver circuit.

5.5.5 Right Leg Driver Circuit

Z_{rl} may be reduced by using the circuit shown in Figure 5.19.

The strategy consists of generating a right leg signal that is proportional to the inverse of the common mode interference V_{CM}^{rld}. This negative feedback reduces the common mode voltage. To simplify the formulas, the capacitor C_g of Figure 5.18 has been omitted. Its effect can be considered by assuming Z_s as the parallel combination of capacitors C_g and C_s. The inputs V_{LA} and V_{RA} of the ECG device are summed together to obtain the common mode interference V_{CM}^{rld}. The signal is introduced in an op-amp in the 'follower' configuration discussed in Figure 5.7 to decouple the stage where the input signals are summed. The V_{CM}^{rld} is at the output of the op-amp U_1. Since the non-inverting input of U_2 is grounded, the inverting input is also at 0 volt when U_2 is not in saturation. This means that for the currents $V_D/Z_f = -V_{CM}/Z_a$, and therefore $V_D = -V_{CM}^{rld} \times Z_f/Z_a$.

The current i generated by the voltage source V_S mainly flows to the right leg since the input impedances are much higher than the impedance of the right leg input. This means that $i = (V_s - V_B)/Z_s \approx (V_B - V_D)/(Z_p + Z_{rl})$ and $V_B \approx V_{CM}^{rld}$. From the previous formulas the common mode voltage reduced by the right leg drive is given by

$$V_{CM}^{rld} \times (Z_p + Z_{rl} + Z_s) = V_s \times (Z_p + Z_{rl}) + V_D \times Z_s, \tag{5.37}$$

and therefore

$$V_{CM}^{rld} \approx V_s \frac{Z_{rl} + Z_p}{Z_{rl} + Z_p + Z_s\left(1 + \frac{Z_f}{Z_a}\right)} \tag{5.38}$$

Z_s is a 200 pF capacitor that at 60 Hz is equivalent to around 13 MΩ impedance and Z_p is at around 56 kΩ. Since $Z_s \gg Z_{rl}$ and $Z_s \gg Z_p$ we have

$$V_{CM}^{rld} \approx V_s \frac{1}{\left(1 + \frac{Z_f}{Z_a}\right) \times \frac{Z_s}{Z_{rl}+Z_p} + 1} \tag{5.39}$$

Comparing V_{CM}^{rld} in equation (5.39) and V_{CM}^{rl} in (5.35), it is clear that the right leg drive reduces the common mode voltage by a factor $\left(1 + \frac{Z_f}{Z_a}\right)$.

Table 5.18 Interference and countermeasures

Interference type	Counter measure
Out-of-band interference	Filtering in analog or digital domain
Transient interference with out-of-range amplitude	Voltage clamping circuits
Electric – electromagnetic interference	Shielding device and cabling with the Faraday cage technique
Magnetic interference	Shielding with high magnetic permeability metal alloy envelope
Interference that can be set as common mode signal	Design device so that interference becomes a common mode signal Increase input impedance Reduce impedance from interference source to device ground Increase isolation impedance between device ground and patient ground Use twisted wires Use active shielding on wires Design symmetric common mode circuits Use instrumentation amplifiers Set negative feedback to common mode interference Improve layout to reduce parasitic effects
Interference reducible through digital post-processing	Eliminate interference through digital filtering when medical diagnostic information is not impacted

Table 5.29 shows that there is a significant decrease of the common mode with active feedback. This value is in line with requirements since the ECG signal must have minimum components in the order of 30 μV_{pp}. In theory, a high value of Z_f/Z_a might be selected but unfortunately this provokes oscillations. Again a tradeoff between CMRR reduction and stability has to be the main focus, with the help of simulations and prototype evaluations as in (Venkatesh, 2011; Hann, 2011). Performance simulations are shown in the implementation section (e.g., Figure 5.67). Table 5.18 summarizes some interference types and the associated countermeasures discussed in this section.

5.6 Analog Filtering

THE NEED Undesired components such as noise or interference may distort the information included in the target signal. Filtering is the process that removes these components by reducing the amplitude of certain specific frequencies where target signal can be neglected.

5.6.1 Frequency Domain

The filtering operation is more easily understood in the frequency domain introduced in Section 5.2.3. Linear time invariant (LTI) filters such as those constituted by LTI components (i.e., resistors, capacitors, inductors and active LTI components) have a complex frequency response $H(f)$ given by equation (5.12). Given an input signal $x(t)$ and its Fourier transform $X(f)$, the output signal is

$$Y(f) = X(f) \times H(f) \tag{5.40}$$

where $Y(f)$ is the Fourier transform of the output signal $y(t)$ given by ($j = \sqrt{-1}$):

$$Y(f) \stackrel{def}{=} \int_{-\infty}^{+\infty} y(t)e^{-j2\pi ft}dt. \tag{5.41}$$

The inverse Fourier transform operator is given by

$$y(t) \stackrel{def}{=} \int_{-\infty}^{+\infty} Y(f)e^{+j2\pi ft}df = \int_{-\infty}^{+\infty} Y(f)(\cos(2\pi ft) + j\sin(2\pi ft))df \tag{5.42}$$

These formulas suggest that signals can be decomposed into sums of sinusoids where $Y(f)$ is the complex function that specifies the magnitude and the phase of each sinusoidal component of frequency f.

The formula $Y(f) = X(f) \times H(f)$ means that the input signal $x(t)$ is directly transformed into $Y(f)$ through the filter with $H(f)$ frequency response. More simply, the component at frequency f of the input signal $X(f)$ will be multiplied by $H(f)$ to determine the output signal at that frequency $Y(f)$. The magnitude of the output $\|Y(f)\|$ is equal to the product of the magnitudes of the filter response $\|H(f)\| = M(f)$ and the input:

$$H(f) = \frac{Y(f)}{X(f)}, \quad H(f) = M(f)e^{j\varphi(f)}, \quad \|Y(f)\| = M(f)\|X(f)\|. \tag{5.43}$$

The phase (or argument) of the output is equal to the sum of the phase of the input and the phase response of the filter $\varphi(f)$ given by (5.11):

$$\arg(Y(f)) = \arg\{X(f)\} + \varphi(f), \quad \varphi(f) \stackrel{def}{=} \arg\{H(f)\}. \tag{5.44}$$

If more filters $H_i(f)$ are cascaded, the overall response of the output $Y(f)$ is the product of each frequency response term with the input $X(f)$

$$Y(f) = X(f) \times H_1(f) \times H_2(f) \ldots \times H_n(f). \tag{5.45a}$$

Filter magnitude transformation is quite intuitive: a signal passing through a filter is attenuated at a specific frequency f by $M(f)$ according to (5.43). The phase transformation, however, deserves some extra explanation.

If we consider an input signal $x(t) = \sin(2\pi ft)$, from eq. (5.40) . . . (5.44), the output signal is given by

$$y(t) = M(f)\sin(2\pi ft + \varphi(f)) = M(f)\sin\big(2\pi f\big(t - \tau_\varphi(f)\big)\big), \tag{5.45b}$$

$$\text{with} \quad \tau_\varphi(f) \stackrel{def}{=} -\frac{\varphi(f)}{2\pi f}. \tag{5.45c}$$

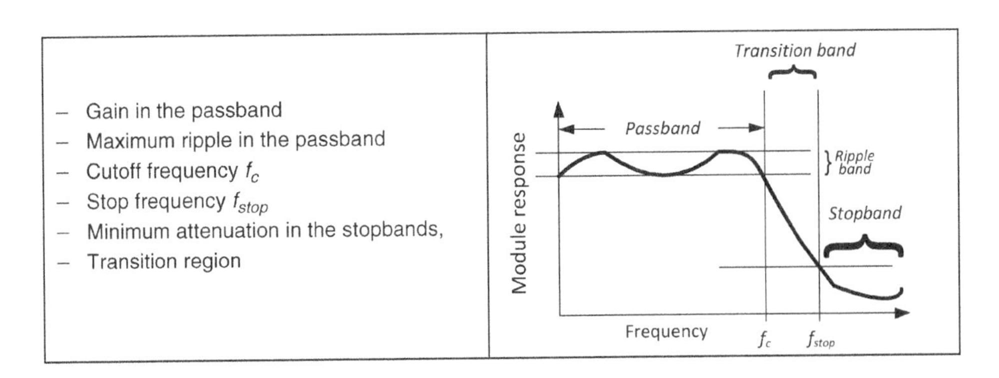

Figure 5.20 Main filter design parameters.

From (5.42), since the output signal $y(t)$ is the integral complex sum of sinusoids and cosinusoids (being $e^{jx} = \cos(x) + j\sin(x)$) if $\tau_\varphi(f)$ is a constant k then all the sinusoids at different frequencies are shifted by the same quantity k in time and therefore the shape of the signal is maintained. Filters where $\tau_\varphi(f) = k$ are called '**linear phase filters**' because from (5.45c) the phase response of the filter is linearly dependent on the frequency ($= kf$). However, if $\tau_\varphi(f)$ is not a constant for different f, then sinusoid components at different frequencies are shifted differently and therefore the shape of the signal changes after filtering.

Linear phase filters are important when signal morphology contains diagnostic information, as in ECG and EEG, and distortions may induce misinterpretation.

Regarding the magnitude, filters attenuate specific group of frequencies and therefore they can be: bandpass, low pass, high pass, bandstop and notch when they reject narrowband frequencies (see Figure 5.20 for a low pass filter response).

Considering the low pass example, the main parameters (as seen in Figure 5.20 left) for a filter design on the magnitude response are therefore as follows. The simple low pass RC filter has the filter response shown in (5.15) here repeated:

$$H(f) = \frac{1}{1 + j2\pi f\,RC}, \quad M(f) = \frac{1}{\sqrt{\left((2\pi f\,RC)^2 + 1\right)}} \tag{5.15}$$

with R and C the resistance and capacitance values of the filter, respectively. The magnitude and the group delay of a first-order low pass filter with 100 Hz cutoff frequency are depicted in Figure 5.21. Based on that figure or alternatively comparing equations (5.15),(5.17) and (5.18), it is clear that the response has an attenuation of 20 dB per 'decade' (i.e. frequency range from a f to $10 \times f$) or 6 dB attenuation per octave (range from a frequency f to $2 \times f$) and 3 dB attenuation at the cutoff frequency 100 Hz see (eq. 5.16). For the steepest attenuation, more RC filter sections have to be used. Unfortunately, simple RC sections cannot generally be cascaded since the low impedance of one section impacts on the following section. So a buffer stage, as discussed in Section 5.3, may be used to decouple impedances. In equation (5.45a), each RC filter section describes a mathematical pole obtained by replacing $j\,2\pi f$ with the complex variable, s. This is called the 'Laplace transform' of the filter transfer function: $H(s) = \frac{1}{1+s\,RC}$ that has a pole in $s = -1/RC$.

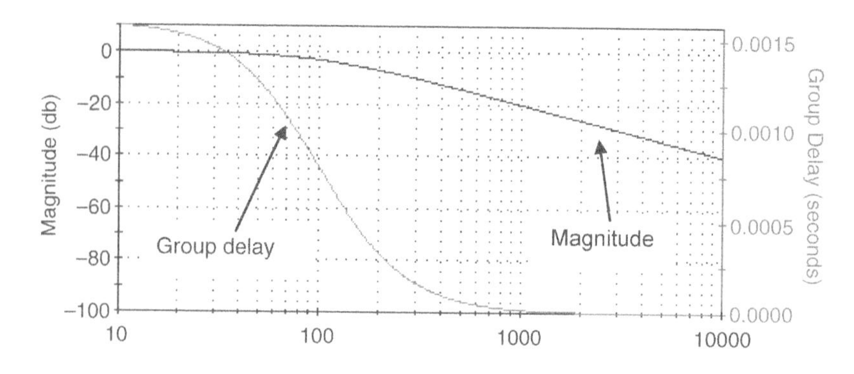

Figure 5.21 Magnitude and group delay of a first-order low pass filter.

Each pole given by an RC section in a transfer function adds 20 dB/decade attenuation so that an eight-pole filter will have 120 dB/decade attenuation. To obtain a filter's parameters, some filters models are available that are optimized in phase or magnitude behavior (e.g., Texas, 2008).

Figure 5.22 shows the behavior of the main **filter design models** for a fourth-order low pass filter with cutoff frequency at 100 Hz.

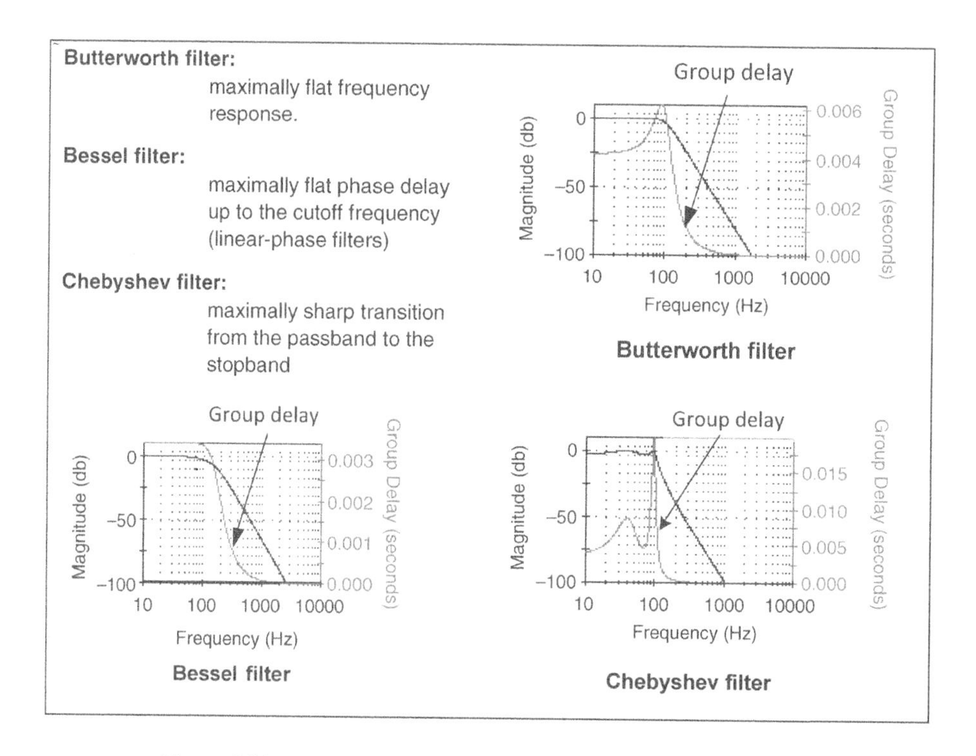

Figure 5.22 Main filter design models (magnitude and group delay).

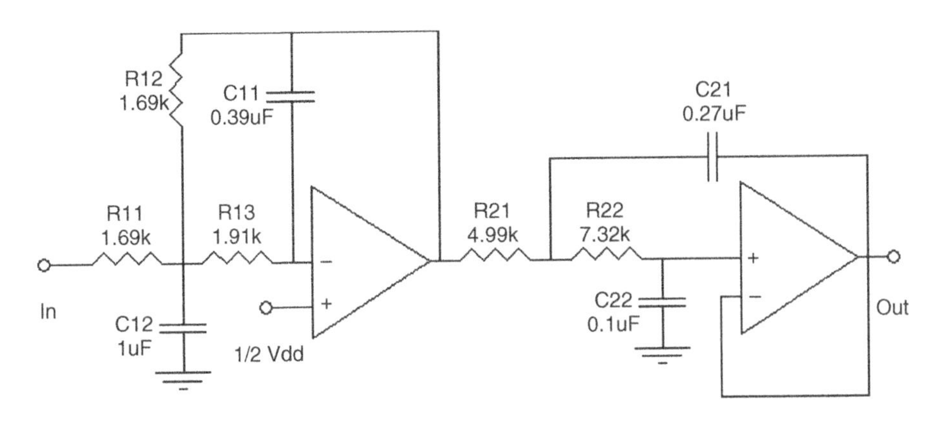

Figure 5.23 Fourth-order Bessel circuit with a first stage MFB and a second stage Sallen–Key topology.

As shown by Figure 5.22, the main tradeoff of the previous models are

- Bessel: limited attenuation over the frequency cutoff
- Butterworth and worse Chebyshev:
  - a peculiar phase response/group delay that provokes so-called 'overshoot', which means that the output signal exceeds its target value at steady state
  - ringing effects in response to step signal response, which means that the output value swings before achieving the target value

The previous filters may be implemented by various circuit topologies. The most popular topologies are

- Multiple Feedback Topology (MFT)
- Sallen–Key – voltage-controlled voltage-source (VCVS) topology.

Figure 5.23 shows an implementation of the Bessel circuit of Figure 5.22 using an MFT topology for the first stage and a Sallen–Key topology in the second stage. Op-amps provide impedance decoupling so that the low impedance of one filter stage does not affect the following stage.

Some problems may negatively influence the **real implementation**. First, resistors and capacitors are commercially available only at specific values, and their real value is different from the nominal value due to the tolerance, temperature dependence and other factors (see Tables 5.8 and 5.13). This produces different frequency response behavior between the ideal and real circuits.

The Sallen–Key unity gain architecture has the benefit of no gain sensitivity with respect to component variations at the cost of a worse attenuation at high frequencies compared to MFB (Karki, 2008).

The values of components are critical in filters. Even small changes in values can generate high performance changes (Texas, 2008). So low tolerance capacitors and 0.1% tolerance resistors are a good choice. It is also recommended that the **resistor** values are between 1 kΩ and 100 kΩ. Lower value resistors may draw too much current from the op-amp that then may not be able to achieve the required dynamic range. Higher resistor values may introduce excessive thermal noise.

Capacitors also have some constraints. Too low capacitors may be close to the parasitic capacitance. In general, free tools are available for filter design, avoiding the calculation of the specific values and performing simulations on frequency responses.

High-order filters can also be implemented using monolithic switched capacitor filters (Horowitz, 1999). These integrated circuit (IC) filters are able to emulate the previous filter models up to the tenth order. IC filters avoid the large number of components required for the active filters (eight resistors, eight capacitors and four op-amp for a Sallen–Key eighth-order architecture). In addition, monolithic switched capacitor filters avoid the sensitivity problem; since op-amp based filters performances are strongly related to component precisions. Unfortunately, IC filters add significant noise at the output, thus reducing their applicability in specific applications. Another approach is to perform filtering in the digital world when possible. Differences on digital and analog processing are discussed in the next section.

5.6.2 Analog versus Digital Filtering

We recall that analog filtering can be equivalently performed in the digital domain (Oppenheim, 2009)

1. when the input signal is band-limited
2. and the sampling rate is high enough to avoid the aliasing effect discussed in Section 5.7
3. and digital filtering is linear and time invariant.

Digital and analog filtering have both pros and cons. Analog filters are faster. Digital processing requires computational power that may be not sufficient for applications with too high signal frequencies, while simple analog components may operate at very high frequencies.

Analog filters also have a better dynamic range. An op-amp can easily have a 10 V dynamic range with an internal noise of around 2.3 μV. This means that there are around 22 equivalent bits for processing bits $= \mathrm{Log}_2\,(10\,\mathrm{V}/2.3\,\mu\mathrm{V})$, still higher than usual digital processing. This benefit is going to be reduced thanks to the increase of the digital resolution.

Another benefit of analog filtering is that signals with very large frequency dynamic ranges may be processed easily while in the digital domain this requires a major effort since computational power is proportional to the maximum frequency due to the Nyquist condition introduced in the next section. For example, an analog circuit can easily process frequencies from 0.01 Hz to 100 kHz, while in the digital domain the 0.01 Hz frequency signal requires 20 million points to be fully captured at 200 kHz sampling frequency (Smith, 1999). Latency (i.e. the time delay) may also be significant in digital filters while in the analog counterpart it is negligible.

On the other hand, digital filtering has superior performance in terms of stability since there is not the problem of parameter changes due to tolerance, temperature dependence or aging of the physical components. In the digital domain, parameters are stored in the memory and they can be changed easily to obtain adaptive filters. Digital filtering can be implemented easily, especially when filters are soft coded on firmware. As a final point, digital filters are not affected by analog circuit noise (thermal, switching, burst, EMI etc.); however, errors may be induced by the limits of the computational system such as: finite length registers, round-off errors and quantization errors.

5.7 ADC Conversion

The analog to digital converters (ADC) are the interface of the analog signal to the digital world. Some basic concepts may help understanding the electronic circuits associated with ADC.

An **analog signal** $x(t)$ is a function where any value of time t in the real domain corresponds to a value in the real/complex domain.

Both the time t and the amplitude $x(t)$ have values that belong to a continuous domain. Referring to Figure 5.24, the **sampling** operation consists of assessing the signal at predefined time intervals. The resulting discrete-time signal is a sequence $x[n] = x(n \times T)$, where n belongs to the set of natural numbers $n \in [-\infty \ldots 0,1,2, \ldots +\infty]$.

The final step toward digital is **quantization**, where the values $x[n]$ are associated with a set of finite numbers. The digital sequence $x_q[n]$ has discrete-time and discrete-amplitude values.

Sampling and quantization reduce the information contained in $x(t)$ so that, in general, it is not possible to reconstruct exactly $x(t)$ from its digital counterpart $x_q[n]$. Under certain circumstances it is possible to reconstruct $x(t)$ from the samples signal $x[n]$, that is to reverse the sampling procedure. In particular, let $x(t)$ be a band-limited signal, that is, with a Fourier transform $X(f)$ (see equation (5.41)) where

$$X(f) = 0, \quad \text{for } \|f\| > f_N/2.$$

According to **Nyquist sampling theorem**, if $x(t)$ is sampled at period $x[n] = x(n \times T_N)$ so that $T_N \leq 1/f_N$ then from $x[n]$ it is possible to reconstruct exactly $x(t)$ with (Oppenheim, 2009)

$$x(t) = \sum_{n=-\infty}^{\infty} x[n] \times \text{sinc}(\pi(t - nT_N)/T_N), \tag{5.46}$$

where the sinc(x) function is defined as

$$\text{sinc}(x) = \frac{\sin(x)}{x}. \tag{5.47}$$

Unfortunately, a real filter that can perform the previous operation does not exist since it would have to perform processing from $-\infty$ to $+\infty$. In any case, since the ideal reconstruction function reduces in time by a factor $1/t$, good approximations can be performed by truncating the sinc(x) function over a specified time window.

The other approximation is that *real band-limited signals do not exist*, since band-limited signals have infinite duration in time being non-null for t spanning from $-\infty$ to $+\infty$.

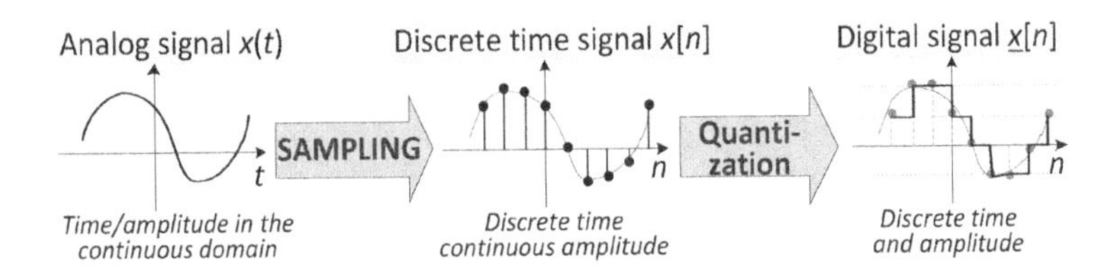

Figure 5.24 ADC conversion.

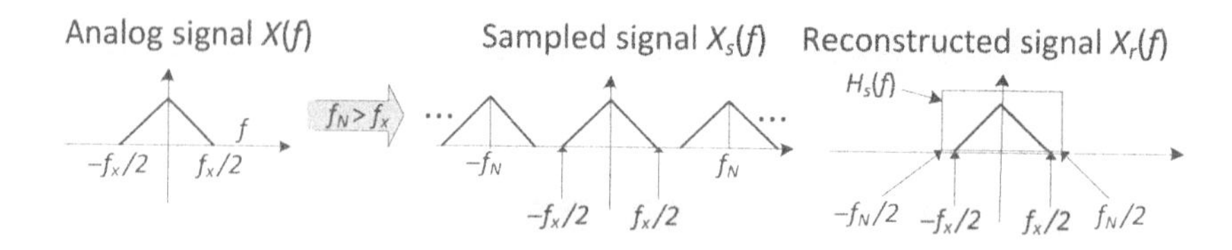

Figure 5.25 Sampling and perfect reconstruction.

When non-band-limited signals are sampled, the **aliasing effect** occurs. This can be easily shown in the frequency domain.

Given $X(f)$ the Fourier transform of the time domain signal $x(t)$, the Fourier transform of $x[n] = x(n \times T_N)$ is given by

$$X_s(f) = \frac{1}{T_N} \sum_{k=-\infty}^{\infty} X\left(f - k\frac{1}{T_N}\right). \tag{5.48}$$

This formula suggests that after sampling for a period T_N, the spectrum of the discrete time signal is obtained by summing infinite replicas of the spectrum of the continuous time version $X(f)$, shifted by the sampling frequency $f_N = 1/T_N$.

If $x(t)$ is band-limited with $X(f) = 0$ for $\|f\| > f_x/2$ and the band of the signal f_x is smaller then the sampling rate f_N (i.e. $f_N > f_x$), from equation (5.46) the different replicas of $X(f)$ do not overlap, as shown in Figure 5.25.

This means that if the sampled signal in the Fourier domain is passed through a filter such as $H_s(f) = \begin{cases} 1, & \|f\| \leq f_N/2 \\ 0, & otherwise \end{cases}$ then the reconstructed signal Fourier transform $X_r(f)$ coincides with $X(f)$:

$$f_N < f_s \Rightarrow X(f) \equiv X_r(f) = X(f)H_s(f) \tag{5.49}$$

This equation reflects the Nyquist criterion for perfect reconstruction of band limited digitally sampled signals. In the time domain, this equation is translated into (5.46) (see also Kester, 2004 for a deeper insight).

If $x(t)$ has frequency components over $f_N/2$ (i.e. $f_N < f_x$), then such components are 'aliased' after sampling. The infinite replicas of the spectrum $X(f)$ produced by the sampling procedure in equation (5.48) overlap and it is no longer possible to reconstruct the original signal as shown in the circles of the right side of Figure 5.26:

$$f_N < f_x \Rightarrow X(f) \neq X_r(f) = X(f)H_s(f).$$

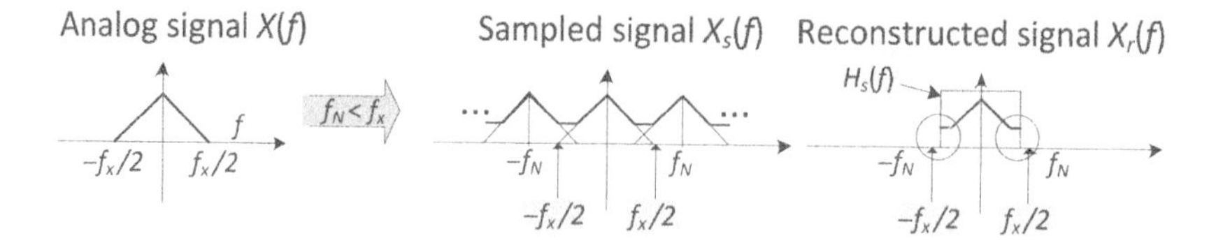

Figure 5.26 Sampling with aliasing effect.

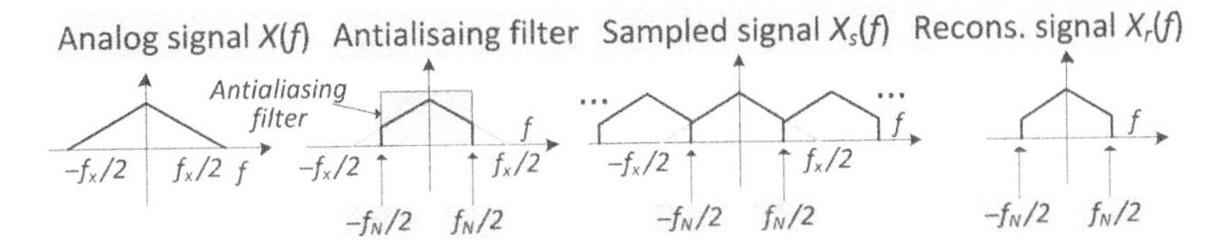

Figure 5.27 Sampling with anti aliasing filter.

Since all real signals are of finite time length, they are not band-limited and some aliasing will always be present. To reduce this distorting effect, an **antialiasing filter** is placed before the sampling as shown in Figure 5.27. This filter reduces the components of the signal $X(f)$ that are out of the allowed band $\pm f_N/2$. This avoids these components being folded back into the $\pm f_N/2$ outer bands after sampling, reducing the aliasing on the reconstructed signal $X_s(f)$. More clearly, with aliasing the signal frequency components f that are over $\pm f_N/2$ frequency (i.e., $|f| > f_N/2$) will be introduced (i.e., folded) into the $\pm f_N/2$ frequency window as shown in Figure 5.26.

The next step to obtain a digital signal is **quantization**. This procedure associates a discrete value (second column of Table 5.19) for all the amplitude values that belong to a specific range (first column of Table 5.19). Such discrete value is binary encoded as for example in the third column of the same table. Further discussions on codes in real ADC can be found in (Kester, 2004). The quantized (i.e. discrete) values may be uniformly distributed within a dynamic input range that we may assume to be from 0 to X_M. Assuming, for example, $X_M = 5$ V and four available values that correspond to a two-bit resolution for an ADC, analog input values of $x[n]$ will be translated into a corresponding code as shown in the Table 5.19.

The minimum resolution of the previous ADC can be considered as the minimum input variation that produces a transition in the output code. This value (1.25 V in the previous case), is called the least significant bit (**LSB**) that for an M-bit ADC is given by LSB $= X_M/2^M$. The maximum error is equal to $\pm$ half an LSB. We may assume that $x[n]$ is statistically uniformly distributed in any quantization range shown in the first column of Table 5.19. This assumption is a fair approximation in most cases of non-DC signals. The error e in any quantization range is statistically equally distributed within the interval $\pm$LSB/2. The probability density function of e, $p(e)$ is therefore equal to 1/LSB in the interval $\pm$LSB/2 and zero otherwise. The **quantization mean square error** σ_e^2 is given by

$$\sigma_e^2 = \int_{\infty}^{\infty} e^2 p(e) de = \int_{-LSB/2}^{+LSB/2} e^2 (1/LSB) de = \frac{1}{LSB} \frac{e^3}{3} \Bigg|_{-LSB/2}^{LSB/2} = \frac{LSB^2}{12}. \tag{5.50}$$

Table 5.19 Transfer function of a two-bit ADC

Analog input value range $x[n]$	Discrete value in reconstruction	Associated digital output code in $\underline{x}[n]$
3.75–5	4.375	11
2.5–3.75	3.125	10
1.25–2.5	1.875	01
0–1.25	0.625	00

$x[n]$ ADC $\underline{x}[n]$
Discrete time signal
Digital signal

It can be shown that the error is roughly uniformly distributed across the sampled frequency band from DC to $f_N/2$.

From eq. 5.50, the root mean square noise is

$$\sigma_{e,RMS} = \frac{LSB}{\sqrt{12}} = \frac{X_M}{2^M\sqrt{12}}, \quad \sigma_{e,\,RMS} = \sqrt{\sigma_e^2}, \tag{5.51}$$

where M is the number of quantization bits. Let us assume that $x(t)$ as a full range $(0\ \ldots\ X_M)$ sine wave signal:

$$x(t) = 0.5\{X_M \sin(2\pi f t) + X_M\}. \tag{5.52}$$

The mean square value of a sine wave is equal to half the squared peak-to-peak value divided by 2, that is

$$\sigma_x^2 = 0.5\left(\frac{X_M}{2}\right)^2 = (X_M)^2/8. \tag{5.53}$$

The signal-to-noise ratio SNR is equal to the ratio of the mean square values of the signal and the noise that in dB is

$$\begin{aligned} SNR_{dB} &= 10\log\left(\frac{\sigma_x^2}{\sigma_e^2}\right) = 10\log\left(\frac{(X_M)^2}{8}\bigg/\frac{LSB^2}{12}\right) \\ &= 10\log\left(\frac{(X_M)^2}{8}\bigg/\frac{(X_M/2^M)^2}{12}\right) \end{aligned} \tag{5.54}$$

which can be simplified into $SNR_{dB} = M \times 20\log(2) + 10\log\left(\frac{4}{3}\right)$ and

$$SNR_{dB} = 6.02 \times M + 1.76\,dB. \tag{5.55}$$

This formula shows the SNR value computed over the whole frequency spectrum from DC to $\pm\, f_N/2$, assuming a full scale sine wave input signal. When the input signal is uniformly distributed in the range $(0\ \ldots\ X_M)$, we have simply $SNR_{dB} = 6.02 \times M$.

For signals that contain frequency components in the whole spectrum $\pm f_N/2$, equation (5.55) suggests the minimum bit resolution to achieve a given SNR.

An analog signal may be sampled at a rate that is higher than its minimum, as shown in Figure 5.25 (oversampling process). In this case, the noise present in the frequencies that are not occupied by the signal itself can be eliminated by digital filtering. SNR can then be computed only on the bandwidth occupied by the signal, and the formula becomes

$$SNR_{dB} = 6.02M + 1.76\,dB + 10\log\left(\frac{f_N}{f_c}\right) \tag{5.56}$$

where f_N is the sampling frequency and f_c is the overall bandwidth occupied by the signal to be sampled.

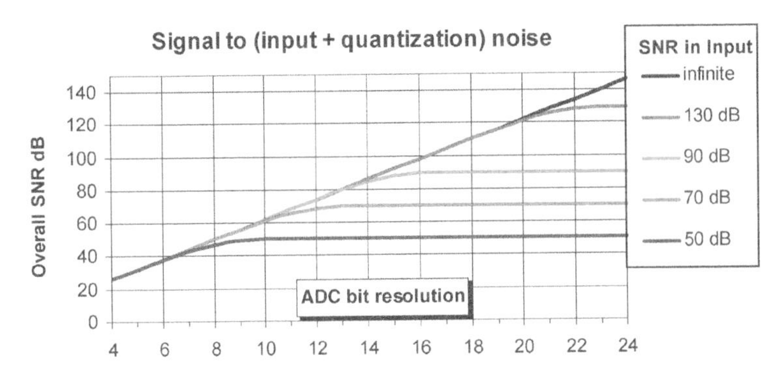

Figure 5.28 SNR vs. ADC bit resolution.

The final term of (5.56) is called the **process gain** and it is of interest especially in some specific ADCs called sigma-delta converters where filtering is greatly simplified thanks to oversampling and noise shaping that shifts noise power out of the signal frequency bandwidth. This allows for a greater reduction of noise.

Equations (5.55) and (5.56) suggest that the quantization process adds noise to the sampled signal and therefore if the signal to be sampled is already affected by noise with variance σ_v^2 (e.g., thermal noise), there is no benefit in increasing the bit resolution over a certain threshold. Figure 5.28 shows the overall $SNR_{dB} = 10\log\left(\frac{\sigma_x^2}{\sigma_e^2+\sigma_v^2}\right)$ with respect to ADC bit resolution for signals that contain noise with an *input* $SNR_{dB} = 10\log\left(\frac{\sigma_x^2}{\sigma_v^2}\right)$ and $\sqrt{\sigma_x^2} = 5$ V. The upper line is the overall SNR with no input noise (*Input* $SNR_{DB} = 0$).

Figure 5.28 shows that if the input signal has an *input* $SNR_{DB} = 50$ dB, there is no benefit in using an ADC with more than 8-bit resolution while with 90 dB input signal, it is worth considering a 16-bit resolution ADC.

Note that a 100 kΩ resistor on 200 Hz bandwidth generates $\sigma_{v,rms} = 0.6$ mV Johnson noise which yields an *input* $SNR_{DB} = 130$ dB. As shown in Figure 5.28, a resolution greater than 22 bits is useless with a similar resistor at the input.

A similar method is applied to designing antialiasing filters. Let us assume that the signal $x(t)$ to be sampled at a sampling frequency of f_N is band-limited with $X(f) = 0$ for $\|f\| > f_x/2$. If there is full-scale interference/noise in the stopband $[f_x/2 \ldots f_N/2]$, then, this interference is introduced into the sampled signal after sampling, if there is not an antialiasing filter. The antialiasing filter must therefore attenuate the stopband signal so that at least the resulting mean square value of sampled interference due to aliasing σ_a^2 is lower than the mean square quantization noise: $\sigma_a^2 < \sigma_v^2$ or from (5.56)

$$SNR_{dB} = 10\log\left(\frac{\sigma_x^2}{\sigma_a^2}\right) > 10\log\left(\frac{\sigma_x^2}{\sigma_v^2}\right) = 6.02 \times M + 1.76. \tag{5.57}$$

However, if the maximum undesired interference in the stopband is lower than the full-scale signal $x(t)$ by k dB, then the antialiasing filter can have a lower attenuation in the stopband by k dB with respect to (5.57). This is a good situation since filters with less attenuation are more simple and less critical.

The choice of the ADC features and the design of the antialiasing filter can therefore be performed considering that: 1) ideal ADC errors are generated only by the sampling procedure that generates aliasing artifacts and by quantization that determines noise 2) real ADCs have other limits that affect performance (see Kester, 2004). In general, there is distortion, noise, cross-channel interference, sampling jitter (i.e., undesired time shift of the clock onset) and intermodulation. For this, datasheets of commercial ADCs include some indicators that evidence the achievable performance. The resolution performance can be disclosed through the ENOB (effective number of bits) that indicates the noise-free bits of resolution, clearly lower than the nominal bit resolution M. More simply, after a certain resolution, individual codes cannot be distinguished due to the noise floor generated by thermal noise and other non-ideal effects. ADCs have to be selected considering their achievable resolution, which is usually 1 or 2 bits lower than the nominal one.

The main features of commercial ADCs are

1. maximum sampling rate (up to over 3 gigasamples/second)
2. nominal bit resolution (2 to 31 bits)
3. number of channels (1 to 32 converters per chip even with 64 inputs)
4. type of analog inputs (single-ended, differential etc.)
5. input dynamic range
6. input noise distortion performance
7. digital interface to communicate digital input signal (serial, parallel, low voltage differential signal (LVDS), I^2C etc.)
8. power and current consumption (from hundreds of μW to watts)
9. packaging.

Demo boards are usually available from vendors for fast prototyping. More often, general-purpose ADCs are also contained within microcontrollers chip as discussed in the next section.

The ADC design process is summarised in Table 5.20.

Table 5.20 ADC design process

1. evaluate the minimum error LSB/2 admissible for the specific application and the maximum input dynamic range X_M at the ADC input
2. the theoretical resolution M is given by $LSB = X_M/2^M$
3. assess the sampling frequency as $f_N > f_x$, where f_x is the bilateral input signal bandwidth
4. design the antialiasing filter so that the unwanted aliased signal is lower than the quantization noise $(\sigma_a^2 < \sigma_v^2)$
5. design the circuit so that the overall noise σ_v^2 (computed with (5.7) for thermal noise) is lower than the quantization error σ_e^2 (equation (5.48))
6. if the overall noise (thermal, aliasing, quantization, . . .) does not achieve specific requirements, oversampling may help reducing noise due to aliasing.

5.8 Programable Devices

To sustain product competitive advantage, there is a continuous effort in electronic device design to reduce costs and development time and to increase performance.

Costs are associated with various factors mainly related to development and production. Time to market and performance are also extremely relevant in competitive markets.

The hardware of devices has a deep impact on cost and development time. So technological progress is now offering more efficient approaches that *replace static hardware*-implemented functions by *soft coded programs*. The use of PC platforms with software-implemented functions or the usage of microcontrollers with specific firmware greatly reduces the cost/time factor over static hardware implementation.

In production, soft coded functions do not increase the unit device cost since software duplication has negligible effect on the product marginal costs. Time required for industrial production of soft coded functions is related to the production of memory support or to the download into programmable devices. Such time is also negligible with respect to hardware production.

At the product development stage, there are other relevant benefits that make soft design less expensive and less time consuming. Soft coded functions are produced by compiling code, while hardware production implies at minimum producing PCBs, buying components and soldering, that is, a relevant cost for each implemented version.

Error fixing is also an issue to be considered: it is immediate in soft coded functions, when revised and debugged code is available, but it requires reworking in hardware.

Flexibility and scalability are a major benefit of soft coded functions since changing functions or increasing functional complexity is easily performed up to the target maximum computational power. Any changes in hardware require extra work.

The suggestion is then to *shift all the functions from hardware to soft programs*. Unfortunately, soft coded solutions cannot implement complex and computationally intensive functions and therefore specialized hardware is usually necessary to perform more involving functions.

Technological improvements are having two effects. From the one hand, programmable device performance is improving and, then, the range of functions that can be soft implemented are increasing in current products. On the other hand, thanks to technological progresses, products may implement more complex functions that require more powerful and specialized hardware.

Figure 5.29 shows the qualitative performance and the convenience of implementing product functions with various hard–soft technologies.

The convenience accounts for factors such as cost of development and production, time to market, flexibility, scalability and after-sales maintenance. At the top left of Figure 5.29, there is the fixed-function hardware chips with specific cost of material and design cost and no programming capability. The next option in Figure 5.29 is based on **ASICs** (application-specific integrated circuit) that are integrated circuits designed for specific purposes. ASICs have low production cost but high design costs to produce the first sample (in the order of millions of dollars). This technology is then suitable for large volume productions. Alternatively, **FPGAs** (field-programable gate array) are reprogrammable components with higher unit cost. With these components, designers may implement logic and some basic analog functions and the result is similar to dedicated hardware with functions performed in parallel.

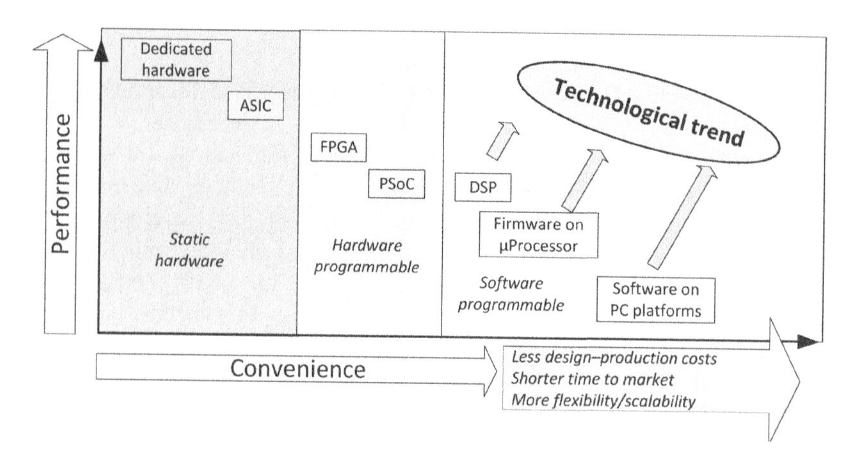

Figure 5.29 Qualitative assessment of component technologies.

FPGAs are composed by parallel blocks that are implemented on the chip surface. A dedicated area is assigned to a specific task that may work without influencing other FPGA blocks. This improves the performance compared with devices that execute programs on the chip such as digital signal processors (**DSP**), microcontrollers, microprocessors). FPGAs contain mostly digital circuits while the programmable system on a chip (**PSoC**) allows a more simple approach to hardware programming with the possibility of using analog functions such as amplifiers, ADC, DAC filters. Microcontrollers (**MCU**) are chips containing in one package simple computers (CPU, memory etc.) plus other valuable analog and digital functions. (timers, ADC, DAC, general purpose I/O, communication interfaces, comparators etc.). Microcontrollers contain soft coded programs (i.e. firmware) that are executed on chip to perform specific functions.

The use of software running on a PC platform is the most convenient approach for implementing a specific function. The reduced cost of software development is compensated by the need of a PC platform and the limited computational power compared to other solutions. PC platforms are generally used for GUI (graphic user interface) implementation: the GUI does not require high computational power and PC platforms have optimized hardware for this task at reduced unit cost. More recently, thanks to the increased computational power, PC software may also perform simple digital processing functions.

In medical devices, PCs may be embedded in the device or may be standalone desktops for large devices. For many tasks requiring limited computational performance and when software/PC platforms are not suited, microcontrollers are a good solution. For embedded systems, **microcontrollers** perform complex tasks with few additional components, little power consumption and the advantage of performing many functions through the internal firmware that can be changed without hardware reworking.

Microcontrollers are extensively used in medical devices since they concentrate, in one chip, many required functions, especially in consumer medical devices. Examples are blood pressure monitors or blood glucose meters, where microcontrollers perform the more relevant functions.

For this reason, in the next section, we will focus on microcontrollers that are also a good example to introduce programmable devices.

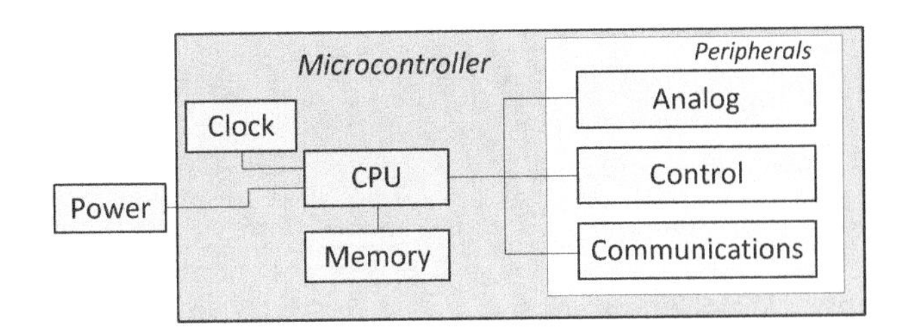

Figure 5.30 Microcontroller simplified architecture.

REQUIREMENTS Microcontrollers (MCUs) are having great market success. The market offers over 20,000 different types of microprocessors from tens of vendors with prices from cents to a hundred dollars. The main benefits of MCU usage are: low cost, scalability, widest choice (performance, packaging, peripherals etc.), limited additional components, simplicity of firmware generation thanks to integrated development environment tools, free libraries, in-circuit programming and debugging and MCU simulation and emulation. All these benefits greatly reduce time to market, risks and costs.

Figure 5.30 shows a simple scheme of a microcontroller. A microcontroller is mainly a CPU with the associated memory, an internal or external clock and peripherals that handle communication, device control and analog functions.

The main requirements for chip selection are on the computational power and the peripheral availability.

Computational power is determined by the core processor type (ARM etc.), the core size (8, 16, 32 bits, with possible dual core capability), clock speed (from MHz to GHz) and memory size and types (EEPROM, RAM, flash etc.). The required **peripherals** determine additional restrictions on MCU choice. **Communication** peripherals allow the MCU to exchange data with external chips/devices. The MCU has specialized peripherals to handle transparently specific protocols such as: USB, Ethernet, SPI, CAN etc.

Control peripherals allow for a simple firmware management of devices such as LCD displays, LEDs and motor control. Other peripherals of this type are timers, configurable I/O ports, interrupts. Analog functions include ADC, DAC, compare and capture modules. **Power supply** is another feature that may suggest specific MCUs models, for example, for low power consumption applications where the voltage supply is in the range of 0.8 1.15 V. An example of a consumer medical device is detailed in Figure 5.31 where an MCU based blood pressure device is shown.

The blood pressure unit is powered by a battery and a DC–DC converter provides regulated power to the MCU and the rest of the circuit as explained in section 5.9. The MCU contains a CPU and the memory where the firmware is downloaded. The CPU can control peripherals by firmware. The upper peripheral in the figure is the pulse wave modulation source (PWM) that generates specific analog signals to drive a buffer that in turn produces audio signals through the speaker. The LCD controller drives the LCD display that shows the values of pressures and other warnings. The I/O pins perform three tasks:

1. power-on LEDs
2. activate the pressure controller that increases the cuff pressure
3. detect keypads pressed by user.

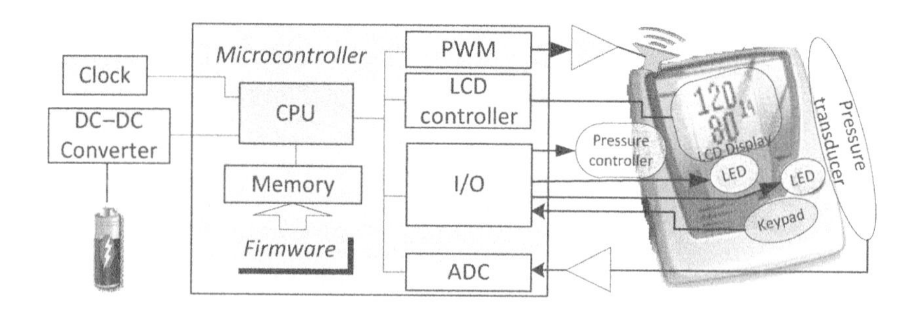

Figure 5.31 Blood pressure architecture.

ADC is connected to the pressure transducer that digitizes the cuff pressure level amplified by an op-amp. From this scheme, it is clear that a blood pressure device and similar devices may have a greatly simplified hardware design, exploiting the capabilities of microcontrollers. Many vendors have organized their products to offer the widest scalability and interchangeability between different MCU types. This simplifies the MCU choice: a simpler MCU may be first tested through evaluation boards. A more powerful MCU may be later selected with limited changes in the developed firmware.

Firmware design remains a major cost. So vendors offer integrated development environments (**IDE**) whose features reduce firmware development time. As outlined in Figure 5.32, an effective IDE should include many components starting from high level programing languages (C etc.) with free available libraries for more common tasks (peripheral usage, memory usage, sensor management etc.).

An IDE will provide easy compilation of the source code with the required libraries, and possibly an 'abstraction layer'. The **abstraction layer** is an included source code file that offers a generic interface to call MCU functions and MCU resources hiding the implementation details of the specific MCU type. For example, an abstraction layer may be composed of files where the physical addresses of peripherals of a specific MCU are associated to a

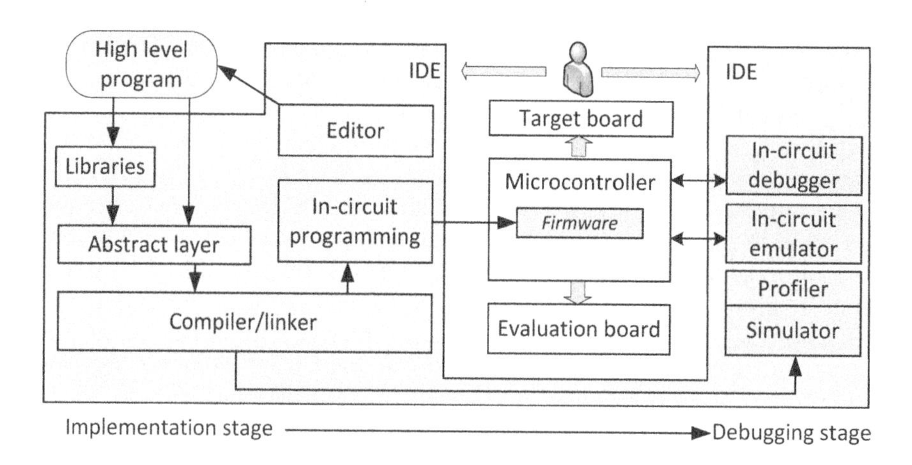

Figure 5.32 Microprocessor work environment.

symbolic label (e.g., *in C #define PIN_A0 40*). Supplying different files for each MCU type will allow the changing of MCUs by simply recompiling the file associated to the selected MCU. Source code written in high-level language with associated libraries and abstraction layers is the input to the compiler and the linker that will generate the corresponding binary code. This code will be downloaded as firmware into the microcontroller memory. This last operation may also be performed by an '**in-circuit programming**' system where firmware is loaded directly through the evaluation/target board without detaching the associated MCU from the PCB. This operation allows the testing of the firmware onboard. Alternatively, the firmware may also be tested through an MCU simulator and performance (memory use, time performance etc.) may be assessed and improved through profiler tools. The testing phase may be greatly reduced through **in-circuit emulation** or debugging. In-circuit emulation allows the replacing of MCU chips with specific hardware that emulates the MCU behavior. This allows extensive debugging capabilities (breakpoints, stopwatch, step-by-step running, etc.).

In most modern environments, MCU boards may offer direct in-circuit debugging, that is, a specific debugging port is available on the circuit for debugging the MCU, and the MCU does not need to be replaced by an emulator. A PC with specific IDE and tools is connected to this debugging port to take control of the MCU and perform the usual debugging. The debugging port may be based on a proprietary approach or may use the JTAG (Joint Test Action Group) standard (IEEE 1149.1). JTAG allows both in-circuit programming and debugging on a variety of devices. Other tools are also offered by vendors to address specific applications (e.g., motor control interface) or to improve the development process (configuration tools, collaboration tools, project managers etc.).

These tools greatly simplify debugging activities. If the IDE allows the easy performing of all these activities, the productivity can be expected to increase significantly, with associated benefits in terms of cost and time to market. For this reason, the effectiveness of the MCU environment is often given more consideration than chip performance in chip selection.

5.9 Power Module

THE NEED Every circuit device needs power to work. Ideally, circuits would need a constant voltage supply *K* for any current consumption. This voltage supply has to be obtained by the available power sources: battery, AC power line or wall plug-in units that transform AC power line into a low voltage supply.

Voltage supply is obtained by a power module circuit that generates V_{out} from a source V_{in} as shown in Figure 5.33. Ideally, V_{out} should not depend on variations of V_{in}. The voltage is '**regulated**' at the output when it does not depend on the current absorbed by the circuit and on the input voltage variations.

Other ideal properties should hold between the input and output pins of the power module. The source should be completely isolated by V_{out}, that is, the impedance between input and output should be infinite. This property is important to reduce interference and increase safety in medical devices when the source is the power line. The power module should also ideally transfer all the input power to the output achieving a 100% efficiency. This feature is important to increase the availability of battery-powered devices. The ideal power module is

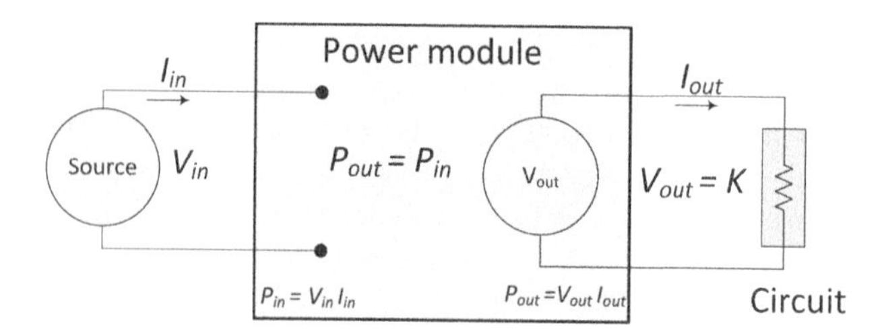

Figure 5.33 Ideal power module.

therefore modeled on the ideal voltage source that provides constant voltage with zero internal impedance.

Real behavior of power modules is constrained by the performance of power sources and the component technology. The designer must therefore find the best circuit design tradeoff, given the constraints of the source technology and the circuit requirements. The usual process is depicted in Figure 5.34. The requirements produced in the requirement analysis stage are the input of the circuit design that has to take into account technological constraints to generate specifications. If the specifications do not satisfy the requirements, further design steps have to be performed. In the specific case, requirements define the sources where the power module is connected and the needs of the circuits to be powered. The power components have greatly evolved in recent years, achieving stringent requirements even considering the limits of the power sources. The following sections will address source characteristics, requirements and circuits design. In addition, Section 5.9.2 will focus on electrical safety and the associated requirements.

5.9.1 Power Sources

The available power sources are the AC power line and the various types of batteries. Table 5.21 summarizes the ideal and real behavior of **AC power lines**. Ideally, the power line supply is simply a sinusoidal voltage source with zero internal impedance, constant nominal voltage V_{rms} and frequency f. Both V_{rms} and f are country-specific in the range of $V_{rms} \in [100 \ldots 240]$ V and $f \in [50,60]$ Hz. Real AC voltage supply sources have two main problems: EMI (ElectroMagnetic Interference) and voltage source variations.

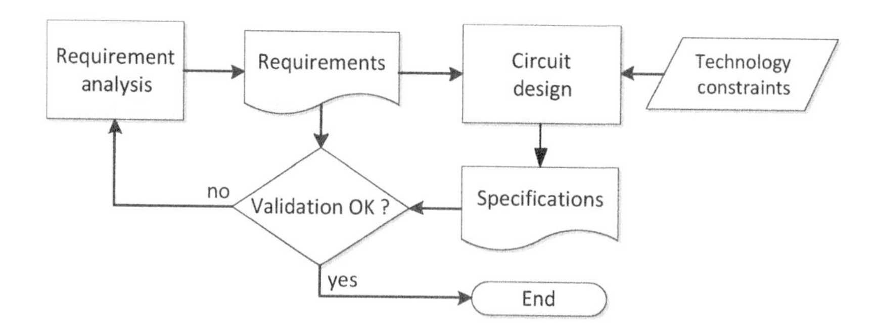

Figure 5.34 Circuit design process.

Table 5.21 AC power ideal versus real behavior

Ideal performance behavior	**AC power**
Voltage output	$V_{out} = \sqrt{2}V_{rms} \times \sin(2\pi ft)$ being $V_{rms} \in [100 \ldots 240]$ V, $f \in [50,60]$ Hz
Current output	$I_{out} =$ any
Resistance	$R_{out} = 0$ – may be derived from previous performances
Real performance behavior	**AC power**
Interferences	ESD, transients, surges are added to power line – see Table 5.7 for details
Variations on voltage	Voltage level experiences dips and interruptions – see Table 5.22 for details

EMI is conducted and radiated into power lines; it is mainly generated by electrostatic discharges (ESD), RF transmitters, switching and transient noise, lighting and low frequency magnetic fields. Expected maximum values for these interferences are included in the IEC/EN 60601-1-2 and associated standards. Medical devices must be immune to the level of interference specified in such standards. Table 5.7 details the sources of interference and the tests that products have to withstand for compliance.

The voltage supplied by the power line is also non-ideal: **dips** and interruptions may be experienced. Table 5.22 shows the expected levels that medical devices must withstand without generating unacceptable risks for patients or operators. For example, the device must withstand 5 seconds of voltage interruption without significant impact on safety. The impact depends on the specific device type and the intended use.

Specific devices may be powered by batteries. **Batteries** are a primary source for low voltage, low consumption devices. Batteries should ideally produce a constant voltage in any condition for an infinite time (Table 5.23). Real components shows expected and less expected behaviors. Batteries have limited capacity and the output voltage decreases with usage. **Capacity**, measured in ampere-hours (Ah) defines the ratio between the current that a

Table 5.22 IEC/EN 60601-1-2 variations on voltage power line (V_{rms})

Test	Reference standard	Description	Tested level
Variations on voltage supply power lines	EN 61000-4-11	Voltage dips, and voltage variations Short interruptions	$V_{rms} < 5\%$ for 0.5 cycle $V_{rms} = 40\%$ for five cycles $V_{rms} = 70\%$ for 25 cycles Interruptions $V_{rms} < 5\%$ for 5 sec

Table 5.23 The battery ideal versus real behavior

Ideal performance behavior	Batteries
Nominal voltage output	$V_{out} = K$, K is constant for any output current and any source variation
Current output	$I_{out} =$ any
Internal resistance	$R_{in} = 0\ \Omega$ (may be derived from previous performances)
Capacity	Infinite (ampere-hours at specific current)

Real performance behavior	Batteries
Voltage output	V_{out} proportional to residual capacity, T, I_{out}, R_{out} ($T =$ temperature)
Current output	$I_{out} <$ maximum load current
Internal resistance	$R_{in} > 0$
Operating temperature range	Capacity decreases at lower temperatures
Capacity	Non-infinite capacity expressed as ampere-hours at a specific drain current
Maximum shelf life	Leakage current discharges batteries continuously; batteries become exhausted even when not used (limited shelf life)
Maximum number of charge cycles	Finite for rechargeable batteries

battery can supply versus the amount of time the battery will sustain the discharge at the specific current. Capacity will decrease when increasing drained current, mostly because power is dissipated into the internal resistance. The capacity also decreases at reduced operating temperatures. The capacity value is specified for a given temperature and for a specific drain current or it is evaluated as the current at which the battery is discharged completely in 20 hours (or other fixed amount of time).

At low operating temperatures, batteries reduce their capacity and therefore discharge rapidly. Under −27°C batteries may have half their normal capacity; at 0°C, around 20% capacity decrease may be expected. This has to be taken into account to determine life-time under different temperature operating conditions.

Batteries may be non-rechargeable (also called primary) or rechargeable (secondary). Secondary batteries can be recharged up to a maximum number of cycles. Recharge conditions are usually expressed in terms of charge current per amount of required time.

Battery usage may provoke hazardous situations. Batteries may leak acid, explode in specific conditions or simply generate risks due to unexpected discharge. This has to be considered when using them in a medical device, and proper countermeasures have to be considered as required by IEC/EN 60601-1.

On the other hand, batteries have an important benefit when used in medical devices since they can power equipment without the risk associated to direct patient connection to a power line. The battery choice also reduces the value of the parasitic capacitance that connects the device with the power line ground, as shown in Figure 5.11. The effect is the reduction of interference from power lines and reduction of risks from excessive leakage current across the patient.

REQUIREMENTS When using electronic medical devices, patient safety depends on two factors: 'basic safety' and 'essential performance'. **Basic safety** guarantees that the device is free 'from unacceptable risk directly caused by physical hazards under normal condition and single fault condition' (IEC/EN 60601-1). The **essential performances** are those that, when absent or degraded, may generate unacceptable risks. These performances depend on the specific device.

The power module may negatively affect both basic safety with electric shock risks and essential performances. As shown in Figure 5.35, the medical device 1 (upper rectangle) and, more specifically its power module, receives radiated EMI from the environment and conducted EMI from power line (lightning, transients etc.). This interference may induce dangerous current values on patients, thus creating a physical hazard. At the same time, such EMI may degrade power module performance that in turn may possibly affect the overall medical device.

If the power module does not sufficiently attenuate conducted EMI from outside device, the EMI may influence other modules such as the ADC and the op-amp high impedance input, where EMI may be significant with respect to the input signal. The power module is also threatened by variations in AC power that again may degrade performance of the device.

A medical device may also affect the performance of other medical devices that are within the same environment. For example, medical device 1 may generate EMI (conducted and radiated) that has a negative impact on the second medical device (lower rectangle in Figure 5.35). Medical devices must respect requirements on emission to avoid impacts on other

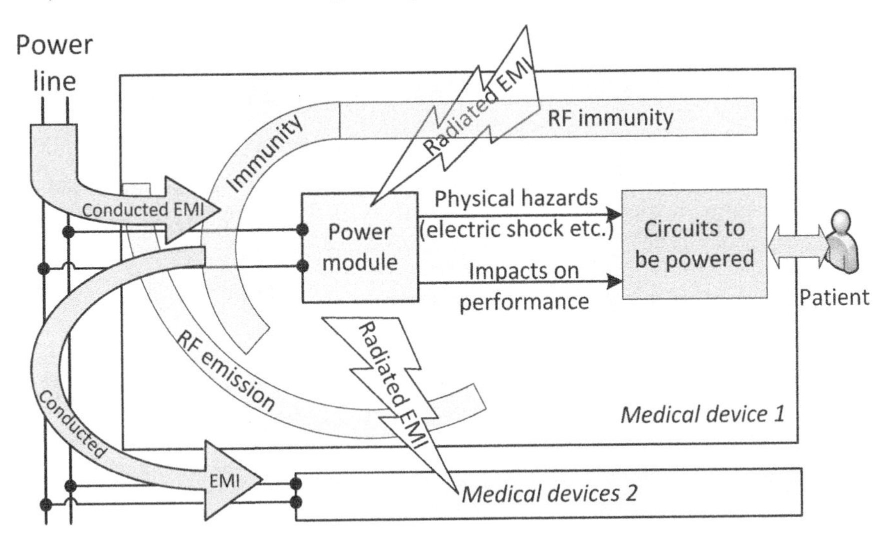

Figure 5.35 Power module safety impacts.

Table 5.24 IEC/EN 60601-1-2 requirements on interference emission

Test	Reference standard	Description
Radiated E-field emissions	EN 55011 (CISPR 11) EN 55014 (CISPR 14)	Product emission constraints on the 30–1000 MHz frequency band
Conducted emissions	EN 55011 (CISPR 11)	Product emission constraints on the 0.15–30 MHz frequency band
Harmonic emissions	IEC/EN 61000-3-2	Limits for harmonic frequency emissions on public mains network for devices with input current lower than 16 A per phase
Flicker emissions	IEC/EN 61000-3-3	Limits for voltage changes, fluctuations and flicker injected into the public main supply network for devices with input current lower than 16 A per phase

medical/non-medical electronic systems and on telecommunications. To avoid unacceptable risks, electrical medical devices have to satisfy the following requirements:

1. immunity to interferences from power line (Table 5.7)
2. immunity to AC variations (Table 5.22)
3. reduced conducted and radiated emissions (Table 5.24).

So the power module has to

1. attenuate the EMI conducted inside and outside the device
2. be immune to EMI that is expected to be radiated by other devices
3. only radiate EMI within prescribed limits
4. provide sufficiently regulated (stable) DC voltage to circuits even in the presence of mains electricity supply input variations.

Power modules have to satisfy additional requirements to avoid unacceptable risks. IEC/EN 60601-1-2 includes many additional prescriptions on specific device components (wiring, flexible cords, layout, power supply chords etc.). These topics will be discussed in chapter 8.

5.9.2 Electrical Safety and Appliance Design

Under specific conditions, the body may become part of an electric circuit. Patients or operators being in contact with an electrical appliance may experience leakage current that enters and exits from the body. According to Ohm's law such current value *i* is:

$$i = V_{applied} \Big/ \sum_i Z_i \tag{5.58}$$

with $V_{applied}$ the applied voltage to the body and Z_i the series impedance encountered between the points where the voltage is applied. The main impedance is usually the dry skin resistance that ranges from 15 kΩ to 1 MΩ. Wet or broken skin may reduce impedance to hundred of ohms.

Physiological effects of current flowing inside a body are proportional to the current value, the duration of the exposure and the body areas affected by current flow. The effects are also dependent to frequency where the range of tens of Hz is the most dangerous. Current may be dangerous at micro to milliampere values (Webster, 2009).

In general, currents may create muscle stimulation interfering with respiratory and cardiac functions. Current also provokes heating, since the body behaves as a resistive load. High currents may cause burns and tissue damage. Also, specific current flow conditions may provoke ventricular fibrillation that, if not addressed, will result in sudden death. The admissible limits for current leakage are in the range of 10–100 μA.

Electric appliances have to be designed to avoid dangerous current leakage, even in the presence of a **single fault condition**, that is, when experiencing a failure of a single element that is employed to reduce risks of electric shock or when a single abnormal condition is present.

Medical equipment may in general supply five types of **leakage current**:

i_1	earth leakage	Current from the device enclosure to ground
i_2	enclosure leakage	Current from the device enclosure to ground via the patient/operator (also known as touch leakage)
i_3	auxiliary leakage	Current between the parts that are physically in contact with the patient in the device normal use (defined as '**applied parts**')
i_4	patient leakage	Current from applied part to ground via the patient
i_5	patient leakage	Current from the power line to the applied part via the patient

The leakage current with the associated numbers are depicted in Figure 5.36.

From equation (5.58), the means of reducing leakage current include the increase of the series impedance and/or the reduction of the applied voltage.

Let us first consider the enclosure. To avoid dangerous currents i_2 from the enclosure, the electric appliances have a 'basic insulation', that is, conductors of power line are not in contact with the enclosure that may be metallic. In case of fault, the conductor may be in contact with the enclosure and users may experience a shock if they touch the enclosure and have a sufficiently low-resistance path to ground.

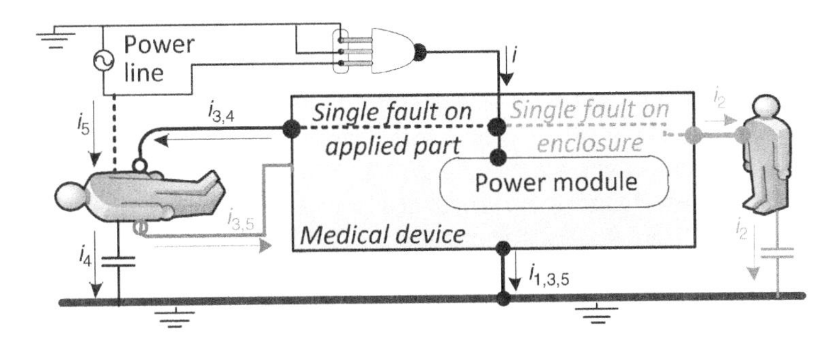

Figure 5.36 Current hazards in single fault conditions.

The single fault hazard may be overcome if the metallic **enclosure** is connected to ground with negligible resistance. In this case, the enclosure never reaches hazardous voltages even in case of unwanted contact with high voltage conductors. If the enclosure is not metallic, then internal metallic parts have to be grounded. The devices with grounded metallic accessible parts are referred to as **Class I** according to IEC/EN 60601-1.

The connection to ground is safe if

1. all accessible conductive parts are connected to a protective terminal
   and
2. the protective terminal is connected to an external protective earthing system
   and
3. the protective earthing system provides a proper low impedance path to earth for each receptacle.

These conditions may be critical, since safety relies also on a proper installation of the power distribution system. An improved approach is followed by **Class II** devices. In this case, a supplementary insulation is provided. If basic insulation fails, the supplementary insulation avoids electric shock, regardless of the grounding system or installation conditions. This scheme is defined as '**double insulation**' and it is indicated with the symbol ⧈. All parts that may be in contact with operators/patients, such as knobs and switches, have to be insulated in double insulation devices. Alternatively, the insulation may be 'reinforced', so that electric shock is avoided, as in the case of double insulation.

The IEC/EN 60601-1 standard requires that devices that are powered by an external powered source must be either Class I or Class II equipment.

Internally powered equipment is generally less critical, thanks to the limited battery voltage: when voltage is low, high leakage currents cannot occur even with wet skin considering (5.58). In addition, the risk of fibrillation is reduced five times for DC currents. It should be noted that lower current values may be sufficient to produce shocks if the device is connected to catheters or electrodes connected to heart. Devices with this type of connection have to limit the current to 10 μA maximum.

Class I and II devices address risks due to excessive leakage current from the enclosure. **Patient and auxiliary leakage** are addressed by specifying current limits on applied parts.

IEC/EN 60601-1 distinguishes different current leakage limits for applied parts in case of

1. direct cardiac application: maximum level of safety '**CF**'
2. patient connection delivering electrical energy or exchange of electrophysiological signals: level of safety '**BF**'
3. patient connections not being BF and CF: level of safety '**B**'.

The rationale is always to guarantee that patient leakage currents are limited to safe levels on normal and single fault conditions.

The CF case is the most critical since even limited current may induce fibrillation. When there is no direct cardiac applications, the current required to induce fibrillation is higher. For the BF case, since the delivery of energy or the signal exchange implies a reduction of skin contact resistance through proper electrodes, a higher risk is still present compared to the third case (B).

To increase security, BF and CF **applied parts** must be **floating**, that is, they must be galvanically isolated. This means that there is no direct DC connection between the power

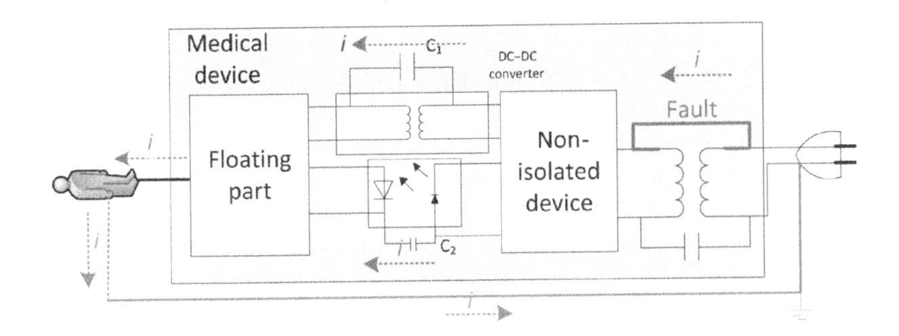

Figure 5.37 Medical device floating parts.

line and applied parts (i.e. parts normally in contact with patient). When the patient is unintentionally connected to the power line as per the case of i_4 current in Figure 5.36, the patient leakage current is still within safe limits since only limited AC current may be exchanged through parasitic capacitances.

Floating parts are obtained, eliminating direct current paths through transformer, capacitor or optocoupler connections. Figure 5.37 shows an example where the patient is unintentionally connected to the power line because of a single fault on the power supply transformer.

Thanks to the transformer of the DC–DC converter and the optocoupler, there is a very high resistance path to ground (in the order of 10^{10}, $10^{12}\,\Omega$, respectively) that does not allow DC current to flow at significant levels. Unfortunately, transformers and optocouplers have non-null parasitic capacitances C_1 and C_2 that conduct AC current. Parasitic capacitances are in the order of tens of pF for DC–DC converters and pF for optocouplers. An overall insulation capacitance of 80 pF at 50 Hz is equivalent a 40 MΩ impedance that with 220 V_{rms} yields 5.5 μA current. This value is within the accepted limits as shown in Table 5.25.

Note also that the optocoupler and the DC–DC converter must withstand a 220 V_{rms} AC continuously. This implies proper component choice so that **insulation** is still guaranteed at these working voltages.

Table 5.25 IEC/EN 60601-1-2 admissible leakage current (values in μA)

		Normal condition		Single fault condition	
i_1	Earth leakage	5,000		10,000	
i_2	Enclosure (touch) leakage	100		500	
		Type B/BF	Type CF	Type B/BF	Type CF
i_3	Auxiliary leakage (either from external voltage to patient either from patient to ground)	DC < 10 AC < 100	DC < 10 AC < 10	DC < 50 AC < 500	DC < 50 AC < 50
i_4 i_5	Patient leakage				

To protect the patient from electrical hazard, the electrical circuit has to provide insulation suitable for the working voltage conditions during intended use and single fault conditions. Inadequate **insulation** may produce current hazards due to electric shocks, sparks generating flames and failures that may impact on essential performance.

Insulation is a means of protecting against electric shocks and the other mentioned risks. Insulation is guaranteed by proper dielectric strengths, creepage and air clearance suitable for working voltages and conditions. IEC/EN 60601-1-2 includes tables to define requirements and tests on these issues. Chapter 8 has additional details.

An insulating solid material can withstand up to a defined electric field strength (i.e., dielectric strength) before breakdown.

Dielectric strength of a device may be expressed as a working voltage that the device can withstand continuously. This condition is tested using higher test voltages for a limited time. IEC/EN 60601-1-2 includes working voltages and the corresponding test voltage to be applied for one minute. For example, a double insulation must withstand an AC voltage of $3\,kV_{rms}$ for one minute to protect the patient and the operator from circuits connected to the main supply at $220\,V_{rms}$.

The other means of protection to avoid voltage breakdown are the **air clearance** and **creepage** defined as the shortest distance between two conductive parts through the air or along the surface of an insulating material.

Clearance and creepage must be proportional to the expected working voltage and the pollution degree to avoid breakdown. With lower working voltages, creepage and clearance distances can be reduced. For working voltages between $125\,V_{rms}$ and $250\,V_{rms}$ air clearance and creepage have to be 3 mm and 1.6 mm respectively.

Battery powered devices have less stringent requirements due to the lower voltage. In turn, battery leakage, reversed battery connections, short-circuiting or inadequate charge may cause unacceptable risk in specific devices. In this case, users have to be informed, and proper countermeasures have to be considered.

This section has shown that power modules have many constraints in medical devices to guarantee safety and essential performance. Additional requirements related to the needs of the circuits to be powered will be addressed in the next section.

5.9.3 Power Module Design

The ideal power module must supply constant zero noise voltage at any current consumption. This guarantees that analog/digital circuits work properly. Real power modules are a compromise with respect to performance required by powered circuits. As outlined in Table 5.26, output voltage accuracy is not infinite and swings with load, and voltage line variations are to be expected at the input of the power module. Output voltage also contains noise that may affect analog circuits.

The power module may derive energy from the AC main supply, batteries or adapters that transform AC main supply to a low voltage AC or DC.

In Figure 5.38, we recall the conceptual scheme of a power module that transforms AC main supply to a DC regulated voltage. The AC supply feeds a first module that protects it from transients, lightning and ESD and reduces EMI in both directions. In the second stage, a transformer reduces the input voltage. The transformer may be also an isolation element since it provides a high resistance path between the main supply and the powered circuit.

Table 5.26 Power module ideal versus real behavior

Ideal performance behavior	**Power module**
Voltage output	$V_{out} = K$ – constant for any output current and any source variation
Current output	$I_{out} = \text{any}$
Power	$P_{out} = P_{in}$ – efficiency = 100%
Isolation	$R_{in\text{-}out} = \infty$ – full isolation between input and output
Impedance	$R_{out} = 0$ $R_{in} = \infty$ – may be derived from previous performances

Real performance behavior	**Power module**
Output voltage accuracy	$Accuracy\,\% = \frac{(\lvert V_{out}\rvert - V_{out,\,nominal})}{V_{out,\,nominal}} \times 100$
Output current	$I_{out} < I_{max}$
Input voltage range	$Input\ Voltage\ range\,\% = \frac{(\lvert V_{in,\,\max}\rvert - V_{in,\,nominal})}{V_{in,\,nominal}} \times 100$
Line voltage regulation	$Line\ regulation\,\% = \frac{(V_{out,\,V in\max} - V_{out,\,V in min})}{V_{out,\,nominal}} \times 100$ $V_{out,V in\,\max}$, $V_{out,\,V in\,min}$ = output voltage at maximum/minimum input voltage
Load voltage regulation	$Load\ regulation\,\% = \frac{(V_{out,\,load\,\max} - V_{out,\,load\,min})}{V_{out,\,no\,\text{minal}}} \times 100$
Output ripple and noise	Ripple and noise introduced by power module
Efficiency	Efficiency (= output power/input power) < 100%

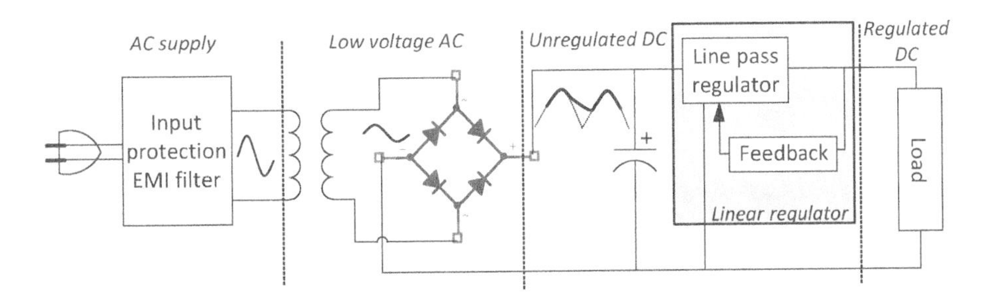

Figure 5.38 Conceptual power module for AC main supply.

The AC voltage from the output of the transformer is rectified by the diode bridge rectifier and by a high value capacitor. From this stage, the voltage is unregulated because it varies with input and with changes in load current. Figure 5.38 shows the ripple in voltage output that is proportional to the supplied current. The final stage is a regulator that makes the output voltage less sensitive to input and load variations.

The circuit of Figure 5.38 may be completely included in the medical device or part of it may be contained in a separate AC adaptor that may supply low voltage AC, unregulated DC or regulated DC supply. Battery powered devices have power modules that include regulators, and possibly chargers and other functions.

In Figure 5.38, a **linear regulator** is used: regulation is performed by sensing the output voltage and by regulating the line pass element voltage. Linear regulators are a good choice, especially for analog systems, since they guarantee low conducted and radiated noise and minimal side effects, albeit at the cost of much lower efficiency. Efficiency is the ratio between the power supplied by the regulator to the circuit P_{out} and the power absorbed by the regulator P_{in}. Linear regulators reduce input voltage to the output level by thermally dissipating the residual energy. This requires thermal dissipaters to avoid an unacceptable rise in temperature. The dissipated energy is given by the voltage drop between input and output pins times the supplied current. This suggests that low current applications are preferred: when voltage drop becomes high across the regulator, low efficiency is achieved because a power equal to (voltage drop) × (current) is dissipated.

More efficiency and flexibility is obtained by AC–DC or DC–DC **switching converters** that transform input voltage into higher/lower output voltage by storing and releasing input energy. Ideally, the energy is simply transformed in these converters and therefore the power remains unchanged between input and output with an ideal 100% efficiency (i.e., $P_{\text{out}} = P_{\text{in}}$.) Output voltage V_{out} is inversely proportional to supplied current i_{out} considering that:

$$V_{in} \times i_{in} = P_{in} = P_{out} = V_{out} \times i_{out} \tag{5.59}$$

where V_{in} and i_{in} are the input voltage and current respectively. The same concept applies to ideal transformers. Energy may be stored as magnetic field, using transformers or inductors, or electric field using capacitors. Switching converters follow various circuit topologies with different performances (Bocock, 2011). Figure 5.39 shows a conceptual scheme applicable to isolated switching converters. The AC or DC voltage is fed into a pulse wave duration (PWD) square wave signal oscillator. The oscillator's duty cycle (i.e. the fraction of the period the square wave is in on state) can be controlled by the feedback. The duty cycle is proportional to the output voltage: this allows the output voltage to be set at the required value. A further regulator is necessary to achieve a stable (regulated) output voltage in the presence of input and load variations.

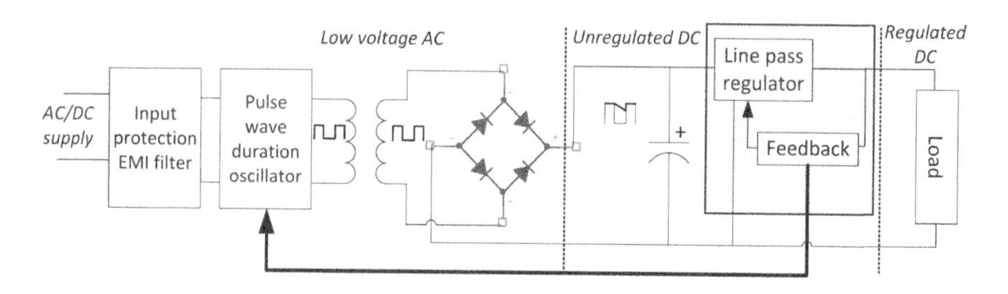

Figure 5.39 Conceptual switching converter.

More recent switching converters may achieve as much as 98% efficiency because power is not dissipated thermally, and inefficient diode bridges are replaced by low resistance FETs. Switching converters may also provide a supplementary insulation if based on transformers. On the other hand, switching converters are more noisy, since oscillators operate at frequencies up to MHz to reduce transformer size. In addition, there is a higher conducted and radiated EMI that has to be addressed to achieve emission requirements.

5.10 Baseband Digital Communication

THE NEED The trend in electronic design is to implement functions in the digital hardware or better in the software domain. Digital hardware is in general to be preferred to analog hardware because of the simpler design, reduced cost and improved robustness to noise. In analog hardware, when noise is added to the signal it is difficult to remove if it is in the same frequency band. In digital hardware, noise can be completely removed when its amplitude is lower than the minimum discrete amplitude that differentiates bits ($= \pm 0.5$ LSB voltage). In digital hardware, signals can therefore be regenerated by eliminating noise. The benefit of the digital world suggests to digitalizing analog signals as close as possible to the analog source. A typical conceptual architecture of digital devices is depicted in Figure 5.40.

More often, actuators and sensors translate analog input/output signals into the digital counterparts. Actuators may be DAC, modules for motor control or power control modules. Sensors may be ADC, pressure or temperature converters. Actuators and sensors are connected to microprocessors or microcontrollers through proper data transmission lines. Such lines connect micros to memories and to the external communication networks. Digital transmission lines become relevant for systems with pervasive digital hardware.

REQUIREMENTS Ideally, transmission lines will allow communication between digital components at the information rate required by the components. For example, an ADC with 12-bit resolution and 1 kbit/s sample rate has to ideally exchange error-free data at 12 kbits/s at minimum. Extra data has to be exchanged for control purposes.

Other parameters have to be considered for design purposes such as: occupied PCB space, cost of components for transmission and generated EMI. To address such topics we recall some of the basic concepts of baseband communications between digital components.

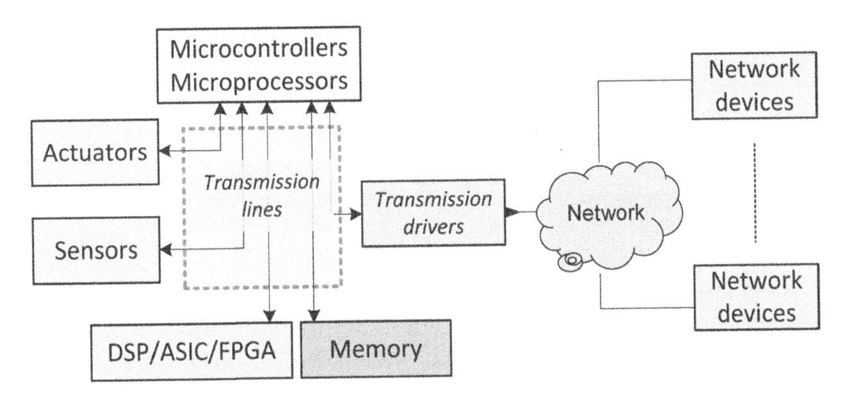

Figure 5.40 System transmission lines.

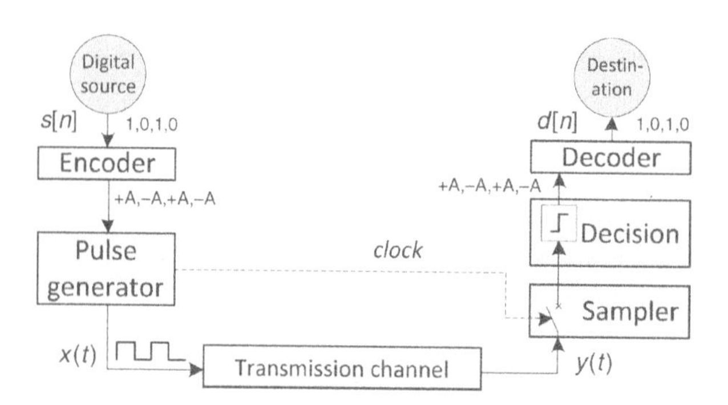

Figure 5.41 Data transmission model.

5.10.1 Data Transmission Elements

In a basic all-digital communication, we consider that a message is transmitted from a digital transmitter (source) to a digital receiver (destination) as shown in Figure 5.41. In baseband all-digital transmission we assume that the transmission channel does not attenuate frequencies from DC to a very high cutoff frequency. This is the usual condition experienced by wireline channels included in the electronic boards or between electronic boards. On the contrary, wireless channels require that the signal is modulated so that its frequency band is suitable for the on-air passband channels (that reduce DC and the lower frequency components).

In the simplest **model of baseband communication**, the message is coded over an electrical signal $x(t)$ which has two possible values $\pm A$.

The message is a sequence of bits $s[n]$ that is coded on $x(t)$ as:

$$nT \leq t < (n+1)T \rightarrow x(t) = \begin{cases} +A : s[n] = 1 \\ -A : s[n] = 0 \end{cases} \tag{5.60}$$

This line coding scheme is called a non-return-to-zero (NRZ) line since there is no state where $x(t) = 0$. This scheme is able to transmit at a **bit rate** $f_{br} = 1/T$ bits/s. If $s[n]$ is a sequence of alternating 0 and 1 (i.e., 0,1,0,1, . . .), then $x(t)$ is a square wave of amplitude $\pm A$ and period $2 \times T$.

Ideally, the signal received by the destination $y(t)$ is equal to the transmitted signal $x(t)$. In practice, $y(t) \neq x(t)$, due to the modifications of the channel transmission and the noise. The message at destination $d[n]$ can be decoded by the following decision rule:

$$Decision\ Rule = \begin{cases} +A : y(nT + T/2) > \lambda \\ -A : y(nT + T/2) < \lambda \end{cases} \tag{5.61}$$

When the probability of transmitting 0 or 1 is the same, it is easily proven that $\lambda = 0$ minimizes the probability of error as shown below. At the end, the decoder simply assigns 1 to the $+A$ value and 0 to $-A$ value.

If the signal $y(t)$ is sampled at the correct time $nT + T/2$ then $d[n] = s[n]$ when the channel does not change the signal and it is noiseless. In these ideal channel conditions,

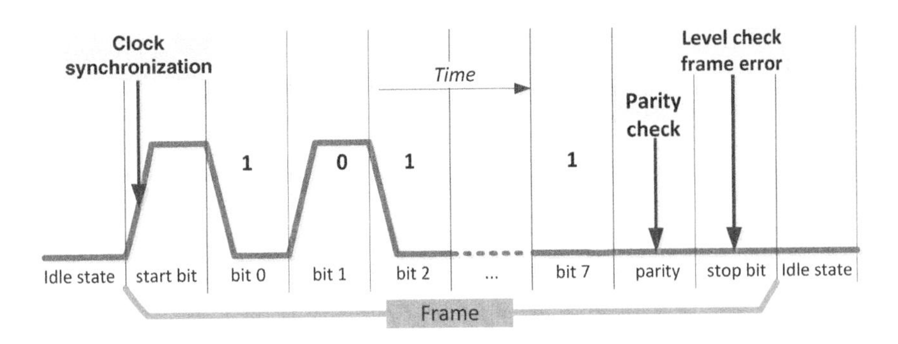

Figure 5.42 RS-232 data transmission.

the **sampling clock** of the source and the destination cannot diverge by an absolute error greater than $\|e_c\| < T/2$. With a greater error, the decision time of a specific bit is shifted on to the following bit. An alignment of source and destination clock may be obtained by also transmitting the clock from the source to the destination (**synchronous** mode) or by regenerating the source clock at the destination with some available information (**asynchronous** mode).

In synchronous mode, the clock is affected by fewer errors than the asynchronous mode and therefore the data rate may be increased considering that the data rate f is limited by $f = 1/T < \frac{1}{2\|e_c\|}$. In the asynchronous mode, clock synchronization may be performed by assuming that some bits are at fixed values.

In the **RS-232** asynchronous mode standard, the transmission is performed by sending a frame of bits as shown in Figure 5.42.

The idle state is at low level. Then, the frame is started by a start bit always at high level. The data bits (5–8) follow the start bit, and then a control bit called parity may be sent. The frame ends with 1 or 2 stop bits always at low level. From the start bit transition, the receiver can synchronize its clock.

Since transmitter and receiver know in advance that the bit rate is $1/T$ the number of start, stop and data bits, the receiver knows when to sample bits and when the stop bit transition should ideally occur. This assumption holds if the clocks of the receiver and transmitter do not diverge too much. Since the stop bit is always transmitted at low level, the receiver will raise a 'frame error' if the sampled value of the stop bit is at high level. This can happen if there is a major difference between the clocks.

To reduce errors at the destination, additional methods have to be implemented. These methods are mainly based on transmitting additional **redundant information** based on data a priori known by the destination. This known data reduce the available bit rate for the message to be transmitted but, in turn, also reduces the probability of errors. For example, the **parity control bit** is an additional method of checking that received data are correct. In the even-parity criterion, the sum of all the one bits plus the parity bit must be even. If this does not occur for a specific frame, there is an odd number of bits in error. Note that if two bits are wrong, the parity check does not reveal the error. In this case, more complex integrity strategies have to be implemented. Such strategies are usually implemented at firmware or software level.

Asynchronous mode is a simpler form of transmission, since the source clock is not sent. Timing is more critical since the clock has to be resynchronized at the destination. This mode is well suited when data is generated in bursts and not on a regular basis. On the other hand, asynchronous communication requires additional control bits to keep error rates at an acceptable level, thus reducing the data rate available for the message to be transmitted.

In the asynchronous mode, errors may be generated by excessive jitter of the receiver clock. Other errors may be generated by the non-ideal behavior of the **transmission channel**. First, consider that active components have a limited rate at which the output can change (slew rate). This generates distortion on the received signal. The **channel is also band limited** due to the parasitic capacitances that produce low pass filtering effects on the signal. On the other hand, **the pulse train signal** $x(t)$ used to transmit bits (equation (5.60)) **has an infinite frequency band** and therefore the attenuation of the highest frequencies of $x(t)$ may produce additional errors.

The proof that $x(t)$ has an infinite frequency band can be simply derived by assessing the square wave signal bandwidth as shown below.

The Fourier transform $X(f)$ of a unity amplitude ± 1 square wave of period $2 \times T$ is given by an infinite periodic train of Dirac impulses $\Delta(f)$ spaced by $2 \times T$ with amplitude shaped by a sync function

$$X(f) = \frac{1}{2} \sum_{\substack{k=-\infty \\ k \neq 0}}^{\infty} sync\left(\pi \frac{k}{2}\right) \times \Delta\left(f - \frac{k}{2T}\right), \tag{5.62}$$

where $sync(x) = sin(x)/x$, k belongs to the natural numbers and $\Delta(f)$ is the Dirac delta impulse defined as

$$\delta(f) = \begin{cases} 0 \quad if \quad f \neq 0 \\ \int_{-\infty}^{\infty} \Delta(f) df = 1 \end{cases} \tag{5.63}$$

Considering the values of $sync(x)$ and that the bit rate $f_{br} = 1/T$, the modulus of equation (5.62) is

$$|X(f)| = \begin{cases} f \neq \frac{k}{2T} \rightarrow 0 \\ f = \frac{k}{2T}, and \begin{cases} k = even,\ k \neq 0 \rightarrow 0 \\ k = 0 \rightarrow \delta(f) \\ k\ odd \rightarrow \frac{1}{k\pi} \times \delta\left(f - \frac{k}{2T}\right) = \frac{f_{br}}{2\pi f} \times \delta\left(f - \frac{k}{2T}\right) \end{cases} \end{cases} \tag{5.64}$$

This equation shows that the Fourier transform is non-null if $f = k/2T$ for any k odd or $k = 0$. When k is odd, the coefficients of the Dirac impulse $\delta(f)$ are inversely proportional to the frequency f.

The signal $x(t)$ can be transmitted only when the channel where the signal is propagated is sufficiently wideband. More specifically, the channel must not attenuate the transmitted signal from DC to a frequency that depends on the bit rate $f_{br} = 1/T$. When f is sufficiently large with respect to f_{br}, from (5.64) the signal spectrum amplitude becomes negligible at frequencies where channel attenuation is significant. More simply, we can say intuitively that the bit rate f_{br}, is limited by the low pass cutoff frequency f_c of the transmission channel since when f_{br} is much higher than f_c the received signal becomes attenuated at high frequencies and this modifies significantly the received signal with impact on error probability.

A more specific indication is obtained by considering the time domain. Assuming for example that **the channel** can be modeled **as a low pass first order filter** that is a simple resistor–capacitor (RC) filter discussed in Section 5.6. After passing through an RC low pass filter, a square wave of maximum value V_{max}, will have an exponential increase/decrease $V_{out\ increase}$, $V_{out\ decrease}$ as (Horowitz, 1999):

$$V_{out,decrease} = V_{\max}\left(e^{-\dfrac{t}{RC}}\right),\ V_{out,increase} = V_{\max}\left(1 - e^{-\dfrac{t}{RC}}\right) \quad (5.65)$$

If we assume that $V_{out,increase}$ must be at least the 90% of V_{max} at the end of one period $T = 1/f_{br}$ to obtain a proper decision, then, from 5.65,

$$f_{br} \leq f_c \times \frac{\ln_e(10)}{2\pi} \approx f_c \times 0.37,\ f_c = \frac{1}{2\pi RC}, \quad (5.66)$$

where f_c is the cutoff frequency of the RC filter. The same formula is obtained for $V_{out,decrease}$ being at least the 10% of V_{max}. Equation (5.66) shows that f_{br} must be less than $0.37 \times f_c$ to have a good confidence in the bit decision at the receiver.

For **active components**, the slew rate parameter indicates the maximum rate at which the output can change. A sinusoid of frequency f is not distorted if the slew rate $\geq 2\pi Af$ where A is the amplitude of the sine wave input signal. This is because the maximum rate (i.e., the maximum of the derivative) of a signal $x(t) = A\sin(2\pi ft)$ is $2\pi Af$. The maximum rate of output variations is obtained for an alternating sequence of A and $-A$ corresponding to an alternating sequence of 0,1 bits.

If we assume that a sinusoid of amplitude A and frequency $f_{br}/2$ corresponding to an alternating sequence of 0,1 bits at f_{br} bit rate can be detected correctly, then the bit rate must be lower than the slew rate as

$$f_{br} \leq Slew\ rate_c \times \frac{1}{\pi A}. \quad (5.67)$$

This means that if the data rate is at f_{br} bits/s then active components must have a minimum value of slew rate as indicated in (5.67).

For example, 400 kbit/s can be processed by components with slew rate greater than 3.6 V/μs.

From what is stated above, the bit rate is limited by the bandwidth of the channel and the associated components. An **upper bit rate limit** can be derived from sampling theory. This offers additional insight on the baseband coding.

At this aim, we define a function $g(t)$ so that

$$x(t) = \sum\nolimits_{n=-\infty}^{\infty} v[n]g(t - nT),\ v[n]\begin{cases} +A : s[n] = 1 \\ -A : s[n] = 0 \end{cases} \quad (5.68)$$

In this formula, each coded bit in the sequence $v[n]$ is multiplied by a function $g(t)$. At the receiver, the signal $x(t)$ will be sampled at $t = nT$ to decode the transmitted bit. To avoid a specific bit $v[n]$ being affected by the value of the other bits in the sequence, $g(t)$ must satisfy the following condition:

$$g(nT) = \begin{cases} 1 : n = 0 \\ 0 : n \neq 0 \end{cases} \tag{5.69}$$

The condition avoids the so-called inter symbol interference (ISI). *We recall that the sequence v[n] for each n may transport binary information (bits) coded with different amplitude levels (e.g., −A, −2A, A, 2A for two bits per symbol coding). The symbol is the information carried by v[n] at a specific n. The symbol coincides with a bit if v[n] has only two values. If v[n] has 2^m values then a symbol conveys information of m bits.*

It can be proved that in the frequency domain, equation (5.69) becomes (Proakis, 2008)

$$f_{br} = 1/T, \ \sum_{k=-\infty}^{\infty} G(f - k/T) = T \tag{5.70}$$

This constraint shows that the Fourier transform $G(f)$ of the signal $g(t)$ cannot have a frequency band narrower than $f_{br}/2$, that is, $G(f) \neq 0 \ \ for \ \ |f| \leq f_{br}/2$ (otherwise eq. 5.70 does not hold).

To avoid inter symbol interference, the maximum bitrate must be lower than the channel bandwidth f_c as $f_{br} \leq 2 \times f_c$. This constraint is related to the Nyquist–Shannon sampling theorem that can be stated as: *a sequence of numbers equally spaced by $1/f_{br}$ can be transmitted by a suitable analog signal $g(t)$, if the bandwidth of $g(t)$, is $f_{br}/2$ at minimum.*

Equation 5.70 also shows that the minimum band ideal function $g(t)$, is given by

$$g(t) = \frac{\sin(\pi f T)}{\pi f T}, \ G(f) = \begin{cases} T : |f| \leq f_{br}/2 \\ 0 : \text{ elsewhere} \end{cases}, f_{br} = 1/T \tag{5.71}$$

These discussions suggest that

1. the transmitted signal has infinite bandwidth, but its frequency components proportionally decrease with frequency f
2. the channel must have a minimum frequency bandwidth of $2 \times f_c$ to transmit at f_{br} bit rate: $f_{br} \leq 2 \times f_c$
3. a better confidence on bit decision is obtained by the more stringent constraint with $f_{br} \leq f_c \times 0.37$
4. the electronic components handling the transmitted signal must have a suitable slew rate for the considered bit rate $f_{br} \leq Slew\ rate \times \frac{1}{\pi A}$ to avoid distortion.

The means of communication considered so far is a baseband channel where frequencies from DC to GHz can be used to convey information. This is the case for digital communication on PCBs , twisted pair wireline channels (up to MHz), coaxial cables (up to GHz) and waveguides (up to 100 GHz). However, wireless channels have a passband behavior.

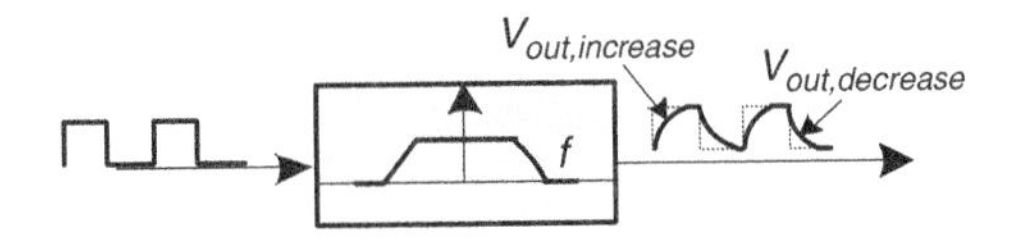

Figure 5.43 Effect of an RC filter on a square wave.

In this case, the baseband signal has to be band-limited and it has to be translated to higher frequencies through a modulator to match the wireless channel passband characteristics. Passband wireless connectivity may be further investigated in specific books such as (Proakis, 2008). The wireless communication frequency band for medical devices is usually in the unlicensed ISM bands that are reserved internationally for industrial, scientific and medical (ISM) purposes.

More recently medical devices have taken advantage of **wireless connectivity**, for example in patient monitoring and home healthcare. The communication technologies are the low cost ZigBee® which is suited for wireless sensors, Bluetooth® and Wi-Fi for PC connection of personal devices, radio frequency identification (RFID) for identification and tracking of entities such as blood bags, medicines, medical supplies or even patient/staff authentication. For such purposes, technology suppliers are adapting these commercial communication technologies to medical usage (Texas, 2010).

Up to now, we have considered two sources of errors in decision that may generate errors on bit decision: the non-ideal channel characteristics and the errors in the decision time due to non-synchronized clocks.

The effects of such distortions are usually assessed through the **eye pattern** diagram. In this diagram, many signals related to different bit values are superimposed to assess if the decision module has the capability to detect the proper bit value. This diagram is practically obtained on the oscilloscope by triggering the horizontal sweep at the symbol rate. The ideal eye pattern for an NRZ coding such as RS-232 of Figure 5.43 is depicted in Figure 5.44.

On the left side of Figure 5.44, two squared waveforms are superimposed. The ideal sampling instant is in the middle of the horizontal extension of the bit. Clock errors provoke a horizontal shift of the sampling instant. This shift does not generate errors in bit decision if the clock error is within the indicated margin for timing errors. The same applies to amplitude errors that do not generate bit errors unless greater of the amplitude margin.

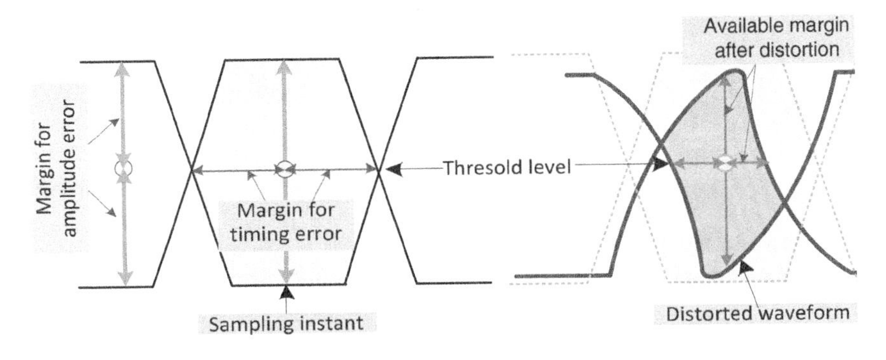

Figure 5.44 The eye pattern diagram.

On the right side of Figure 5.44, the square waveform is distorted after transmission over a low pass channel. Time and amplitude margins, evidenced by the filled eye area, are strongly reduced. The left image can be considered as an open eye patten sufficiently resistant to signal degradation. The right image shows a more closed eye due to the limited slew rate capability that may induce a wrong bit decision. Note that when the bit rate increases and the slew rate is not changed, the opened area of the eye pattern reduces proportionally. Thus, an increased error rate is expected when increasing the bit rate.

Clock and channel effects on bit decision are worsened by the **noise** generated by resistors and electronic components. Since noise is mainly generated by thermal sources, it can be modeled as an additive Gaussian process. As shown in Figure 5.46, we assume that the baseband transmission channel can be modeled by a low pass filter and additive white gaussian noise (AWGN) is superimposed on the signal. In this case, the **probability of bit error** P_e can be simply derived. The received signal $y(t)$ is $y(t) = z(t) + w(t)$, where $w(t)$ belongs to a white Gaussian stationary ergodic process with unilateral power spectral density N_0. To have a better understanding of the following discussion, Table 5.27 summarizes some basic concepts of the stochastic processes (Papoulis, 1993). For thermal noise $N_0 = KT_k$, where T_k is the temperature in kelvin and K the Boltzmann constant (see Section 5.2.1).

In practical systems, white noise with power spectral density $= N_0$ for any value of frequency f, does not exist. It is then of interest, considering a white band-limited noise whose spectrum is constant only for a limited bandwidth.

The received signal $y(t)$ is $y(t) = z(t) + w(t)$ where $w(t)$ belongs to a white Gaussian stationary ergodic process with bilateral power spectral density $N/2$.

The signal $y(t)$ is sampled at T-spaced time intervals. Assuming that the additive white Gaussian noise is band-limited with band $B = 1/T$, the noise $w(t)$ has an autocorrelation function $R_w(\tau) = N_0 B \cdot sinc(B\tau)$. Since the autocorrelation is null for $\tau = k/B$, with k integer, the sample values at k/B distance are not correlated. It can be proved that in Gaussian processes uncorrelated variables are also statistically independent and therefore, sampling $y(t)$ at T equally spaced positions will yield the signal sample $z(kT)$ added to a Gaussian noise sample statistically independent of all the other noise samples located at different time instants. Considering a binary transmission where only the two values $+A$ and $-A$ are transmitted for the two bit values (1,0) and such values are left unchanged at sampling instants kT, there may have two conditions according to the transmitted bit: $y(kT) = \pm A + w(kT)$ where $w(kT)$ belongs to a zero mean Gaussian process with variance:

$$\sigma_w^2 = R_w(\tau = 0) = N_0 B = N_0/T. \tag{5.72}$$

When a bit $s[k] = 1$ is transmitted, the signal is $y(kT) = +A + w(kT)$, and there is an error in bit decision when $w(kT) < -A + \lambda$, where λ is the decision threshold introduced in (5.46).

Conversely, when a bit $s[k] = 0$ is transmitted, the signal is $y(kT) = -A + w(kT)$ and there is an error in bit decision when $w(kT) > A + \lambda$.

The probability of having an error P_e is then

$$P_e = P_s(s[k] = 1) \times P_w(w(kT) < -A + \lambda) + P_s(s[k] = 0) \times P_w(w(kT) > A + \lambda), \tag{5.73}$$

where $P_s(s[k] = 1)$ and $P_s(s[k] = 0)$ are the probabilities of generating the bit 0 or 1. When the probabilities to have 0 or 1 are equal, $(P_s(s[k] = 1) = P_s(s[k] = 0))$, it is easily proven that

Table 5.27 Stochastic processes: main properties

Definition	Meaning
Stocastic (random) process	Set of $\{t, w(t), P_w(w(t),\ t)\}$ where for any given time instant t, the values of the function $w(t)$ belongs to a probability function $P_w(w(t),\ t)$
Strong stationary process	Statistical properties of random processes do not depend on time, that is: $-P_w\{w(t_m)\} = P_w\{w(t_n)\}$ for any m, n $-P_w\{w(t_m),\ w(t_n),\ t_m,\ t_n\} = P_w\{w(t_m),\ w(t_n),\ t_m - t_n\}$ for any m, n
Expected value of a function $g(w)$	$\mathrm{E}\{g(w)\} \overset{def}{=} \int_{-\infty}^{\infty} g(w)P_w(w)dw$, where w is a random variable
Ensemble values of strong stationary processes.	Mean value:$\eta_w \overset{def}{=} \mathrm{E}\{w(t)\}$ Square mean value: $m_w^{(2)} \overset{def}{=} \mathrm{E}\left\{(w(t))^2\right\}$ Variance: $\sigma_w^2 \overset{def}{=} m_w^{(2)} - (\eta_w)^2$ Moment: $m_w^{(1,1)}(\tau) \overset{def}{=} \mathrm{E}\{w(t)w(t+\tau)\}$ *Ensemble values do not change over time t*
Process time averages of $w(t)$	Mean value (= DC): $M_w \overset{def}{=} \lim_{T\to\infty} \frac{1}{T}\int_{-T/2}^{T/2} w(t)dt$ Mean square value (= power):$\wp_w \overset{def}{=} \lim_{T\to\infty} \frac{1}{T}\int_{-T/2}^{T/2} \lvert w(t)\rvert^2 dt$ Autocorrelation: $R_w(\tau) \overset{def}{=} \lim_{T\to\infty} \frac{1}{T}\int_{-T/2}^{T/2} w(t)w(t-\tau)dt,$
Property of the ergodic processes	Time average values equal to ensemble average values: , $M_w = \eta_w, m_w^{(2)} = \wp_w, m_w^{(1,1)}(\tau) = R_w(\tau)\ldots$
Wiener theorem	The power density spectrum $\wp(f)$ of a stochastic process is equal to the inverse Fourier transform ($\mathrm{F}^{-1}\{\}$ defined in (5.42)) of the autocorrelation function: $\wp(f) = \mathrm{F}^{-1}\{R_w(\tau)\}$
Statistically uncorrelated variables	Two stochastic variables x, w are uncorrelated if their expected value is null:$\mathrm{E}\{w, x\} = 0$
Independent random variables	Two stochastic variables x, w are statistically independent if the joint probability function is equal to the product of the single probability functions: $P_{w,x}\{w,\ x\} = P_w\{w\} \times P_x\{x\}$
White noise process	Stochastic process with $\wp(f) = N_0$ constant for any f or equivalently $R_w(\tau) = N_0\Delta(\tau)$ where $\delta(\tau)$ is the Dirac impulse defined in (5.62); from the previous equation, the autocorrelation is non-null only for $\tau = 0$, that is, sample values at different times are not correlated
White noise band-limited process	Stochastic process with $\wp(f) = N_0$ constant for any $\lvert f\rvert < B/2$ or, equivalently, $R_w(\tau) = N_0 B \cdot \mathrm{sinc}\,(B\tau)$ where sinc(x) is defined in (5.47); from the previous equation, the autocorrelation is null only for $\tau = k/B$, with k integer, that is, sample values at k/B distance are not correlated
White stationary Gaussian noise	Probability density function: $P_w\{w\} = \frac{1}{\sqrt{2\pi}\cdot\sigma} \cdot e^{-\frac{1}{2}\left(\frac{w-\eta_w}{\sigma}\right)^2}$ Power density spectrum: $\wp(f) = N_0 = \sigma^2$

the probability of error P_e is minimized for $\lambda = 0$. In this case, since $P(w(kT) < -A) = P(w(kT) > A)$ and $P(s[k] = 1) + P(s[\mathrm{k}] = 0) = 1$:

$$P_e = P_w(w(kT) < -A) = P_w(w(kT) > A) \tag{5.74}$$

For Gaussian processes we have

$$P_e = P_w(w(kT) < -A) = \frac{1}{\sqrt{2\pi} \cdot \sigma_w} \cdot \int_{-\infty}^{-A} e^{-\frac{1}{2}\left(\frac{w}{\sigma_w}\right)^2} dw. \tag{5.75}$$

We now introduce the error function $erf(x)$ and its complementary $erfc(x)$:

$$erf(x) = \frac{2}{\sqrt{\pi}} \int_0^x \cdot e^{-w^2} dw, \quad erfc(x) = 1 - erf(x). \tag{5.76}$$

It is easily shown that $erf\left(\frac{A}{\sqrt{2} \cdot \sigma_w}\right)$ is the probability that a random variable w belonging to a Gaussian process with zero mean and variance σ_w^2 is inside the $-A$, $+A$ interval. From (5.76), $0.5 \times erfc\left(\frac{A}{\sqrt{2} \cdot \sigma_w}\right)$ is the probability that the Gaussian random variable is greater than A. From (5.72) the probability of error can be then written as

$$P_e = 0.5 \times erfc\left(\frac{A}{\sqrt{2} \cdot \sigma_w}\right) = 0.5 \times erfc\left(\sqrt{\frac{E_b}{2 \cdot N_0}}\right), \tag{5.77}$$

where $E_b = A^2 \times T$ is the energy of the signal per bit on the assumption that the signal is simply coded as non-zero return (NRZ) as in (5.60). The complementary error function is depicted in Figure 5.45 with respect to the abscissa value: $\mathrm{SNR(dB)} = 10\log_{10}(\mathrm{SNR})$, $\mathrm{SNR} = \frac{E_b}{N_0} = \frac{A^2 T}{N_0}$. The probability of error in Figure 5.45 increases with the increase of the signal amplitude A, while it decreases when the noise N_0 and the bit rate $1/T$ increases.

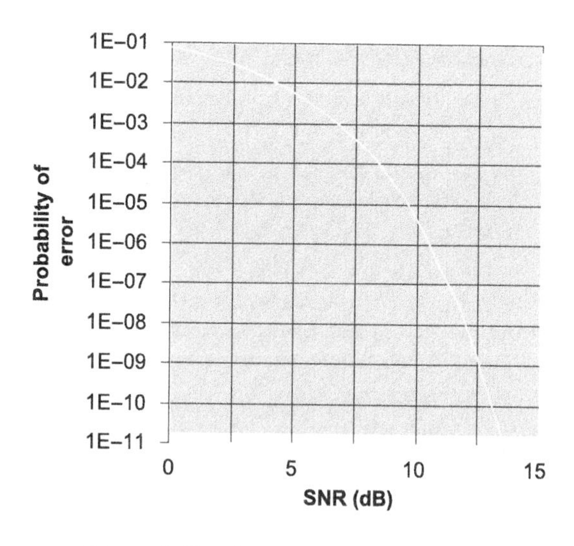

Figure 5.45 The Gaussian error function.

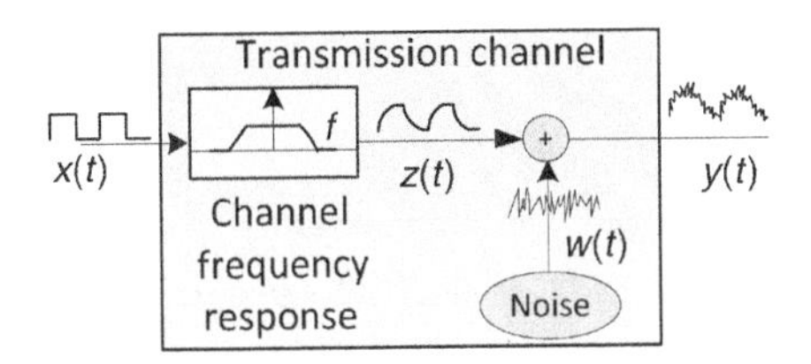

Figure 5.46 The transmission channel model.

We note also that the received amplitude A_r is lower than the nominal value A due to the low pass effects of the transmission channel. In addition, the nominal value of A is limited by the maximum supply voltage of the digital circuit: to reduce consumption, circuit voltage supply is continuously reducing. For example, when A has the typical maximum values of CMOS-TTL logic family ($+A = 4.7$ V, $-A = 0$ V), a 10Mbit transmission with typical noise may have $P_e = 5 \times 10^{-12}$. The amplitude of the signal may be lower than the nominal value A due to the effect of the transmission channel. As shown in Figure 5.46 the received signal will be distorted by the transmission channel, which reduces the amplitude of the signal itself at the time of decision. This increases the error probability due to the additive Gaussian noise. The CMOS-TTL logic family can usually accept worse threshold levels ($A = 2$ V, $-A = 0.8$ V) that reduce the probability errors to $P_e = 3 \times 10^{-1}$, which is equivalent to three errors over ten transmitted bits. To obtain the same error probability $P_e = 5 \times 10^{-12}$, the bit rate has to be reduced to 650 kbits/s. To obtain better performance and lower costs, more recent logic families may reduce the supply voltage to 3.3 V or even down to 1.0 V. In this case, noise may increase probability of error significantly. It should be noted that in real electronic boards, EMI may also generate noise with negative impact on error probability. In turn, high bit-rate baseband transmissions may determine how the EMI has to be limited in order to comply with regulations.

From an implementation point of view, digital components are increasing in electronic devices. These components need data exchange techniques with the following requirements:

1. economic in space, cost and consumption
2. simple and fast in design and time to market
3. performing at the required data transfer rate.

Therefore, the electronic industry has defined standard interfaces using parallel or serial buses. **Parallel buses** (e.g., ISA, SCSI, PCI in personal computers) are used for high-speed connections but require more space and often they are more expensive solutions. Alternatively, data may be serialized so that information is transmitted sequentially.

Serial buses reduce the required physical lines and the number of IC pins, thus lowering the overall device cost. The serial bus overall transfer rate may also be as good as parallel bus performance, since the transmission clock can be increased because no synchronization is required as it is with parallel data transmission. Well-known serial buses are RS-232, RS-422, SPI, I^2C, USB and Ethernet. Both serial and parallel buses are based on transmission lines experiencing the probability of error as discussed above. Medical devices may also have the stringent requirement of galvanic isolation, that is, the floating parts connected to the patient must not have a direct DC connection to the power line in order to avoid electric

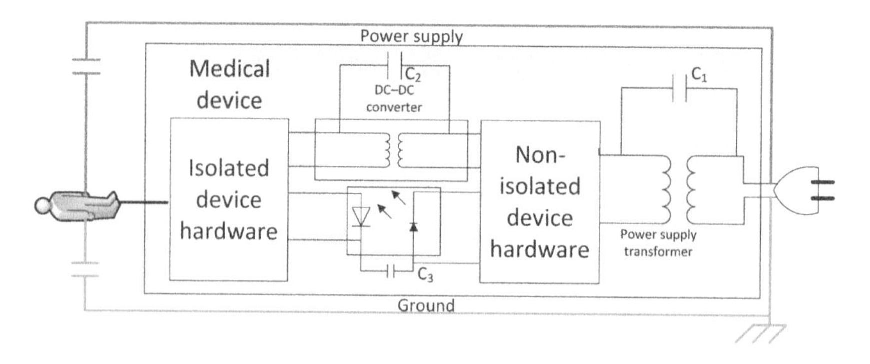

Figure 5.47 Medical device isolation model.

shock. Galvanic isolation of data transmission can be nicely performed using optical devices. Industry is now producing optocouplers compliant to the necessary voltage isolation.

Figure 5.47 shows a basic scheme for galvanic isolation of a medical device. Data transmission is isolated through optocouplers to be selected with proper maximum bit rate and voltage isolation. The power supply may be isolated through transformers, as explained in the section 5.9. The outcome of this discussion is

1. digital components are increasing in electronic boards
2. digital components communicate through transmission lines
3. transmission lines have limited bandwidth and are affected by additive noise and EMI
4. the bit rate between digital components is limited by the available bandwidth of transmission lines, noise and other EMI
5. additional error correction bits (parity etc.) may reduce error probability at the cost of a reduced effective bit rate
6. the increase in bit rate creates higher frequency EMI which in any case has to be limited to the allowed valued
7. standard serial and parallel data interfaces reduce design time and costs for data transmission; standards also reduce problems related to baseband data transmission (probability of error, EMI etc.) and, in general, to medical certification
8. galvanic isolation of transmission lines, required in medical device, may be performed through optocouplers.

Part II: Implementation

This section shows how electronic theory is realised in the design of marketable medical products. This is done by disclosing the hardware of the product **Gamma Cardio CG**. The implementation details are provided for this book by the Gamma Cardio Soft (www.gamma cardiosoft.it), who has the vision of supplying low-cost high-performance devices, showing the device implementation details with the aim of increasing the world diffusion of EGGs. This is a means of reducing the effects of the cardiovascular disease that are the leading cause of deaths and disabilities in the world (WHO, 2011). The sections of this implementation part follow an organization similar to the theory part introducing: the need for the specific electronic stage, the associated requirements and the implemented circuit. The sections of the implementation part shown in Figure 5.48 have the same unit numbers as the corresponding sections in the theory part.

5.20 Gamma Cardio CG Architecture

Figure 5.48 shows the basic Gamma Cardio CG architecture and the associated electronic stages. Circled numbers specify the section where the block is discussed.

The Gamma Cardio CG unit is connected to the patient electrodes through a patient cable. The PC connection is through a USB plug. The patient is connected through a shielded cable with ten terminations, which are connected to the six precordial and the four peripheral electrodes. The ten **protection circuit stages** eliminate mains interference and transients that are superimposed to the ECG signal that may damage the electronic circuits. The ten **buffer** stages offer high input impedance so that the patient impedance does not affect the measurements, as explained in Section 5.23. The input signals have to be combined together to obtain the target measurement signals. The **lead generation** network determines the proper

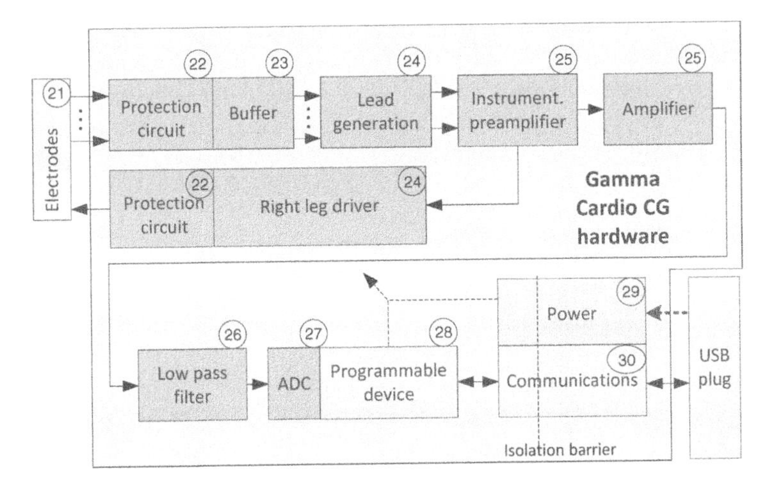

Figure 5.48 Gamma Cardio CG acquisition unit architecture.

combination of input signals to obtain the lead signals as described in Chapter 2. The **instrumentation preamplifier** performs the subtraction operation between the two input signals, eliminating common mode interference, that is the undesired signal equally added to both the input signals (Section 5.25). The module also generates the sum of the undesired input signals (common mode signal) that is used by the **right leg driver** to reduce interference at input (Section 5.24). An **amplifier** increases the ECG amplitude, and the **low pass filter** significantly reduces out-of-band interference so that the signal is suitable for analog to digital conversion (**ADC**) that may be performed directly inside the **programmable device**. The programmable device controls the analog circuit part through the control lines and sends the input signal to the PC. This device is connected to the PC through a **communication** stage that handles serial and USB communication through an isolation barrier. The **power** stage provides a supply to all the acquisition unit stages through an additional isolation barrier which prevents a direct connection from the patient to the power distribution system, even in presence of faults. In figure 5.48, the power supply connections are omitted for clarity. The power of the board is obtained through the USB connection.

5.20.1 ECG Design Choices

The improvements in the electronic components allow the choice of different ECG design schemes suitable for handling the ECG signal. According to the AAMI EC 11 standard (AAMI, 1991), an ECG signal is a differential mode signal of maximum ± 5 mV in the range of 0.05–150 Hz added to a differential mode DC component of up to ± 300 mV and a common mode interference mainly due to power line electric fields that may rise to tens of volts. The ECG system noise must not add noise greater than 30 mV_{pp}.

A first design choice is the so-called **single channel 12 lead ECG** where the 12 lead signals are recorded one by one and not simultaneously. This scheme is depicted in Figure 5.49.

After circuit protection, a lead selector circuit defines the appropriate combination of input signals to obtain the required lead signal. Then, the instrumentation amplifier reduces common mode interference and amplifies input around ten times so that the output signal is in the ± 3.05 V range, as $\pm 3.05\ \text{V} = (\pm 300\ \text{mV DC} \pm 5\ \text{mV ECG}) \times 10$. This is suitable for a power supply at $\pm$ 5 V. In addition the instrumentation amplifier provides the common mode signal to the right leg drive (RLD) stage that eliminates additional common mode interference. The high pass filter eliminates the DC offset signal. After additional amplification, a low pass filter reduces out-of-band components (> 150 Hz) and avoids minimizing aliasing after sampling process. Thanks to the overall gain of $10 \times 50 = 500$, at the ADC input we

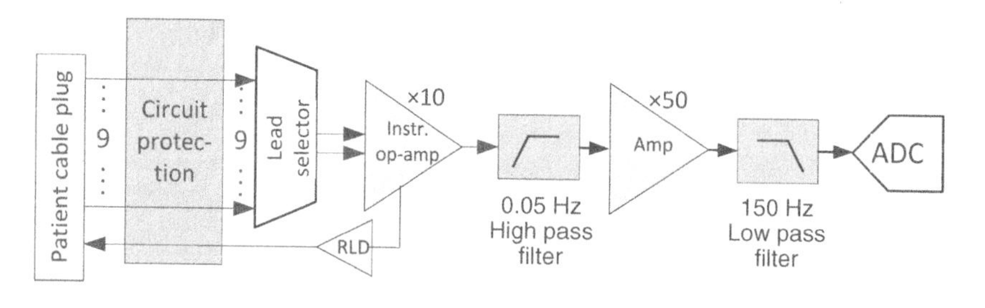

Figure 5.49 ECG single channel architecture.

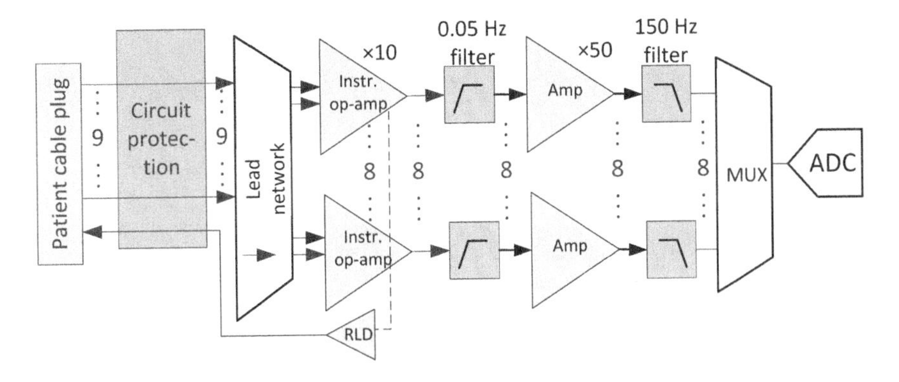

Figure 5.50 ECG basic multichannel architecture.

have an ECG signal in the range of ± 2.5 V with reduced common mode interference and no DC offset. This range may be level shifted by 2.5 V so that the typical ADC input dynamic range of 0–5 V is achieved.

The single channel may be generalized to a **multichannel** scheme as shown in Figure 5.50 where eight single channel sections are implemented and the other four lead signals can be digitally derived from the previous eight channels, as explained in the previous chapters.

In this scheme, the eight high pass filters have to be built with a 1 μF large capacitor which may be critical. The number of components can be reduced thanks to the use of high resolution (> 22 bit) ADC devices such as sigma delta multichannel ADCs. Using these chips, the ECG scheme shown in Figure 5.51 is simplified, since the ± 300 mV DC offset signal may be sampled and digitally eliminated. This avoids using a high pass analog filter. When a sigma delta oversampling ADC is used, the low pass filters required to prevent aliasing can also be simplified to an RC network as outlined by equation (5.56) where the final term allows for increasing SNR value.

More recently, some vendors have developed chips with specialized ECG front-ends (Texas, 2011b; Analog, 2011b). This has reduced the number of components, and also the design risks, but at generally increased costs. In addition, this choice implies the single supplier risk: if production of the component is phased out, or the chip cannot be used because of cost escalation, the design has to be completely redone with respect to a design with generic components. In general, ECG design has many challenges, including safety

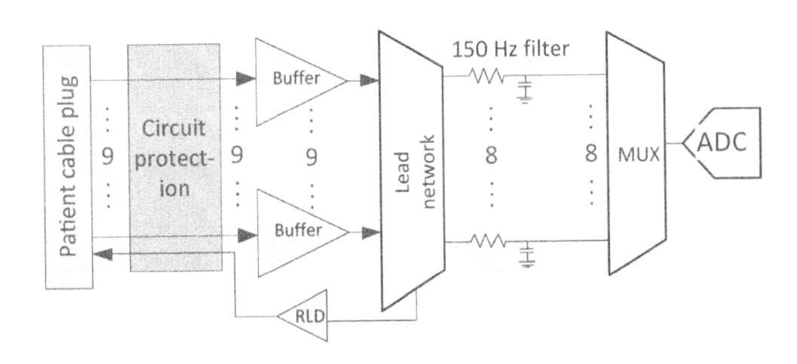

Figure 5.51 ECG improved multichannel architecture.

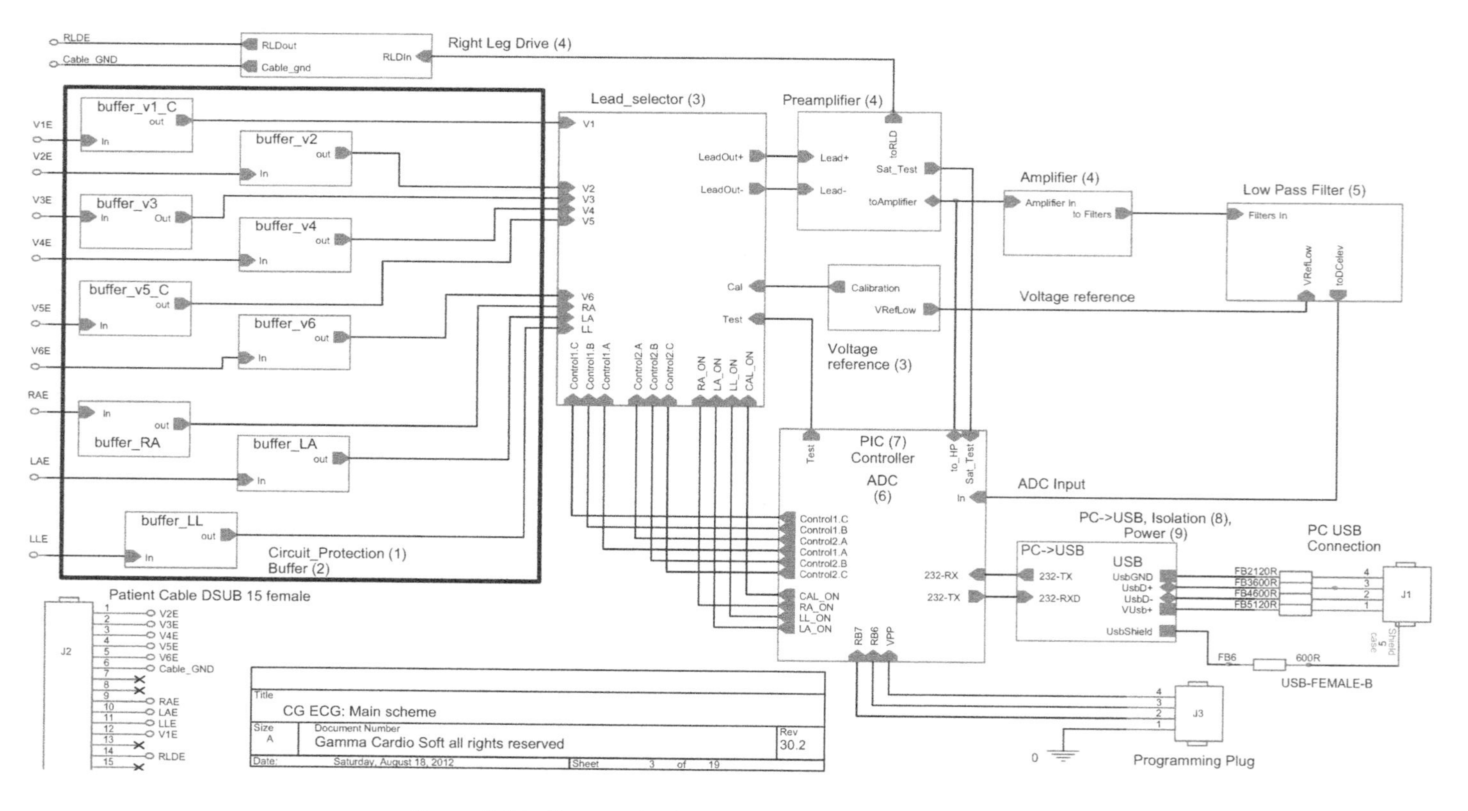

Figure 5.52 Gamma Cardio CG acquisition unit general scheme.

assurance, EMI/RFI interference reduction, correct dynamic range, device reliability – protection and noise minimization (Crone, 2011).

The single-chip analog front-end may not solve this challenges: without a deep comprehension of the theory behind ECG, the main design problems remain. So the Gamma Cardio CG is a good example for understanding the ECG implementation choices and also for academic purposes, since it addresses the main problems of a typical ECGs, regardless of the specific architecture that may be single channel as in Figure 5.49, multichannel as in Figure 5.50, improved multichannel as in Figure 5.51 or single chip. The Gamma Cardio CG is a high-resolution single channel device that may be the base for the multichannel variants addressed in Figure 5.50 and Figure 5.51.

5.20.2 Gamma Cardio CG Complete Scheme

The **complete main schematic** diagram of the Gamma Cardio CG is shown in Figure 5.52. Each block of the figure is a specific schematic diagram discussed in the section whose number is indicated in parenthesis.

Referring to Figure 5.52, the patient cable is connected to the ECG acquisition unit through a female 15-pin (D-Sub) standard connector (**J2** at bottom left) which is connected to the 9 buffer stages that contain the protection circuits. The buffer stage outputs feed the lead selector stage that is controlled through ten digital lines of the Pic controller ADC stage. The lead selector receives the calibration signal from the voltage reference stage and a test signal from the PIC.

These signals allow performance tests to be executed on the acquisition unit. The lead selector outputs the ECG signal to the instrumentation preamplifier via two balanced differential signals (LeadOut+, LeadOut−). A voltage reference block is required to provide a calibration signal. This stage also has a pin (Sat_Test) that evaluates if the signal is in saturation. The signal is then amplified, filtered and sent to the ADC contained in the PIC. The PIC stage has a plug for downloading the firmware (J3) and it has a serial data interface connection with the PC−>USB stage that translates serial data into a standard USB interface with two input–output data lines, ground and power supply (+5 V) provided by the PC. The PC−>USB stage contains components for galvanic isolation of power as discussed in section 5.29 and data transmission and the components that provide power supply for all the active devices.

5.21 ECG Sensors

THE NEED The ECG measures the electrical potentials (i.e., biopotentials) generated by the human heart. More generally, biopotentials are generated by the ionic activity of the cells in the body. The heart's excitable cells produce specific voltage potentials associated with heart contractions (systole) and with the following relaxing resting state (diastole), as explained in Chapter 2. Since the human body can conduct ionic currents, the potentials generated by the heart may be measured at the body's surface. The current flowing in the body is a ionic current consisting of negative and positive charges. There is therefore the need for sensors that convert the body's ionic currents into an electron current. This is obtained by a chemical reaction at the interface of a sensor (i.e., the electrode) made up of a

Table 5.28 ECG performances impacting electrodes selection

Ecg performance	Requirements
ECG maximum allowed voltage offset (EC 11)	$\geq \pm 300\,\text{mV}$
ECG Input impedance (EC 11)	$\geq 2.5\,\text{M}\Omega$ at 10 Hz
Skin–electrode overall impedance (all lead) (EC 11)	$< 250\,\text{k}\Omega$ to have error $< 10\%$
Recovery time after defibrillator pulses (IEC/EN 60601-2-25)	<10 s after defibrillator discharge (test condition: ECG sinal readable in the screen-paper)
Biocompatibility (ISO 10993-1).	Electrodes must not be a cytotoxic irritant or a sensitizer, in compliance with the requirement of ISO 10993-1.
Certification (EEC, 2007)	Class I Medical device according to the EU Medical Device Directive

conductive metal connected to the device and an electrolyte that is located between the electrode and the skin.

REQUIREMENTS Electrodes constitute the interface between the patient body and the ECG device. Requirements and usage have already been discussed in Chapter 2, while design specifications are addressed in Chapter 3. Electrodes must satisfy requirements on:

1. electrical performance
2. defibrillation recovery performance
3. biocompatibility
4. certification.

Table 5.28 outlines the ECG performances that affects electrode selection. The main electrical ECG performances relate to voltage offset and impedance. Regarding **voltage offset**, the electrodes rely on a chemical reaction to transform ionic current into electronic current. An ideal electrode should completely convert the ionic current into the electronic counterpart. Unfortunately, the skin–electrode interface has a non-ideal behavior that can be modeled as an electronic passive impedance (i.e., a RC network circuit) with a DC source generator modeling the offset potential.

More specifically, the skin is mainly a dry dielectric surface that behaves as an impedance for ion–electron current transfer. A chloride gel or saline solution may be used to reduce the skin–electrode impedance. In addition to the skin impedance, the electrode–skin interface includes the impedance from the electrolytic gel and the electrode–electrolyte interface. The DC voltage offset is due to the half-cell potentials caused by different energies of the electrode, electrolyte and skin.

The electrical performance of electrodes varies greatly according to patients, test conditions and type of electrodes (e.g., Baba, 2008). This may affect the ECG input stages that may be saturated by large electrode offset potentials experienced especially with suction chest electrodes that have a reduced contact surface. The AAMI EC11 standard (AAMI, 1991) requires that the ECG must withstand at least $\pm$ 300 mV offset potential, and (AAMI, 1884) requires that disposable electrodes have an offset potential lower than 100 mV.

Skin–electrode impedance is another source of error when its value is not significantly lower than the input impedance of the ECG device. This is clarified by equation (5.19) and Figure 5.5 in Section 5.3. Unfortunately, the impedance value depends on various elements such as the electrode placement, the position, the type of electrode paste and the type of electrode. Skin preparation is also a major factor affecting offset potentials and impedance.

This suggests imposing a requirement on minimum ECG input impedance, so that, from a statistical point of view, a significant sample of acquisitions will be affected by an acceptable error. More specifically, with an ECG input stage with a minimum impedance of 2.5 MΩ at 10 Hz per lead, as required by AAMI EC 11 (Table 5.28), a 50 kΩ skin-to-electrode impedance will yield a error of $<$2%, while a 250 kΩ skin-to-electrode impedance will yield an error of $<$10% as per eq. (5.19).

According to some statistical studies, this requirement will ensure that the 98% of ECG acquisitions with usual skin preparation will be affected by an error $<$10% (AAMI, 1991). It is also argued that the remaining 2% of ECG acquisitions with higher skin-to-electrode impedances will be detected by physicians due to the excessive artifacts on the ECG plots.

The **defibrillation recovery performance** is related to the capability of the ECG to show the correct signal after defibrillation. A defibrillator may often be used with an ECG. A defibrillator is a medical device that can supply energy to a patient. This energy usually reestablishes a normal sinus rhythm from life-threatening conditions such as ventricular fibrillation and ventricular tachycardia. The ECG and the associated electrodes have to withstand the energy that is delivered by the defibrillator. In addition, after the defibrillation discharge, the ECG has to be operative within 10 seconds in order to asses if a normal sinus rhythm has been reestablished. For the tests of the ECG performance, the circuit shown in Figure 5.53 is depicted. The defibrillator behaviour is modelled by a 5000 V DC generator that supplies energy to a series combination of a 500 μH inductance (L1) and a 32 μF capacitor (C1). The capacitor and the inductance are then discharged into a 100 Ω load (R1) in parallel with the ECG inputs simulated by R4 and R5 (IEC/EN 60601-2-25). Relays S1 and

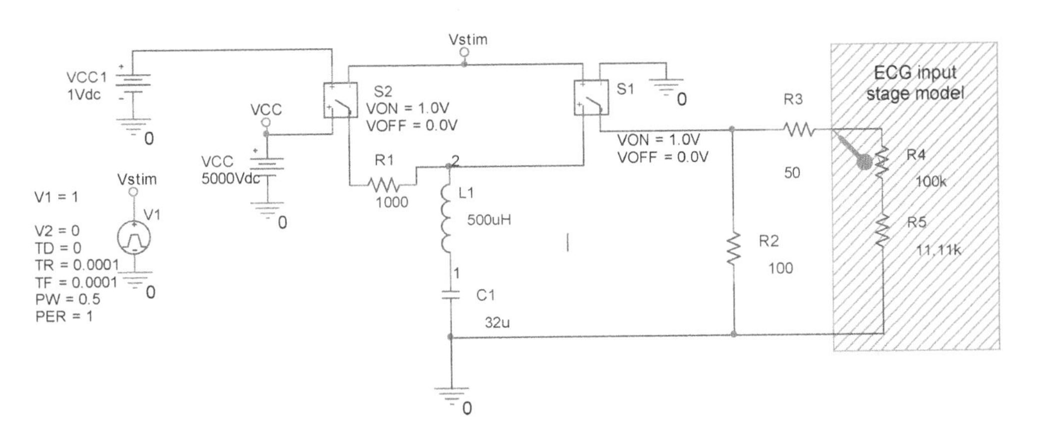

Figure 5.53 Conceptual defibrillator simulator scheme.

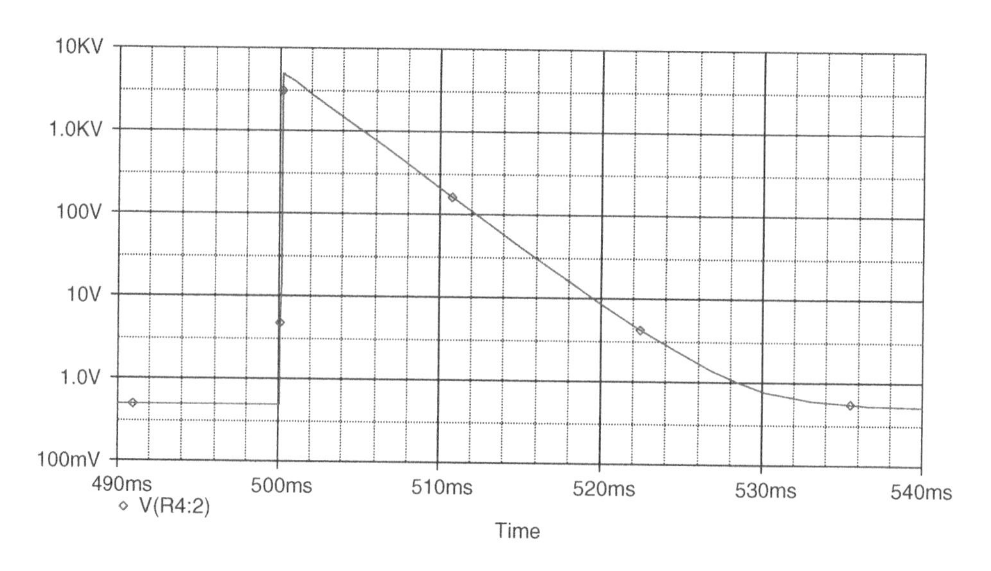

Figure 5.54 Defibrillator discharge against time.

S2, the voltage generator V_{stim} and the DC generator VCC1 are used in the Spice circuit simulation to charge L1 and C1 ($V_{stim} = 0V$) and to trigger the discharge on the load R2 at 500 ms ($V_{stim} = 1V$).

The defibrillator discharge produces a large voltage for a limited time period at the ECG inputs, as shown in Figure 5.54. The defibrillator discharge may then produce large DC offset potentials by charging the electrode impedance with high voltage value that takes some time to be discharged. These offset potentials may saturate the ECG for a long time. The effect on the ECG display may even be a line that indicates the total absence of heart activity, thus suggesting the wrong counteractions to physicians.

Following IEC 60601-2-25, electrodes must be suitable for defibrillation procedures, with signal recovery time lower than 10 s.

ECG performance is strictly related to the electrodes performance. The AAMI EC11 standard (AAMI, 1991) requires that the ECG manufacturers specify the electrodes type to be used to achieve compliance with the requirements of the standard.

THE SOLUTIONS The most used electrodes for ECGs and similar applications are based on silver/silver chloride (i.e. Ag/AgCl). These electrodes consist of a silver foil coated with silver chloride (AgCl). The electrode's reversible chemical reaction that transforms the ionic current into an electron current is

$$Ag + Cl^- \Leftrightarrow AgCl + e^- \tag{5.78}$$

The Ag/AgCl electrode gives a better performance on offset, impedance and frequency dependence. For example, Ag/AgCl may have a DC voltage offset around 0.1–5 mV, around ten times higher than stainless steel electrodes. Ag/AgCl impedance is in the order of 70–300 Ω w.r.t. 1–2 kΩ of stainless steel.

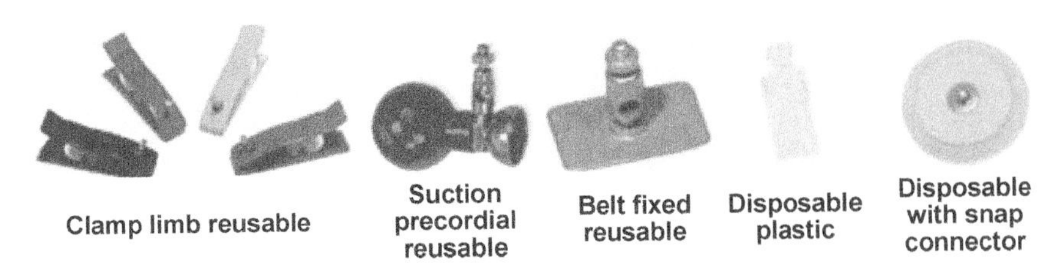

Figure 5.55 ECG electrodes.

Electrodes may be disposable or reusable. Examples are shown in Figure 5.55. For the four limbs, reusable electrodes are based on Ag/AgCl clamps kept tight by an elastic mechanism.

Alternatively Ag/AgCl peripheral plates electrodes may be fixed to the limbs through rubber belts. For chest applications, Ag/AgCl suction chest electrodes provided with rubber balls may be suitable for precordial recording at rest. Rubber balls must be squeezed after positioning to improve adherence. Alternatively, disposable electrodes are now available at reasonable cost with the advantage of reducing risks associated with cross infections more frequent in reusable electrodes.

5.22 Gamma Cardio CG Protection

The ECG acquisition unit must withstand signals originated by

1. the patient
2. external interference transients (conducted or radiated as in Table 5.7)
3. high voltage transients from a defibrillator (AAMI, 1991).

The protection circuit reduces interference to values that do not cause damage to the hardware unit and do not affect patient safety. The specific circuit requirements discussed in Chapter 3 are summarized in Table 5.29; the most challenging requirement is the protection against defibrillator discharge. The associated over-voltage, shown in Figure 5.54, is usually handled by a gas discharge tube or a metal oxide varistor (MOV). Pros and cons of MOVs are shown in Table 5.30 (adapted from Wikipedia). MOVs are worst for leakage and capacitance but thanks to the reduced switching time, they avoid dangerous transients to electronic circuit.

Two approaches are usually considered for protecting devices: introduce over-voltage protection directly on the acquisition unit or use patient cables with embedded resistors and voltage clampers. This latter approach has the benefit of relaxing stringent defibrillation requirements on the acquisition unit, because most of the overload is handled within the cable itself.

Many hardware failurs are due to transients propagated into circuits by the external connections. In the Gamma Cardio CG device protected cables have been used. This option reduces the probability of failure, since overvoltage is usually reduced by the cable itself with minimal propagation into the device.

Table 5.29 Requirements for protection circuit

Requirement	Performance	Value
R 9: IEC/EN 60601-1 and IEC/EN 60601-2-25 compliance	Protection against defibrillator discharge	$\leq 5\,\text{kV}$
R 12 Class II CF type device	IEC/EN 60601 classification	Class II type CF device
R9: IEC/EN 60601-1 and IEC/EN 60601-2-25 compliance	DC current on input leads	See Table 5.31 for Class II type CF devices
R 3: AAMI EC 11 compliance	Input dynamic range (R 3.3)	$\geq \pm 300\,\text{mV} \pm 5\,\text{mV}$
	System noise(R 3.12)	$\leq 30\,\mu\text{V}$
	Input impedance (R 3.9)	$>2.5\,\text{m}\Omega$
	Overload protection (R 3.14)	$\geq 50\,\text{Hz}$, 1 V, 10 s
R 10 IEC/EN 60601-1-2 compliance	EMC	see Table 5.7

Assuming an MOV with 65 V DC clamping voltage, the acquisition unit only has to withstand these 65 V transients after defibrillation. Figure 5.56 shows the overall circuit including defibrillator test model, patient cable and the acquisition unit. The defibrillator is applied to the two terminals of the patient cable that may have resistors and varistors arranged as in Figure 5.56. The resistors R1 and R2 must be specifically designed to survive the peak power during the discharges that have a maximum of $(5000\text{-}130)^2/(4 \times 10000) = 593$ W (see eq. 5.6) although for a very limited time as shown in Figure 5.57, where the dissipated power of one resistor is plotted on a log y axis over time.

The ECG cable construction must also avoid internal arcs occurring during defibrillator discharge.

Thanks to the varistors, the input pins of the acquisition unit will experience only a maximum of $65 + 65 = 130\,\text{V}$ which is easily absorbed by the resistors and transistors. Considering that during defibrillation operations, the transistors clamp the voltage at 0.7 V, there will be a current of $(130 - 1.4)/(R3 + R4) = 0.6\,\text{mA}$ with 41 mW dissipated in each resistor. Such values are easily handled by the transistors and resistors.

The specific configuration of transistors has been selected instead of simple diode protection because voltage clamping is higher and no saturation occurs for the expected differential input range ($\geq \pm 305\,\text{mV}$). The 2N3904 transistor is a usual choice because of the reduced

Table 5.30 Typical performance of MOVs and gas dischargers

Device	Switching time	Capacitance	Leakage current
Metal oxide varistor	<nS	100–1000 pF	$10\,\mu\text{A}$
Gas discharge tube	$<5\,\mu\text{s}$	<1 pF	$\approx$pA

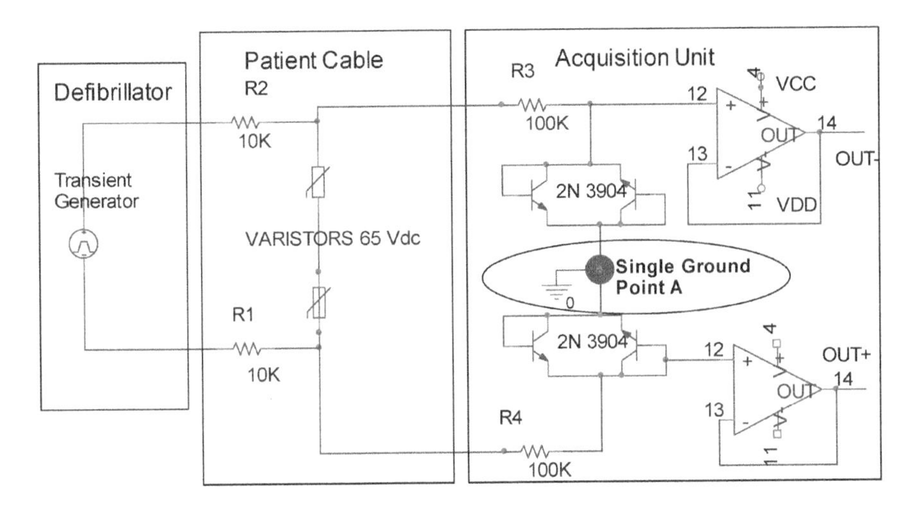

Figure 5.56 Patient cable and protection circuit.

leakage and the input capacitance in off-state (less than 100 pA with tens of pF capacitance) (Sommerville, 1994).

Note that the circuit design is very conservative with respect to transient protection, especially for the high value of R3 and R4 (100 kΩ) that is a tradeoff in terms of system noise. This value is in any case acceptable since the AAMI EC 11 standard (AAMI, 1991) allows for a maximum of $V_{pp} = 30\,\mu V$ noise at the input. The two 100 kΩ resistors at the input will give $V_{pp} = 3\sqrt{2}\,\mu V = 4.2\,\mu V$ (see Section 5.2.1 and Table 5.44 with noise analysis at system level) which is still an acceptable value, while with 10 MΩ, the noise value would rise to an unacceptable $V_{pp} = 42\,\mu V$ which would not satisfy the AAMI requirement and would hide a significant amount of signal information.

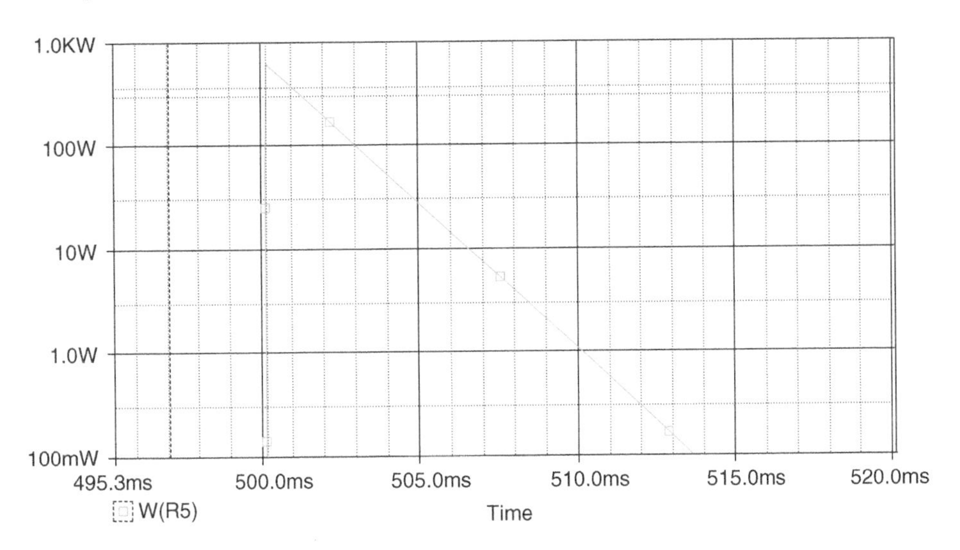

Figure 5.57 Dissipated power after defibrillation charge.

Table 5.31 Admissible leakage current values from IEC/EN 60601-1-2 3rd edition

	Normal Condition (NC) (mA)	**Single Fault Condition (SFC) (mA)**
Earth leakage current (i.e. current between all patient leads connected together and earth)(limits from the 60601-1 third edition, note that in the second edition the limits were 0.5 mA NC, 1 mA SFC)	5	10
Allowable values of **touch current** (i.e. current between all patient leads connected together and a metal envelope embracing the device enclosure, in the second edition named 'enclosure leakage')	0.1	0.5
Patient leakage current and **patient auxiliary current** (i.e. current between one patient lead and all the others connected together)	(B/BF) 0.1 AC 0.01 DC (CF) 0.01 AC 0.01DC	(B/BF) 0.5 AC 0.05 DC (CF) 0.05 AC 0.05 DC
Total Patient leakage current with maximum mains voltage on applied part (i.e. current between all patient leads and maximum mains voltage or with all applied parts of the same type connected together)	(B/BF) 0.5 AC 0.05 DC (CF) 0.05 AC 0.05 DC	(B/BF) 1.0 AC 0.1 DC (CF) 0.1 AC 0.1 DC

Transients from the conducted EMI detailed in Table 5.7 are also easily handled by the previous circuit. The maximum voltage at the op-amps inputs will be always less than 0.8 V. This is a good design choice to limit the input voltage as required by most of op-amps.

Regarding the **leakage current requirements**, limits are outlined in Table 5.31, please refer to the IEC 60601-1 standard for details. The input circuit does not draw current apart from op-amp bias current which is very limited, therefore leakage is minimized at least at the input leads. Other factors affect leakage and therefore these requirements are tested at system level in Section 5.29. The Gamma Cardio CG outperforms requirements since leakage current is always $\leq 1.1\ \mu A$. Regarding transient immunity, there are also some PCB design specifications. The general rule consists of avoiding transients propagating within the acquisition unit since components withstanding the defibrillator transient produce interfering signals within the ground plane that in turn may negatively affect other circuits. The rules in Table 5.32 help in reducing interference.

Table 5.32 PCB guidelines to improve transient immunity

1.	Keep resistors and associated voltage clampers in a compact area, separate from the other devices. Do not fold the defibrillators resistors and associated voltage clampers over the top of other circuits, to prevent coupling of interference into other circuits.
2.	Connect all the ground pins of clampers together and connect them to ground at one point only, to prevent transient propagation between PCB tracks.
3.	Surround the ground area almost completely by a void space
4.	Locate the connection ground point as far as possible away from all digital circuitry.

Table 5.33 Requirements for buffer stage

Requirement	Performance	Value
R 3 AAMI EC 11 compliance	Input dynamic range (R 3.3)	$\geq \pm 300$ mV DC ± 5 mV
	Input impedance (R 3.9)	>2.5 mΩ at 10 Hz
R 3 AAMI EC 11 compliance	System noise (R 3.12)	$\leq 30\ \mu V_{pp}$
Circuit protection maximum voltage	Input maximum voltage	at least 0.8 mV

5.23 Gamma Cardio CG Buffer Stage

The buffer stage is required in the ECG to increase the input ECG's impedance, because the source impedance may be significant and may negatively affect the measured signal.

An op-amp based buffer stage (Figure 5.7) has its output voltage equal to its input, and has very high input impedance that minimizes errors due to skin–electrode impedance. The buffer stage also has a low output impedance to drive easily the following stages. The main requirements for this stage are summarized in Table 5.33.

Figure 5.58 shows both the protection and the buffer circuit that are repeated nine times in the Gamma Cardio CG general scheme shown in Figure 5.52. This scheme has already been discussed in Section 5.3.

We focus here to assess the **input dynamic range** and the input impedance requirements of Table 5.33 can be easily assessed. System noise and CMRR requirements will be assessed at system level. Referring to Figure 5.56, two buffer stages are used to obtain a differential signal. Each buffer stage has unity gain and clamps the signal amplitude to within ±0.7 V. This means that the input differential signal will not be clamped until a peak-to-peak value of 1.4 V of input is reached. This value is much higher than the maximum expected ECG

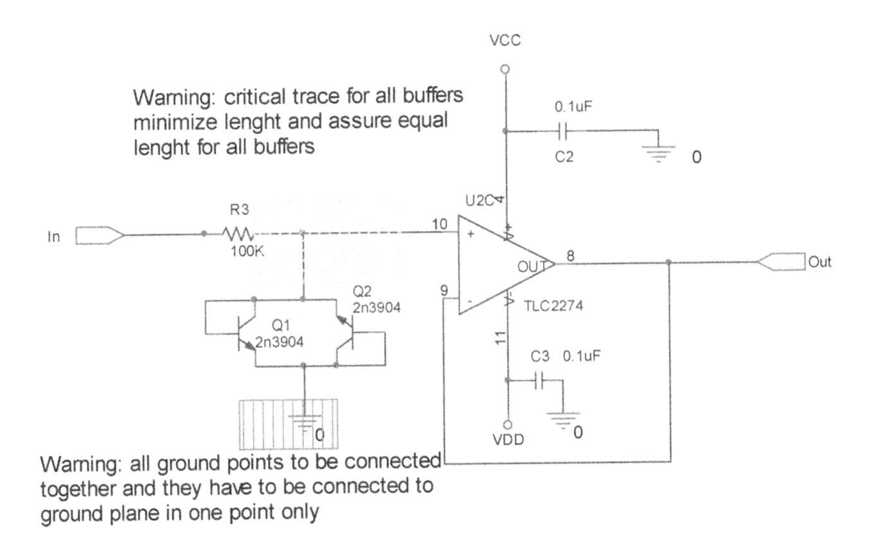

Figure 5.58 Buffer and protection stage.

Table 5.34 Effects on ECG performance from real op-amp behavior

Real op-amp behavior (TLC 2274)	Effects on performances
Input impedance $< \infty$	The input impedance of 663 MΩ is expected, which is sufficient for the ECG application.
Input bias current > 0	The op-amp draws current from the input which creates voltage drops across the input resistance (R3 in Figure 5.58). Since the input offset current is picoamperes (10^{-12} A) and R3 plus the source resistance is in the order of 10^5 Ω, the voltage drop is in the order of 0.1 μV. This value is lower than the minimum required resolution.
4.7 V< Maximum input voltage < 5 V	The transistors limit the input voltage range to ± 0.7 V. This value is much lower than maximum ratings −4.7 V to + 5 V for ± 5 V power supply.
Input offset voltage $V_{IO} > 0$ (required to have zero output)	In the worst case, according to the datasheet, two TLC 2274 may have a V_{IO} difference of 2.4 mV which is propagated to the output due to unity gain. This will add a DC value of 0.0024 V that will be summed to the ECG maximum dynamic range (± 0.305 V). The overall worst DC value (± 0.329 V) is lower than the maximum allowed DC value that can be present at the next stage without producing saturation. This higher value (± 0.352 V) is computed in Section 5.25.2.
Common mode rejection ratio (CMRR)	This aspect does not influence the performance since the op-amp is in unity gain configuration.
PSRR = (variation in power supply)/(variation in input that causes this experienced variation in output)	Capacitors are included to reduce variations in the power supply.
Negative rail $\leq V_{out} \leq$ positive rail	Voltage output has to be included within the positive to negative rails to avoid output saturation. This condition is guaranteed for the ECG signal input dynamic specifications (± 0.305 V).
Slew rate (rate at which output can change at maximum) < 1.7 V/μs	The slew rate at unity gain is 1.7 V/μs which is much higher than the target performance of 320 mV/s
Maximum output current < 50 mA	The buffer stage is connected to a 1.2 kΩ in the worst case. Considering that the maximum output voltage is 0.7 V, the output current (0.5 mA) is much lower that maximum rating (50 mA). According to the datasheet, this value will enable the required maximum output voltage (± 0.7 V).
Frequency/phase response dependent on frequency	The bandwidth is much higher than required (i.e. 150 Hz)
Temperature dependence	Parameters are not significantly affected by temperature changes.
Noise power > 0	The device has a peak-to-peak equivalent input noise voltage of 1.4 mV_{pp} from f = 0.1 Hz to 10 Hz at 25°C. This value is low considering with maximum noise value (30 μV). In any case, the system noise analysis has to be done considering all noise sources as in Table 5.44.

input = ± 305 mV. A single-ended input impedance of at least 2.5 MΩ at 10 Hz will be needed to meet the **input impedance** requirement (AAMI, 1991). Referring to Figure 5.58 and to the datasheets, the leakage current of the transistor is in the order of nanoamperes (10^{-9} A), and therefore it is negligible, and the maximum capacitance of the junction is 8 pF.

The op-amp has an input bias current in the order of picoamperes (10^{-12} A) and an input resistance in the order of teraohms (10^{12} Ω) and therefore both phenomena are negligible. The input capacitance is 8 pF, so the input circuit is thus represented by a 100 kΩ resistor in series with a capacitance of 24 pF (two transistors and the op-amp in parallel). The impedance of this circuit at 10 Hz is therefore equal to 663 MΩ (see formulas in Table 5.13). The next step consists of evaluating how real component behavior might affect overall performance. Table 5.34 summarizes such analysis.

5.24 The Lead Selector

THE NEED — As discussed in Chapter 3, the ECG signals captured by the electrodes positioned on the patient, have to be combined together to obtain the ECG leads.

Referring to Figures 5.48 and 5.52, the ECG signals are available at the buffer output as a low impedance source. They can therefore be combined to obtain ECG leads as specified by the AAMI standard (AAMI, 1991). This function is performed by the lead selector, which will also select test signals and calibration signals to test the acquisition unit.

REQUIREMENTS — The main requirements of this stage are summarized in Table 5.35.

Table 5.36 shows that the nine input signals (LA, RA, LL, $V_{1...6}$) have to be recombined to obtain the required 12 lead signals.

Figure 5.59 derived from the schemes discussed in the theory section (Section 5.4) is suitable for this purpose. Leads I, II and III are obtained easily. For example, in lead I the control lines of the MUX select LA input, and the switch associated to RA is the only one closed in the analog switch component. Golberger leads are obtained by closing two switches. For example, in aV_R, LA and LL, the switches will be closed.

Since LL, LA and RA are the exit of a buffer, they can be considered as voltage sources with a low resistance R_S in series. At point A of Figure 5.59 we will have

$$V_A = (LA + LL) \times (R + R_S)/(R + R_S + R + R_S) = (LA + LL)/2 \quad (5.79)$$

Thus, selecting RA in the MUX of Figure 5.59, the aV_R will be obtained.

Table 5.35 Requirements for lead selector

Requirement	Performance	Value
R 3 AAMI EC 11 compliance	**Lead definition** (R 3.3)	See Table 5.36 and Chapter 3 for specific discussion
	Accuracy of input signal reproduction: (R 3.7) error in lead weighting factors	≤5%

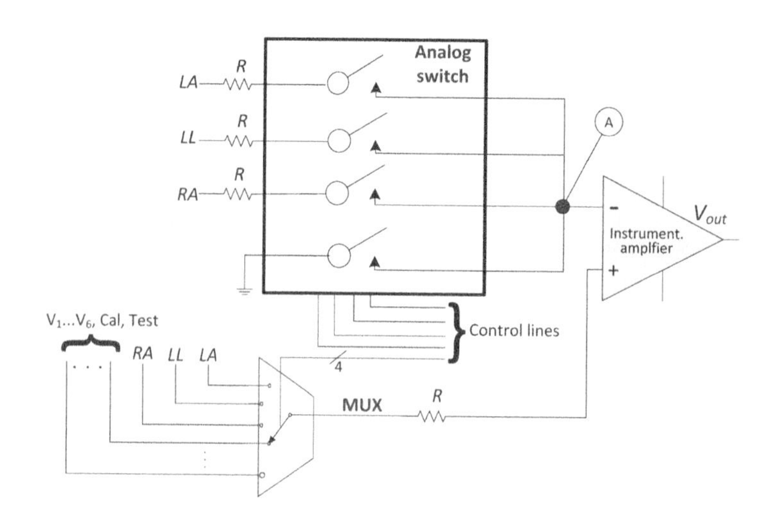

Figure 5.59 The lead selector conceptual scheme.

A similar consideration applies for the unipolar chest leads. In this case, all the three switches associated with LA, LL, RA inputs are closed.

Calibration and test signals are simply obtained by opening all the analog switches apart from ground connected switch. These signals are useful for testing performance of the acquisition board.

Figure 5.60 shows the detailed scheme that uses the two multiplexers U8, U9 to select all the 12 lines.

The 6 MUX control lines indicated as Control 1.A to Control 2.C and the 3 analog switch control lines are connected to the PIC. LeadOut− and LeadOut+ are connected to the input of the instrumentation operational amplifier. In the real scheme, capacitors are used to bypass interference on the negative and positive supplies VCC and VDD set to ± 5 V. The test signal

Table 5.36 Requirements for lead definition

Lead	Formula	Name of lead
I	LA – RA (left arm – right arm)	Bipolar limb leads
II	LL – RA (left leg – right arm)	
III	LL – LA	Einthoven
aV_R	RA – (LA + LL)/2	
aV_L	LA – (RA + LL)/2	Augmented leads (Goldberger)
aV_F	LL – (RA + LA)/2	
V_N, $N \in [1 \ldots 6]$ (V_N = chest lead)	V_N – (RA + LA + LL)/3	Unipolar chest leads (Wilson)

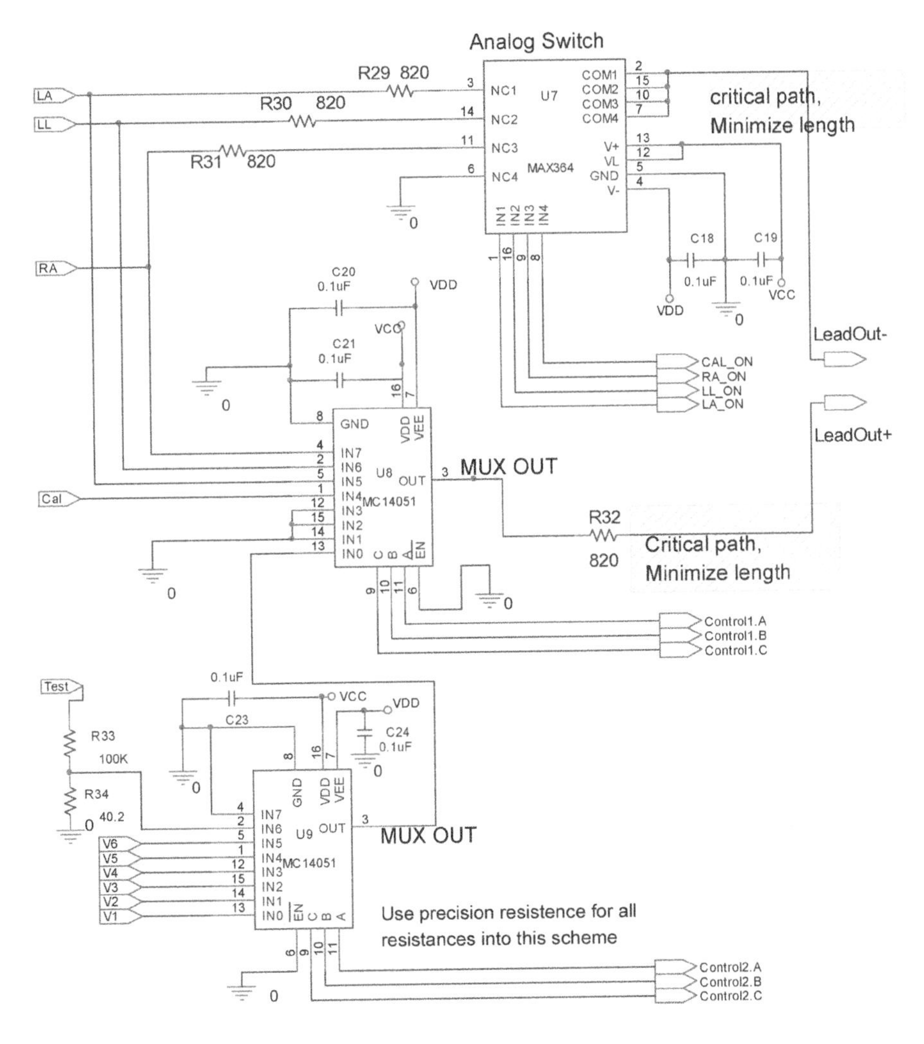

Figure 5.60 Gamma Cardio CG lead selector circuit.

(Test) is connected to a PIC pulse width modulation (PWM) output. This output can produce square waves with different frequencies and duty cycles. Some PCB traces are considered as critical since they are connected to the high impedance inputs of the op-amp. Track length minimization is required to reduce EMI effects.

The resistors in Figure 5.60 must be of precision type to reduce signals errors. The requirement of Table 5.35 states that lead weighting factors must be accurate to within ±5% error. The verification of this requirement is outlined in the following.

In leads I, II, III, RA, LA, LL, the signals captured by the patient electrodes are decoupled by the unity gain buffers and passed respectively through

1. a precision 820 Ω resistor and an analog switch (U7 Max 364)
2. a precision 820 Ω resistor and a mux (U 8 MC14051)

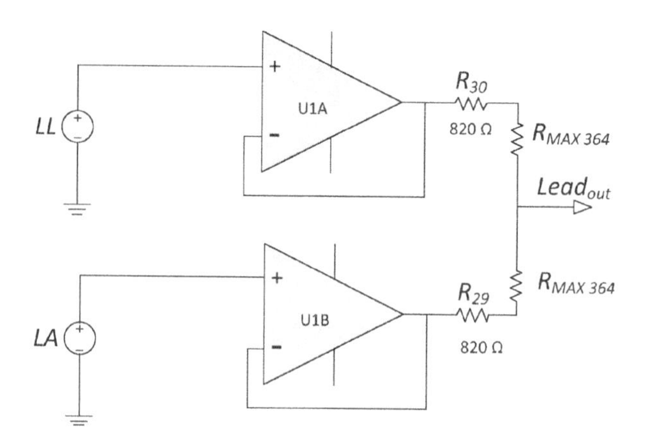

Figure 5.61 Goldberg leads conceptual scheme.

and finally they arrive at the input of the instrumental preamplifier (pins $\text{Lead}_{\text{Out}}-$ and $\text{Lead}_{\text{Out}}+$).

Since the preamplifier addressed in Section 5.25.1 has a very high input impedance, the the Lead_{Out} pins maybe considered as disconnected.

In Goldberg's lead, two terms have to be summed and halved as shown in Figure 5.61, where $R_{MAX\ 364}$ represents the resistance of the analog switch.

In the ideal case, the resistances of the first upper circuit and second lower circuit are equal and therefore since the op-amp output can be modeled with a resistor in series with a voltage source connected to ground from (5.79):

$$\text{Lead}_{\text{Out}} = \text{LL}/2 + \text{LA}/2 \tag{5.80}$$

From the TLC2274 op-amp datasheet, the output impedance at unity gain is less than 1 ohm and therefore possible variations can be neglected considering the small currents involved. Considering that the input signals at Max 364 are less than ± 0.7 V, the two $R_{MAX\ 364}$ resistances are at around 100 Ω and may differ by a maximum of 4 Ω. Since resistors are 1% tolerance, the maximum error occurs in the worst case where R29 = 820 + 1%, R30 = 820 − 1% and max difference between the two $R_{MAX\ 364} = 4\ \Omega$. In this case, we have

$$\text{LA at Lead}_{\text{OUT}} = \text{LA} \times \frac{(820 + 1\%) + (100 + 2)}{(820 + 1\%) + (100 + 2) + (820 - 1\%) + (100 - 2)} = 0.5055$$

A value of 0.5055 over 0.5 target represents a 1.11% error. The same applies for LL. The error is less than the target maximum value of 5%.

In the case of the unipolar chest leads V_N, referring to Figure 5.60 and considering that NC1, NC2, NC3 are closed, the scale factor for LA is given by

$$\text{LA Lead}_{\text{Out}} = (\text{R30}/\!/\text{R31})/(\text{R29} + \text{R30}/\!/\text{R31}) \tag{5.81}$$

where // in (R30//R31) indicates the value of resistors in parallel. The scaling factor is equal to 1/3, if all the resistors have equal value.

Table 5.37 AAMI requirement for standardizing voltage

AAMI requirement	Value
Nominal value	1.0 ± 0.01 mV
Rise time	<1 ms
Decay time	≥100 s.
ECG amplitude error	$\leq \pm 5\%$

Assuming the worst case where R29 and R31 and the associated switch resistance have the lower values $= 820 - 1\% + 100 - 2$ and R30 has the upper value $= 820 + 1\% + 100 + 2$ we have

$$\text{La at Lead}_{\text{OUT}} = \text{LA} \times 0.328 = -1.48\% \text{ error} \tag{5.82}$$

Similar calculations yield the error for LL and RA. The error is thus below the required upper threshold (5%).

5.24.1 Calibration

Instrumentation devices should have embedded test functions to assess the proper functioning of the device, especially when operational problems may have consequences on patient health. For an ECG, the AAMI standard includes a standardized voltage step signal for assessing the accuracy of gain and frequency response. This signal, also known as a calibration pulse, has features indicated in Table 5.37.

The requirement is satisfied when the standardizing voltage step is within $\pm 5\%$ of an external step signal with amplitude $= 1.00 \pm 0.01$ mV applied at the input. The circuit in Figure 5.62 generates the proper amplitude value. The signal step is generated through the switching of MUX (U8) from ground to calibration input and again to ground. The circuit uses a voltage reference device that has the characteristic of producing a sufficiently constant voltage, regardless of other circuit and environmental parameters (e.g., loading of device, temperature, power supply).

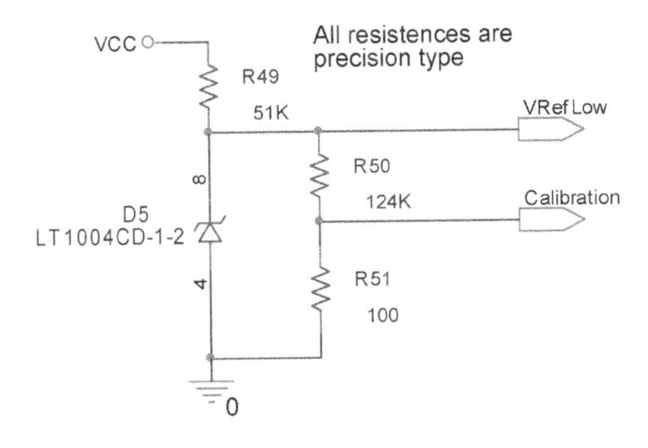

Figure 5.62 Calibration circuit.

Table 5.38 Calibration circuit performance

	Ideal case	Worst case 1	Worst case 2	Remarks
V_{diode} (V)	1.235	1.239	1.231	value of reference diode and initial accuracy
R50 (ohm)	124,000	122,760	126250	tolerance ± 1%
R51 (ohm)	100	101	99	tolerance ± 1%
$V_{calibration}$ (mV)	0.995	1.019	0.965	
Error (%)	−0.5	1.9	−3.5	

Voltage references is probably included in most of the medical instrumentation since there is the need to assess accuracy over time of the measurements (temperature, pressure, potentials, speed, length etc.) by using a constant standard value. Voltage references are also used in voltage regulators, AD converters and for measurement comparison. There are basically two types of voltage references: Zener based (diodes or IC) and bandgap reference. This latter type is used in the circuit of Figure 5.62 to generate a calibration voltage and to generate a voltage reference used for other circuits. Again ideal and real devices have different behavior: voltage values depends on temperature, supply voltage and the load on the device. The voltage value of commercial devices is specified within a specific tolerance, long-term stability is not guaranteed and noise is an extra component.

Real behavior has to be considered when designing the circuit. A simple analysis taking into account initial accuracy and resistance tolerances is included in Table 5.38. The analysis shows that calibration achieves a maximum of 3.5% error, that is compliant to the requirement (< 5%).

The table shows that also the error of the standardizing signal may be greater than 1%. This is to be considered if this standardizing signal has to test the other requirement of 'ECG amplitude error < ± 5%'. Since the −3.5% to +1.9% error of the calibration signal may hide the ECG amplitude error, when using the internal calibration signal, the ECG amplitude error must be included into the −3.1% ... + 1.5% range. If the error is not included in this range, there may be an error on the standardized signal or on the ECG system.

5.25 ECG Amplification

THE NEED The overall ECG amplification stage must amplify the ECG signal to a level suitable for the ADC input, reducing significantly the common mode interference. The stage must avoid signal distortion and must introduce noise at acceptable level.

As discussed in the theory section (Section 5.5), **interference** is a main problem for instrumentation, and some general countermeasures have been developed as summarized in Table 5.18. Referring to Figure 5.52, the ECG amplification includes a preamplifier based on an instrumentation op-amp, an amplifier and a specific circuit (the right leg driver RLD) that allows a further reduction of common mode interference.

The source differential signal contains the ECG signal (±5 mV, in the 0.05–150 Hz frequency range) added to a **DC offset voltage** in the range of +300 mV. The DC offset voltage

Table 5.39 Requirements for ECG amplification

Requirement	Performance	Value
R 3 AAMI EC 11 compliance	Input dynamic range (3)	±5 mV varying at a rate up to 320 mV/s added to a DC offset voltage in the range of ±300 mV
R 3 AAMI EC 11 compliance	Gain (4)	Gain error ≤5%
R 3 AAMI EC 11 compliance	Common mode rejection (11)	CMRR ≥95 dB also with electrode imbalance
R 9 IEC/EN 60601-2-25 compliance	Saturation signal	Presence of saturation indication. Compliance checked with 10 Hz, 1 mV sinusoidal signal superimposed on a DC supply voltage variable from −5 V to +5 V.
R 12 Class II CF type device	Maximum leakage current	See Table 5.31

does not contain information and therefore has to be eliminated by a high pass filter with 0.05 Hz cutoff frequency. A further amplifier stage has to be added after the high pass filter because the preamplifier must have a limited gain to avoid signal saturation. A specific circuit must monitor the saturation according to the requirement of Table 5.39. This indication shows that when the ECG signal is too high in amplitude, it may be distorted and therefore it may suggest wrong medical information.

Since the input signal at preamplifier swings in the range of ±305 mV, with a power supply of ±5 V, a maximum gain $G_{max} < 5/0.305 = 16.39$ can be achieved by the preamplifier. This is because op-amps clamp output signals at a value equal to or lower than the power supply. A further stage is required to obtain the range required by the AD converter.

REQUIREMENTS > The main requirements of this stage are summarized in Table 5.39.

The proofs related to input dynamic range and gain will be given in Sections 5.25.2 and 5.25.3 respectively. The other requirements have to be verified through specific demonstrations on the ECG equipment.

5.25.1 ECG Circuits

The lead selector block is connected to the preamplifier that is based on an instrumentation amplifier U10 as shown in Figure 5.63.

The internal scheme of this component is given in Figure 5.15. The op-amp gain is set by the two resistances (R36 and R37) connected at the pins GS1 and GS2 as defined by equation (5.31). These resistances are the equivalent of RG in Figure 5.15. The connection point of R36 and R37 contains the common mode voltage signal that is sent to the RLD stage. The signal is equivalent to the V_{CM} in the RLD circuit of Figure 5.19. The input ECG signal may have a DC component in the range ±300 mV that has to be eliminated.

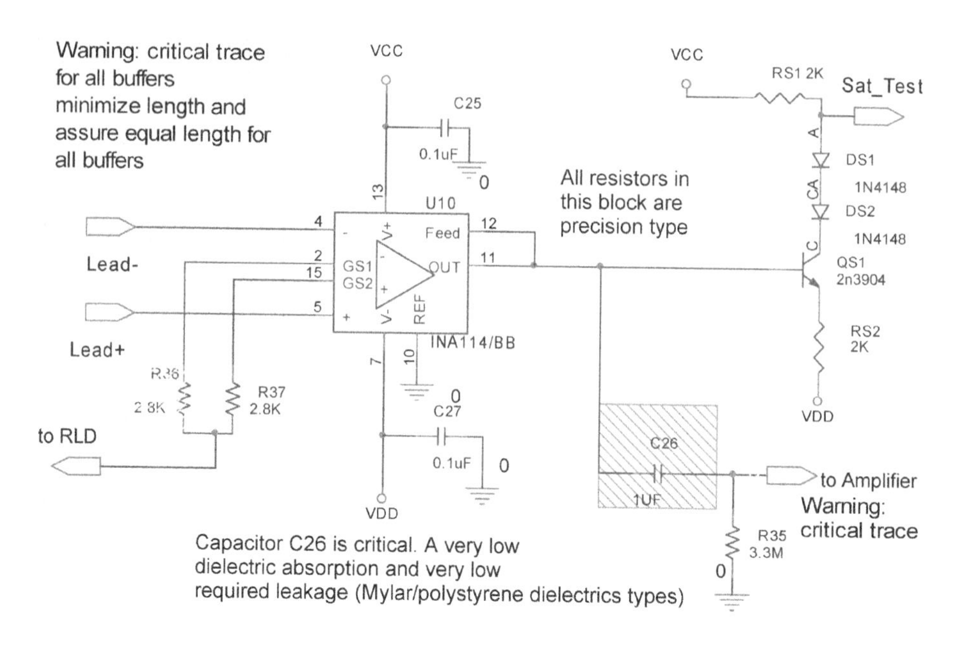

Figure 5.63 Preamplifier circuit.

The output of U10 is sent to a **high pass first-order** circuit whose cutoff frequency is 0.048 Hz. This value is easily computed by equation (5.16). The filter eliminates the DC while preserving the ECG information. Since the capacitor has a high value, it has to be discharged to quickly adapt to newer DC voltage offsets. This is done by a discharging circuit that is connected to the capacitor (see 'toHP' pin in PIC block of Figure 5.52).

The output of the high pass filter is sent to a non-inverting op-amp amplifier shown in Figure 5.65. Gain of the amplifier is easily computed through the rules (5.21) and (5.23).

The output of U10 is also sent to the base of the transistor 2N3904 that forms a circuit that monitors **saturation** of the input differential stage as indicated by the requirement R 9 (see Table 5.39). The circuit behavior is clarified in Figure 5.64 where the Spice simulation of the Sat_Test pin (upper signal) is drawn with respect to a 1 V_{PP} input differential sinusoidal signal. The circuit generates a positive signal that corresponds to specific values of the input signal. The threshold of 3.83 V of Sat_Test corresponds to an input of around 0.32 V for the positive half-wave and −0.34 V for the negative half-wave. These values are still in the non-saturation area since the preamplifier saturates for a 0.35 V input value. The Sat_Test pin is converted into a digital value by an ADC contained in the PIC. If this value is over 3.83 V, the system will display the saturation indication since the input signal is going to saturate.

The right leg drive (RLD) circuit of Figure 5.66 is similar to the conceptual circuit shown in Figure 5.19 of Section 5.5.5.

U2D is a voltage follower (buffer) op-amp that decouples the exit of the instrumentation amplifier where the common mode voltage V_{cm} is available. The output of U2D drives the cable shielding to V_{cm}. This technique, also called '**signal guarding**' (Horowitz, 1999), reduces the effects of external interference on the shielded cable by reducing parasitic input capacitance and leakage. The external interference is driven to ground thanks to the low impedance of the op-amp output. The capacitance between the inner conductor and the

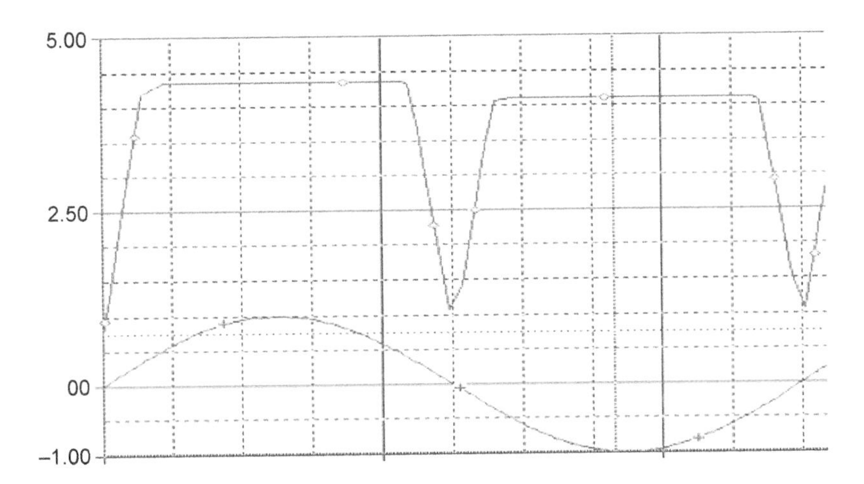

Figure 5.64 Saturation circuit behavior.

shield creates a parasitic effect that may reduce input mpedance. Since the shield is driven to V_{cm}, the inner cable and its shield achieve similar potential and therefore the effect of current leakage due to the capacitance is reduced. The U2A op-amp is an inverting amplifier that provides common mode negative feedback. The RLDout pin of Figure 5.66 is connected to the right leg of the patient. Note that the two transistors clamp the voltage to a maximum of 0.7 V even in case of failure of the op-amp. This limits the current that can flow within the patient.

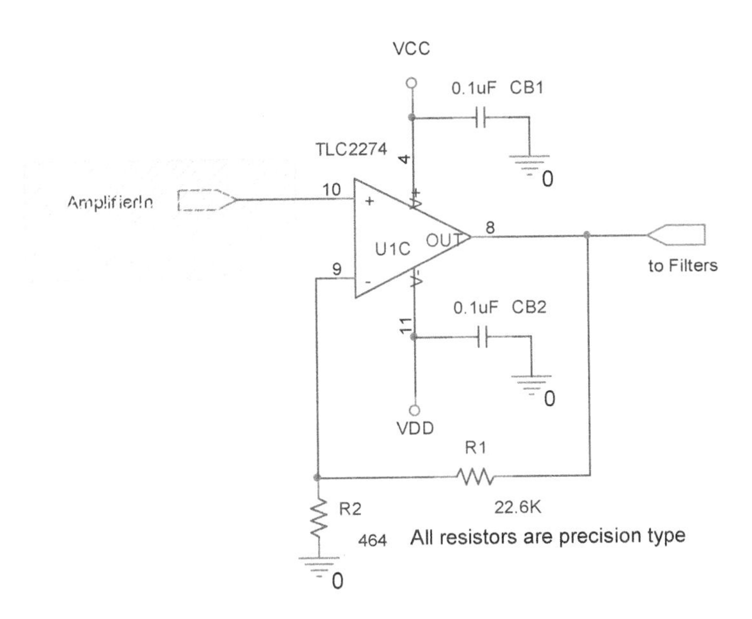

Figure 5.65 Amplifier circuit.

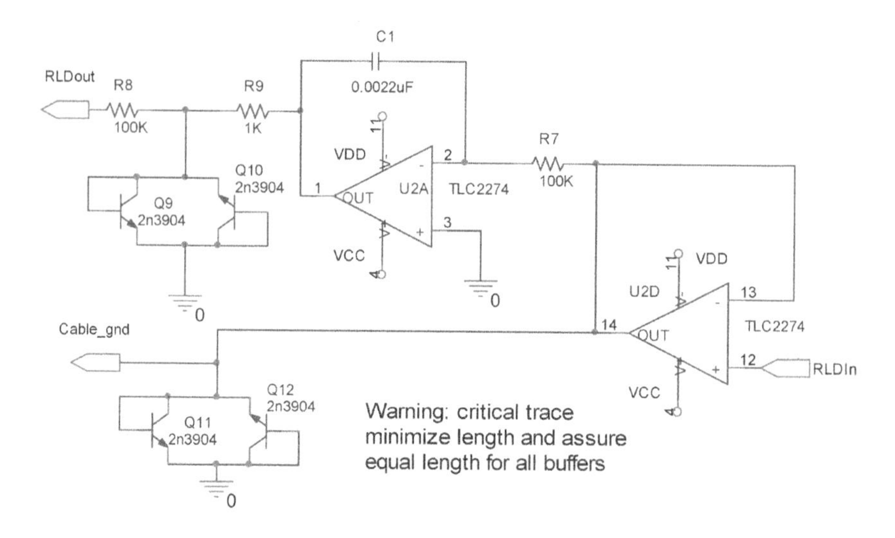

Figure 5.66 Right leg driver circuit.

The performance of the overall amplification stage in terms of CMRR can be easily simulated with Spice, which reveals that in the AAMI EC 11 test conditions, with a 20 V_{rms} input common mode signal interference, around 2 μV_{pp} signal is obtained at the ECG input. This yields an ideal CMRR > 140 dB. Real circuits have asymmetries that create imbalance. This may reduce CMRR to 110 dB.

The frequency performance with RLD circuit is depicted in Figure 5.67. The lower plot shows the common mode potential at different frequencies and the upper plot shows the

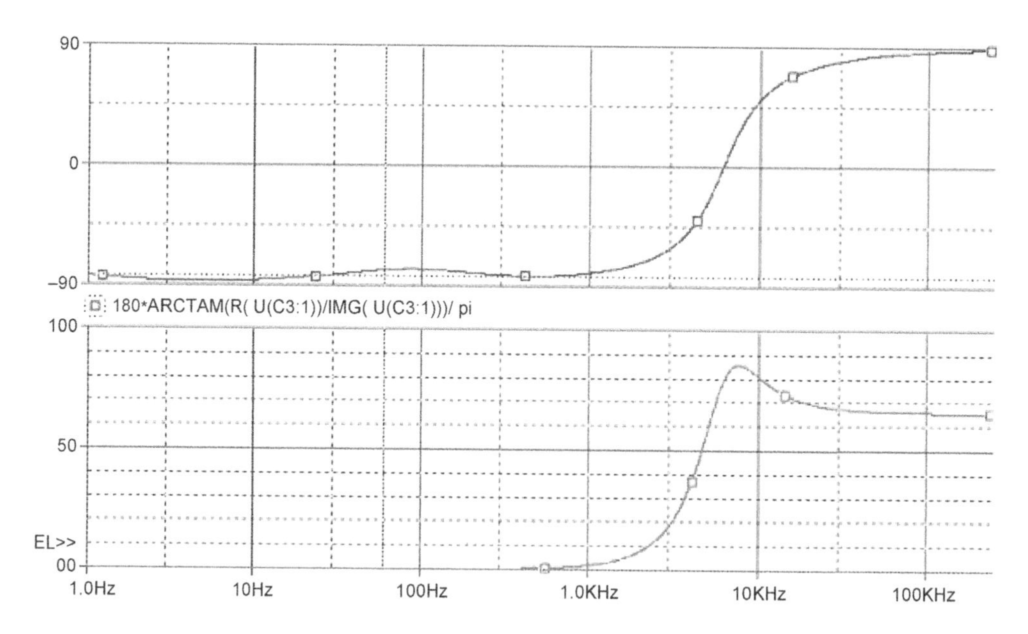

Figure 5.67 Right leg driver frequency performance.

phase in degrees. The ECG input is stable below 4 kHz, where common mode signal is deeply reduced. Above 10 kHz the RLD circuit has no effect since there is no attenuation of the common mode signal. At around 8 kHz the negative feedback of the RLD circuit becomes positive and there is a limited oscillation.

The previous circuits are designed to comply with the specific requirements of Table 5.39. The demonstration of input dynamic range and gain error requirements gives further insight on design choices.

5.25.2 *Input Dynamic Range: Requirement Demonstrations*

REQUIREMENTS The input dynamic range must be at a maximum differential voltage of ±5 mV with a slew rate up to 320 mV/s. The slew rate is the maximum limit at which the output can change. This is equivalent to saying that, at minimum, a sine wave of frequency $f = 10.2$ Hz and amplitude $A = 5$ mV must pass undistorted. The values are given by the formula valid for sine wave signals: the slew rate $\geq 2\pi Af$ for sinusoids to be unchanged. The ECG must tolerate a DC differential voltage in the range of ±300 mV without changing more than ±5% over ±300 mV DC.

DEMOSTRATION For this demonstration we refer to the main scheme of Figure 5.52 and the detailed schemes of the stages where the input signal is processed. The first stage (the buffer of Figure 5.58) does not amplify the input signal, but limits its excursion to $\pm 2 \times 0.7\ V_{pp}$ because of the transistors.

From the Ina 114 datasheet, we obtain the following information:

- $G = 1 + 50\,k\Omega/R_G$ with $R_G = R_{36} + R_{37}$
- the preamplifier amplifies the signal with G = 9.928 times. The maximum allowed op-amp output voltage is $V_{out} \leq 3.5$ V with ±5 V power supply.

The maximum ECG input signal to avoid saturation of the preamplifier is thus:

$$V_s = 3.5/9.928 = \pm 0.352\ V > \pm 300\ mV\ \pm\ 5\ mV \quad (5.83)$$

The preamplifier stage contains a 0.05 Hz high pass filter that removes the DC component (±300 mV). The ECG signal (Vs = ±5 mV) is amplified by the amplifier stage of $G_a = 49.707 = (1 + 22.6\,k\Omega/464\,\Omega)$, for a total amplification of

$$G_T = G_a \times G = 493.51 \quad (5.84)$$

Given an input ECG signal $V_s = \pm 5$ mV, the amplifier output signal has a value $V_a = \pm 2.468$. The amplifier stage does not clamp the input signal since the maximum swing of the output signal with ±5 V power supply is ±4.99 V (see datasheet TLC 2274, parameter $V_{OM+} V_{OM-}$ with current $I_{OH}\ I_{OL} < 1$ mA).

The low pass filter stage, discussed in Section 5.6, does not change the signal dynamic range in the frequency range of interest, that is, the ECG signal can only reach ±2.468 V at maximum. This stage shifts the level of the signal up by +2.47 V. The output signal of the filter therefore has a dynamic range of [±2.468 V + 2.47 V] = [+4.938, +0.002 V].

The output current I_{OH} I_{OL} of the final level shifter stage U_{11D} is $I_O < 50$ mA ($I_O =$ 25.6 μA = 4.938/R with R = 96.3 k + 96.3 k). From the TLC 2274 datasheet, the maximum output voltage reaches 4.99 V, and therefore the output signal is not attenuated even at the maximum positive value. The output signal of the filter stage is sent to the AD converter that has a dynamic range of 0–5 V.

The software system records and stores the samples without degradation and it is capable of displaying 95 mm amplitude, much greater than the 10 mm = 10 mV required. In the end, no ECG circuit stage clamps or distorts an input signal $V_S = \pm 5$ mV, even if added to a differential component of ±300 mV.

5.25.3 Gain Error: Requirement Demonstrations

REQUIREMENTS The maximum gain error must not exceed 5%.

DEMONSTRATION The gain error is mainly due to resistors that determine the amplification of the amplifier and the preamplifier stages.

The other stages where the signal is managed are the voltage follower or FET switches that do not change the gain and, according to the datasheet, have negligible change in gain. The total gain is therefore

$$G_{tot} = (1 + 50\,\text{k}\Omega/(R_{36} + R_{37})) \times (1 + R_1/R_2) \tag{5.85}$$

The resistances provide a precision temperature coefficient of resistance (TCR) at 100 ppm/C (variation of 100 part per million every celsius degree of charge), being the resistance value R_T at temperature T:

$$R_T = R_0 \times (1 + \text{TCR} \times (T - T_0)) \tag{5.86}$$

where R_0 is the value at temperature $T_0 = 27°$C. The change in value due to temperature variations is therefore negligible. According to the datasheet of the instrumentation amplifier, its gain error is also negligible. The resistors are 1% tolerance and then, in the worst case, there are errors of the order:

$$G_{tot,\,max} = (1 + 50\,\text{k}\Omega/(R_{36} + R_{37} - 2\%)) \times (1 + (R_1 + 1\%)/(R_2 - 1\%))$$
$$G_{tot,\,min} = (1 + 50\,\text{k}\Omega/(R_{36} + R_{37} + 2\%)) \times (1 + (R_1 - 1\%)/(R_2 + 1\%))$$

Substituting the values we have

$$G_{tot,max} = 507.86 \text{ (equivalent to 2.91\% error)}$$

$$G_{tot,min} = 479.63 \text{ (equivalent to } -2.81\% \text{ error)}$$

This shows that the gain error is less than 3%.

Note that the resistances have in general a mean value equal to their nominal value and an actual value that is distributed according to a Gaussian distribution with standard deviation =

tolerance/3. This comes from the reasonable approximation that the tolerance is three times the standard deviation. Since resistor values are statistically independent, the probability of having four resistors with the worst case values is given by the product of the single probabilities. This worst case has therefore a negligible probability.

5.26 Analog Filtering

THE NEED — Analog filtering is needed in the ECG before the AD conversion to avoid aliasing out of band interference and to fulfill the specific requirements of AAMI EC 11 on the ECG signal.

As discussed in Sections 5.6 and 5.7, the design of this filter depends on various factors such as: required resolution, out-of-band interference, amplitude and the selected sampling rate.

The analog filter also has to shape the frequency response so that the bandwidth is within 0.05–150 Hz and the out-of-band response degrades according to the AAMI EC 11 standard requirements.

The analog filter stage must also shift the level of the signal so that it is suitable for the input dynamic range accepted by the ADC (0–5 V)

REQUIREMENTS — The main requirements of this stage are summarized in Table 5.40.

High pass filter can have a time constant >3.3 s which is equivalent to a first-order high pass filter with corner frequency ≤0.05 Hz.

Since the ECG signal shape must not be changed, to avoid ECG misinterpretation, as explained in Section 5.6.1, the linear phase requirement is added for both of the filters (low and high pass).

The low pass filter must reduce the aliasing as discussed in Section 5.7. The antialiasing requirements are derived by the following features:

- Resolution = 12 bit (see next section for details)
- out-of-band interference amplitude = full scale
- selected sampling rate = 1200 Hz standard, or 1000 Hz optional

From the previous data, at 1200 Hz sampling rate, the components at 1200 − 150 = 1050 Hz will be aliased at 150 Hz. If we assume that at 1050 Hz there may be a full scale interference, then 74 dB attenuation has to be achieved at this frequency to obtain a noise lower than the quantization error as per equation (5.55). The choice of a Bessel filter among

Table 5.40 Requirements for ECG filtering and resolution

Requirement	Performance	Value
R 3 AAMI EC 11 compliance	Frequency and impulse response (3)	Time constant >3.3 s for high pass – see Table 5.41 for low pass. Linear phase response
R 15 Horizontal resolution	Horizontal resolution	1200 samples per second

Table 5.41 Requirements for ECG low pass filtering

Frequency range (Hz)	Min % amplitude increment versus amplitude at 10 Hz	Max % amplitude increment versus amplitude at 10 Hz
1–40	−10%	+10%
40–150	−30%	+10%
150–∞	−100%	+10%

Table 5.42 Bessel filter attenuation

Order	Frequency = 850	Frequency = 1050
6	62.6	73
7	69.5	87.7
8	76	90

the filter models of Figure 5.22 is compulsory since the filter must be linear phase. From Table 5.42, it is clear that a sixth order filter is sufficient at 1200 Hz sampling rate. Since the sampling rate may also be set at 1000 Hz, then a eighth order filter is needed to achieve attenuation at the aliased component, that is 1000 − 150 = 850 Hz.

Note that from sixth to eighth order there are four resistors, four capacitors and two op-amps more. Since the TLC 2274 op-amps are in a four op-amp package, and the two saved op-amps would not need to be used in the Gamma Cardio CG, then the conservative choice to implement an eighth order Bessel filter does not overcomplicate the circuit. Values, performances and topology of the filter circuit may be easily computed using the free tools available from chip manufacturers. For example, Figure 5.68 shows the frequency response starting from 0 dB and the group delay that is constant for the band of interest: 0–150 Hz.

CIRCUITS As discussed in Section 5.6, two main topologies are available: multiple feedback topology (MFT) and the Sallen–Key configuration. Sallen–Key is simpler, with no gain sensitivity with respect to errors in component values, but with worse performance at

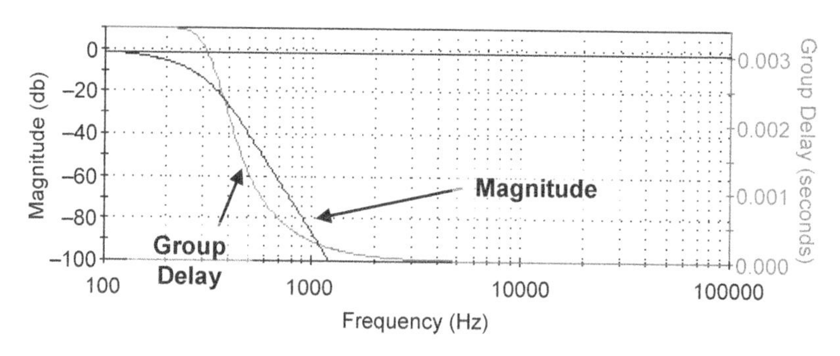

Figure 5.68 Theoretical low pass filter response.

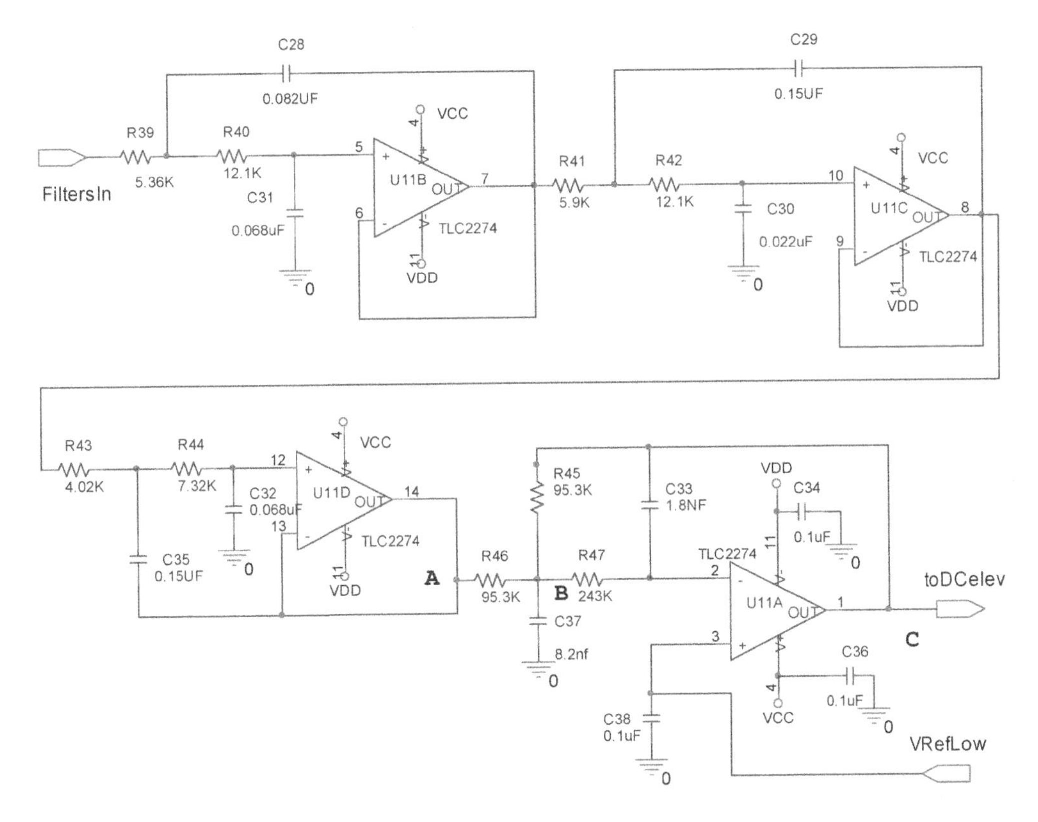

Figure 5.69 Low pass Gamma Cardio CG filter.

the highest frequencies. This latter issue is not a problem considering the conservative choice of an eighth order filter. The MFT has to be selected for the final stage since it can perform level shifting. Figure 5.69 shows the implemented scheme. Apart from the bypass capacitors (C34, C36, C38), all resistors and capacitors must be precision type (1% or better) to avoid performance degradation.

The final stage (U11A) shifts the level by 2.47 V since V_{RefLow} is at a stabilized voltage of 1.235 V as per Figure 5.62. This also means that pin 2 of the final op-amp U11A is at 1.235 V and since no DC current flows through R47, the voltage divider made up of R46 and R47 has 0 V DC at the output of U11D (point A), 1.235 V at point B and therefore 2.47 V DC at the U11A output (point C). This proves that the final module shifts the DC level from 0 V to 2.47 V.

The circuit design and validation are performed through the usual steps. First, the circuit performance is assessed theoretically. This can be easily done, also using the various available free tools. Then the filter is assessed through a circuit simulator such as Spice. Finally, the circuit is tested on a demonstrator, on a prototype and on the final target. Figure 5.70 shows the performance as obtained from a theoretical tool, from Spice and from the real equipment. Theoretical and simulated performance are quite indistinguishable, while the real performance shows some differences, especially at low frequencies. This performance is in any case in line with requirements in Table 5.41.

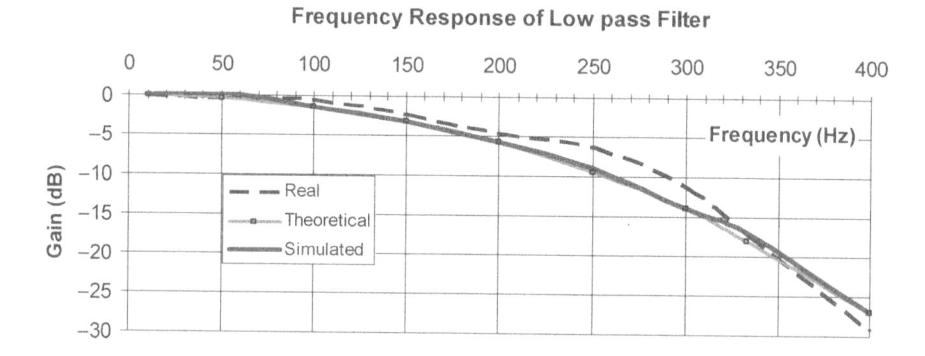

Figure 5.70 Low pass filter performance.

5.27 The ADC Circuit

THE NEED The ADC converter samples and quantizes the analog signal processed by the low pass filter. The digital signal is then sent to the microprocessor which encodes the digital signal into the protocol exchanged with the PC. The ADC basically has to preserve the features of the analog signal into the digital version: proper vertical and horizontal resolutions have to be guaranteed, by designing the ADC circuit with suitable quantization error and sampling rate. The ADC circuit depends on the chosen ECG architecture. For multichannel ECGs, a eight-to-one multiplexer connected to the ADC can be chosen. We recall that the 12 leads can be derived from the eight sampled channels, as explained in Chapter 2. Multiplexers and ADCs, also available in single chip, allow the sampling of eight channels sequentially. This approach requires that the time skew between the AD conversions of different channels has somehow to be addressed, either with a sample-and-hold circuit or by digital processing the correction so that the other four additional signals can be derived properly.

Alternatively, eight ADCs may be used to synchronously digitalize the eight different ECG channels. Another design decision influencing ADC selection is the use of the high pass filter. Referring to Figures 5.50 and 5.51, if high pass is not used, an ADC with increased resolution is required because the input dynamic range changes from $\pm$ 5 mV to $\pm$305 mV, given that the $\pm$300 mV DC component remains at the ADC input without the high pass filter. In this latter case the DC component must be then digitized and eliminated by digital filtering. Since the input dynamics increase by a factor of 61 (610 mV/10 mV) then ADCs with an additional six bits nominal ($2^6 > 61$) have to be selected to obtain the same LSB as the case of $\pm$5 mV input dynamics.

Low pass filter design is also related to ADC selection. If oversampling, or better, sigma-delta ADCs are used, then a simplified low pass filter can be used. With appropriate values of sampling rate, simple first-order filters are sufficient as in Figure 5.51.

REQUIREMENTS Table 5.43 summarizes requirements related to the ADC.

ADC selection implies requirements on vertical and horizontal resolution. We consider the single channel Gamma Cardio CG case where the input signal at the ADC does not contain the DC component that is eliminated by a high pass filter.

Table 5.43 Requirements for ECG amplification

Requirement	Performance	Value
R 3 AAMI EC 11 compliance	Frequency and impulse response (R 3.3)	High pass: see Table 5.41 for frequency response Low pass: time constant >3.3 s Linear phase response
R 3 AAMI EC 11 compliance	System noise (R 3.7)	$<\pm 40\,\mu$V
R 3 AAMI EC 11 compliance	Response to minimum signal (R 3.7)	20 μV_{pp} sinusoidal signal at 10 Hz, visible with time base selection = 25 mm/s and gain setting = 10 mm/mV.
Design requirement	Vertical resolution	$<10\,\mu$V
R 15 Horizontal resolution	Horizontal resolution	1200 samples per second

The **vertical resolution** requirement can be assessed considering that the human eye can hardly perceive differences less than 0.5 mm on the pre-ruled ECG paper. With the standard maximum vertical resolution (20 mm/mV) this translates to a 25 μV (=0.5/20) resolution for a $\pm$5 mV input signal. The required minimum resolution is nine bits > $\log_2$ (10 mV/25 μV). This resolution corresponds to a root mean square quantization noise $\sigma_{e,rms} = 7.21\,\mu$V (see equation (5.51)).

This constraint is in line with the $\pm 40\,\mu V_{pp}$ requirement for maximum system noise specification, equivalent to a root mean square $\sigma_{a,rms} \approx 13.3\,\mu$V (=40/3). $\sigma_{a,rms}$ is obtained considering that, for white noise, the ratio between the peak value and the rms of a signal may be approximated to 3. Selecting nine-bit resolution corresponding to 25 μV LSB, we have $\sigma_{e,rms} < \sigma_{a,rms}$ and therefore the quantization error allows us to achieve the system noise target.

The AAMI EC 11 standard also requires that a sinusoidal signal with 20 μV_{pp} must be visible. This suggests a conservative resolution of 10 μV LSB which can be achieved by $\log_2$ (10 mV/10 μV) equivalent to ten bits. A ten-bit effective resolution can be achieved by ADCs with nominal resolution of 11–12 bits as discussed in Section 5.7.

To satisfy AAMI EC 11 requirements, a sampling rate greater than 500 Hz is required. The proposed sampling rate is much higher, to allow better signal representation.

COMPONENTS The selected ADC in the Gamma Cardio CG is contained in the microprocessor PIC 16c773. The performances are

1. maximum sampling rate $\gg$ 1200 Hz
2. nominal resolution = 12 bit
3. number of channels >1
4. input dynamic range = 0–5 V.

PERFORMANCE After digitization, the ECG signal does not suffer further degradation, apart from numerical errors that are negligible. The overall hardware system performance

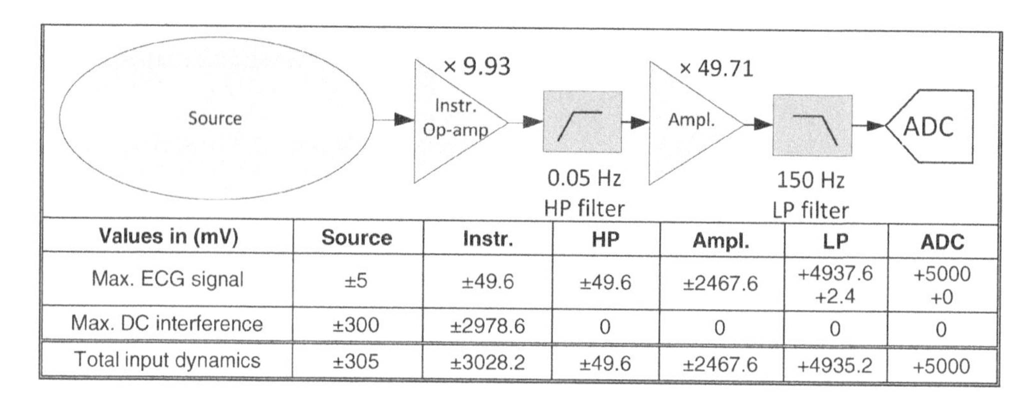

Values in (mV)	Source	Instr.	HP	Ampl.	LP	ADC
Max. ECG signal	±5	±49.6	±49.6	±2467.6	+4937.6 +2.4	+5000 +0
Max. DC interference	±300	±2978.6	0	0	0	0
Total input dynamics	±305	±3028.2	±49.6	±2467.6	+4935.2	+5000

Figure 5.71 ECG gain analysis.

can then be computed at the ADC input considering the cascade of analog stages and the ADC performance.

First, the **gain** of the overall analog system has to be verified, checking that each stage does not saturate or experience damage when the maximum dynamic range is applied at the input. Power supply rails, excessive common mode differential input dynamic range and output voltage swing have to be considered. The final performance is also dependent on the maximum current generated by the output stage of the op-amps.

The resolution and the associated quantization error can then be computed given the gain and the input dynamic range at the ADC. Finally, the system noise has to be evaluated and compared to the quantization error to check if an unnecessarily high resolution has been selected.

The gain of the overall analog system can be analyzed in Figure 5.71, where the voltage of the ECG and the DC interference is computed across the various stages. All the stages must avoid saturation over the maximum dynamics that may occur at the input or at the output. The input signal dynamic range must be suitable for the ADC stage.

A more detailed analysis of the gain has been performed in Section 5.25.2. The Figure 5.71 also shows that the ADC resolution will be worse at the input since the input dynamic range of the ADC (0–5V) has not been fully exploited by the ECG signal, which ranges from +2.4 mV to +4937.6 mV. Assuming 11 bits effective resolution, then, at the ADC, the LSB is around $2.44\,\text{mV} = 5\,\text{V}/2^{11}$. Considering that the overall gain is $G_{tot} = 9.93 \times 49.71 = 493.51$, the LSB at input is $2.44\,\text{mV}/G_{tot} = 4.96\,\mu\text{V}$ instead of $4.9\,\mu\text{V}$ obtained if the full ADC dynamics were exploited. This value is still less than the maximum target of $10\,\mu\text{V}$.

The final analysis is related to the overall **system noise**. Recalling that T is the temperature in kelvin and K *is* the Boltzmann constant and Δf is the frequency bandwidth, the main rules are (see also Horowitz, 1999)

1. uncorrelated noise signals must be added in the squared amplitude. (rms values do not add) as per equation (5.8))
2. resistors have noise rms voltage $= e_{rms} = \sqrt{4 \cdot K \cdot R \cdot T_K \cdot \Delta f}$ V 25.2
3. op-amps have voltage and current noise generators; the overall noise has to be computed according to the circuit configuration (see Texas 2007)

4. the noise value has to be computed as if it were generated at the input (therefore it must be divided by the gain)
5. simplifications can occur, considering that uncorrelated noise adds in a squared way; noise components can be neglected when their amplitude differs by one order with respect to the highest noise term, since in a squared sum they differ by two orders
6. computations have to be performed in the original units and not in their multiples (i.e., in volts and not in nV) to avoid errors.

The overall scheme for the noise analysis is shown in Figure 5.72.

In general, noise-related worst components are those generated in circuits where the signal has not yet been amplified, that is, the buffers, the lead selector and the input of the preamplifier. The noise generated in the other stages has to be divided by the overall input gain performed by the previous stages so that it can be evaluated as if it were generated at the input. Noise analysis is usually performed by representing the noise value at a given stage as if it were generated at the input stage. In this approach the noise is expressed as RTI (RTI = return to input), and it is calculated as: noise value (RTI) = noise value/(gain achieved by the signal from the input stage to the point where the noise is measured).

A simplified analysis of the more relevant components of noise is outlined in Table 5.44. For each component, the noise spectral density has been computed in the second column of the table. This value has to be divided by the overall gain to evaluate the noise at the input (third column with RTI value). The fourth and the fifth columns are the rms and the mean square noise, computed over the whole bandwidth (150 Hz). The final column is the peak-to-peak noise amplitude assuming a crest factor of 6. Note that, according to (5.8), only the mean square noise column has been added and the total mean square noise has been used to compute the total peak-to-peak noise voltage and rms noise value.

As expected, the main relevant noise is due to the two 100 kΩ resistors at the input and to the buffer op-amps. Resistors generate noise with a mean square amplitude:

$$e^2 = 4 \cdot K \cdot R \cdot T \cdot B = 1.65 \cdot 10^{-4} \cdot R \cdot 150 \qquad (5.87)$$

where R is the resistance value, T the temperature in kelvin and K the Boltzmann constant. Op-amps in the voltage follower configuration generate

$$e^2 = \left(e_{rms,opamp}\right)^2 \cdot \Delta f + \left(i_{rms,opamp} \times R\right)^2 \qquad (5.88)$$

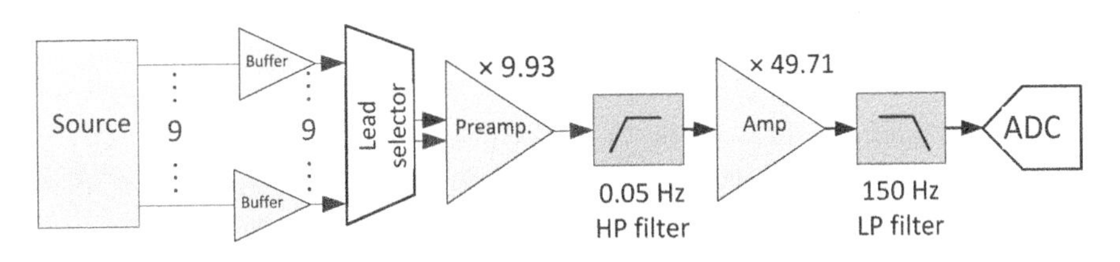

Figure 5.72 ECG Noise analysis.

Table 5.44 ECG noise analysis

	Spectral density noise	Spectral density noise RTI	Root mean square noise		Mean square noise		Peak-to-peak voltage value
	$e/\sqrt{B}$	$e/\sqrt{B_{RTI}} = e/\sqrt{B}/G$	$e_{rms} = e/\sqrt{B_{RTI}} \times \sqrt{B}$		$e^2 = (e_{rms})^2$		$V_{pp} = 6 \times e_{rms}$
Buffer resistances		5.7E−08	7.0E−07		5.0E−13		4.2E−06
Buffer op-amps		3.1E−08	3.8E−07		1.5E−13		2.3E−06
Lead selector		5.2E−09	6.4E−08		4.1E−15		3.8E−07
Preamplifier		1.4E−08	1.7E−07		2.8E−14		1.0E−06
HP filter	2.3E−07	2.4E−08	2.9E−07		8.3E−14		1.7E−06
Amplifier	2.2E−08	4.6E−10	5.7E−09		3.2E−17		3.4E−08
LP filter	2.2E−08	4.7E−11	5.8E−10		3.3E−19		3.5E−09
Total without ADC			**8.7E−07**	←	**7.6E−13**	→	**5.2E−06**
ADC			1.5E−06		2.2E−12		9.0E−06
Total noise			**1.7E−06**	←	**3.0E−12**	→	**1.0E−05**

where $e_{rms,\ op\text{-}amp}$ and $i_{rms,\ op\text{-}amp}$ are the op-amp noise contribution in terms of voltage and current, and Δf is the frequency bandwidth (see Horowitz, 1999). The lead selector contribution is negligible since resistors are low values, and active components do not introduce significant noise. Preamplifier noise is also quite negligible. The noise from the following stages is negligible since their contribution is divided by the gain. From Table 5.44, the overall noise including quantization noise is equal to $10\,\mu V_{pp}$ (bottom right value).

This value is in line with experimental results, and is compatible with the requirements which impose a system noise lower than $40\,\mu V_{pp}$.

Note that the ADC resolution cannot be improved further since the total rms noise of the signal (=0.87 μV) is comparable with the quantization rms error (=1.5 μV). Adding one more bit will yield 0.75 μV quantization rms error which is lower than the signal noise. This shows that the resolution is at the correct value.

5.28 Programable Devices

THE NEED In ECGs, as in general in any embedded systems, designers tend to allocate product functions on soft coded programs to reduce design/product costs and time to market. This is possible for functions that do not require intensive computational power.

For example, in ECGs, the control logic of the system may be implemented more effectively through firmware running on programmable device (MCU, FPGA etc.) and, when possible, on software for PC platforms.

Table 5.45 Requirements for ECG amplification

Requirement	Performance	Value
R 3 AAMI EC 11 compliance	Lead definition (R 3.3)	See Table 5.36 and Chapter 3 for specific discussion
R 3 AAMI EC 11 compliance	Standardizing voltage: (R 3.8)	See Table 5.37
Design requirement	Vertical resolution	$<10\,\mu V$
R 15 Horizontal resolution	Horizontal resolution	1200 samples per second
R 9 IEC/EN 60601-2-25 compliance	Indication of inoperable electrocardiograph (Section 51.103* IEC/EN 60601-2-25)	Fully operative when signal is within intended input dynamic, and indication when excessive signal dynamic makes the ECG inoperable
R 9 IEC/EN 60601-2-25 compliance	Unblocking after defibrillation (Section 51.101* IEC/EN 60601-2-25)	ECG trace readable in less than 5 s, after defibrillator discharge

In addition, the programmable devices and the associated firmware may conveniently perform other functions such as

1. AD conversion
2. data transmission
3. generation of analog control signals
4. digital control of some specific components.

REQUIREMENTS Table 5.45 summarizes the requirements related to the controller and the programmable device.

The functions related to these requirements are mainly implemented in firmware. The hardware has to guarantee connections and minimal additional circuits to perform the electrical interfaces.

5.28.1 Circuit Design

Figure 5.73 outlines how microcontroller unit (MCU) peripherals are connected to the ECG board to satisfy the requirements of Table 5.45.

The **pulse wave modulation** (PWM) module generates the control signals used to test the analog chain. An output line is used to generate the calibration signal. The ADCs are used to digitize the ECG signal and the output of the instrumentation amplifier. This signal, represented in the upper part of Figure 5.63 (Sat_Test), is used to check if the amplifier saturates, thus making the ECG inoperable as per the associated requirement in Table 5.45. Nine output lines of the **I/O** module are connected to the lead selector module to obtain the appropriate signal combinations as explained in Section 5.23.

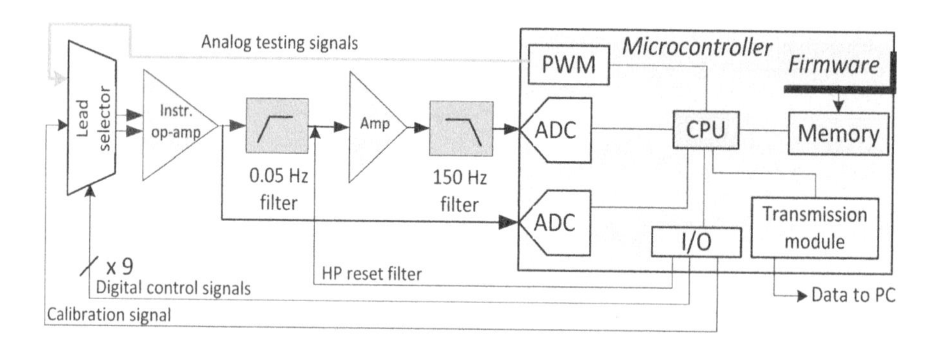

Figure 5.73 MCU functions in the ECG.

In defibrillation, a high DC voltage charges the capacitor of the HP filter. The **HP filter** would take 7.58 seconds to discharge to 10% of its initial value. This value is computed considering the formula

$$V(t) = V(t = 0) \times e^{\frac{t}{RC}} \tag{5.89}$$

where $V(t)$ is the value of voltage at time t for a capacitor C discharging through resistor R. This suggests using another output line to discharge quickly the 1 μF capacitor to satisfy the associated requirement. The transmission module of the MCU allows the MCU to exchange data with the PC.

The MCU circuit shown in Figure 5.74 demonstrates the flexibility and the convenience of the microcontrollers: few additional hardware components are needed to implement the circuit that is outlined in Figure 5.73.

More in detail, the bypass capacitor (C15) eliminates AC ripple from the DC. The diodes D3 and D4 complete the **in-circuit programming** port that allows the downloading of firmware and the performing in-circuit debugging.

The MOSFET M1 on the right side of Figure 5.74 is used as a switch to discharge the energy accumulated by the 1 μF capacitor C26 that is part of the high pass filter included in Figure 5.63. The MOSFET, controlled by one of the microcontroller output ports, has been used as a switch since it has low resistance in active mode, and low current leakage when it is in the off state. The communication between the MCU and the PC is performed using the internal universal asynchronous receiver transmitter (UART) that allows data exchange according to the RS-232 communication standard. R27, C16 and C17 are the vendor's recommended circuit for ensuring proper **clock** generation with a crystal.

5.28.2 The Clock

MCUs need a digital clock signal to work. This signal can be obtained by connecting an RC network to the oscillator circuit inside the MCU. This is a cost-saving option suitable for time insensitive applications, since the RC oscillator frequency will depend on many factors (R and C values, power, temperature, parasitic capacitances and tolerance).

In contrast, Figure 5.73 has functions that are heavily time dependent such as ADC, PWD and the calibration. Analog to digital conversion requires precise timing to trigger sampling periodically at time 1/(sampling rate). A non-periodic clock might trigger the sampling at a

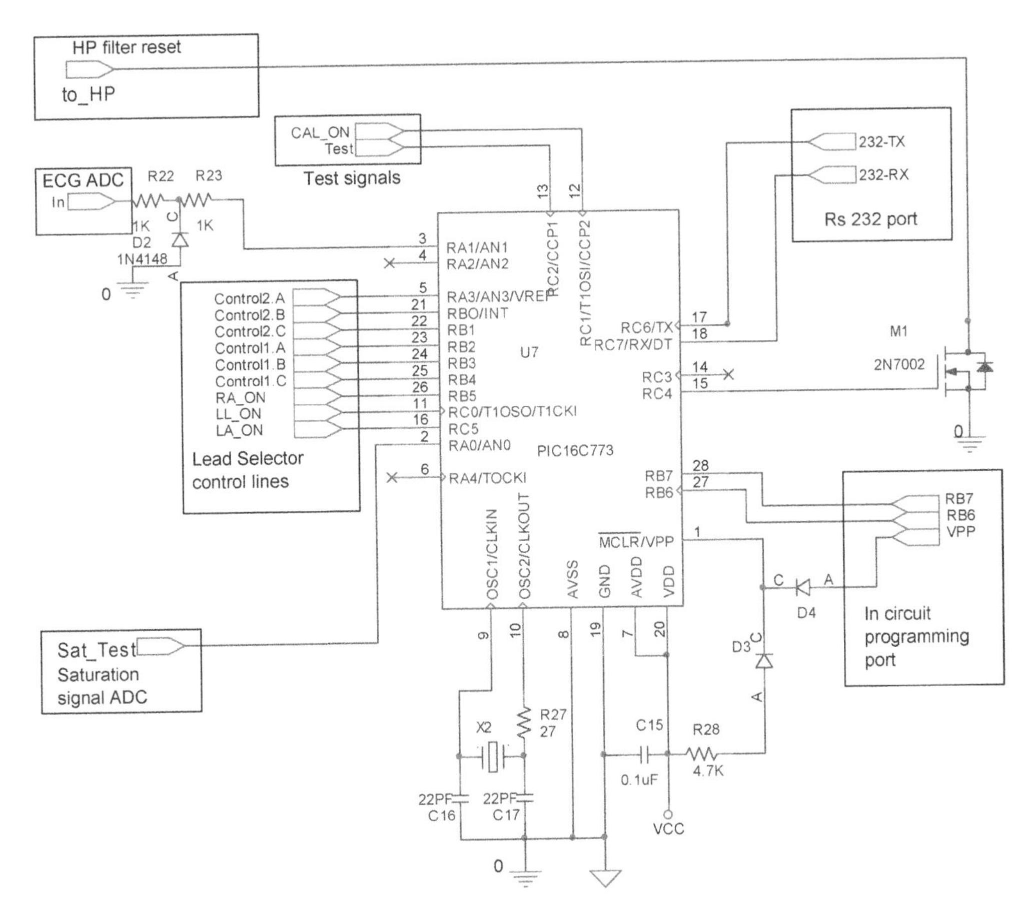

Figure 5.74 Microcontroller ECG circuit.

rate different from expected, thus distorting the sampled signal. For example, the PWD module generates a square wave signal to test the analog chain. The square wave can be changed in wave period and duty cycle: duty cycle is the ratio between the time that the square signal is on and the time that the signal is in off state. The period of the square test wave has to be exact, since frequency analysis of the analog part is based on the frequency of the test signal. The calibration signal also has to have specific time constraints detailed in Table 5.37.

To guarantee exact timing, good clock stability and precision are required. This is obtained by using more precise clock generators. Better clock generators are mainly based on crystals, ceramic resonators and IC oscillators. Crystals are quartz piezoelectric devices that are cut so as to vibrate mechanically at a specific frequency. In a piezoelectric material, a strain generates a voltage and vice versa: a voltage applied to the crystal generates acoustic waves that in turn generate a signal at a specific frequency. Crystals basically emulate an RLC circuit as outlined in Table 5.46. The clock may also be generated by ceramic resonators that rely on a similar mechanism with reduced performance and cost. Crystals and ceramic resonators have good accuracy and low cost but they are sensitive to many factors: vibration, temperature,

Table 5.46 Crystal ideal versus real behavior

Ideal behavior	**Crystal quartz** X2
Resonator at frequency f	$f = 1/2\pi\sqrt{LC}$ INDUCTOR CAPACITOR L, C ideal components
Q factor	Infinite Q factor ($=f/bw$, where f is the resonant frequency and bw the bandwidth, that is, the width of frequencies where the filter gain is halved)
Resonant frequency f	f does not depend on temperature, aging, shock, EMI, vibration conditions

Real behavior	**Crystal quartz** X2
Resonator at frequency f	$f = \frac{1}{2\pi\sqrt{L_1C_1}} \times \sqrt{1+\frac{C_1}{C_2}}$ C1 L1 R ESR C2 C_2 is the capacitance of the crystal electrodes and R_{ESR} is the equivalent series resistance.
Frequency Tolerance	$f_{real} = f_{Nominal} \times (1 \pm \text{tolerance}\%)$
Frequency Stability	Frequency depends on temperature, aging, power supply shock vibration
Temperature range	Operating conditions are guaranteed only within the specified temperature range
Mode of oscillation	Quartz has to be used in specific configurations (parallel/serial mode)

humidity and interference. Such disadvantages may be overcome by specific integrated circuits that, at increased cost, include a resonator component and circuits that are guaranteed to oscillate at a stable frequency. On the other hand, quartz may be a problematic component since, in some circumstances, it may not oscillate. When quartz does not oscillate, the MCU will not start.

AT this aim, MCU vendors indicate the specific crystal quartz models and vendors that are guaranteed to oscillate. Manufacturers' datasheets have to be followed carefully to avoid such problems. For example, low effective series resistance (R_{ESR}) indicated in Table 5.46 reduces problems at start up. Stray capacitance at the PCB negatively influences performance and therefore quartz must always be close to the MCU. Quartz may not oscillate at specific temperatures or voltage supply levels. So electronic medical devices are tested at all temperature/humidity conditions for their intended use (IEC/EN 60601-1).

5.29 Power Module

THE NEED Electronic circuits require constant DC voltages with proper current power supply. The power module generates the required DC voltage from the available power sources (AC main power line or battery). Power modules have specific requirements for safe usage of the electrical devices. Medical devices have additional requirements to avoid unacceptable risks for patients and operators in normal and single fault conditions. For example, two means of protections have to be considered to avoid electrical shock hazards. If one fails, the other guarantees protection. The means of protection are given by the basic insulation and the protective earth (**Class I type**) or by a supplementary or double insulation (**Class II types**) as indicated in the IEC/EN 60601-1-2 standard.

Electrical medical devices also have '**applied parts**', that are the physical parts that in normal usage are in contact with the patient. Such parts have to be designed so that acceptable leakage current level is supplied to the patient. In addition, ECGs have to withstand defibrillator discharges. This implies additional constraints for the power module.

REQUIREMENTS Chapter 3 introduces the main requirements for the Gamma Cardio CG acquisition unit. Requirements applicable to the power module specify compliance to the following standards:

1. basic safety and essential performance (IEC/EN 60601-1)
2. electromagnetic compatibility (IEC/EN 60601-1-2)
3. safety of electrocardiographs (IEC/EN 60601-2-25).

These standards, addressed in Section 5.9, are applied to the Gamma Cardio CG and the power module, as outlined in Table 5.47. Compliance tests are performed on the whole device, but performance compliance is significantly dependent on the power module behavior.

Emission and immunity requirements affect the EMI filter included in the power module. Floating part condition must be guaranteed by the modules that interface the patient with PC: the power module and the communication module. These modules must have a galvanic isolation so that parts in contact with patient are floating as shown in Figure 5.75.

The isolation is obtained by optocouplers and the isolated DC–DC converter. CF applied type devices have low leakage currents (Table 5.25) that are obtained by selecting isolating components (optocouplers, DC–DC converter) with high isolation impedance. Isolating components have to withstand test voltages associated with the working voltage in single fault condition. Air clearance/creepage requires that PCB metal plates are at a safe distance (>4 mm) across the isolation barrier.

The same applies for the facing pins of the isolation components that cross the isolation barrier. This is because, since left part is floating, the maximum voltage is expected between the two facing conductive parts across the isolation barrier. The overall system must withstand the defibrillator test (5 kV discharge). Most of the protection is performed at patient cable level. In any case, the power module has to withstand the residual voltage.

Table 5.47 Main requirements of the Gamma Cardio CG power modules

Topic	Reference	Outcome compliance level
Radiated emission	Table 5.24: CISPR 11	Group 1 Class B compliant
Immunity to ESD	Table 5.7: EN 61000-4-2	±6 kV contact ±8 kV air
Immunity to electrical burst	Table 5.7: EN 61000-4-4	Compliant: ±2 kV
RF conducted immunity	Table 5.7: EN 61000-4-6	Compliant: 3 V_{eff}/m from 150 kHz to 80 MHz
RF radiated immunity	Table 5.7: EN 61000-4-3	Compliant: 3 V/m from 80 MHz to 2.5 GHz.
Power frequency H-field immunity	Table 5.7: EN 61000-4-8	Compliant: 3 A/m
Electrical equipment class	IEC/EN 60601-1	Class II (double insulation)
Applied part type	IEC/	CF type
Current leakage	Table 5.25: IEC/EN 60601-1	Compliant: current leakage ≤0.1 μA, apart From current leakage caused by main supply on applied parts: ≤11 μA (requirement ≤50 mA)
Working voltage (dielectric strength,)	IEC/EN 60601-2-25	1500 V test voltage
Air clearance creepage	IEC/EN 60601-2-25	4 mm minimum
Battery	Table 5.23	Not applicable
Defibrillation proof	IEC/EN 60601-2-25	5 kV pulse test
Power module voltage	Table 5.26	USB $V_{in} = 5$ V, $V_{out} = \pm 5$ V, Isolated at 220 $V_{ac,rms}$ for single fault condition

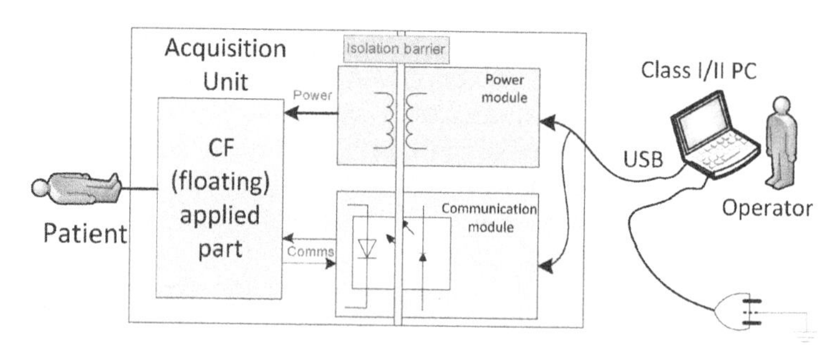

Figure 5.75 Medical device floating parts.

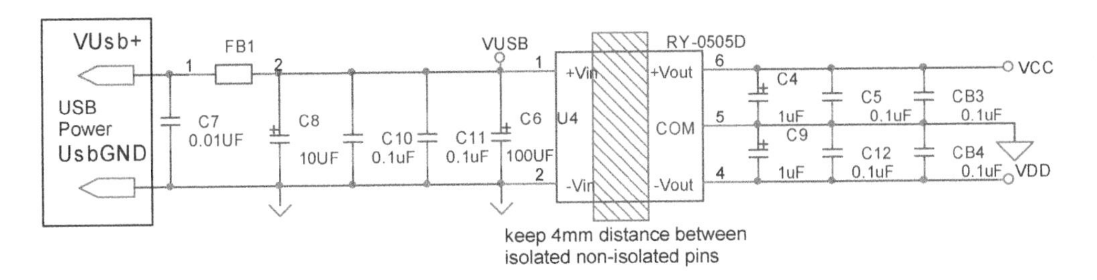

Figure 5.76 The power module circuit.

In addition to the previous requirements, the power module has to provide a regulated voltage supply for the digital and analog floating circuits. These are powered at +5 V and ±5 V respectively. The source is the USB power which provides 5 V.

5.29.1 Power Module Circuit

Figure 5.76 shows the complete power module circuit. Power is obtained from the USB plug. A first filter is made up of the ferrite bead and the two capacitors C7 and C8. The filter reduces noise interference from the device being radiated down the USB cable to the host PC and vice versa.

C10 and C11 are the **decoupling capacitors** of the communication module. In addition, a large capacitor (C6) is provided to reduce the effects of input voltage dips. The core of the power module is a DC–DC isolated converter that provides ±5 V to the floating part through a galvanic isolation guaranteed by an internal transformer.

AC bypass capacitors are always to be included in the circuit, as suggested by datasheets of active components, to reduce power noise especially when high resolution ADCs are powered. The bypass capacitor must be placed at the power source (1 μF and 0.1 μF) and at every active analog and digital component. Ideally, a capacitor should have reducing impedance at increasing frequency. Real capacitors offer this behavior only at specific frequencies, due to parasitic effects. So two capacitors are set in parallel where the 1 μF value handles the lower frequencies while the 0.1 μF capacitors reduce higher frequencies.

5.30 Communication Module

THE NEED ➤ Medical electronic devices almost always have digital components that require data exchange. So digital components (sensors, processors, actuators etc.) often embed standard interface modules for communication. But these modules do not eliminate design problems related to communication lines (high unexpected bit error rate, EMI etc.). In addition, medical devices have stringent requirements for the parts physically connected to the patient (applied parts) that must not have a direct DC connection to power lines, to avoid electric shock. Since a communication path has to be present between applied parts and non-applied parts (e.g., the medical device with its GUI) galvanic isolation has to be implemented. This section shows the simple example of the Gamma Cardio CG where standard communication interfaces (UART-USB) are used for data exchange between the PC and the Gamma Cardio CG through a galvanic isolation.

REQUIREMENTS The communication module requirements are similar to the power module insulation requirements, since both the modules have to isolate the applied part and the non-applied parts as shown in Figure 5.75. In addition, the transmission module has to guarantee the data exchange functionality between the microprocessor and the PC: required data rate, minimal error rate and so on.

Regarding **insulation**, the power module requirements shown in Table 5.47 that apply to the communication module are

1. immunity to EMI
2. current leakage
3. air clearance between CF applied parts and non-applied parts.

These requirements have to be satisfied with proper electronic components and PCB layout.

Regarding **data exchange rate** requirements, the communication module has to guarantee a minimum capacity that is higher than the maximum data exchange between the microprocessor and the PC.

The maximum exchange of data holds when the Gamma Cardio CG is in acquisition mode and transmits data to the PC. In this case, we have

$$\text{Communication module data rate capacity} \gg \text{Maximum expected data rate}$$
$$\text{Maximum expected data rate} = \text{ECG bits/second} \times \text{bit redundancy factor} \times \text{packet redundancy factor} \times \text{activity time factor.}$$

ECG data are transmitted at 12 bit resolution × 1200 samples/s = 14,400 bits/s. Assuming no parity bit, there are two extra bits for start and stop as shown in Figure 5.42. This increases the bit rate by a bit redundancy factor = 10/8 = 1.25. ECG data bytes are transmitted within a packet according to a protocol. Each packet includes control bytes and traffic (ECG) data. The data rate is therefore greater by a factor = (number of bytes in a packet)/(ECG data bytes in a packet).

Assuming 120 ECG bytes per packet plus eight bytes of redundancy, we have:

$$\text{Packet redundancy factor} = 128/120 = 1.067. \quad (5.90)$$

The activity time factor accounts for the fact that data flow is not continuous. There are time intervals where the microprocessor is not able to transmit because it is busy on other tasks. Even if the processor has a hardware UART that does not occupy processor time when transmitting bits, the firmware itself may not be ready to feed the UART continuously at the maximum speed.

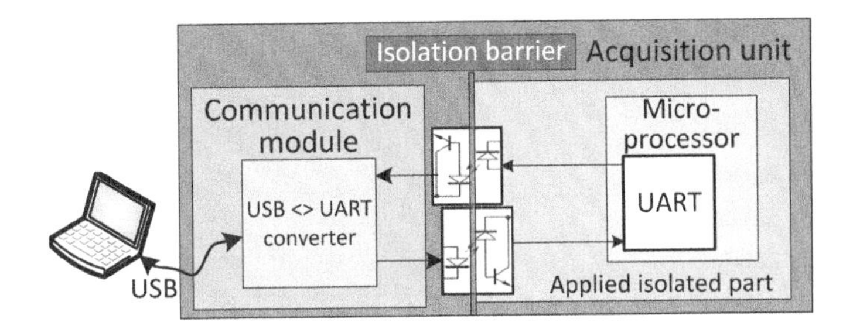

Figure 5.77 Gamma Cardio CG conceptual communication scheme.

It can be considered that 20% of time cannot be used to transmit bits over the serial line. In this case we have: activity time factor $= 1/(1-0.2) = 1.25$ and so

$$\text{Maximum expected data rate} = 14,400 \times 1.25 \times 1.067 \times 1.25 \cong 24,000 \text{ bits/second}$$

THE CIRCUIT Figure 5.77 shows the conceptual data flow scheme between the PC and the Gamma Cardio CG acquisition unit. The PC uses the standard USB for data communication. Since the microprocessor is equipped with a UART, a USB to UART converter is needed.

Apart from the voltage levels, the UART implements the RS-232 communication standard that needs, at minimum, two unidirectional communication lines: one for transmission and one for reception. This simplifies the galvanic isolation barrier that can be implemented through simple optocouplers, which are unidirectional devices. Other options are available: more expensive components are now available to guarantee USB galvanic isolation directly. These have the drawback of requiring more powerful DC–DC converters, since the USB–UART converter would be located in the isolated applied part. Another solution would be to consider a microprocessor with onboard USB at extra cost.

Figure 5.78 shows the hardware scheme. From the application/software point of view, the PC will address the UART of the microprocessor as if it were its own (virtual) RS-232 port, thanks to the specific software drivers. Figure 5.78 shows the USB–UART converter circuit. Newer versions of the USB–UART converter chip reduce external components to a single capacitor. However, updating components implies performing certification and functional tests again. This suggests planning component updates with care. The USB converter chip handles the USB PC port and translates it into an RS-232 data flow with TTL voltage levels through the pins 24 and 25. These pins are connected to the optocouplers that are connected to the floating part of the microprocessor as shown in Figure 5.79.

Galvanic isolation can be achieved using optocouplers, capacitors, transformers or even relays for certain applications. **Optocouplers** are gaining more attention for their improved isolation, common mode noise elimination and limited cost and space, especially for I/O interfaces on PC boards. Optocouplers (also known as optoisolators or photocouplers) are electronic devices where the electric signal is transferred from input to output through light waves. More specifically, the input signal is transformed into a proportional light signal by a specific device (light emitting diode).

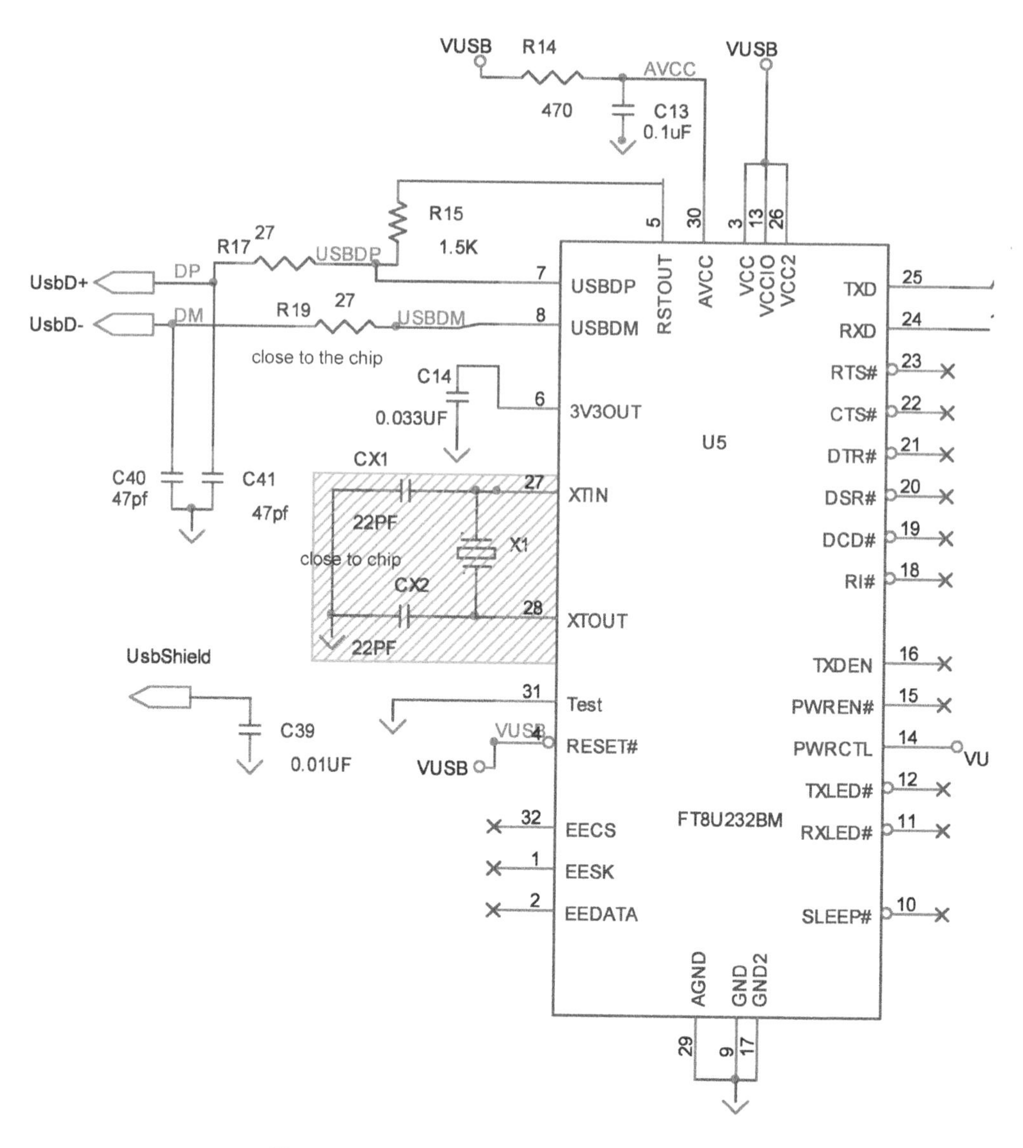

Figure 5.78 Gamma Cardio CG USB converter.

The light is then retransformed into an electronic signal by a light sensor (e.g., a phototransistor). With this approach, there is no electronic connection between the LED input and output photosensor pins. The current between input and output can be propagated only through the material that insulates the emitter and the sensor.

The leakage current is generally around a microampere with high voltage applied (thousands of volts). Other typical isolation characteristics are: resistance $>10^{12}$, capacitance <1 pF, withstand insulation test voltage >1000 V. These features make optocouplers ideal for medical applications.

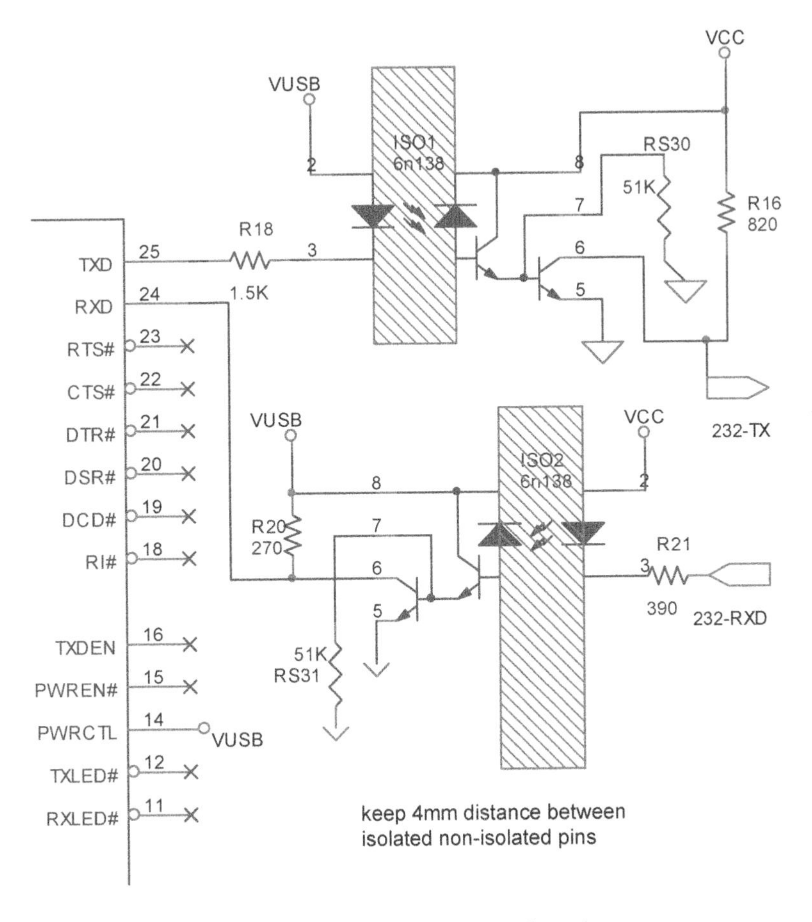

Figure 5.79 Optocouplers: connection scheme.

Conclusion

In this chapter, we have addressed the electronics behind medical instruments. The topics of this chapter are generally applicable to other fields where electronic devices are implemented. The theory section has outlined the behavior of the main electronic components, the typical circuits and the background required to understand the implemented functions (e.g., signal processing, telecommunication and information theory). The implementation part has shown a market-delivered realization with a discussion of more common practical problems.

The current industrial trend shows that electronics is always more integrated with software, firmware, signal-information theory and telecommunications. Hardware is complemented by software and firmware functions that can now be efficiently implemented thanks to the advances in development tools, programmable devices and PC platforms. Software and firmware are often implementing complex algorithms that employ signal/information theory concepts more deeply. Medical services on space-distributed telemedicine platforms are increasing. This requires knowledge of telecommunications.

Notably, the current trend suggests then that electronic design must be faced with a proper background on software-firmware engineering, signal/information theory and telecommunications. This chapter is an effort to offer this comprehensive 'system' approach to designing medical electronic devices. The next two chapters will enforce this approach, focusing on software and telecommunications with ICT-enabled E-Health medical services. The final chapter will combine electronics, software and telecommunication subcomponents to certify the overall medical device system.

References

AAMI (1991), Association for the Advancement of Medical Instrumentation, American National Standard, Diagnostic Electrocardiographic Devices, ANSI/AAMI EC11-1991.

Analogue Device (2011a), Technologies for High Performance Portable Healthcare Devices, AH09519-0-11/11(A).

Analogue Device (2011b), ADAS 1000 datasheet.

Baba A. *et al.* (2008), Measurement of the electrical properties of ungelled ECG electrodes, International Journal of Biology and Biomedical Engineering.

Barnes J. (2001), Designing Electronic Equipment For ESD Immunity, Circuit Design, July 2001. Available on Internet.

Bocock G. (2011), Power Supply Technical Guide Issue 3, XP Power.

Bronzino J. D. editor (2006), The Biomedical Engineering Handbook, Taylor & Francis.

Bryant J. *et al.* (2000), Protecting Instrumentation Amplifiers, Sensors 2000.

Crone B. (2011), Mitigation Strategies for ECG Design Challenges, Technical Article MS-2160, Analogue Device.

EEC (2007), Council Directive 93/42/EEC of 14 June 1993 concerning medical devices available at http://ec.europa.eu/health/medical-devices/index_en.htm amended by 2007/47/EC into in force on 21 March 2010

Freescale (2010), High Volume Sensor for Low Pressure Applications.

Hann M. (2011), Signal chain basics: Analyzing RL drive in ECG front-end with SPICE, EE Times-Asia.

Horowitz P. *et al.* (1999), The Art of Electronics, 1999, Cambridge press.

Karki (2008), Active Low pass Filter Design, Application Report, SLOA049, Texas Instruments.

Kester W. (2004), Analog-Digital Conversion, Analogue Devices.

Kester W. (2005), The Data Conversion Handbook, Analogue Devices Technical Handbooks.

Oppenheim A., Schafer R. (2009), Discrete-Time Signal Processing, Prentice Hall.

Papoulis A. (1993), Probability, Random Variables and Stochastic Processes, McGraw-Hill.

Proakis J. G. (2008), Digital Communications, McGraw-Hill Book Co.

Scott W. (2000), Finding the Needle in a Haystack, Analog Dialogue, 34.

Smith S. (1999), The Scientist and Engineer's Guide to Digital Signal Processing, California Technical Publishing.

Sommerville T. (1994), Isolation amps hike accuracy and reliability, Application Bulletin, Burr-Brown.

Texas (2008), Active Filter Design Techniques, Application Report, SLOA088, Texas Instruments.

Texas (2010), Medical Applications Guide, Texas Instruments.

Texas (2011), Analogue Switch Guide, Texas Instruments.

Texas (2007), Noise Analysis in Operational Amplifier Circuits, Application Report, SLVA043B, Texas Instruments.

Texas (2011b), ADS1198 datasheet, Texas Instruments.

Venkatesh A. (2011), Improving Common-Mode Rejection Using the Right-Leg Drive Amplifier, Texas Application Report SBAA188.

Webster J. G. (2009), Medical instrumentation: application and design, John Wiley & Sons.

WHO (2011), World Health Organization, The top 10 causes of death, Fact sheet no. 310, 2008 updated June 2011.

Zumbahlen H. (2008), Linear circuit design handbook, Analogue Devices, inc, Elsevier.

6

Medical Software

The remaining 10% of the code accounts for 90% of the development time

Chapter Organization

Software is receiving increased attention in the medical device area due to its pervasive usage in newer medical devices and its specific nature generating risks that have produced serious injuries in recent years. These reasons have drawn attention from regulatory and standards bodies that have produced specific additional regulations and standards to reduce incidents. This chapter is organized following the organization of the standard IEC/EN 62304 concerning medical device software and software life cycle. As shown in Figure 6.1, software production process is divided into specific stages that are discussed in the sections indicated in the circled numbers.

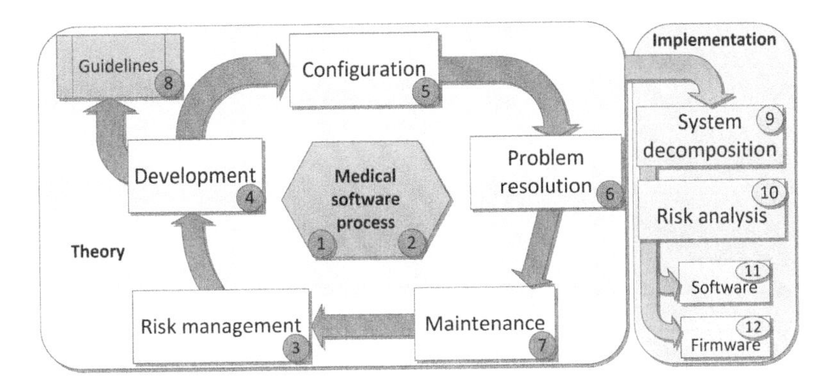

Figure 6.1 Medical software design model.

Medical Instrument Design and Development: from Requirements to Market Placements, First Edition.
Claudio Becchetti and Alessandro Neri.

Risk management is the driver for all safety/mission-critical activities such as medical device software. In the end, software must not generate unacceptable risks as discussed in Section 6.3 and in more in details in Chapter 8. Risk management drives the successive **development,** which includes all the activities from requirements to integration and testing (Section 6.4). Software implementation may be greatly improved based on **design guidelines** (Section 6.8). **Configuration** is another critical aspect; software is made up of thousands or even millions of lines of code. This complexity may be driven only by a specific process assisted by dedicated tools. With a configuration process (Section 6.5), it can be reasonably ensured that the right piece of software among the thousands/millions of code lines is in the right place with the right version. The growing number of code lines makes software more prone to errors which, again, may generate risks. So a specific process (**problem resolution)** introduced in Section 6.6 is a control measure for such risks. When software has already been released on the field, additional modifications may be required. This suggests introducing a simpler process (**maintenance**) to change and update the software with respect to the development process (Section 6.7). The implementation part outlines how theory is translated into the Gamma Cardio CG product whose disclosed details are addressed in Sections 6.11 for software and 6.12 for firmware. This chapter attempts to explain the complexity of software design with the risks that may be generated in medical devices. Such complexity is often underestimated. For example, in terms of effort required to finalize software applications, it is sometimes evident that: '*the remaining 10% of the code accounts for 90% of the development time*'.

Part I: Theory

6.1 Introduction

A software product is a set of programs containing instructions and data that computers and programable devices (e.g., microcontrollers, FPGAs etc.) read to perform specific functions. In this **definition of software**, we include the firmware which is the software mainly devoted to embedded systems and programable devices. From this software definition, digital data files that do not contain programing instructions are excluded.

In the general **medical instrument architecture** introduced in Section 1.4 (see Figure 6.2), the patient is connected to a device through sensors that transform diagnostic information into electrical signals. These signals are processed by specific hardware boards that are managed by programable components such as microcontrollers. Programable components such as DSPs and FPGAs also have an important role in processing the electrical signals that are usually digitized within the hardware board. Programable devices are governed by firmware that can usually be downloaded and updated for error-fixing or upgrading. The custom hardware may be connected to a PC based platform also through a network connection. PC platforms are increasingly used in medical instruments since they implement complex functions via software at lower development and manufacturing costs.

The firmware and the software running on PC based platforms share the same problems and they are often referred to generically as (medical) software in standards (e.g., IEC/EN 60601-1) and regulatory frameworks (e.g., EEC, 2007).

In general, **a medical software** is

- a stand alone software used for medical purposes (diagnostic or therapeutic)
- a software component of a medical device (i.e., embedded software)
- a software which drives a device or influences the use of a device
- a software that is an accessory of a medical device.

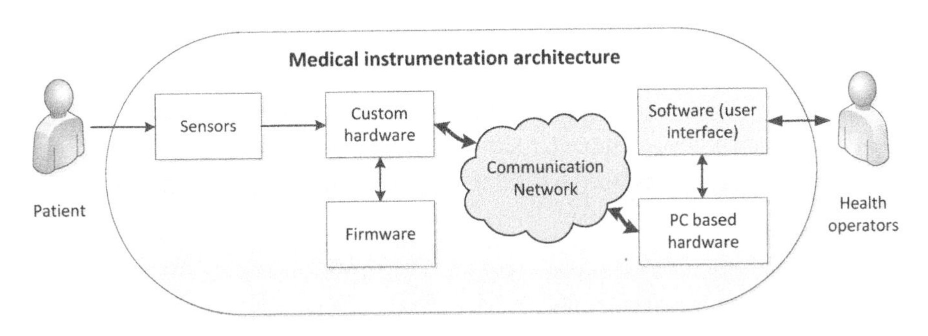

Figure 6.2 Medical instrument architecture.

It may be also considered medical software a product that is used

- in the production of medical devices
- in the implementation of a quality system focused on medical device management.

Medical software is a very significant branch of software engineering that is frequently regulated and must comply with national and regional laws. In the European Union (EU), software requirements are included in the Medical Devices Directive amended in 2007 (EEC, 2007). In the USA, the Food and Drug Administration (FDA) has increased its involvement in reviewing the development of medical device software starting in the mid 1980s, when coding errors in a radiation therapy device resulted in the overdose of patients (Leveson, 1993). In the USA, in the years 2010 and 2011, 250 medical device recalls were due to software and five recalls to firmware (FDA, 2012).

According to (EEC, 2007), software used for a human being for diagnosis, prevention, monitoring, treatment or alleviation of disease is a **medical device** in itself. A similar approach holds in the US market where medical device software may be simply a piece of software on its own (FDA, 2002).

Medical device software and software incorporated in medical devices have therefore to incorporate all the controls required in the regulatory framework. Not all software that is used in the context of healthcare is considered to be a medical device. For example, software used for storage, or lossless compression is not covered by the EU MDD (MEDDEV, 2012). On the other side, software may also perform a lossy compression; for example, it may decrease the size of a file containing medical images with an algorithm that reduces the quality of the image itself. In this case, the information loss may affect diagnosis and the associated treatment thus impacting on safety. For this reason, this latter software has to be classified as a medical device.

6.1.1 Intrinsic Risks and Software Engineering

Medical devices and instruments are now usually embedded by software which plays a significant and increasing role in safety-critical functions. Unfortunately, the absence of errors and therefore **safety cannot be ensured** deterministically (i.e., for 100% of the operative time) in non-trivial software programs.

Between 2002 and 2010, software-based medical devices resulted in millions of recalled devices where more than 10% was directly due to software failures. This recall rate is nearly double that for the period of 1983–1997 (Fu, 2011). Despite recalls, software failures are still frequent with impact on patients' health. Software is a critical component with respect to other medical device technologies. This is mainly due to the nature of software itself which is easily changed and produced through a simple compilation. This feature is generating risks for the devices.

According to (FDA, 2002), an analysis on 3140 medical device recalls, in the years 1992–1998, showed that 7.7% of the cases were due to software failures. It is interesting to note that software failures in the 79% of the cases were due to changes at the maintenance stage, that is, after software release (FDA, 2002).

To reduce costs and risks of software development, a systematic approach to the analysis, design, assessment, implementation, testing, maintenance and reengineering of software has been defined, with the application of engineering to software. This approach gave birth to a new discipline known as 'software engineering'.

Software engineering (SE) may be defined as the application of a systematic, disciplined, quantifiable approach to the development, operation and maintenance of software, and the study of these approaches (Abran, 2004). Software engineering is the application of engineering to software, since it integrates significant mathematics, computer science and engineering methods. A proper application of software engineering may reduce risks of failure, which may be particularly critical in medical devices. This approach has been followed by regulatory frameworks and standards that have set requirements over the whole process of software development.

6.1.2 Main Concepts in Software Development

In the medical industry, as in many other safety-critical environments, software implies additional relevant activities extra to writing the code. A software product must comply with regulatory requirements, it must satisfy the user needs, it must be reliable with acceptable unrecoverable errors or wrong results and it must be easily installed and maintained. In a broader sense, all the steps included between the software product conception and its deployment and maintenance must be properly managed. Software development process must encompass research, new development, prototyping, modification, reuse, re-engineering, maintenance and any other activities ideally in a planned and organized process.

There are different **approaches to software development**. Some methods follow a more structured, engineering-based approach for developing solutions, whereas others may implement a more incremental approach, where software programs evolve incrementally and iteratively. Such topics have already been discussed in Chapter 1. In any case, most methodologies share a combination of the following stages of development:

- gathering requirements for the proposed solution
- analyzing the problem
- defining plans and design specifications for the software-based solution
- implementing (coding) the software
- testing the software
- deploying programs on the target environment
- performing maintenance, change management and error fixing.

Different approaches to software development may organize these stages in specific orders, or devote more or less time to different stages. The level of detail of the documentation produced at each stage of software development may also vary. These stages may also be carried out in sequence (the so called 'waterfall' approach introduced in Chapter 1), or they may be repeated over various cycles or iterations (a more 'iterative' approach). Iterative approaches promote testing throughout the development **life cycle**, as well as having a working (or error-free) product at all times.

More structured and waterfall-based approaches attempt to assess the main risks at the beginning. A detailed plan is developed before any software implementation takes place; this avoids significant design changes and recoding at later stages of the software development life cycle. Regardless of the the adopted process, the sequence of different stages is

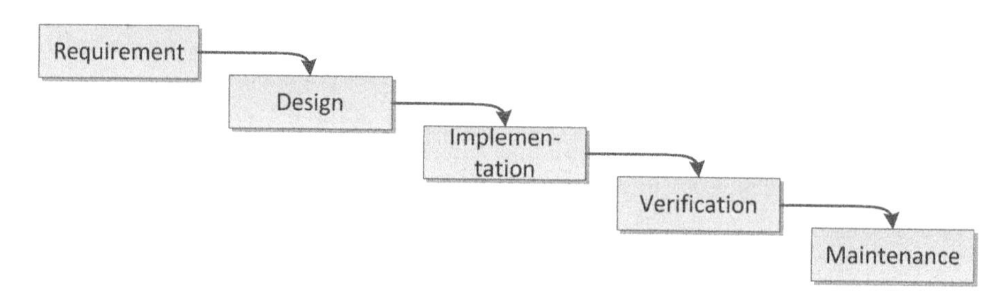

Figure 6.3 Waterfall software design process.

usually named '**software development life cycle**' while the 'style' is referred to as '**development methodology**'.

For example, Figure 6.3 shows the widespread waterfall model. This model implies a sequential development strategy where each stage starts at the end of the previous one: first, requirement analysis, then, design and so on.

Other life cycles derive from the waterfall, as discussed in Chapter 1. When considering the success of a project, the organizational, managerial and support aspects of software development are at least as important as the technical aspects. Software engineering management usually also encompasses planning, control, people management, configuration management and tools management.

It is worth noticing that without a well-balanced method suited to the context and the objectives, a software project is likely to fail, even with the most proficient developers.

Without the hope of being exhaustive, this chapter will outline the medical software development process from the regulatory framework to the principles of software management processes and the software life cycle based on the specific standards for medical device software.

6.1.3 Regulatory Requirements for Software

In recent years, regulatory bodies have increased their attention to medical device software since software is a growing component in medical devices. Its presence has amplified the device complexity, and the associated systematic failures are difficult to manage by conventional device testing approaches.

It has been seen that *safety and performance of medical devices may be ensured only through a controlled software development process.* The earlier European Medical Device Directive (EEC, 1993) introduced requirements over the design of **electronic programable systems**, (i.e., devices with processing units, software and interfaces):

> *12.1 Devices incorporating electronic programable systems must be* ***designed to ensure*** *the*
>
> 1. *repeatability*
> 2. *reliability and*
> 3. *performance of these systems according to the intended use.*
>
> *In the event of a single fault condition (in the system) appropriate means should be adopted to eliminate or reduce as far as possible consequent risks.*

In the new version of the Directive (EEC, 2007), software can be a medical device on its own (*'medical device' means any instrument, apparatus, appliance, software, material or other article, whether used alone or in combination'*). In addition, the new directive added that

> *For devices which incorporate software or which are medical software in themselves, the software must be* ***validated*** *according to the state of the art taking into account the principles of: 1) development life cycle, 2) risk management, 3) validation and verification.*

The standard IEC/EN 60601-1, focusing on safety and essential performance for medical electrical equipment, has introduced additional requirements on programable systems. (For an explanation of standard codes IEC, ISO, EN, ANSI and so on, see Chapter 8). Basically, the standard prescribes that software can achieve adequate safety and performance level only if specific process areas are controlled. Such areas, which are historically suggested by safety-critical software development, do not change significantly between EU, US (e.g., FDA, 2005) or other regulatory frameworks.

IEC/EN 60601-1 defines broad process areas: risk management, development life cycle, problem resolution, change management, requirement specification, design and implementation, verification and validation. The standard also outlines the additional hazards related to network connected devices. Specific requirements on these areas are left to vertical standards such as ISO/EN 14971 for risk management and, more specifically, to IEC/EN 62304 for medical device software – software life cycle processes (IEC/EN 62304, 2006). The following sections will outline the main software-related processes based on the IEC/EN 62304 organization.

6.2 The Process: a Standard for Medical Software

Until recently, safety regulations for medical device software were not rigorous. In addition, software was not formally classified as a medical product by the national authorities. With the rise of a new generation of software-empowered devices, a new regulation is in force governing medical device software development for all classes of equipment. Previous software safety standards were best suited to medical devices with low risk levels, as opposed to recent products where software failure could result in death. With the growing importance of embedded systems, the focus has shifted to the reliability of software and to the associated device usage risk. As a result, the new **IEC/EN 62304** has emerged as a global benchmark for the management of the software development life cycle. IEC 62304 is a standard for software design in medical products recognized by the European Union (i.e., harmonized) and by the US FDA. Since the standard is harmonized, medical devices complying to this standard are presumed to satisfy essential requirements contained in Medical Devices Directive 93/42/EEC (MDD) with amendment M5 (2007/47/EC) as related to software development. This may be a simple route to ensuring compliance with the MDD. The US FDA has also recognized the standard (ANSI/AAMI/IEC 62304:2006) as evidence that medical device software has been designed to an acceptable quality level. This standard is identical to the EN/ISO variant in all essential details. Designing according to IEC/EN 62304 ensures that software is produced by means of a defined and controlled process.

6.2.1 IEC/EN 62304 Overview

Medical device regulatory frameworks and standards currently have a similar approach to software management. As shown in Figure 6.4, a medical device (and therefore a medical

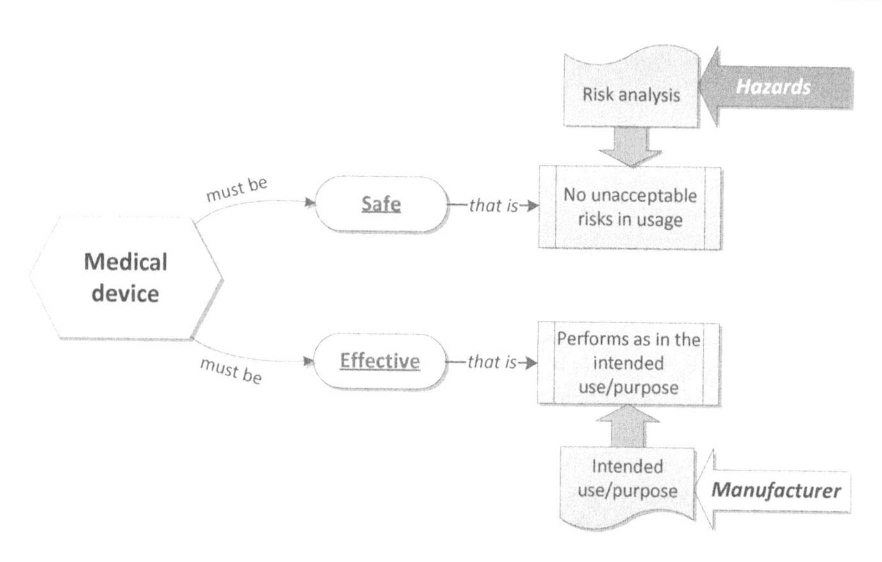

Figure 6.4 Medical device objectives.

instrument) must be safe and effective. ***Safety*** *is ensured through a risk management process that excludes the existence of unacceptable risks with respect to the benefits* derived by the use of the medical device itself. **Effectiveness** means that the medical device performs as described by the manufacturer in the intended use/purpose.

This approach is shared for example in the US and EU regulatory frameworks and it is considered in the IEC/EN 60601-1 that is the main standard for safety and performance of medical electrical equipment. The IEC/EN 62304 on software development is based on the approach depicted in Figure 6.4, since software is either a stand-alone component that is therefore considered as a medical device, or it is an integral part of the medical device that conditions the safety and effectiveness requirements.

IEC/EN 62304 provides a framework of life cycle processes needed to safely design and maintain medical device software. This standard provides requirements for each life cycle step and process. Each process is further divided into a set of activities and tasks. The underlying **assumption** is that software testing alone cannot ensure safety and effectiveness. In the software area, only a controlled management of the processes and the critical activities of development can ensure that software is safe (IEC/EN 60601-1).

A first requirement is that medical device software must be developed and maintained following a **risk management system** and a **quality management system** compliant with the requirements imposed by the regulatory framework (e.g., the Medical Device Directive in Europe or the applicable Code of Federal Regulations in the USA). Figure 6.5 shows the relations between IEC/EN 62304 and the device life cycle. IEC/EN 62304 relies on ISO 14971 which addresses the application of risk management to medical devices. Quality management may be addressed by ISO 13485 or equivalent approaches.

The medical device is affected by IEC/EN 60601 (safety and performance of medical electrical equipment) and other horizontal or vertical (i.e., specific to a class of devices) standards. The main sources for development requirements are always the applicable

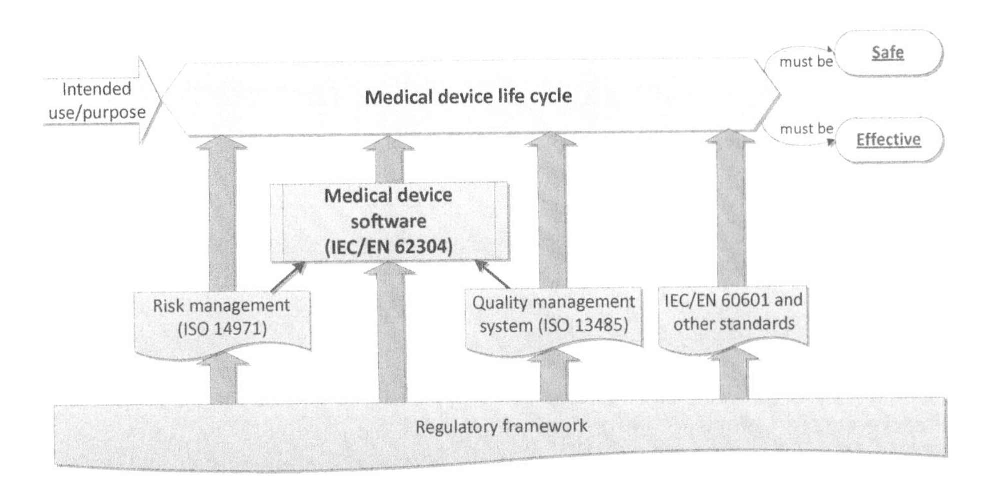

Figure 6.5 IEC 62304 relations with standards.

regulatory framework and the intended use/purpose specified by the manufacturer that impacts on the final objective of ensuring safety and effectiveness for each produced device.

The **software process** consists of the activities shown Figure 6.6. **Risk management** is the driver of software development. It has to ensure that all unacceptable risks are mitigated by proper control measures: in no circumstance must software generate unacceptable risks.

The **development process** includes the usual activities of software production discussed in Section 6.4. The standard does not recommend any specific software life cycle. The choice is left to the manufacturer who has to select the most appropriate approach according to the context.

Because most incidents are related to service or maintenance of medical devices, including inappropriate software updates and upgrades, the **software maintenance process** is

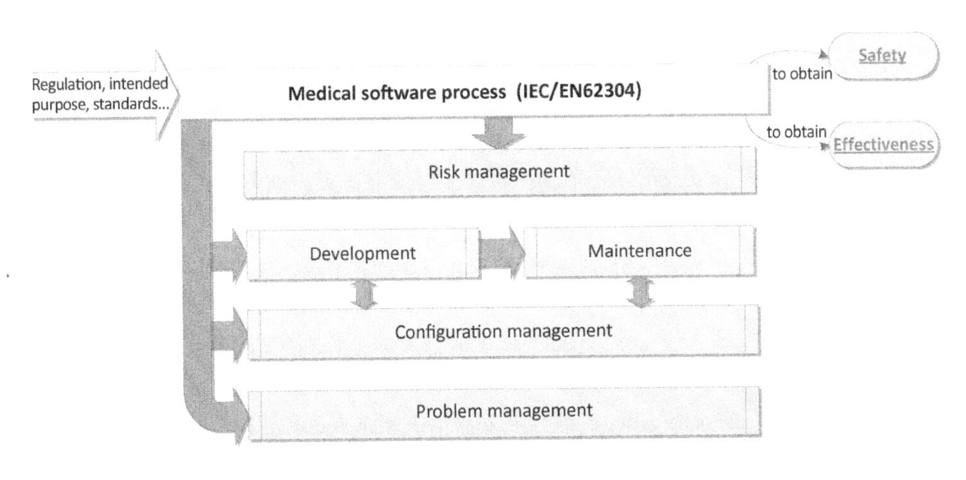

Figure 6.6 Software process.

considered as important as the software development itself for ensuring safety. The software maintenance process is very similar to the software development process but with a reduced burden.

The standard identifies two additional processes considered essential for developing safe medical software. They are configuration management and software problem resolution.

Configuration management process has to define procedures to identify and manage software items (including documentation), throughout the product life cycle. In general, configuration management has to be considered in broader terms, thus applying to the whole medical device system components, including for example hardware, accessories, system documentation and marking.

Configuration management not only ensures safety and effectiveness, but it is also a valuable core process for the manufacturing industry to ensure producability of the marketed products. Configuration specifies which items (hardware software, documentation, accessories etc.) are required to produce a specific version of a specific product. Configuration management has an increased benefit as the number of items required to reproduce and manufacture the product increases. More simply, more complex products in terms of the number of subcomponents require a more stringent configuration management process. This explains why, in any industrial software production, configuration is compulsory. Software consists of thousands/millions of code lines that may be changed between different product versions and types. The complexity in terms of numerousness is so high that even simple software items require stringent configuration to avoid significant inefficiencies. Collaborative software production increases complexity and again requires more advanced and strict configuration management processes.

Problem resolution is another critical process included in the standard. This process, in turn is strictly guided by the requirements set by the regulatory framework since discovered problems may impact on safety and often specific national procedures have been introduced to handle filures. In the worst problem cases, products have to be withdrawn from the market.

Similarly to other software standards that have been in place from decades, IEC/EN 62304 does not specify an organizational structure for the manufacturer or specific roles for process executors; it requires only that the prescribed processes are performed with a documented and effective approach that satisfies the standard requirements. The standard does not prescribe formats and/or contents for the documentation to be produced; the decision is left to the manufacturers. This standard also does not prescribe a specific life cycle model. The manufacturers are responsible for selecting a life cycle model for the software project. The lack of requirements on life cycle is consistent with the criterion that the effectiveness of a specific software approach depends on many factors (people, procedure in place, context, environment etc.), and there is no method to determine the best approach.

In the following some basic concepts will be introduced. The five processes constituting the core of the standard will be discussed in the sections from 6.3 to 6.7.

6.2.2 Risk Analysis for Hardware and Software Design

Risk management is used to reduce risks associated with medical device usage. ISO/EN 14971 has been used as the standard for risk management for medical devices. The latest version of this standard (ISO 14971:2007: harmonized in the EU with EN 14971:2009) has been considerably extended, and the contents are now focused to both software and hardware components.

The approach to risk management is similar to the system–subsystem decomposition described in Chapter 1. To reduce an unmanageable system complexity, the system is decomposed into smaller subsystems whose complexity can more easily be dealt managing one subsystem at time. In the risk management context, the system (i.e., the medical device), has to be analyzed as whole; then, risks have to be associated and detailed to the subsystems that are part of the medical device in a top-down process. Again, subsystems may be hardware boards, packages, pieces of software or accessories. Risk management can be performed separately for each subsystem, including software modules to define the safety requirements. Chapter 1 includes examples of ECG risks managed in the Gamma Cardio CG product. In particular, the following project risks have been discussed:

1. distorted visualization of ECG signal
2. expensive error-fixing in the overall system integration
3. non-stable or unfeasible requirements.

For example, the first risk is clearly safety critical since it may induce a wrong diagnosis that may lead to an incorrect treatment. This risk is also associated to the software/firmware component.

According to ISO 14971, the software risk management process has to be considered within the system risk management process involving the device as a whole. Risk management is extensively discussed in Chapter 8. Here, we will focus on specific topics related to software.

During the **hazard** identification activity of the risk management process, harms eventually generated by software modules are identified and assessed. Hazards that could be indirectly caused by software (e.g., by providing misleading information that could lead to inappropriate treatment) need to be also considered when determining whether the software is a source of specific risks.

Due to its flexibility and its insignificant marginal production cost, software may also be used effectively to mitigate unacceptable risks through additional software-based control measures. This decision is made during the risk control activity of the risk management process. Chapter 8 includes a description of all the activities of risk management.

It is worth **comparing hardware and software in risk generation**. In software programs, hazards are due to a specific cause – the software errors – while in hardware a common reason may be not defined so clearly. On the other hand, localization of software errors is generally time-consuming since errors may be in any of the thousands/millions of lines of code, and the effect of the error may be far distant in time and space. More specifically, in software programs, an error in a specific line of code may produce effects in an entirely different part of the program at a time instant again far from the time when the error has occurred. This makes error fixing a complex and time-consuming activity. In some cases of poor quality software, errors may be fixed only with a major redesign spanning over the whole cide. In the worst cases, this activity may be so onerous that the whole software application has to be abandoned.

Since software includes a huge number of internal states that, for different values, may generate different output, classical risk analysis may be difficult. On the other hand, since risk is associated to errors, error fixing generally completely eliminates the risk, while in hardware, risks are only mitigated. These simple concepts show that risk management for software has to be faced with a different approach from hardware.

6.2.3 Software Safety Classification

According to IEC/EN 62304, in the first steps of software design, the manufacturer has to assign a safety class to the software system as a whole. This classification is based on the potential to create a hazard for patients, healthcare operators or other people involved.

The software is classified into three classes according to Table 6.1, The right-hand column includes the level of concern used in US medical device classification (FDA, 2008). For EU classification, stand-alone software, if covered by the Medical Device Directive, is considered as an active device that may be considered as Class IIa, unless it is intended to control or monitor the performance of active therapeutic devices in Class IIb, or intended directly to influence the performance of such Class IIb devices. Software that drives a device or influences the use of a device, falls automatically into the same class as the device (EEC, 2007).

Defining 'serious injury,' 'non-serious injury,' 'injury' and 'damage to health' is important to apply this classification effectively. The concept of injury may appear obvious at first glance, but this can be a far more complex problem when the context of the device is taken into account. The standard defines 'serious injury' as follows:

Injury or illness that directly or indirectly

a. *is life threatening,*
b. *results in permanent impairment of a body function or permanent damage to a body structure, or*
c. *requires medical or surgical intervention to prevent permanent impairment of a body function or permanent damage to a body structure.*

   Note: Permanent impairment means an irreversible impairment or damage to a body structure or function excluding trivial impairment or damage.

Manufacturers have the responsibility of deciding the safety classification of the software. Unfortunately, the previous definitions are not strict enough to determine an unquestionable

Table 6.1 Software safety classification and US level of concern

Software Safety Classification IEC 62304		**Level of Concern** *Guidance for the content of premarket submissions for software contained in medical devices*
Class A	No injury or damage to health is possible	Minor: Failures or latent design flaws are unlikely to cause any injury
Class B	Non-serious injury is possible	Moderate: Failure or latent possible design flaw could directly or indirectly result in minor injury
Class C	Death or serious injury is possible	Major: Failure or flaw could directly or indirectly result in death or serious injury

classification. Manufacturer assumptions are then required during the certification stage. This may be critical since higher classes imply greater development effort in terms of cost and time, as discussed in Section 6.2.5. It is recommend that Class B is the minimum standard to be applied for medical software, since Class A safety classification does not prescribe sufficiently rigorous software development processes.

6.2.4 *System Decomposition and Risks*

The software safety classification determines the procedures required for software development. The next step consists of decomposing the system into its subsystems: hardware, software....

This allows associating each risk at system level to the subsystems that are affected by the risk. Then, the risk can be managed at subsystem level through proper counter measures. The architectural diagram of the system will show the hierarchical tree based diagram consisting of intermediate nodes and terminal nodes usually referred to as leaves (of the tree).

To explain system decomposition and risks, let us consider the case of the Gamma Cardio CG product whose details are provided for this book by the Gamma Cardio Soft Company. The Gamma Cardio CG is a marketed 12-lead high-resolution ECG that is connected to a PC platform. Figure 6.7, introduced in Section 3.7.1, shows the functional scheme of the product with its interfaces indicated by the circled numbers. The acquisition unit with its firmware records the ECG signal and digitally transmits data to a PC through a USB connection. The PC is equipped with a software application that manages the ECG samples to obtain the visual ECG traces.

Figure 6.8 shows the decomposition of the Gamma Cardio CG product at system level. Risk analysis has to associate system risks at the lower subsystems (i.e., items) of the tree.

We need to look at the risk of producing a distorted ECG signal that in turn may induce an erroneous diagnosis and treatment. Among the subsystems (leaves of the diagram of Figure 6.8), we can consider that the risk is carried only by the hardware platform, the firmware and the application software. Electrodes and cabling are assumed to be already certified and to have no unacceptable risk. The plastic case of the acquisition unit does not affect the ECG signal, and the PC with its operating system (OS) does not perform operations that may change the signal. This has to be assumed since the PC and OS are commercial elements.

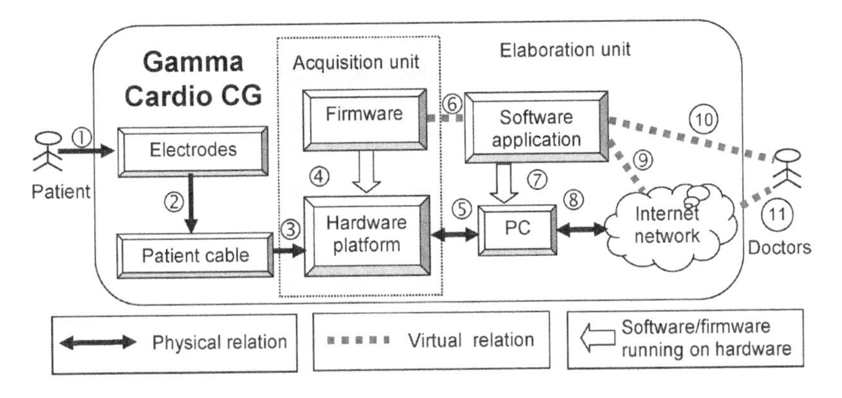

Figure 6.7 Gamma Cardio CG system–subsystem architecture.

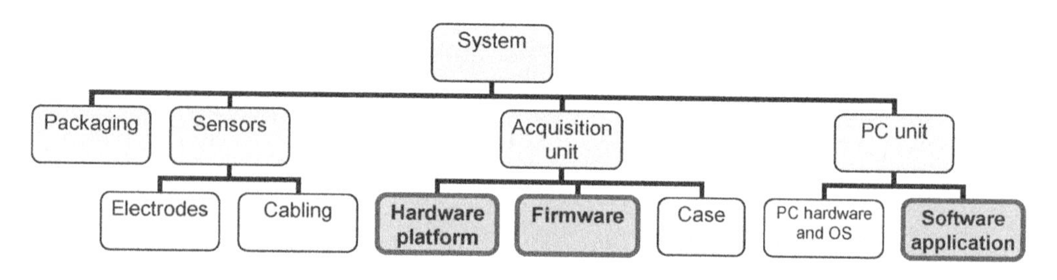

Figure 6.8 Gamma Cardio CG system architecture.

Thus, the ECG has to be designed so that the PC hardware with its operating system does not impact on safety. The risk of distorting the ECG signal is therefore associated to the dark colored bold items of Figure 6.8: application software, hardware platform and firmware. The same approach is applied for all the other risk.

6.2.5 *Impact of Safety Classification*

Safety classification has a significant impact on the software development process. Medical software manufacturers have to consider with care the correct classification to avoid expensive and time-consuming reworking at later stages of the development project.

Table 6.2 outlines the activities to be performed according to safety classification. Unit verification, integration and system testing have to be performed on all classes, but for Class A, formal detailed documentation does not need to be produced. Class A software thus requires less effort. On the other hand, only higher classes ensure an adequate control level for mission-safety-critical software.

6.2.6 *SOUP*

Software Of Unknown Provenance (or software of uncertain pedigree) is the derivation of the term SOUP; this refers to any code where the development quality process or the applicable quality standard is not known. This software may have inadequate accompanying documentation or may be developed by third parties that do not disclose the development process.

SOUP is often used in safety-critical contexts for pragmatic reasons since it reduces cost and development time. However, SOUP, especially when not sufficiently tested, may have a significant impact on medical device safety. This suggests that a software risk analysis has to be carefully performed on any SOUP code, and a clear justification has to be provided to verify that software safety is not jeopardized by the SOAP.

SOUP management is affected by the safety classification of the medical software. If software is included in Class A, SOUP is suitable for integration into the medical device software without specific safety analysis.

As the class increases, the risks increase and the safety analysis becomes more complex. This means that only simple functions that cannot affect safety of the overall medical software can be performed by SOAP components.

Notably, **open source software** is more suited to being used in safety-critical software since, in open source, the software code is accessible unlike with conventional third-party

Table 6.2 Requirements for the different software classes

Ref. IEC 62304	Software document	Class A	Class B	Class C
5.1	**Development plan**	Plan required, minor contents	Plan required	Plan required, major contents
5.2	**Requirements specification**	Document required with minor contents	Document required	Document required with major contents
5.3	**Architecture**	Not required	Software architecture	Software architecture refined to software unit level
5.4	**Detailed design**	Not required		Document detailed design for software units
5.5	**Unit implementation**	All implemented units are documented and source controlled		
5.5	**Unit verification**	Not required	Tests and acceptance criteria Unit verification	Process, additional tests and acceptance criteria Unit verification
5.6	**Integration and Integration Testing**	Not required	Integration testing	
5.7	**System Testing**	Not required	System testing	
5.8	**Release**	Documented version of the released software	List of remaining software anomalies, annotated with an explanation of the impact on safety or effectiveness, including operator usage and human factors	

modules. The availability of the source code allows the performing of activities that are compulsory for the higher safety classes, as shown in Table 6.2. In addition, it should be noted that errors occurrence is inversely proportional to the effort already spent in analyzing and testing the code, and open source software is generally jointly developed by a community, which implies a higher number of developers and therefore a greater control on the software itself. Open source software is therefore a promising approach for reducing development time and cost while increasing safety. Such benefits are counterbalanced by industrial strategies on intellectual property right and in some cases by regulatory frameworks (Economist, 2012). For example, as discussed in Chapter 1, open software under specific licenses (e.g., GPL), can hardly be used in commercial applications since 'all works derivative of code under the GPL must themselves be under the GPL'.

6.3 Risk Management Process

A key improvement of IEC/EN 62304 over earlier standards and policies is the integration of risk management into the development process. In the past, such processes were managed by non-integrated standards. Software cannot ensure deterministically the absence of errors that, in turn, may guarantee the safety in all conditions. Safety can be ensured only on statistic basis designing the product so that the risk R given by joint effect (e.g., the product) of the severity S of harm and the probability of occurrence P is acceptable:

$$R = S \times P < \textit{safety threshold}. \tag{6.1}$$

More simply, referring to Figure 6.9, harms with high severity must have reduced probability of occurrence, while only low severity harms are allowed to have high probability of occurrence. If these conditions do not hold, an unacceptable risk is in place and control measures have to be implemented. These control measures either reduce the probability of occurrence either reduce the severity of harm so that the newer risk level R' becomes acceptable with respect to the benefit of the medical product:

$$R' = R \times E' = S' \times P' < \textit{safety threshold} \tag{6.2}$$

In the equation (6.2), E' is the 'effectiveness level' of the control measure that represents the quantifiable value of the risk reduction when the control measure is in place. These concepts justify why software risk management must be an integral part of the software development process: software safety is ensured by the criterion that **software does not imply foreseeable unacceptable risks**.

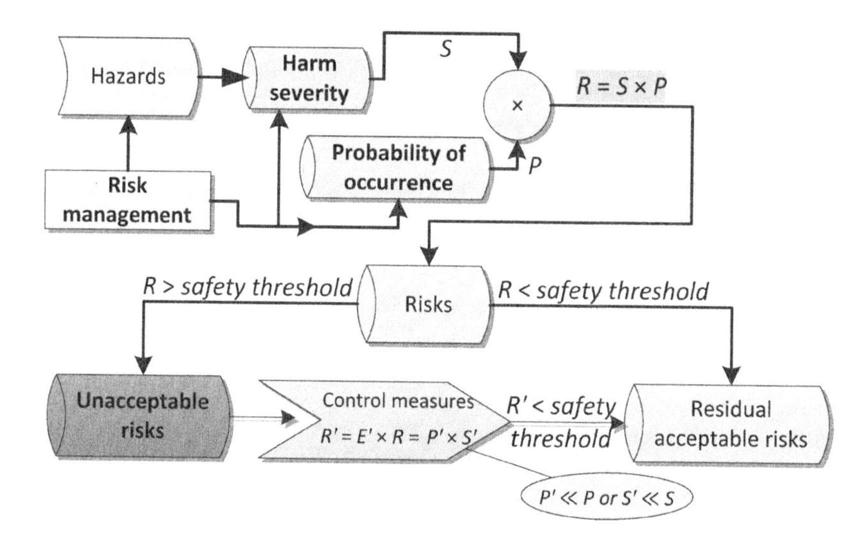

Figure 6.9 Risk model.

The IEC/EN 62304 standard integrates the risk management standard (ISO/EN 14971). In addition, since ISO/EN 14971 does not focus explicitly on software development, IEC/EN 62304 introduces additional requirements focused on medical software:

1. a development process compliant with ISO 14971
2. a documented software risk management plan
3. a hazard analysis (identification of hazardous situations, probability of occurrence, harm severity level and identification of risk control measures as depicted in Figure 6.9
4. a documented implementation of applicable software control measures.

Figure 6.10 shows the four activities of the risk management process. **Risk identification** reveals hazardous situations that the following step – **risk analysis** – quantifies in terms of harm severity and probability of occurrence. We assume here that the risk level is obtained as the product of severity and probability of occurrence $R = S \times P$.

In the **risk evaluation** stage, risks are assessed on a quantitative and qualitative basis. If risks are considered as acceptable with respect to the benefits expected from the use of the medical device, no additional activity is performed. For unacceptable risks, an additional step is required – **risk control** – that introduces a control measure that reduces either the probability of occurrence or the severity of the harm. This outcome is quantitatively accounted for by the effectiveness level E'. If the residual risk, $R' = E' \times R$, is evaluated as not acceptable, additional control measures have to be implemented.

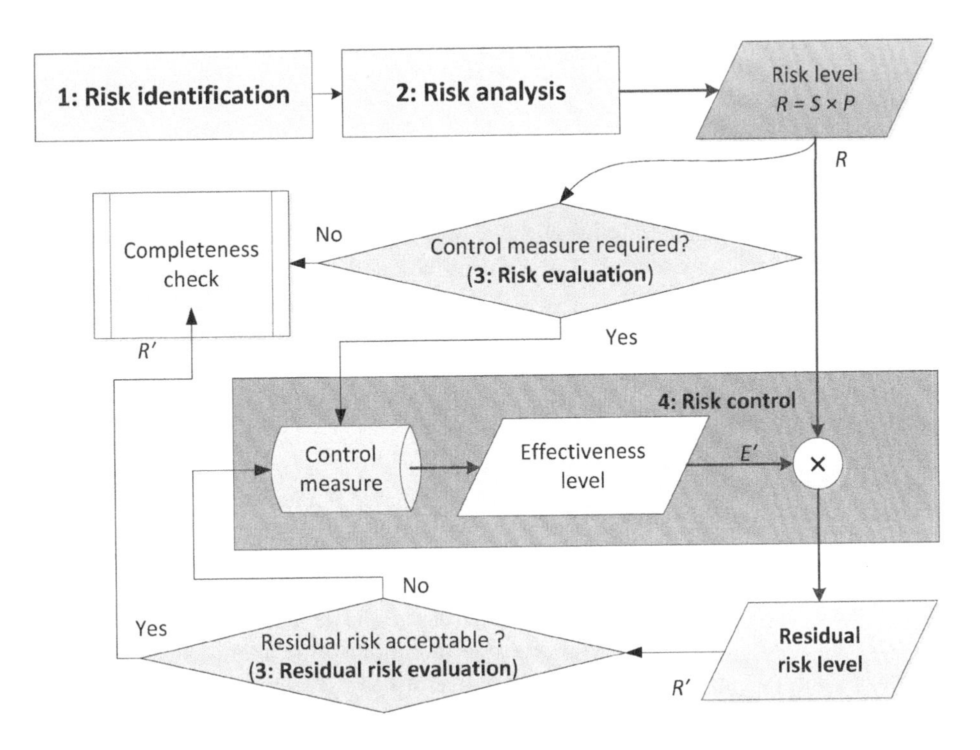

Figure 6.10 Risk management process.

6.3.1 Risk Management in Software

The previous discussion has outlined the main concepts of risk management. In the following, we will only be dealing with risk management for software, but a deeper discussion on risks is included in Chapter 8.

The **risk identification** stage of the risk management process determines the hazardous situations defining the scenarios and the sequence of events 'triggering' the hazard. For each hazard, software items that contribute to the hazardous situation have to be identified.

We have also to focus on **SOUP** since it may be the source of hazardous situations. This analysis may take into account possible published SOUP anomaly lists. Such information has to be assessed carefully on the assumption that SOUP and associated documentation can not normally be trusted.

The final product of the risk identification will be the complete and formal documentation of the potential causes of hazards from software items contributing to a hazardous situation and the sequences of events that could result in hazardous situations. **Risk analysis** will introduce the harm severity and the probability of occurrence. These two activities imply that the software safety class has been assigned to software items, based on the hazard analysis.

For each unacceptable risk generated by a software item and documented in the risk management file, the manufacturer must define and document **risk control measures** to eliminate risk or at least mitigate its level. From a practical point of view:

- the risk control measure shall be inserted in the software requirements
- the software safety class shall be re-evaluated, taking into account the control measure implementation
- the risk management documentation shall formally record the risk control measure, identifying the associated hazards.

After defining and documenting the risk control measures, the manufacturer must verify the effectiveness of such measures and document the verification procedures. In particular, the risk control measures must be assessed to verify possible new hazardous events. An example of a software control measure may be associated with the temperature of the device. More specifically, a device may have temperature sensors whose value is controlled by software. When specific thresholds are exceeded, software may display alarms and reduce the power to circuits that are overheating. The action performed by the software is critical considering the risk when this action is not performed. A formal test is then required to verify the correct behavior. The test may include a procedure where the device is heated in a climatic room.

Risk management is performed in parallel with the other processes, such as the software changes. Changes will have to be analyzed not only on the usual cost and performance axes, but also the risk impact has to be managed according to the defined process. Changes have to be analyzed from a safety point of view, considering also the additional impact of the implemented risk control measures. Following this analysis, an updated hazard list has to be available; such a list will be the input of the risk management process as depicted in Figures 6.9 and 6.10.

6.3.2 *Risk Management for Medical Instrument Software*

The previous section has focused on the main concepts related to risk management for software. This section will address risks from an **industrial perspective**. In the following, we first recall some important points.

Software is pervasive in the healthcare industry. Software is going to perform more and more functions of a medical device due to its benefits with respect to hardware implementation (e.g., minimum cost of production and higher flexibility).

Software is not a risk in itself. Software generates risks if its failure can lead to hazardous situations at system level through a defined sequence of events. This suggests that software risk management has to be performed in the context of the medical device 'system' where risks affecting safety can be identified and analyzed. (IEC/TR 80002-1).

Among the drawbacks of soft implementation is that **software has a more complex error detection**. Non-trivial software applications have an increased number of states given by the values that may be stored in its internal variables and the possible branching that the software program may execute. This adds an additional complexity to risk management, since risk has to be assessed with regard to the sequence of events that may generate hazardous situations. The analysis of the **complete sequence of events** may be practically unmanageable due to the number of events and states. The higher complexity also affects the estimation of the probability of occurrence of a software failure. Again, considering the number of states, the estimation of this probability is a challenging task. This suggests focusing with great care on software in risk identification and risk analysis. Some techniques may help.

Software risk analysis may be performed with a top-down and/or bottom-up strategy. Specific techniques such as preliminary hazard analysis (PHA), fault tree analysis (FTA), failure mode and effects analysis (FMEA), hazard and operability study (HAZOP) and hazard analysis and critical control point (HACCP) may help in the various stages of development (see IEC/EN 14971). In general, the manufacturer performs risk analysis at system level and then assesses if a system risk may occur because of a software failure (top-down approach). For this, the FTA techniques may help. Briefly, in **FTA** (IEC 61025 standard) from an undesired top event, the possible consequences are identified, going down to the subsystems until reaching component unit level.

The bottom-up approach is also useful for identifying system risks due to the software. In this case, the designer must consider the consequences at system level in the case of subsystem failure. We then consider a software subsystem and the consequences in terms of harm that it may have on a medical device in normal and fault conditions. This approach recalls the failure modes and effect analysis (**FMEA**) technique (IEC 60812 standard) where possible failures at subsystem level are identified assessing

1. the probability of occurrence P
2. the severity of the effect at system level in terms of harm or injuries S
3. the probability of detecting the failure D.

The risk priority number (RPN) is given by the product of the previous three terms: $RPN = P \times S \times D$. This accounts for the priority that a risk has to be addressed. The probability of detecting a failure D is a critical number in software where failure detection may be particularly cumbersome. Chapter 8 addresses FMEA-like techniques more deeply.

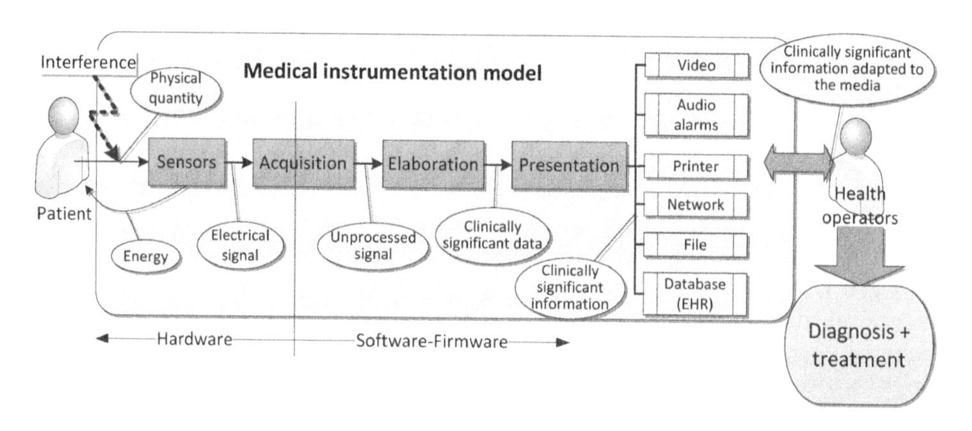

Figure 6.11 Medical instrument model for risk management.

To identify the main risks of **medical instruments** we rely on the simple functional model shown in Figure 6.11. The patient is connected to a medical instrument through sensors that can also supply energy to the patient itself to perform measurements. This happens in imaging medical instruments, but it is also common in biopotential instruments that energy is supply to reduce interference, as addressed in Chapter 5 (i.e., right leg drive stage). Sensors measure physical quantities that are translated into electrical signals. These signals are recorded by an acquisition stage that outputs an unprocessed signal to an elaboration stage. This stage transforms the input signals into clinically relevant data that are adapted by the later presentation stage for the media that is available for operator usage. The healthcare operator may view this data through a printer or on video with audio alarms. The presentation stage must adapt clinically relevant data to such media. This stage also has to adapt the data when they are stored in files, through the network and via shared databases such as the Electronic Health Records addressed in Chapter 7.

Regardless of the media on which the data are presented, the operator will perform a diagnosis based on this data that will be used for treatment. Software and firmware may have a functional role from the acquisition stage. We can now define some **main general risks** for medical instruments software based on the model of Figure 6.11. Such risks, adapted from (IEC/TR 80002, 2009), are outlined in Table 6.3.

Referring to Table 6.3, first the medical instrument may be unavailable for diagnosis due to internal software problems. Unrecoverable errors may be possible reasons. Such errors may be generated by various causes, such as wrong memory accesses, numeric errors, abnormal terminations and hang-ups. Regardless of the cause, the severity of the risk has to be assessed with regard to the intended use. Instruments may monitor vital physiological parameters, where variations could result in immediate danger (e.g., some European Class IIb devices). In this case, the severity of the risk is high, since unavailability may result in injuries. In contrast, instruments used for routine diagnostics may not consider **unavailability** as a severe risk.

Starting from left-hand side of Figure 6.11, software may control the amount of energy, such as radiation, supplied by the hardware to the patient. Errors may have a

Table 6.3 Medical instrument software: general risks

ID	Software function area	Risk	Hazard
1	**Availability**	Diagnostic service unavailable	Delayed treatment
2	**Hardware wrong control**	Wrong energy applied	Injuries, wrong clinical data, wrong diagnosis
3	**Clinical data display**	Wrong display (artifacts, distortions, lossy compressions etc.)	Wrong diagnosis
4	**Alarms**	Not perceived	Delayed treatment
5	**User Interface**	Error prone	Wrong diagnosis
6	**Patient data**	Wrong access, data unavailability, corruption	Misplaced or wrong diagnosis,
7	**Security**	Security breach	Sensitive data violation, modification, injuries

dramatic result as happened in the Therac-25, when coding errors in a radiation therapy device resulted in overdosing patients with X-rays (Leveson, 1993). Going on with Table 6.3 to risk ID 3, clinical data may be **wrongly processed** and displayed by the data processing chain: acquisition, elaboration or presentation in Figure 6.11. Modifications of clinically relevant data may induce wrong diagnosis and associated treatments with the harm severity dependent on the context. **Alarms** (risk ID 4) usually managed by software, are a critical issue, especially in intensive care units where ignored alarms may result in delayed treatments. To reduce such risks, the IEC 60601-1-8 standard proposes some guidelines. The **user interface** may increase risks of unwanted operations with possibly serious injuries or death in medical devices. This has been addressed in Chapter 3. Medical instruments have usually network capability to share their data. These functions are mainly performed by software. Failures in managing this data may have safety-critical consequences.The last two risks of Table 6.3 are instruments that share data with networked healthcare databases, usually referred to as electronic health records (EHR). Many safety and security events may generate harms. Increasingly, healthcare practitioners rely on shared databases for their diagnosis and treatments. Corruption of these databases may have dramatic consequences.

Table 6.3 and Figure 6.11 may be used as a framework for analyzing risks and evaluating control measures for medical instruments. We will use this approach to determine a design.

6.4 Software Development Process

The software development process includes the usual activities that, in Figure 6.12, are depicted in a waterfall life cycle model.

It is worth noting that the IEC/EN 62304 standard does not recommend specific life cycles outlined in Chapter 1. The manufacturer has the freedom to select the more appropriate life cycle, documenting and implementing the process related to the activities depicted in

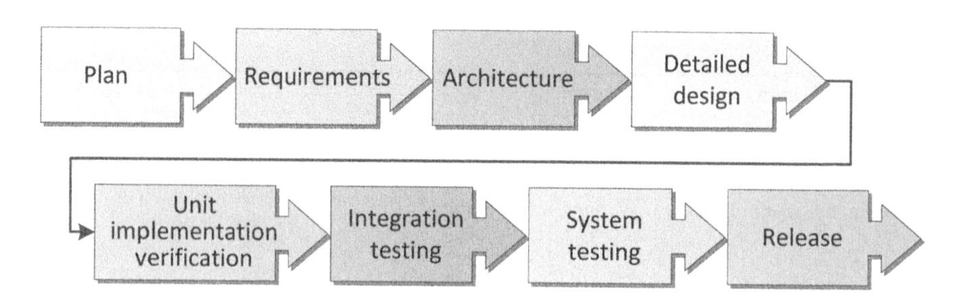

Figure 6.12 Software development process.

Figure 6.12 with the requirements outlined in Table 6.2 and in the related sections of the standard. Below, we will detail the development activities.

6.4.1 Software Development Planning

Software development planning is the activity where the whole software development process is analyzed to create a document (or a set of documents) that describes the events to be performed during the software life cycle. Chapter 1 has already introduced project management and the **development plan** that has to be available at the start of the project and it has to be monitored and updated during the whole development process. The plan should include at least the so-called 'who does what, when, how and why'. More formally, the plan should set the goals (the why) of the project and define tasks (the what) required to achieve the goal. Tasks are generally defined in a hierarchical form under a work breakdown structure (WBS). The project's deliveries and the associated milestones with delivery time (when) are also to be specified. Finally, the plan should include the procedures (how) and the people (who) involved in the project.

There are many template plans that focus on specific areas. Chapter 1 focuses on the MIL-STD-498 standard (MIL, 1994) which still includes valuable templates for product development. In the same chapter, the development plan of the Gamma Cardio CG product is disclosed and discussed in detail.

The planning process is particularly important since it coordinates software development with the quality management system; it makes risks manageable and, in general, it differentiates a consistent development at industrial quality level from the so-called 'code and fix' approach where minimal design is performed at the cost of reduced productivity and increased risk level.

The manufactures have to produce plans for managing the activities required by the standard and by other requirements that may be present in other applicable sources such as contracts, standards and regulatory frameworks. The plan has to be consistent from system-level to software level activities and must include all applicable items. The overall plan may include other plans related to specific activities such as integration, verification, risk management, quality assurance and configuration management.

The systematic planning and recording of engineering project information is an intrinsic part of the software development process and, in general, of any non-trivial project. This process should be performed, regardless of whether a deliverable is formally required. In addition, if this process is performed through specific procedures and output templates that

Table 6.4 Project management checklist ☑

1.	do the main project development plan (who does what, when, how and why)
2.	check if a project plan template exists for the specific type of project (e.g., software development plan) to be sure to address all the issues/potential problems of project management
3.	address at the beginning what can go wrong (risks) with risk mitigation and contingency actions; allocate extra budget to address consequences
4.	allocate realistic consistent effort of time and resources
5.	monitor and control the project regularly in terms of cost, time and performance, updating the plan
6.	remember human factors (choose proper resources and monitor motivation).

are derived from company recommendations, the procedures/templates should be maintained and possibly integrated, since consolidated processes are more effective and less risky than new processes to be implemented. Planning is a critical activity in every project that involves non-trivial activities. Table 6.4, derived from the analogous checklist in Chapter 1, summarizes the main planning tasks.

6.4.2 Software Requirements Analysis

Insufficient requirements are a first reason for failure in medical device design (Kelleher, 2003), in software design (Standish Group, 1995) and probably in any complex human venture. Errors in requirements produce a cost for fixing that is hundreds of time higher when discovered at later development stages. As suggested by common sense, to reduce consequences, errors, especially in requirements, have to be fixed as early as possible. This has been also discussed in Chapter 1.

Software requirements have to define ***what*** *the software has to do rather than* ***how*** *the software will perform specific functions.* This is a main error of software analysts who can tend to define how the programs have to work. Such errors reduce the degree of freedom of the developers who inherit additional unnecessary constraints that may in turn increase the required effort. Requirements are extensively discussed in Chapter 2 from theory to the implementation part, where the requirements of the Gamma Cardio CG product are disclosed.

The **Software requirements analysis** is the process of formalizing needs implemented in software components, from a technology independent viewpoint, creating analytical models and documenting the results of the analysis through the list of requirements (IEEE, 1993). During software requirements analysis, analysts define and record the software requirements. Figure 6.13 from Chapter 1 shows typical classes of requirements for medical devices. This requirements may be generated from various sources and may generate different type of requirements.

Software requirements are documented in the software requirements specification (SRS) that includes the requirements and the methods to be used to ensure that each requirement has been met. Software requirement documentation should be validated through internal and

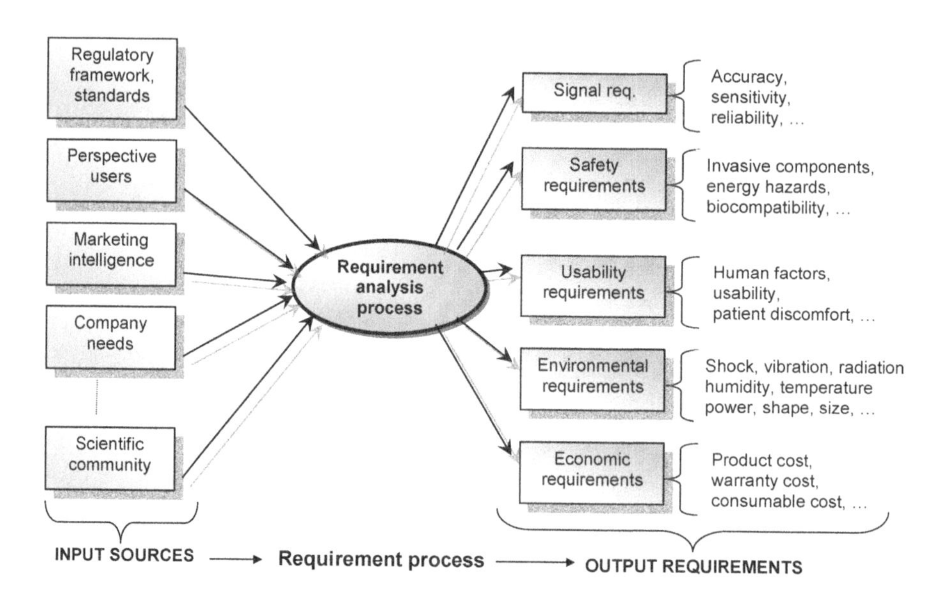

Figure 6.13 Requirement process model.

external review techniques such as inspections held by system engineers, software engineers, software analysts and software testers.

Quality assurance should perform audits to verify execution of the activities within the software requirements process, as well as compliance to document standards on a requirement-by-requirement basis.

6.4.3 Software Architectural Design

Software architectural design is conceptually equivalent to system architectural design which is the process of decomposing a **system** into subsystems, defining interfaces between the subsystems. When dealing with decomposition subsystems may also be referred to with the similar terms, such as modules, components, packages, configuration items, boards or units according to the context. This process was explained in Chapter 1, where system–subsystem decomposition is discussed in detail from a theory point of view, and showing the implementation of the Gamma Cardio CG product.

The decomposition process is used because any non-trivial product design has a high level of **complexity** given by the set of the design details that have to be faced and solved coherently. Unfortunately, the human mind is only able to face a very limited number of problems at once and with restricted details. This activity can therefore be handled through the use of the '**divide and conquer**' paradigm, that mainly recommends to divide enemy forces into small components until the component size is small enough to be handled by the existing capabilities. Subsystem components should be designed independently as possible so that, when fighting against one component, the other components do not 'arrive and change the balance of power'. The strategy then consists of fighting one component at a time hoping, that the other components do not break into the battle.

This strategy is the core of system/software architectural design. Project design forces have limited capacity to simultaneously solve all the design details (the enemy). The strategy then consists of **breaking the system down** into smaller subsystems until subsystem sizes can be faced within the project designers' capabilities. Architectural design is generally limited to the first top level hierarchies and the associated interfaces.

More formally, adapting IEEE 14764, system/software architectural design is the process of analyzing design alternatives and defining architecture, components, interfaces for a computer software configuration item (**CSCI**) of a system or a segment of a system. Software architectural design only consists of the initial or first level decomposition that identifies a CSCI. Software architectural design is composed of a hierarchy of top-level software components, including their identification, type, purpose and function. Software architectural designs also identify the subordinates (i.e. the software units in a component), dependencies (relationships with other components) and interfaces. Finally, software architectural design identifies the resources (external elements used), processing (rules used by the component to achieve its function) and data elements internal to the component itself. Figure 6.8 shows an example of a very top level system functional architecture design applicable to medical instruments.

The system (i.e., the medical instrument) is decomposed into four subsystems with three internal interfaces and two interfaces with the external actors (i.e., the doctor and the patient). The architecture may be customized into the architecture of the Gamma Cardio CG as shown in Figure 6.7. The conceptual Gamma Cardio CG architecture has 11 interfaces, two actors and the items that are part of the product. The items to be developed are the firmware, the hardware and the software application while sensors and the patient cable may be considered as commercial-off-the-shelf (COTS). The related architecture is depicted in a breakdown hierarchical tree in Figure 6.8.

A core criterion of system architecture design for the selection of the subsystems is the **independence of the subsystems** themselves. This criterion referred to as 'maximum cohesion and minimum coupling' (see Chapter 1) recommends selecting subsystems with maximum independence with respect to the others, thus minimizing the contents of the interface. The complete independence is obviously obtained by the absence of the interface. Various benefits stem from this paradigm. Maximally independent subsystems may be designed separately, thus reducing the complexity of the design. The approach improves the reuse of software. A real benefit is also in the development and testing of the software itself, since reducing the interfaces minimizes the side-effects that one subsystem induces on others. This dramatically reduces testing that in architecture with highly coupled subsystems may become particularly onerous. This is due to the fact that errors whose effects are discovered in one subsystem may be caused by any code line spread in all the other connected subsystems. The independence of subsystems also has a deep impact on safety and risks. If one subsystem is completely independent of the others, then its risk analysis may be handled separately and then simplified. Independence is also related to the **segregation** principle in the IEC/EN 61304 standard which offers extreme example of segregation: software processes running on different processors with no resources in common. By contrast, in coupled systems, risks and safety of one subsystem may have impact on the other coupled subsystems. Therefore, *risks have to be analyzed considering the joint effect of all the subsystems that are coupled with the subsystem under analysis.*

Architectural design choices are then guided by reducing complexity and defining self-contained modules with minimal interfaces with the external subsystems. This has a positive impact on the overall safety, since risk analysis can be mainly focused at the level of smaller and simpler subsystems.

Software architectural designs are documented in a software architectural design description (SADD), for each CSCI. The SADD describes

- CSCI-wide design decisions
- the CSCI architectural design with CSCI identification
- the system decomposition
- the interface characteristics of one or more hardware configuration items (HWCI), CSCIs, manual operations or software components
- segregation and other choices to maximize risk control
- functional and performance requirements of external software (SOUP)
- hardware and software required by external software (SOUP).

In addition, the software architectural design for a CSCI must show the link between architecture and software requirements contained in the SRS and must explain the requirements implementation.

In practical architectural design, experience suggests that, in the first stages, graphical design tools may actually reduce productivity, since designers take their attention off the main task (defining systems, subsystems and their relations) and focus instead on tool management, graphical aspects and other distracting details. Design and technology details are also a distracting factor at the top level which diverts attention from the main problem. Details have to be considered at later steps when lower hierarchical design layers are available and coherence has to be ensured between layers. The architecture may then be upgraded so that coherence is jointly ensured at all levels. Table 6.5, adapted from Chapter 1, summarizes the main steps of architectural design (SADD).

Table 6.5 Architectural design checklist ☑

1. to reduce complexity and risks, apply system–subsystem decomposition process as follows:
    a. at first level define actors, interfaces and the system with limited detail
    b. create a second level by decomposing system into subsystems and interfaces
    c. define roles and responsibilities of subsystems
    d. define subsystems, minimizing coupling between subsystems and maximizing cohesion within each subsystem
    e. analyze risks and introduce required control measures
    f. iterate the process, defining the lower layers and introducing improved options in terms of risks and complexity
2. at the end of the process, there should be a hierarchical structure of subsystems and interfaces with an acceptable risk set

6.4.4 *Detailed Software Design*

The detailed software design activity is the process of defining implementable software units and interfaces based on the software architecture. Taking into consideration the more general system–subsystem decomposition process focused on in Chapter 1, the detailed software design activity is the process that logically follows the definition of the architecture. The decomposition divides the system into successive layers until the lower layers are sufficiently large to be tested separately and with sufficient design detail to allow subsequent coding into a programing language. The detailed design should include all the critical decisions, especially those affecting safety, so that the risk analysis is complete at this level and programers do not add additional elements that may affect risks.

At the end of this process, the software units and their interfaces will be detailed and the overall system–subsystem decomposition from the overall product to the software units will not contain any contradictions.

Detailed design is critical for later code implementation, since defects at this level will produce code errors that will require additional time for fixing in order to eliminate errors that were made at the design stages. This suggests performing careful design reviews to detect errors at this stage. Detailed unit design usually includes (MIL 498, 1994)

1. identification, type, purpose and function (a statement of what the unit does)
2. subordinate units, dependencies and interfaces for each unit
3. resources, processing and data
4. concept of execution, showing the dynamic relationship of the software units
5. interface design
6. traceability of specifications versus CSCI requirements.

There are many **tools** and techniques used to describe the details of software units. Program design languages can be used to describe inputs, outputs, local data and algorithms. Other common techniques for describing logic include metacode or structured English or graphical approaches such as structured, real-time, relational, object oriented or formal notations. These approaches use entity relationship diagrams, data flow diagrams, state transition diagrams and other diagrams to describe the software functions and data. Tools are selected according to the software and the development context which includes maintenance and compatibility issues with the customer and the development company environment. Detailed software designs are documented in a software design description (SDD). In addition, the detailed software design must describe specifically how it achieves its software architectural design-level requirements, as stated in the software architectural design description.

6.4.5 *Software Unit Implementation and Verification*

Software unit implementation and verification is the process of translating design specifications into instructions expressed in a programing language. The process includes the execution of unit test cases that have to verify the correct implementation of the software-level detailed design.

An individual unit test may target an individual software procedure: an object oriented programing language class, a file containing an ad hoc collection of source code statements, or any grouping of procedures, classes or files that collectively meet the requirements of the SDD. Examples of software that may be the target of an individual unit test include a C, Java or Visual Basic subroutine, a C++ class, a hyper-text markup language (HTML) file or a combination of different programing language instructions that are associated with a single detailed software design unit in the SDD.

In medical device software, the **unit-testing** process is a major step to ensuring safety and effectiveness since these features have to be validated and verified even at the lower level of software unit components. The documentation and the implemented code is usually tested by internal review and external review techniques. Unit testing is usually performed by software programers who use sample data to verify the unit correctness and compare the actual and expected unit behavior. **Code inspection** is also effective in reducing defects and may be performed based on coding and testing standards or guidelines.

Unit test cases validate the correctness of the addressed requirements, with respect to prerequisite conditions, test inputs, expected test results, assumptions and constraints.

Quality assurance resources have to perform audits to verify the execution of activities related to software unit implementation and verification, as well as compliance to document standards on a requirement-by-requirement basis.

More specifically, quality assurance has to verify the execution of software implementation, preparing for unit testing, performing unit testing, revision and retesting, and analyzing and recording unit test result activities, as well as the proper execution of reviews. The acceptance of software units has to be performed according to the criteria defined by manufacturer procedures. Such criteria may include

1. compliance to requirements
2. compliance to programing standards and procedures
3. correct implementation of critical programing issues: memory management, resource allocation, fault management, variable initialization, self diagnostics, defensive contract checks, data flow and event sequence
4. positive results of unit testing.

Acceptance test reports must include at least

- test results
- tested version
- hardware and software configuration
- test procedures and tools
- identification of the name of the tester and date of test
- list of defects.

If defects are found, and changes are implemented on the code, the manufacturer must

1. test the changes and repeat the relevant tests
2. perform regression testing to ensure that changes have not created side effects
3. activate the risk management process.

6.4.6 *Software Integration and Integration Testing*

Software integration and integration testing is the process of combining and jointly executing all the software units that are part of a CSCI. The activity includes the testing of the overall CSCI with the analysis and evaluation of the results, comparing actual and expected behavior.

Experience recommends performing such a final step only after a deep unit testing, since detecting errors in a wider software aggregate usually requires more effort then **single unit integration**. In principle, unit testing should detect all the defects inherent in a specific unit, leaving to integration and testing stage only the errors that are generated by the joint execution of all the components that are part of the CSCI. This process is usually performed incrementally, aggregating one unit after another and testing the aggregate until all the CSCIs are integrated.

Software integration and testing is a systematic technique for building the program structure while performing tests to discover errors associated with the integration. The process generally includes the following steps:

- define a set of tests that must be executed (possibly automatically) whenever CSCI is changed (**regression test set**)
- incrementally integrate the software units
- incrementally test the integrated software
- analyze results and correct any defects
- execute the regression test on the integrated software.

Testing should focus on the behavior of the CSCI and the data exchanged at the internal and external interfaces of the CSCI. At this aim, the so-called black box and white box testing are employed. In **black box testing**, the module (e.g., system, CSCI, software aggregate or unit) is tested, analyzing the output for specific module inputs. The correct input–output response is derived from the sole knowledge of the requirements associated with the module. No knowledge of the internal implementation is considered in deriving the correct input–output sequences. By contrast, **in white box testing**, the testing data input is derived based on the knowledge of the internal structure of the module. Table 6.6 outlines the main goals of test procedures.

Table 6.6 Software integration testing

Test goals	Description
CSCI build-up	Incremental tests performed during integration. Modules are gradually integrated into the application at each integration. At each integration, the application is tested to evidence errors generated by the modules integrated at each stage.
Interface	Interfaces are more likely to be tested completely and a systematic test approach may be applied. Data transfer and protocol exchange is tested across interfaces.
Effectiveness	Tests on software that shall behave as intended by the manufacturer
Safety	tests on software that shall not generate that are unacceptable with respect to the expected benefits

Test effort should be proportional to the level of risk of the device and to the risks that the CSCI might generate at system level. Safety-critical CSCIs and functions obviously have to be tested by specific detailed test cases that should be introduced in the regression test set. External review techniques should include a test readiness review (TRR), assessing the unit integration and testing activities. This review should be conducted for each CSCI to determine whether the software test procedures are complete and to ensure that the manufacturer is prepared for formal CSCI testing. Since errors discovered after software delivery requires much more cost for fixing, investments in test activities is generally well repaid. On the other side, since extensive testing may cause delays in program scheduling, testing is often reduced.

6.4.7 Software System Testing

In software system testing all the software system is tested as a whole with the aim of verifying if the system may be operationally used, that is, if it behaves as described by the manufacturer in the intended use and is safe for all intended users. This process must then

1. evaluate completeness and effectiveness of test procedures that ensure that
   a. the software behaves correctly
   b. the software behaves as specified by all the associated requirements related to safety and effectiveness
2. perform all test procedures in a simulated or actual environment
3. evaluate the test results
4. proceed with error correction and reworking and then re-execute the test session until software does not show any defects.

Defects may also remain in the final release, but, in this case, a proper rationale has to be documented. Residual risk also has to be assessed in terms of acceptability. Test procedures have to be traced to requirements, and all the requirements must have at least one test to validate that requirement's content.

The system qualification test must demonstrate that the product is ready to be released onto the market. This qualification is performed under the manufacturer's responsibility.

6.4.8 Software Release

The software release activity is the process of preparing the whole software package for the final deployable release. The software package includes executable software, version descriptions, user manuals, installation procedures and so on. The software release activity must ensure that

- the released version is identified and documented with all test reports and possible remaining defects
- the manufacturer's quality system has been applied throughout the software life cycle
- the software is reproducible, that is
  - the software environment for code design and implementation and reproducibility is fully archived for future use
  - the procedures for software release are documented
  - the software itself (source, executable, libraries, installation packs etc.) is archived

- the configuration history is documented
- the overall release package is archived properly to ensure that it can be retrieved for a time at least equal to the device lifetime (i.e., a minimum 10 years after the final device is sold).

During the software release activity, the manufacturer prepares the executable software, including any other software files needed to install and operate on the target, and records the version of the available software. Every element that is included in the release must be identified in terms of part and version number. The manufacturer also identifies and records installation procedures needed for end users or IT personnel. Finally, the manufacturer installs and checks out the executable software, provides training to users and provides other assistance as specified in the contract.

Software release activity documentation is validated through internal and external reviews. Quality assurance has to perform audits to verify the correct execution of activities for the software release, as well as conformance to document standards on a requirement-by-requirement basis. Summarizing, during the software release activity the manufacturer must ensure the following:

- software validation and verification completed
- residual defects documented with rationale
- all components of the released version identified and archived in long-term storage (components include the development environment, development and installation procedures etc.)
- design plan activities and tasks completed
- software release repeatability verified.

6.5 Software Configuration Management Process

Compared to hardware design, software is easy to modify and release. Software also consists of a huge number of elements (i.e., code lines) that may be subject to change with possible side effects that may be detected in code areas far away from the modifications. **Uncontrolled and unplanned modifications** can result in software code that can be hardly managed due to the increasing number of discovered defects. Software configuration helps to limit this nightmare condition with an overall significant reduction in the effort of managing the code and the associated files.

Configuration management focuses on establishing and **maintaining consistency between all deliveries of the system** (documents, software, hardware schemes, requirements, specifications and operational information) throughout the product life. This process cannot be limited to software; it must be an integrated activity that embraces all the components that are required to manufacture the medical device. This also includes the development environment needed to reproduce and manufacture the product.

Configuration management (CM) is a valuable core process for any manufacturing industry to ensure reproducibility of the marketed products, since it represents the **'backbone' of the enterprise structure**. It instills discipline and keeps the product attributes and documentation consistent. 'CM enables all stakeholders in the technical effort, at any given time in the life of a product, to use identical data for development activities and decision making'.

Inadequate configuration management may produce confusion, inaccuracies, low productivity and unmanageable configuration data (Nasa, 2007).

Safety-critical software may also generate unacceptable risks when it is not properly configured.

In the medical device sector, configuration management is critical to **ensure safety**. Medical device design history and documents have to be available in case of incidents to reduce incidental negative effects and to avoid additional risks related to the other product samples placed on the market. During the product life cycle, configuration management is necessary to ensure that changes in the software/system do not jeopardize the risk analysis by introducing critical functionalities not dealt with in accordance with the risk management process. In medical electronic equipment including software, inadequate configuration management is considered a source of risk on its own (IEC/EN 60601-1). For this reason, manufacturers need a process that manages the evolution of a system/software product, retrieves the history of the changes, recovers earlier versions and controls the team collaboration efforts to avoid side effects. Configuration also has to manage the different product items released in different product versions for different customers.

Figure 6.14 shows the main steps of a **configuration management process**.

The first step (**planning**) is the definition of an approach with procedures and a tool that will practically manage the workflow and store all items required to reproduce the product in all its historical versions. The tool includes a database usually referred to as the 'revision control system'. The configuration management plan describes the items to be configured, the activities to be performed along with their scheduling, the associated procedures and the roles and responsibilities within the process. In **identification**, the manufacturer defines rules for the unique identification of a configuration item in its specific version. Configuration items also include documentation needed by the process and third-party (e.g. SOAP) components.

Change control is the process of monitoring, documenting, authorizing and executing changes to a software item. This activity can have different timescales at different stages within the project. In the development phase, the frequent changes do not need specific management, while after release on the market, changes have to be evaluated carefully and formal processes of assessment and approval have to be performed to fulfil the required modifications. In particular, when the product has been released, changes have to be performed through the problem report process or the change request process.

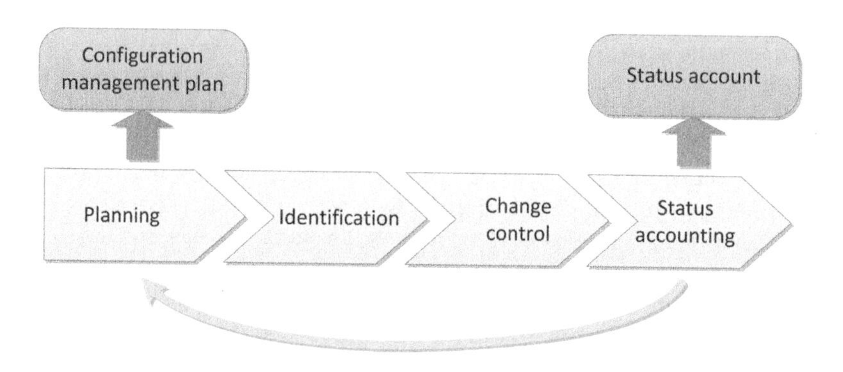

Figure 6.14 Configuration management process.

Different levels of configuration management can exist for different process phases. The manufacturer will have more formal configuration management for the released versions, and configuration working areas for products under development with reduced procedural management effort.

Configuration status accounting is the process of documenting and reporting the data necessary to manage a configuration item. This includes the set of the current and all the past versions, with the history of the changes related to the versions and requirements.

The configuration management process has to be verified regularly by inspecting items and testing the actual capability of the configuration tool and the stored data to reproduce specific versions, to document changes and to show an item's history.

6.6 Software Problem Resolution Process

The software problem resolution process aims to assess and fix defects following a documented approach that has impact on the configuration control, on the risk management and possibly on other procedures included in the applicable regulatory framework. This activity has to follow a formalized and structured process. Possible impacts also have to be assessed with respect to safety and effectiveness of the medical product. Figure 6.15 shows the main tasks of the process.

Problem reports are first **analyzed** and classified, considering the type of error, scope and level of criticality. This allows for a correct planning of the investigation and the resolution in terms of human resources (the 'right' problem to the 'right' people), resources (i.e., broader scope problems will imply more resources for investigation and resolution) and time (i.e., criticality will govern the urgency in term of calendar and committed resources).

It is then required to maintain **records** of problem reports life cycle, updating the history whenever important changes occur. The manufacturer must **investigate** the problems and identify the causes and their impact on safety. The manufacturer will then record the results of the investigation and the evaluation generating a change request or a rationale for taking no action.

Communication is critical in managing problems, especially considering the safety impacts of medical devices. In more serious cases, the manufacturer must **warn applicable stakeholders** as requested by the regulatory framework.

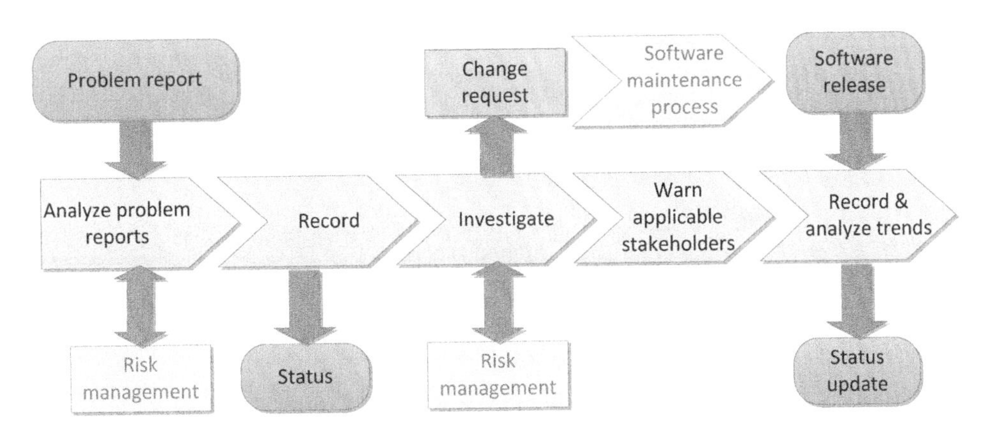

Figure 6.15 Problem resolution process.

Records of problem reports have to be stored and analyzed to discover potentially critical trends. In this phase, a periodical trend report should be performed to identify the general behavior of the software development and to spot specific problem patterns that have to be solved at a general level.

6.7 Software Maintenance Process

Software may experience several upgrades throughout its life cycle for a number of reasons: problem fixing, introduction of improvements, compliance with changed regulations, adaptation to new customer needs, upgrading due to new releases of libraries, operating systems or underlying technologies. This aspect becomes particularly important in safety-critical medical software that is characterized by a long life period after the first commercialization.

The software maintenance process includes all the activities related to the software upgrades after the first software release. Successive upgrades have to be managed with a consistent and formalized approach to avoid impact on safety and to ensure effectiveness. However, a smaller maintenance process within the development process is preferred for updating software promptly on critical problems. Figure 6.16 shows the typical process of software maintenance.

The maintenance activities have to be performed according to a software maintenance **plan** that should address what, who, when and how maintenance issues will be managed. The IEC/EN 62304 standard recommends that the plan should include

- procedures for managing feedback after the release of the software
- criteria for classifying feedback into problem reports
- management of the other processes (risk management, problem resolution, configuration management)
- management of upgrades, patches, bug fixes, obsolescence and upgrades of SOUP.

Depending on the impact of the changes, the implementation of a modification may be performed using the maintenance process or the usual development process.

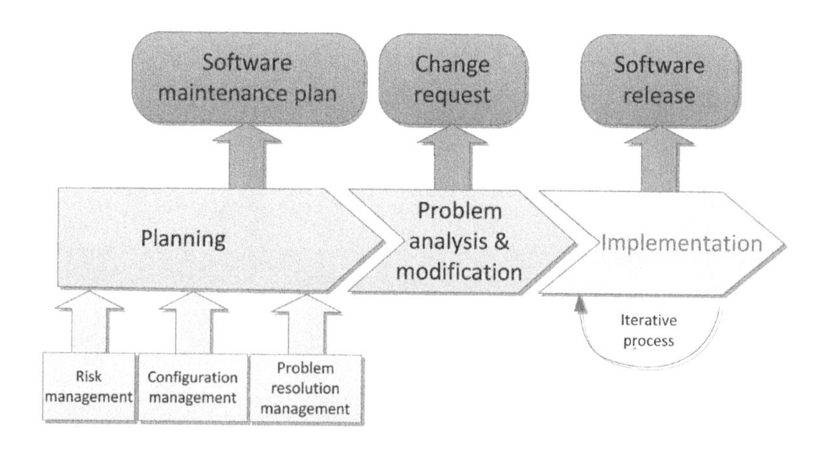

Figure 6.16 Maintenance process.

Problem analysis and modification is focused on the evaluation and the approval of change requests which modify released software products. In addition, a number of actions will be performed to officially communicate the release of a new version to regulatory bodies, especially when changes are due to safety-critical problems. In this case, the analysis must identify side effects, propose corrections and investigate possible similar problems affecting other products. The new release must be revalidated and a reasonable regression test set must be re-executed to prove that additional problems have not been inadvertently introduced after the changes. The activities related to the changes have to be recorded in terms of change request, change report, test report for validation or verification, acceptances test.

6.8 Guidelines on Software Design

Quite often, software programs implementing new features increase in their level of complexity, so that bug fixing, testing and maintenance become critical. Frequently, software can become so complex that it is more convenient to do a new development from scratch than to continue working on the existing software. This situation is well represented by the expression included at the beginning of this chapter: '*The remaining 10% of the code accounts for 90% of the development time*'. In more general terms, hardware has been reducing costs and doubling performance every 18 months, according to the revised version of Moore's Law (Moore 1965). Software implementation has not experienced such rapid improvements. 'Software remains in large part a craft industry' (Dyson, 1998) where the human skill is critical and no dramatic automation can be implemented apart from an extensive reuse which allows us not to write code at all. A classic analysis suggests that software has an **essential complexity** due to the activities that have to be performed by the software itself (Brooks, 1986). This complexity cannot be reduced and therefore no 'silver bullet' exists that can even reduce efforts by one order of magnitude as experienced in hardware.

In medical devices, software is not just a question of performance because safety is much more critical. This suggests focusing on the elements that may guarantee these characteristics: safety and performance. IEC/EN 62304 includes some elements that may ensure these conditions, but other elements should also be considered. As suggested by (Dyson, 1998) and (Brooks, 1986) **software is a creative and craft activity** and therefore, human resources for designing and implementing software may have a great difference in productivity compared to more conventional jobs. Although it is difficult to measure, some authors argue that a tenfold difference can exist between different designers and programers (Brooks, 1986). Experience in complex mission-critical or safety-critical software may suggest greater differences that may rise to 30 or even be infinite in the worst cases where only the best programers can implement complex software features that in medical devices may be related to safety.

In addition to the proper selection of resources, there are other factors that may **improve software** quality. The adoption of specific **guidelines** in computer programing is critical for reducing software development efforts especially in debugging, testing and maintenance stages. These later stages account for most of the software project efforts which can rise to 80% of the overall time and cost budget (Sun, 1997).

Table 6.7 shows typical **quality factors** for software, adapted from (Hetzel, 1984) and ISO/IEC 25010, 2011).

Table 6.7 Software quality factors

Category	Quality factor	Description
Exterior qualities impacting on safety and effectiveness	**Correctness**	Capability to behave as specified in the intended use and purpose
	Reliability	Frequency of ineffective/unsafe conditions due to programing errors
	Robustness	Frequency of ineffective/unsafe conditions due to user errors
	Usability	Ease of using software for the intended use
	Security	Degree of protection against unauthorized uses
Interior qualities	Efficiency	Amount of computational resource use
	Testability	Ease of testing to ensure that software behaves as intended
	Readability	Ease of reading code and associated documentation
	Modularity	Degree to which the system's components are independent
Qualities impacting future	Flexibility	Degree to which the system can be adapted to external changes
	Reusability	Ability of an item to be used repeatedly in different contexts
	Maintainability	Ease of program modification

Factors are grouped into three categories: exterior, interior and future qualities. Exterior qualities at the top of Table 6.7 affect both safety and effectiveness when related to medical device software. Qualities that affect safety and effectiveness in a medical context are written in bold. Interior qualities refer to the structural features of software. Future qualities address aspects that help in the life cycle stages after deployment. Maintainability can also be considered as critical in response to errors experienced in the device that may require prompt fixing. In the following sections, we propose an example of design and implementation guidelines adapted from (Becchetti, 1999) and used in the development of the Gamma Cardio CG. Design and implementation guidelines help in improving reliability, robustness and maintainability quality factors that directly affect safety and effectiveness. In addition, they also have a positive impact on readability and modularity with an overall benefit in the life cycle costs.

The proposed guidelines are based on a conservative programing approach that is suited to safety critical projects as for medical device software projects. Such guidelines may be a simple example to understand the guidelines and coding conventions usually found for many contexts and programing languages. The reason for guideline distribution is that around 80%

of the overall life cycle software effort is focused on maintenance, and software is rarely managed by the same author for long periods (Sun, 1997). Guidelines on code help in understanding the code more quickly and thus enabling errors to be found more easily. In software, 'cleanness' of source code is usually proportional to reliability and correctness. These simple guidelines are outlined by first introducing definitions (Section 6.8.1) and recommendations (Section 6.8.2). A hierarchical approach to software development, where classes are all derived from a few ancestors, is now common in many software frameworks that have core services already implemented by specific available classes. This theme is addressed in Section 6.8.3. In mission-critical software, design and programing should be driven by prudence to reduce risks. To this end, a conservative and defensive approach is outlined in 6.8.4.

6.8.1 Definitions

6.8.1.1 Modules and Interfaces

Modular programing is a design technique where software is decomposed into interchangeable components by breaking down the program functions. Modules interact with the external environment through interfaces. The interface practically contains all the functions used to access the services that a module offers. An interface should be the only approach to interacting with modules. Each module should basically accomplish one or more defined and complete functions (i.e., maximum internal cohesion) while minimized interaction with other modules (i.e., minimum coupling). Modules designed according to maximum cohesion and minimal coupling, as discussed in Chapter 1 and Section 6.4.3, improve maintainability and produce logically interchangeable components when interfaces are compatible regardless of internal module implementation.

6.8.1.2 Classes and Objects

In object oriented programing, a class is the general construct that is used as a template to generate the associated instance usually referred to as 'objects'. A class `PatientClass` may look like:

```
class PatientClass  {
    String surname;
    String name,
    ...
    public:
    Create_Patient(String specific_surname,String specific_name)};
```

The class `PatientClass` contains data and functions (i.e. `Create_Patient(...)`) to manage the patient. Within a software program, the class is 'instantiated' into an object, that is, objects are created and refer to specific patients:

```
PatientClass my_object_patient;
my_object_patient.Create_Patient("Neri", "Alessandro");
```

Every class publishes an interface of functions (possibly not data) that allows access to the class services. In the specific case, the interface is constituted by `Create_Patient (...))` which is the only published function that allows access to the class services. A module encompasses a set of collaborating objects that may be accessible only through a unique possibly minimized interface.

6.8.2 Basic Recommendations

Readability is the first recommendation since it is a necessary condition for most of the other quality factors. In principle, comments should not be necessary since the code itself should tell how the application works. Unclear, unclean or unreadable codes hide errors and, in general, they are an indication that the designers/programers have not understood the application logic. When a programer produces unclear code, errors are likely. These concepts are particularly addressed by newer software development methodologies such as extreme programing (Beck, 1999).

Readability is a primary quality of code, as it has many positive effects, such as

- reduction of the development process: fewer errors are generated from the beginning because some of them are self-evident from inspection
- reduction of debugging efforts
- decrease in maintenance effort by the author or by other programers: any author tends to forget details of a program especially over the years' life span as in the case of medical software; a readable code reduces this problem.

The main recommendation for improving readability is that the code should be written 'in plain English'. In principle, even those who are not accustomed to the specific programing language should be able to understand what a fragment of a code does simply by reading the code. This means that all the names of the data and functions should be self-explanatory and abbreviations should be avoided when possible. In addition, the definition of the variables should be close to their use, and one statement per line is mandatory. Some programers tend to adopt a cryptic style and structure, claiming a better performance especially in embedded software. On the contrary, modern compilers are extremely proficient in code optimization, so a number of practices which were common few years ago are now totally obsolete and may even be less effective since they cannot be optimized by compilers.

A unit test should also be embedded in the software. Test routines with an automatic result check are a strategy to reduce the overall effort since they dramatically reduce error detection when repeatedly executed in regression tests. In medical software, they are also valuable risk protection measures if associated to safety-critical performances. For example, in the Gamma Cardio CG, test routines with automatic check results have been implemented to verify the system performance required by the ECG standard AAMI EC 11.

6.8.3 Software Core Services

The previous recommendations suggest an architectural approach as evidenced by Figure 6.17. Application software written in high-level or natural language requires low level

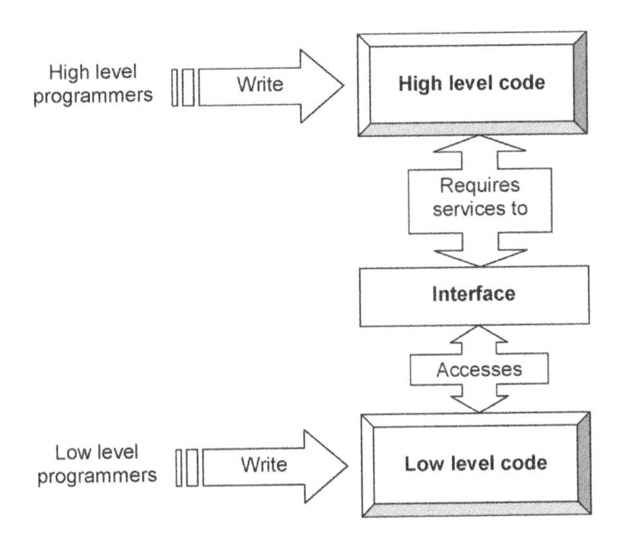

Figure 6.17 Software systems layering.

services through predefined interfaces that are the access point to low level services provided by operating system or by foundation classes common to many projects. Examples of core services that should not be developed are

- safe memory management and data container structures
- diagnostics and defensive programing procedures
- services involving the operating system (file management, user interfaces, networks, etc.)
- user program configuration management
- text data management.

Such low level services may be provided by specific libraries, foundation framework libraries or, better, directly offered by the programing language, as in the case of more modern programing languages.

In any case, it is important that most of the software is written in natural language style while low level services are either provided by existing facilities written by expert programers and tested accurately. This allows critical software to be written in a style that is close to the way the problem is represented in the analysis and design phases.

In the Gamma Cardio CG, high and low level codes have been clearly separated. Low level features of the language are used for the implementation of lower level services and they are not visible to the high level programers. These programers can avoid any low-level features of the language using the services supplied.

Thanks to the availability of libraries, most of the developed code can be exclusively high level. By this, we mean for example that programers may never need low level characteristics of the C/C++ language such as pointers that advantageously are hidden in more recent languages. The approach underlying this design style implies that the high level programers must follow strict rules that stem from the following 'conservative programing law', highly recommended in the risk-driven medical software.

> If there are many techniques to implement the same functionality, and some are safer than others, only the safest must be used; **the others are forbidden**

6.8.4 Defensive Programing

The debugging operation is performed in two steps: localization of the cause of the error and modification of the code. The first step is much more time-consuming. In a 'defensive programing' approach, code should be written so that errors are localized more spatially and temporally close to the generating code lines. That is, errors should evidence effects as soon as they occur and in the fragment of the code that is close to the statement that has caused the error. This is implemented by considering that a module can receive any input and it must work properly even with the wrong input. No **assumption** should be made with wrong data. This suggests that writing code with the (wrong) assumption that input data are correct will result in errors and unpredictable behaviors. Code should be written so that any input data inconsistency is handled, taking into account that even the least probable input may arrive as described by Murphy's law: '*Anything that can go wrong will go wrong*'.

More specifically, unrecoverable errors should never occur within a module, especially in a safety-critical context. *The module should report detailed information of the unexpected input while still trying to recover from the problem.* Specific 'stress' test are proven to be effective for detecting unrecoverable errors. Such tests call modules at a frequency much higher than in normal operation, with a random approach and random input data.

This style is similar to the 'design by contract' style where any function must obey a contract where benefits are expected and obligations are accepted in return (Meyer, 1998). In this book, the term 'defensive programing' will be used in a wider sense indicating strategies to prevent and/or easily debug errors. Defensive programing is performed by inserting proper **checks** on the data in the code. Three questions arise:

- Which are the data to be checked?
- When must these data be checked?
- What is the action to take in case of fault?

For the first question, in general, all data must be considered. In particular, the arguments passed to a function and its return value should be checked. These checks that are usually called 'preconditions' and 'post-conditions' are particularly required in object oriented environments where classes are largely reused (Meyer, 1998).

The strategy behind these checks, called 'design by contract', consists of thinking that any module must obey a contract that specifies benefits and obligations like the common contracts between 'two parties when one of them (the *supplier*) performs some task for the other (the *client*)'. The benefits are the preconditions that the module may assume, while the obligations are the post-conditions that the module must ensure. The module must implement preconditions and post-condition checks *reporting detailed information of unexpected input while still trying to recover the problems.* Lack of these controls may result in huge disasters such as the crash of the European Ariane 5 launcher on 4 June 1996 (Jézéquel and Meyer, 1997). Similar problems occur in medical software. In the years 2010–2011, they were 244 Class II recalls in the USA related to software and five related to firmware. In Class II recalls, the product may cause temporary or medically reversible adverse health consequences with remote probability of serious adverse health consequences (FDA, 2012). In the years 2000–2011, there were 19 Class I recalls in the USA related to software and five related to firmware. A Class I recall is a situation in which there is a reasonable probability that the use of or exposure to a violative product will cause serious adverse health consequences or death (FDA, 2012).

Part II: Implementation

This section shows how medical device software theory and processes are realised in the design of medical products. This is done by revealing software and firmware details of the **Gamma Cardio CG** product supplied by the Gamma Cardio Soft (www.gammacardiosoft .it). The sections are organized following the design process. In particular, in Section 6.9, the Gamma Cardio CG's **system decomposition** is outlined, identifying the developed items. In the same section, a concise use-case recalls the main ECG functions. **Risk management** at system level is introduced in Section 6.10 to identify risks affecting software/firmware and introducing requirements with the associated countermeasures. Then, the focus is on the two configuration items: **software** in Section 6.11 and **firmware** in Section 6.12. For each configuration item, hazards are analyzed and specific requirements are generated to mitigate risks. Other requirements are generated from system requirements and from the functions that the items have to perform. The **architectural design** is then derived to identify the modules and the units with the associated requirements. A generalized model suitable for any biomedical instrument is used to broaden the scope of the discussion.

The software development process from requirements to the release is a quite complex process for the Gamma Cardio CG, as it is for any non-trivial medical device. A detailed description of this activity cannot practically be included in the scope and the size of this book. We will therefore focus on two significant aspects for medical instruments: the **signal processing chain** and the automatic test capability.

The ECG, as with any biomedical instrumentation, transforms signals obtained by sensors into clinically significant data that have a format suitable for the available output device and the specific user. Most of this processing is performed by software. Section 6.11.3 describes the mathematical transformation from input ECG samples to clinical data shown to the users.

A valued-added capability for reducing risks and development/manufacturing costs and improving product quality is the **automatic test capability**. This capability is used to automatically validate the hardware at the factory facility, thus reducing the risk of using human validation. The automatic test capability, which is also useful for performing regression tests under development, is outlined in Section 6.12.3.

6.9 System Decomposition

6.9.1 Gamma Cardio CG Use Case

The Gamma Cardio Soft ECG system (Gamma Cardio CG) is composed of an acquisition unit with firmware (FW-CG) which communicates with management software running on a PC platform. The management software is called the 'multifunctional ECG software' (MES).

The operator uses MES to show and record the ECG signal. The ECG is recorded and stored by the MES, which can show, print and transmit the examination results to a competent physician for diagnosis. Transmission is performed through the internal telemedicine system. The operator who records the ECG can be a doctor, a nurse or, in general, any properly trained personnel.

The specialist physician can determine the ECG interpretation through

- the ECG signal shown on the PC screen

- a printed copy of the ECG
- a printed fax sent directly from an operator through the MES
- a file containing the ECG.

The file can be e-mailed directly from the MES or it can be stored on removable or shared media (USB flash drive, cloud network etc.).

The MES offers all the basic ECG functionalities plus additional functions:

1. configuration and management of the acquisition unit
2. ECG acquisition and signal elaboration
3. visualization of the ECG in on-line (during acquisition) and off-line mode
4. creation of reports with associated patient data
5. report preview, print, fax and save
6. storage of ECGs reports in files and in the internal database
7. creation of reports in Acrobat® PDF format
8. telemedicine capability through the internal mail system
9. help system management
10. autotest system management
11. step by step procedures for fast ECG acquisition and interpretation.

6.9.2 System Decomposition

The Gamma Cardio CG system is decomposed into subsystems as shown in Figure 6.18. The system–subsystem breakdown structure shown in Figure 6.8 has three subsystems to be developed at the second level: the hardware of the acquisition unit, the firmware of the acquisition unit (FW-CG) and the software application (MES).

As shown in Chapter 2, system level requirements are obtained from applicable frameworks and standards, risk analysis and the intended use/purpose of the device. In Chapter 3, all system requirements are associated with subsystems, and associated subsystem requirements are generated. For a complete design, it is important that all system requirements are associated to some subsystem requirement at all levels of the decomposition hierarchy, that is, at first level on the subsystems indicated in Figure 6.18 and at all other lower levels on modules and units. More simply, at each level of detail, there must be a subsystem

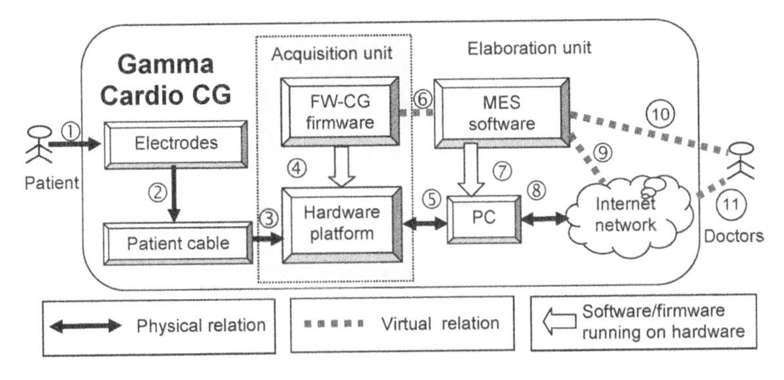

Figure 6.18 The Gamma Cardio CG subsystem identification.

requirement that can be traced back to a system requirement. In addition to requirements generated by system requirements, other specifications may be introduced at subsystem level to realize the concept of executions and use cases.

6.10 Risk Management

At software/firmware subsystem level we consider two sources of risks. The system risks that have some impact also at (subsystem) software level and the new risks that are identified while designing the software. System risks have to be assessed considering the risk effect at subsystem level. For both these risks, it has to be evaluated as to whether countermeasures are required. If so, proper requirements have to be generated. Table 6.8 shows risks that are identified according to the intended use, to the usual medical practice and to the model outlined in Table 6.3. The second column shows whether some software risk is already

Table 6.8 Gamma Cardio CG software: risks

ID	System requirement	Software function risk area	Software applicability	Level of concern	Countermeasure on software/firmware
1	R 11	**Availability**	Diagnostic service unavailable	minor	[RS_0] Defensive programing style (Section 6.8), software validation (MES and FW-CG) watchdog on firmware
2		**Hardware wrong control**	Not applicable for software		(countermeasure on hardware)
3		**Clinical data display**	Wrong display (artifacts, distortions, wrong filters etc.)	moderate	See Table 6.9
4		**Alarms**	Not applicable		Not applicable
5	R 15	**User interface**	Error prone	minor	[RS_8] MES reproducing conventional ECG interface
6		**Patient data (on shared database)**	No access to remote database	minor	Local database
7		**Security**	System is equipped with telemedicine functions through e-mail functions	minor	Not applicable

associated to system level countermeasures. If so, the associated requirement ID is indicated. Table 6.8 is a simple outline of the risk analysis, risks managed at a more formal and complete level are addressed in Chapter 8.

The degree of risk (i.e., level of concern), shown in the fifth column, is defined according to (FDA, 2005). The level is assessed without taking into account the requirements for countermeasures shown in the final column. Such additional requirements imposed on the system, identified with label [RS_XX], allow a further reduction of the probability of occurrence of the risk and/or of the harm severity.

In Table 6.8 two requirements are introduced to ensure better system availability and a reduced probability of usability errors.

Table 6.9 details the countermeasures to overcome the risk of a distorted ECG signal. The analysis is performed following the scheme of Figure 6.11. The final column contains the countermeasure requirement to be set on the software or firmware to reduce risk probability or severity.

Seven safety requirements are defined to reduce such risks, which span from automatic test capabilities to the performance of the whole signal processing chain and communication protocol controls. The next two sections introduce additional subsystem requirements for firmware and software.

6.11 Software Application

6.11.1 Software Requirements

MES application development has to be performed based on requirements at software subsystem level.

At subsystem level, we consider two sources of requirements. There are the system requirements that have to generate subsystem requirements when applicable. These requirements are not sufficient to design the subsystem; additional requirements are then needed to realize the use cases and the concepts of execution of the subsystem. Subsystem requirements may be

- countermeasures to mitigate
- functional requirements
- standard compliance requirements.

The first set of safety requirements is included in Tables 6.8 and 6.9 with the label [RS_XX]. The second set of requirements identified by the label [RF_XX] is generated from use cases such as the one outlined in Section 6.3.2 (Table 6.10). The third type of requirements arises from compliance to AAMI EC 11 standard which ensures performance suitable for clinical usage. These requirements, included in Table 6.11, are identified with the label [R_AA_XX].

Table 6.10 collects the **functional requirements**. The first column shows the system requirement number associated to software requirements. All system requirements that have an impact on software must generate an appropriate requirement that explains what the software application must do to accomplish the system requirement. The final column indicates the software application module (unit) as detailed in the architectural design in the next section.

Table 6.9 Gamma Cardio CG software: risks related to wrong clinical data display

ID	Reason	Level of concern	Scenario discussion	Countermeasure on software requirements
8	Due to incorrect designed or implemented signal processing	moderate	Degradation may lead to wrong diagnosis	[RS_1] signal processing in MES and FW-CG in accordance with AAMI EC 11 standard [RS_2] MES and FW-CG with built-in automatic test system to validate all signal processing chain
9	Due to degraded operations in hardware acquisition unit	minor	Degradation may occur for saturation of the preamplifier stage	[RS_3] MES with self-test function [RS_ 4] MES with saturation indication on screen
10	Due to incorrect positioning of the electrodes	minor	Easily identifiable electrode position by specialists (usually described in ECG medical books)	[RS_5] MES with on-line help showing electrode positioning.
11	Due to mains supply interferences or artifacts	minor	Easily identifiable by specialists (usually described in ECG medical books)	[RS_6] The MES will allow the operators to add or remove filters even on recorded ECGs
12	Due to communication problems with the acquisition unit or to problems of storage medium	minor	USB standard ensures high reliability of transport. Storage media with internal integrity mechanisms. Probability of occurrence is low and due to the high sampling rate artifacts would be hardly noticeable	[RS_7] MES and FW-CG with redundant communication protocols (CRC) able to detect errors

From the acquisition module, the raw signal is stored in the internal memory and it is then processed by the **elaboration module**. This module performs all the filtering configured by the user and the scaling on time and amplitude axes as explained in Section 6.11.3.

Functional requirements formalize the capabilities already introduced in the previous chapters.

Table 6.11 contains the **AAMI EC 11** IEC/EN 60601-2-25 standard requirements applicable to the software. The final table column indicates the requirement number. Software is

Table 6.10 Gamma Cardio CG software: functional requirements

System req.	ID	MES Requirements	Addressed module
R 22	RF_1	Management of the acquisition unit with ECG recording and hardware unit test capability. Capability to manage different transport layers (USB, RS-232, Ethernet).	Acquisition
	RF_2	Capability to simulate hardware acquisition unit by MES software for test mode	Acquisition (board simulator)
R 3	RF_3	Generation of patient and synthetic signals as per EC 11 to test software	Acquisition (signal generator)
	RF_4	ECG saved on hard disk or removable media without degradation of the signal with patient and report additional data	Presentation (file)
	RF_5	Patient data reports and ECGs stored in a local database	Presentation (database)
R 4, R 14	RF_6	Measuring units: mm/mV e mm/ms, Sampling rate resolution at 1200 Hz	Elaboration
	RF_7	ECG sharing and telemedicine capability via internal mailer system	Mailer
	RF_8	ECG display in on-line and off-line mode with supporting tools for reporting	Presentation (Video)
R 5, R 6, R 7	RF_9	Implemented filters: noise net filter, low pass filter and isoelectric line stability filter	Elaboration
R 19	RF_10	ECG PDF data output available	Presentation (file)
R 20	RF_11	Print and print preview of ECG and reports with zoom capability	Presentation (printer)
R 21	RF_12	Touch screen like interface with customization capability	Presentation (video)
	RF_13	Help manager	Controller (help)

involved in the elaboration and the presentation of the ECG signal. The first requirements are related to vertical and horizontal scale; the other requirements are related to the visual presentation (R 3.6), the overall accuracy and the upper cutoff frequency that is also set by the digital filter implemented in software.

Tables 6.8–6.11 contain all the requirements needed to design the software application. The next step consists of performing software decomposition to identify modules and units.

Table 6.11 Gamma Cardio CG software: performance requirements

System requirement	Subsystem requirement	Requirement description	Standard performance	Gamma Cardio CG performance	Validation method
R 3.2	R_AA_1	**Lead definition**	Standard 12 leads	Compliant (Standard 12 leads recorded one channel at time, see Section 2.13.3)	Inspection
R 3.3		**Input dynamic range**			
R 3.3.1	R_AA_2	Maximum input dynamic range that ensures linear response from ECG	$\geq \pm 5$ mV	Compliant	Demo
R 3.4		**Gain**			
R 3.4.1	R_AA_3	Gain selections	Selectable minimum at 20, 10, 5 mm/mV	Compliant	Inspection
R 3.4.2	R_AA_4	Gain error	$\leq 5\%$	$\leq \pm 3\%$	Demo
R 3.5		**Time**			
R 3.5.1	R_AA_5	time base selections	Min. 25, 50 mm/s	5,25,50 mm/s	Inspection
R 3.5.2	R_AA_6	time base error	$\leq \pm 5\%$	Compliant	Test
R 3.7		**Accuracy** of input signal reproduction			
R 3.7.1	R_AA_7	overall error for signals up to ± 5 mV amplitude and 125 mV/s	$\leq 5\%$ and $\leq \pm 40\ \mu$V	Compliant	Test
R 3.7.2	R_AA_8	Frequency response	See Chapter 5	Compliant	Demo
R 3.6		**Output display**			

Table 6.11 (*Continued*)

System requirement	Subsystem requirement	Requirement description	Standard performance	Gamma Cardio CG performance	Validation method
R 3.6.1	R_AA_9	Visibility of signals with minimum amplitude	$\geq \pm 5$ mV	Compliant	Inspection
R 3.6.2	R_AA_10	Vertical size of the display per channel	>40 mm	Compliant	Inspection
R 3.6.3	R_AA_11	trace visibility of 25 Hz sinusoidal test signal 20 mm p-p amplitude	Sinusoid peaks distinguish-able	Compliant	Test
R 3.6.4	R_AA_12	Trace width after (R 3.6.3) test signal switch off	<1 mm	Compliant	Inspection
R 3.6.5	R_AA_13	Pre-ruled paper division	10 div/cm	Compliant	Inspection
R 3.6.6	R_AA_14	Error of rulings	±2%	Compliant	Inspection
R 3.6.7	R_AA_15	Time marker error	±2%	Compliant	Inspection
R 3.6.8	R_AA_16	Indication for degraded mode	Indication of signal saturation of input stage	Compliant	Test

6.11.2 Architectural Design

Figure 6.19 shows the decomposition of software into modules. Continuous lines indicate ECG data exchange while dotted lines show control paths. This scheme recalls Figure 6.11; software is basically composed of an acquisition module, an elaboration module and a presentation module. The acquisition module manages the transmission of data with the acquisition unit. The elaboration module performs all the signal processing functions to extract clinically relevant data suitable for the specific output media (video, printer, file etc.). The presentation module manages the specific output media and the controller monitors and configures the other modules based on the operator inputs. This scheme is applicable to any medical instrument.

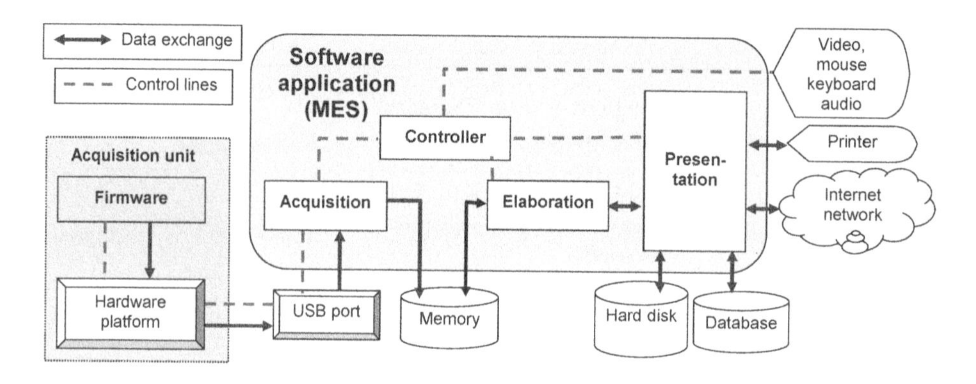

Figure 6.19 Gamma Cardio CG software application architecture.

The software architecture can then be further decomposed in the scheme of Figure 6.20. Again, continuous lines indicate data exchange while dotted lines show control data transfer.

From the bottom left of the figure, the **acquisition module** manages the transmission of data with the acquisition unit. The *communication driver* handles the different communication channels (USB, RS-232, Ethernet) so that the other modules manage a data exchange regardless the transport media.

The *protocol manager* handles all the message exchange and the data formats adding redundancy mechanisms, that is cyclic redundancy check (CRC) and synchronization data for the detection of transmission errors.

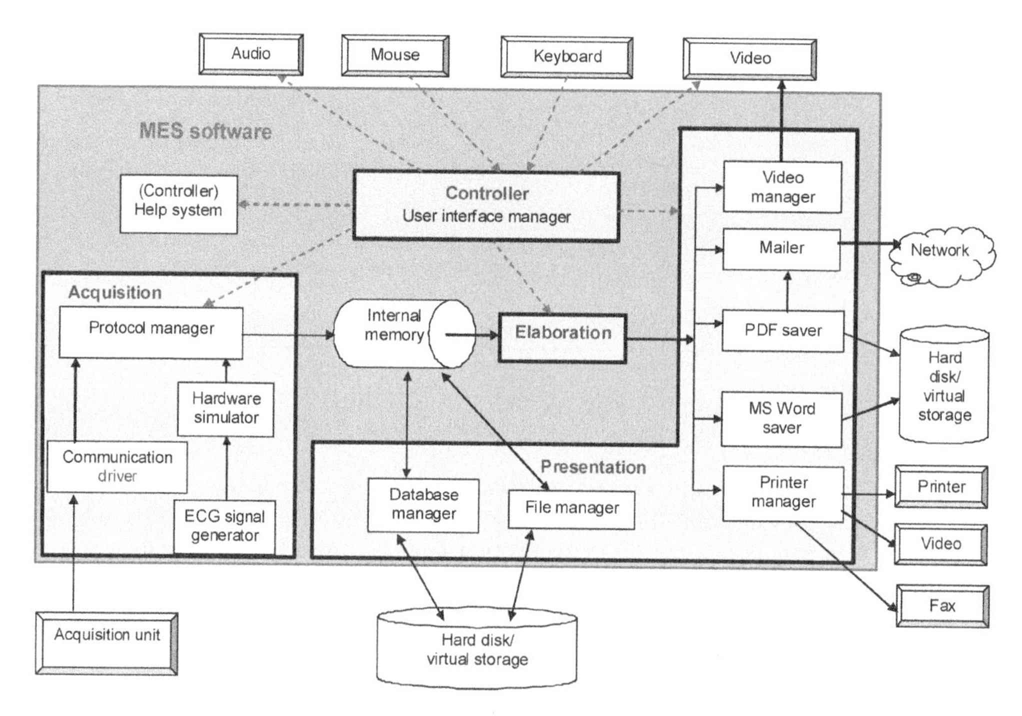

Figure 6.20 Gamma Cardio CG software: unit level.

The module may be connected to the hardware acquisition unit or to a simulator that responds to protocol manager messages with the same behavior as the hardware acquisition unit. The *hardware simulator* is connected to a *signal generator* that can produce the ECG signal or certain synthetic signals required for the tests. The generator may produce a signal, summing various sources (sinusoids, rectangular waveforms, white noise, ECG sources) varying the associated parameters: amplitude, frequency and duty cycle, as applicable. These tests are used to validate the software application and, in particular, the response of the digital filters implemented in software. The output signal is then managed by the **presentation** module that includes various drivers to handle the different output media (video, files, mail, PDF format, patient database with electronic health records).

The **controller** module supervises and synchronizes all the other modules in the MES and provides help messages when required.

The description of all the modules is derived by the requirements introduced in the previous sections. It is important that all system requirements are associated with the units defined in the architectural design so that units are properly implemented.

6.11.3 Elaboration Module

The elaboration module provides an output signal that is suited for the resolution of the visualization media (e.g., 96 dots per inch as for Microsoft Windows operating system video). The module receives an input sequence of numbers sampled at a specific rate (e.g., 1200 samples/s) and quantized at a defined resolution (e.g., 12 bits). The module performs the following operations on the input sequence:

- adaptation and interpolation of the vertical scale as a function of the gain required and the characteristics of the output medium
- adaptation and interpolation of the horizontal scale as a function of the required scrolling speed and the characteristics of the output medium
- low pass filtering for the elimination of the out-of-band noise
- filtering according to user requirements (muscle noise, baseline wander and noise net filters).

Let us, for example, consider as output a video frame of 600×400 pixels where the vertical pixel index starts from 0 up to 399 and the horizontal pixel index ranges from 0 to 599. The input ECG sequence belongs to an X space where values are sampled at 1200 Hz and 12 bits. The output sequence of the elaboration module $y[n]$ in the Y space (e.g., the video frame) is given by

$$\begin{aligned} y[n] = {} & out\_zero\_value + filter\_overall\_response * \\ & (scale\_constant \times (x[n \times time\_constant] - input\_zero\_value) \end{aligned} \quad (6.3)$$

where *out_zero_value* is the vertical displacement that the zero value must have on the y axis, that is, 200 considering that y is in the [0,399] range. *input_zero_value* is the value corresponding to zero in the X domain For a 12 bit sampled sequence, it is expected that sample x values are in the range [0,4095] and, therefore, 2048 is the x value corresponding to zero in the analog signal domain. Symbol $*$ denotes the convolution operation introduced in Chapter 4, and the *filter_overall_response* is the finite impulse response (FIR) of the joint effect of all the filters applied to the signal (muscle noise, baseline wander and noise net filters). This response is obtained by the convolution of the required filter impulse responses. Since the convolution may be computationally too complex in the time domain, the operation is performed in the frequency domain as explained in Chapter 4.

scale_constant and *time_constant* are the values of vertical and horizontal units in the Y space (i.e., pixels) corresponding to the unit in the X space (i.e., samples). The vertical scale constant is given by

$$\begin{aligned} scale\_constant &= Lsb\_Value\_In\_Millivolt \times gain \\ &\times V\_Ratio\_Dpi / Mm\_Per\_Inch / V\_Zoom \end{aligned} \tag{6.4}$$

Lsb_Value_In_Millivolt is the constant that translates the low significant bit (LSB) in millivolt equal to: $Lsb\_Value\_In\_Millivolt = (input\ dynamic/2^{BIT\_RESOLUTION})$. With a 10 mV input dynamic, as per standard, and 12 bit resolution, *Lsb_Value_In_Millivolt* is equal to 2.44 μV. *Gain* may have the value of 20, 10, 5 mm/s as per the requirement R_AA_3 (i.e., gain selections) in Table 6.11. *V_Ratio_Dpi* is the dot per inch horizontal value equal to 96 in Microsoft Windows operating systems. *Mm_Per_Inch* (i.e., millimeters per inch) is equal to 25.4 and *V_Zoom* is the required vertical zoom factor (e.g., 1 for no zoom required). For the horizontal axis we have

$$\begin{aligned} time\_constant &= Mm\_Per\_Inch \times Sample\_Frequency/ \\ &/speed/H\_Ratio\_Dpi/H\_Zoom \end{aligned} \tag{6.5}$$

Sample_Frequency is in this case assumed to be 1200 Hz; *speed* is determined by one of the values of requirement 3.2.05 (i.e., time base selections: 25 or 50 mm/s), *H_Zoom* is the horizontal required zoom factor and $H\_Ratio\_Dpi = V\_Ratio\_Dpi$.

To avoid numeric errors, the input samples have to be converted to **double-precision** data (15-digit precision). All the numeric operations are performed with *doubles* to reduce precision errors. With the exception of the anti-drift filter, all filtering operations are linear, time invariant and minimum phase. MES signal filtering has a constant group delay and a substantially flat frequency response in the frequencies of interest ($f < 200$ Hz). This avoids distortions on the ECG signal waveforms as detailed in Chapters 3 and 5.

In equation (6.3) $x[.]$ sequence is taken at **non-integer positions** since *time_constant* is not usually an integer. Using the previous parameters, $time\_constant = 12.7$. This means that between two adjacent pixels of the $y[.]$ output sequence shown on the screen, there are around 12 samples of the input sequence $x[\cdot]$. The samples at non-integer positions may be evaluated through the adjacent samples at integer positions. Since the input signal can be considered as band-limited, the Whittaker–Shannon interpolation formula (eq. 4.18) may be used, taking into consideration of the Nyquist–Shannon sampling theorem detailed in Chapter 4. Unfortunately, this formula computes the sample based on a high number (theoretically infinite) of adjacent samples. Since the input sequence is already oversampled. Much simpler interpolation schemes may be evaluated using only the neighborhood samples.

With reference to Figure 6.20, note also that the elaboration module obtains an output sequence always from the original sampled signal without changing its value. This avoids that successive filtering procedures degrading the output signal.

6.12 Firmware

The firmware of the acquisition unit (FW-CG) manages hardware and exchanges data and control signals with the software application. Figure 6.21, discussed in Chapter 5, shows the hardware modules that the firmware has to manage to perform the ECG acquisition.

The following functions are included in firmware with, in parenthesis, the hardware module of the microcontroller that is used to perform the function:

1. communication with the software application (transmission module)
2. analog to digital converter control for ECG acquisition (ADC1)
3. management of control lines for lead selection and calibration signal (I/O)
4. monitoring of the signal saturation (ADC 2)
5. reset baseline automatic management (I/O HP reset)
6. management of the internal signal generator (PWM)
7. management of the internal autotest function
8. management of the internal memory registers through the software application.

These functions were discussed in Chapter 5.

6.12.1 *Firmware Requirements*

The firmware requirements are generated from the relevant system requirements and risks. Other requirements are generated to perform the functions included in the system concept of execution. Firmware requirements are outlined in the rows of Table 6.12. The first column of each row indicates the associated system requirement, while the second column shows the label of the firmware requirement.

As in the software case, labels change when requirements are generated by security issues (RS_XX), functional requirements (RF_XX) or from the performances specified by the AAMI EC 11 standard (R_AA_XX). The final column specifies the firmware module where the requirement is implemented. From Chapter 3, **system requirements** that are applicable to firmware are R 3, 9, 11, 14, 15 and 22. More specifically, software validation (R 11), AAMI EC 11 compliance (R 3), horizontal resolution and sampling (R 14), (R 15) saturation indication and the usability functions are applicable also to firmware as indicated in Table 6.12. Regarding requirement R_AA_6 and RF_15, a correct sampling of the input signal is a

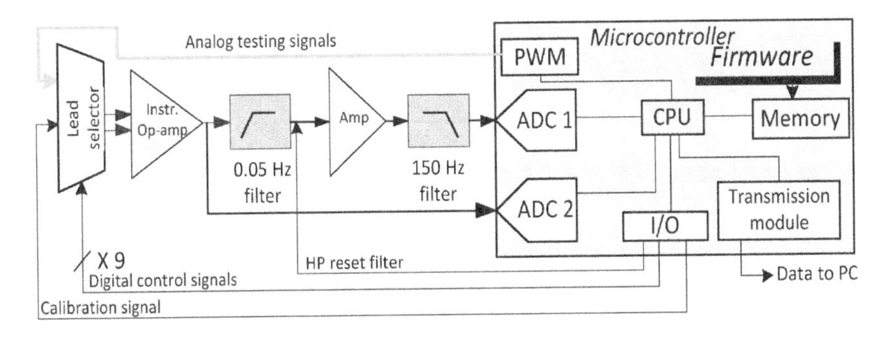

Figure 6.21 Firmware-controlled hardware functions.

Table 6.12 Gamma Cardio CG firmware requirements

System req.	ID	Requirement FW-CG	Firmware module
R 11	RS_0	Software written according to defensive programing style (Section 6.8), software validation	All
	RS_0	Implementation of watchdog functionality	FW controller
	RF_14	Management of the other modules of the system diagnostics and initial configuration	FW controller
R 3	RS_1	Signal processing in accordance with AAMI EC 11 standard (detailed in R_AA_17, R_AA_18, R_AA_19)	FW sampler
R 3.5.2	R_AA_17	Jitter control system to ensure maximum time base error $< \pm 5\%$	FW sampler FW controller
R 14	RF_15	Acquisition of the ECG signal at sampling rate of 1200 Hz and 12 bit resolution.	FW sampler
R 22	RS_2	Built-in automatic test system to validate the complete signal processing chain	FW controller, signal generator Memory manager
	RS_7	Implementation of redundant communication protocols (CRC) able to detect transmission errors.	FW communication
	RF_16	Management of the communication with the software application (SEM) hosted on the PC	FW communication
R 9 R 15	R_AA_18	Acquisition of the signal monitoring the saturation of the preamplifier stage (indication for degraded mode)	FW monitor
	RF_17	Management of the system reset of the high pass filter on software request	FW reset
R 3.13	R_AA_19	Automatic activation of the baseline reset in case of lead change or signal overload Return time to baseline ≤ 3 s after overload, ≤ 1 s after lead switch	FW reset

key factor to avoid distortion on the sampled sequence.The firmware has to trigger the AD converter at exact time steps given by the inverse of the sampling rate (i.e., =0.83 μs ≅ 1/1200 Hz). Jitter on the sampling period may cause errors in ECG interpretation. In addition, the firmware has to support sufficient data transfer to allow 1200 samples with the required resolution, packet redundancy and control messages to be transferred to the application unit. This issue has been discussed in Chapter 5. The production cost (R 22) may be reduced if an automatic test capability with the widest control of hardware is implemented. In this way, the human effort of validating each device sample is strongly reduced and, consequently, so is the associated manufacturing cost. The system requirements associated with compliance to IEC/EN 60601-2-25 and to the reduction of training for usage (R 9, R 15) affect the firmware for the indication of degraded mode which both reduces risks of wrong diagnosis and improves usability. So, the firmware has to monitor and communicate the saturation indication (degraded mode).

6.12.2 *Architectural Design*

To understand the CG-firmware architecture, we introduce first the typical model of a firmware program that is similar across various application fields. We will use the software model of Arduino which is a widely used open source electronics prototyping platform (Arduino, 2012). A firmware program is, mainly composed by three functions:

1. setup()
2. loop()
3. ISR() (interrupt service routine).

The **setup() function** is executed at startup to perform initialization of the software and hardware. Variables are initialized and hardware components of the microcontroller are configured to implement the required functions. For example, I/O ports have to be configured as input or output, AD converters require a sampling rate, communication ports need to be configured to the right data format and communication rate.

loop() is a function executed repeatedly until the board is powered off. This function contains the core of the program with all the code that is executed sequentially.

However, some activities have to be executed immediately after some specific events occur. For such activities it is not possible to rely on the loop function since, for this function the execution of a specific part of the code occurs only when the microprocessor passes over the code. In the case of event-triggered functions, the loop function has to be interrupted to execute the part of the code associated with the event. This mechanism is handled by the interrupts. Interrupts are signals that, when triggered, suspend current processor activity and execute the function associated with the interrupt (i.e. the interrupt service routine **ISR()**). When ISR() terminates, the program returns to the point where it was suspended immediately before the interrupt occurrence. The typical structure of a C firmware program may be written as in Figure 6.22.

We can now describe the firmware logic within the three functions (setup(), loop(), ISR ()) with the help of Figure 6.23, which shows the architectural decomposition of the firmware with all the modules indicated in Table 6.12.

The loop() function shown in Figure 6.24 contains the core of the program.

The setup() function has to initialize all the modules included in Figure 6.23. In particular,

- FW communication: UART, buffer structure to transmit and receive bytes
- FW sampler: ADC 1 configuration, timer configuration at sampling rate
- FW monitor: ADC 2 configuration
- Lead selector, filter reset, calibration: I/O configuration
- FW control: activate watchdog.

```
void main ()   {
        setup ();

        while (true) {
                        loop (); } //end while (true)
        return;
        }

void setup () { // initialization code}
void loop () { // core code iteratively executed   }
void ISR () { // code executed on interrupt   }
```

Figure 6.22 Firmware program: typical structure.

The setup() also calls the **Set_Status(required_status)** function that changes the status of the acquisition board. Firmware starts in the ECG_STOP status and it may change into ECG_ACQUISITION or ECG_TEST.

When the status is ECG_ACQUISITION or ECG_TEST, the lead selector is configured according to the required lead, the buffer is initialized and the interrupt timer is activated. This allows the associated **interrupt service routine** function ISR() to be invoked every 0.83 μs, when the ECG signal has to be sampled. The same routine monitors the time lag between to two successive ISR() invocations so as to monitor the time differences with respect to the value of 0.83 μs. Significant distortions may create artifacts.

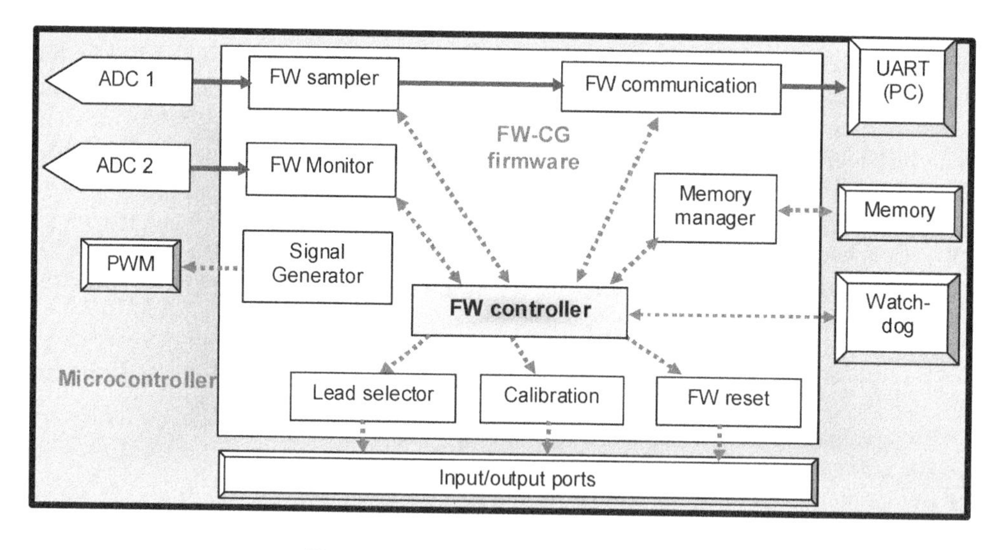

Figure 6.23 CG Firmware architecture.

```
void loop (){
      Restart_Watchdog();
      Analyse_Input();
      if (status!=ECG_STOP)  {
            if (sampled_data_available)  {
                  Load_Value_In_TX_Buffer(acquired_sample);
                  Manage_Signal_Monitoring();
                  sampled_data_available=false;
                  }
            Send_Byte_From_TX_Buffer_To_PC();
            Generate_Test_Signal();
            }
      return;
      }
```

Figure 6.24 FW-CG: loop function.

The setup() function also activates the **watchdog** timer. Table 6.12 includes the requirement RS_0 specifying the implementation of the watchdog to increase the availability and reliability of software. A watchdog is a particular timer that activates the microcontroller reset when the timer value reaches zero. In the loop() function, this timer is set to a time value (say x) that is much greater than the time required to execute again the instruction that sets the timer to x. If the processor stops and the loop function is not re-executed after time x, the microprocessor is reset to the power-up condition, allowing for the firmware to restart. This technique, widely used in embedded systems, avoids processor hang-up for software or hardware reasons. The first instruction sets the watchdog timer to the predefined time lapse. The FW communication module analyzes the packets sent by the PC to the firmware through the Analyse_Input() function. Referring to Figure 6.24, the PC may send the following application **commands**:

ACQUI_START_CMD: start ECG acquisition STOP_CMD: stop ECG acquisition	CONFIGURE_CMD: configure sampling rate CONFIGURE_TEST_SIGNAL
GETSTATE_CMD: receive the status	SET_MEMORY: to write memory of the μc GET_MEMORY: to get memory of the μc

The firmware answers with an acknowledge message to confirm that the message has been received. In the case of acquisition, the firmware starts to send data_messages with ECG samples. In the case of GET_MEMORY, the firmware sends the Memory_Status(memory_value) message. The board status is sent after receiving GETSTATE_CMD from the PC.

When the acquisition starts after the ACQUI_START_CMD is received, the interrupt service routine is invoked asynchronously to record the ECG signal samples. Referring to Figure 6.24, when sampled_data_available is true, the ECG samples are packed in the transmission buffer through the function Load_Value_In_TX_Buffer. The successive function Send_Byte_From_TX_Buffer_To_PC() sends the bytes into the transmission buffer used to communicate with the PC. In Figure 6.24, there are two other functions: Manage_Signal_Monitoring() checks if the signal saturates the input stage and sends a warning to the software application accordingly. Generate_Test_Signal() produces the signals required to test the hardware board.

6.12.3 Automatic Test Capability

A valued-added function to reduce risks and development/manufacturing costs and to improve product quality is the **automatic test capability**. This function is used to automatically validate the hardware at the factory facility, thus reducing the risk of human validation. The automatic test capability is also useful for performing regression tests under development. An automatic test capability is effective when firmware, hardware and software are designed from the beginning using a design-for-testing approach. This means that the hardware has to have a proper configuration so that tests validate most of the functions. Table 6.13 shows the results of an automatic test session. The session tests transmission, firmware, calibration function, **HighPassFilterReset**, FET reset module, multiplexers of the lead selector and the overall frequency response. Readers will find details of these functions in Chapter 5. A major piece of information of a test report is the identification of the device under tests. This is done by reporting the serial number and the version number of the software, hardware and firmware in the first lines of Table 6.13. Considering the *Calibration* test section as an example, there is a line (e.g., calibration_high_value) for each test, indicating the result (OK, NOK), the measured value (e.g., 0.979) and the admissible value or range (e.g.,:=<0.95,1.05>)). The calibration_high_value is equal to 0.979 mV with an error of −2.1% with respect to the expected value (1 mV), which is in line with the maximum permissible error $< \pm 5\%$ (i.e., from 0.95 to 1.05 mV) according to the AAMI EC 11 standard.

The first test session is on the transmission. As shown in Table 6.13, 100 bytes are exchanged with no error. In the **checkfirmare** session, specific internal hardware registries and firmware configurations are checked. The calibration session generates a calibration signal and measures the error with respect to the ideal value. **CheckHighPassFilter** verifies whether the response of the filter is correct. Since the filter is a simple first-order RC network, the filter is charged and then the exponential decay of the signal is checked at different times. Chapter 5 describes RC filters and their exponential decay characteristic. The TestFet session performs a check on the FET component that discharges the high value capacitor of the high pass filter. **TestMux** tests the lead selector lines in terms the functionalities of the multiplexers and of the measured noise. Finally, in the **TestFrequencyResponse** square wave signals are generated at various frequencies to check the frequency response amplitude and the frequency position of the peaks. The control ranges are set to the limits imposed by the standard. This allows the performance required by the standards to be validated.

At the beginning of this chapter we included the sentence: '*The remaining 10% of code accounts for 90% of the development time*', which accounts for the possible problems in the software development where, due to continuously generating defects, software estimated as finished at the 90% level, may even be impossible to finalize. A suitable software engineering process can mitigate this risk, but the value of the human resources employed in design and programing remain a compulsory ingredient, much more so than in other fields (Brooks, 1986).

Table 6.13 Gamma Cardio CG automatic test report

Test Report Gamma Cardio CG

Apr 15 11:19:54 2012

Software version v. 5.0.1.34b,10/4/09
build Apr 15 2009: 10:46:47
Firmware version: V.19.3,11/12/05
Serial Number = 15

Test: CheckTransmission

num_of_wrong_chars = OK, (0 := 0)
num_of_received_chars = OK, (100 := 100)
num_error_message = OK, (1 := 1)
last_error_string = OK, (1 := 1)
number_of_ferr_errors = OK, (0 := 0)
number_of_oerr_errors = OK, (0 := 0)
Test: CheckTrasmission: OK

Test: CheckFirmware

num_error_message = OK, (1 := 1)
last_error_string = OK, (1 := 1)
number_of_ferr_errors = OK, (0 := 0)
number_of_oerr_errors = OK, (0 := 0)
sampling_rate_lag = OK, (186 := 186)
fw_version = OK, (86 := 86)
TRISA = OK, (3 := 3)
. . .
PORTA = OK, (8 := 8)
. . .
Test: CheckFirmware: OK

Test: Calibration

sampling_rate = OK, (1176.47 := <900,1210>)
recorded_samples = OK, (3600 := 3600)
seconds_to_be_recorded = OK, (3 := 3)
calibration_high_value = OK, (0.979 := <.95,1.05>)
zero_average_value = OK, (0.026 := <−0.14,0.14>)
noise_standard_deviation=OK, (0.0067 := <0,0.015>)
Saturation_system_failure = OK, (1 := 1)
Test: Calibration: OK

Test: CheckHighPassFilter

average_offset = OK, (0.0572607 := <−0.1,0.1>)
max_value_millivolt = OK, (1.99352 := <1.8,2.3>)
max_value_1 = OK, (1.45397 := <1.4,1.65>)
max_value_2 = OK, (1.07311 := <1.0,1.25>)
max_value_3 = OK, (0.792349 := <0.7,0.95>)
min_value = OK, (−1.4098 := <−1.6,−1.3>)
sampling_period_error = OK, (186 := <100,190>)
Test: CheckHp: OK

Test: TestFet

average_offset = OK, (0.042627 := <−0.1,0.1>)
max_value_millivolt = OK, (1.98374 := <1.8,2.2>)
fet_plus_average_offset = OK,(0.0054 := <−0.1,0.1>)
min_value_millivolt = OK, (−1.99331 := <−2.2,−1.8>)
fet_minus_average_offset =OK, (0.005 := <−0.1,0.1>)
sampling_period_error = OK, (186 := <100,190>)
Test: TestFet: OK

Test: TestMux

average_signal_LA = OK, (0.055 := <−0.1,0.1>)
average_noise_LA = OK (1.93e−06 := <0,0.001>)
average_signal_LL = OK, (0.0465 := <−0.1,0.1>)
average_noise_LL = OK, (1.82e−06 := <0,0.001>)
average_signal_RA = OK, (0.03970 := <−0.1,0.1>)
average_noise_RA=OK,(1.365e−05 := <0,0.001>)
sampling_period_error = OK, (186 := <100,190>)
Test: TestMux: OK

Test: **TestFrequencyResponse**

peak_freq_10 = OK, (10.59 := <8,12>)
freq_mod_10 = OK, (2.04 := <1.90,2.2>)
peak_freq_37 = OK, (37.5 := <36,39>)
freq_mod_dB_37% = OK, (−1.17 := <−10,+10>)
. . . . . .
peak_freq_400 = OK, (399.609 := <398,402>)
freq_mod_dB_400% = OK, (−90.71 := <−100, + 10>)

Gamma Cardio CG overall result: OK

Test completed: Apr 15 11:19:57 2012

References

Abran A. *et al.* (2004), Guide to the Software Engineering Body of Knowledge, IEEE Computer Society.
Arduino (2012), Open source electronics prototyping platform: http://www.arduino.cc/
Becchetti (1999), Speech Recognition: Theory and C++ Implementation, John Wiley and Sons.
Beck K. (1999), Extreme programming explained: Embrace change, Addison-Wesley.
Brooks F. (1986), No Silver Bullet – Essence and Accident in Software Engineering. Proceedings of the IFIP Tenth World Computing Conference.
Dyson F. J. (1998), Science as a Craft Industry, Essays On Science and Society.
Economist (2012), Medical technology: Applying the 'open source' model to the design of medical devices promises to increase safety and spur innovation.
EEC (1993), Council Directive 93/42/EEC 14 June 1993.
EEC (2007), Council Directive 93/42/EEC 14 June 1993 concerning medical devices available at http://ec.europa .eu/health/medical-devices/index_en.htm amended by 2007/47/EC 21 March 2010.
FDA (2002), General Principles of Software Validation; Final Guidance for Industry and FDA Staff, Jan 2002.
FDA (2005), Guidance for the Content of Premarket Submissions for Software Contained in Medical Devices.
FDA (2008), Murray, J., FDA Training-Module on IEC 62304.
FDA (2012), Search Medical & Radiation Emitting Device Recalls, US Food and Drug Admin., website facility.
Fu, Kevin (2011) – 'Trustworthy Medical Device Software'. In Public Health Effectiveness of the FDA 510(k) Clearance Process: Workshop Report, Washington, DC.
Hetzel B. (1984), The Complete Guide to Software Testing, John Wiley and Son.
IEEE (1993), IEEE Recommended Practice for Software Requirements Specifications, IEEE Std 830-1993.
IEC/TR 80002 (2009), Medical device software. Part 1: Guidance on the application of ISO 14971.
ISO/IEC/EN 62304 (2006), Medical device software – Software life-cycle processes (EN 62304:2006/AC:2008).
Jézéquel J. M. and Meyer B. (1997), 'Design by contract: the lesson of Ariane', IEEE Computer.
Kelleher B. (2003), The seven deadly sins of medical device development. Medical Device & Industry Magazine 2003: http://www.devicelink.com/mddi/archive/01/09/002.html
Leveson N. *et al.* (1993), An investigation of the Therac-25 Accidents, IEEE Computer.
MEDDEV (2012), European Commission, Medical Devices: Guidance document: Qualification and Classification of stand-alone software.
Meyer B. (1998), 'Building bug-free O-O software: An introduction to Design by Contract'.
MIL-STD-498 (1994), Software Development and Documentation Standard, AMSC NO. N7069, US Department of Defense.
Moore G. E. (1965), Cramming more components onto integrated circuits, Electronics Magazine.
Nasa (2007), Systems Engineering Handbook.
Standish Group (1995), The Chaos Report.
Sun (1997), Sun Microsystems, Java Code Convention, September 1997.

7

c-Health

Medical Informatics

Supply medical services anytime and anywhere

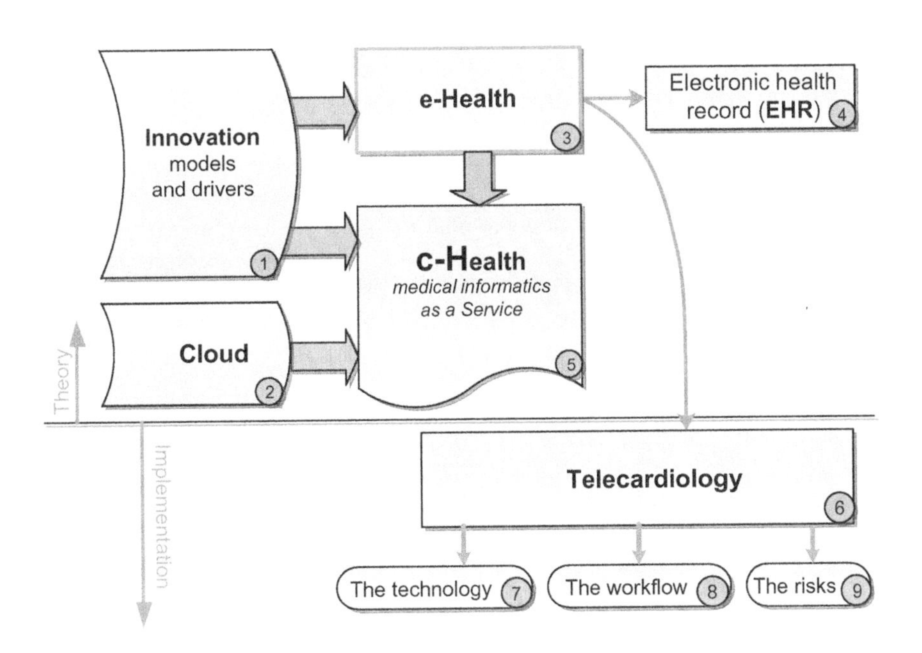

Figure 7.1 Chapter organization.

Medical Instrument Design and Development: from Requirements to Market Placements, First Edition.
Claudio Becchetti and Alessandro Neri.

Chapter Organization

This chapter deals with the innovation that information and communication technology (**ICT**) is bringing to medical devices and, in particular, to medical instruments.

Medical instruments are expected to be part of integrated health system environments, exchanging seamless continuous flows of clinical and administrative data. This is driving innovative products in this sector usually referred as '**e-Health**'.

This chapter is organized as shown in Figure 7.1 where boxes represent chapter sections whose numbers are included in a circle. Innovation in medical devices and medical instruments is a key factor for improving health services, increasing patient accessibility and enhancing disease-control effectiveness.

Unfortunately, innovative products can often fail in the health market. Section 7.1 will therefore discuss innovation in medical devices, introducing the drivers for change and an **evaluation framework** for assessing the feasibility of implementing innovative products. This model may also help in evaluating emerging technologies with their impact on medical devices.

Then Section 7.2 will describe a promising technology, **cloud computing**, which is expected to significantly change medical products, especially in the e-Health area.

e-Health, introduced in Section 7.3, is a novel area that encompasses products and services, where electronic processes and ICT have a major impact. e-Health is having a strong impact on medical instruments, which are expected to be designed to suit a service model where data will be available 'anytime, anywhere' on the network.

Electronic health records (**EHR**), discussed in Section 7.4, is a widespread example of e-Health, where medical instruments have an increasing role in supplying diagnostic data.

These approaches are expected to evolve into the concept of **c-Health** (Section 7.5) where e-Health solutions evolve through cloud computing to offer e-Health as a service. Cloud computing is an emerging ICT technology that will have a deep impact on the e-Health market with an estimated growth to $5.4 billion by 2017. Cloud computing is expected to achieve an average growth rate of 20.5% per year from 2012 to 2017 (Marketsandmarkets, 2012). For this reason, we define the term c-Health as '*the set of healthcare services supported by cloud computing*'.

The implementation section includes an example of e-Health: **telecardiology** (Section 7.6) and its associated **technologies** that are evolving, especially on the patient side with domestic and mobile devices (Section 7.7). The products in these areas are already quite mature; the critical factor is the **process** and the model, which needs to ensure safety, effectiveness, compatibility with approved clinical procedures and advantages for all stakeholders (i.e., patients, professional operators, insurers, healthcare facilities etc.). This topic is discussed in Section 7.8.

The final chapter outlines the **risks** of telecardiology services; this risks have to be evaluated and managed, by implementing proper countermeasures to achieve an acceptable risk level.

In this chapter, readers may consider the role of ICT in medical instruments, focusing on a promising technology (i.e., cloud computing) and assessing the key factors necessary for successful products. By comparison of the theory and implementation parts of the chapter, readers can understand how e-Health is implemented in one of the most impactful areas (cardiology).

Part I: Theory

7.1 Introduction

Current societal needs and challenges are powerful drivers for innovation in the health sector. In addition, current low-cost consolidated technologies may be a key success factor that enables the realization of **innovative medical solutions**. This chapter deals with innovative products that may be conceived when driven by new societal needs and technological innovations. This trend has been experiencing in the medical device sector and, more significantly, in medical instruments in the context of the e-Health.

Among the **societal driving needs**, ageing will be overwhelming in the health sector, especially in certain areas of the world, such as Europe (EC, 2011). The increase of the ageing population will prompt new sustainable models to ensure health rights. At the same time, budget restrictions at national and citizen level put further pressure for more efficient healthcare, suggesting the concentration of the health services on reducing costs and increasing effectiveness. This is contrasts with the necessity to ensure health rights with the widest universal coverage of health services, that is, where and when they are needed.

Technology is now offering enabling components that allow innovative products to be conceived. Electronics is greatly improving in performance and in cost, enabling the pervasive use of software that in turn reduces costs and facilitates the realization of smarter devices. These aspects have been addressed in Chapters 5 and 6. Telecommunications and associated technologies are another innovative driver that allows the virtualization of medical products and services as is happening in telemedicine and in electronic health records. This trend is generating new products in the area of the e-Health, which include health solutions relying on electronic processes and communications. More recently, **cloud computing** promises additional innovative solutions that may improve health and quality of life for patients and families, mitigating the shortage of healthcare professionals and making the national healthcare systems more sustainable.

Technology is not sufficient; there are other factors that are necessary for success in the medical device sector. New products must address strong societal needs, they must comply regulation overcoming preconceptions of patients/operators who are usually diffident about innovative medical products. Product marketing based only on technology, may be misleading with respect to the success of a specific device.

In the following sections, we introduce an evaluation model to assess innovative products, highlighting their strengths and weaknesses. We then discuss major drivers in the medical sector from technology enhancements and societal-political needs.

7.1.1 The Assessment Framework

Figure 7.2 shows a simplified model for assessing sustainability of innovative products considering macro-environmental factors. This model is derived from the usual marketing frameworks. For example, in the PESTEL model, Gillespie (2007) suggests categorizing various macro elements that influence firms under six factors: Political, Economics, Societal, Technological, Environmental and Legal. These factors are included in the model of Figure 7.2.

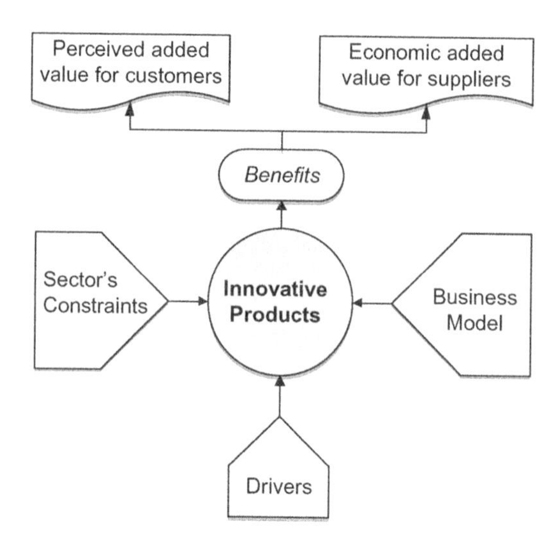

Figure 7.2 An assessment framework for innovative products.

Political and societal factors may be strong **drivers** that trigger product conception. Strong social needs may suggest innovative products, especially when technology progress improves performance and reduces cost. Political decisions may also be strong drivers, for example, by promoting specific market sectors and supporting the associated research.

Products that are based only on competitive technology and strong marketing needs may fail on the market because other necessary conditions have to be analyzed. This suggests extending the analysis to other factors such as the 'business model' and the **constraints** that rule the specific market sector.

For example, the legal activities related to a specific product market may be significant. The regulatory entities may impose stringent requirements and processes on product design, certification, manufacturing and usage.

A **business model** has to specify why and how the company will generate value for all stakeholders, describing information such as the market dynamics, the cost structure, the customer context and the distribution channel. Finally, all this analysis must identify clear **benefits** for all the stakeholders that offer the innovative products/services and for customers–patients who have to perceive specific added value that they are willing to pay for.

7.1.2 Assessment Framework for the Health Sector

The previous model may be adapted to the health sector, taking into consideration the specific factors affecting innovation (CEU, 2011). Table 7.1 shows how this model is translated into a template useful for a specific product assessment. The final column has to be filled in with the reference to a specific product in terms of the classical SWOT analysis – strengths, weaknesses, opportunities and threats/risks (Kotler, 2011). The key factors on the rows of Table 7.1 are: drivers, constraints, business and benefits. These general terms are translated into specific factors considering also (EC, 2011b).

Table 7.1 Evaluation template for innovative health products

Evaluation of innovative health products		
Key factors	**Specific factors**	**Strengths, weaknesses, opportunities, threats**
Benefits	Economics for suppliers	to be completed with product data
	Added value for patients	
Business	Patient perspective	
	Operator perspective	
	Sustainability (market, revenues, cost model etc.)	
Constraints	Workflow	
	Regulation and legal aspects	
	Effectiveness	
	Safety	
Drivers	Technology	
	Politics (R&D support etc.)	
	Society (diseases impacts, socio-cultural, ethical aspects)	

The **drivers** may be political, societal or technological factors. National and regional regulatory entities such as US FDA or the European Commission recognize the importance of innovation for making medical devices safer and more effective (FDA, 2011).

Regulatory and government authorities are therefore committed to encourage innovation by implementing specific measures, such as funding programs for research and development and improvement of the medical device market approval processes so as to reduce time to market and certification costs. In particular, funded themes are related to the main political and social challenges envisaged by the local authorities. Such grants may help companies to develop innovative products with economically sustainable models. For example, in Europe, a major challenge is the growing ageing population, taking into account that the number of Europeans aged over 65 is expected to rise from 87 million in 2010 to 124 million by 2030 (EC, 2011).

The European Union has implemented initiatives, with associated budget, to support innovative health products in the following **disease areas** (EC, 2012a):

- brain research and related diseases
- human development and ageing
- major diseases and disorders, including cancer, cardiovascular diseases, diabetes and obesity, rare diseases and severe chronic diseases.

Brain research is related to diseases of the brain itself, of the spinal cord and of the peripheral nerves. These disorders account for more than 35% of Europe's total disease burden.

Ageing is a major economic and social issue considering that in EU by 2050, the over-65 population is expected to increase by 70% and the over-80 population by 170%.

Regarding the other major diseases, despite the huge research efforts, cancer will probably remain one of the main causes of the death in the coming decades. It is estimated that in the EU in 2006, 1.5 million people died of cancer and three million new cases were diagnosed.

Cardiovascular diseases are the world-leading cause of death, with 30% of deaths worldwide and 42% of deaths in the EU.

Diabetes, whose effects are abnormal blood sugar levels, is an increasingly prevalent disease, with a huge incidence. Diabetes currently affects around 250 million people worldwide with an estimated increase to over 380 million by 2025, according to International Diabetes Federation (EC, 2012a).

All these diseases areas imply the use of diagnostics, and there are therefore strong social, political and economic drivers to develop innovative medical instruments.

The availability of low cost, high performance electronics and ICT is also a major driver to improve products and to introduce innovative e-Health solutions that may help in reducing costs for national healthcare systems. ICT is now a pervasive technology used by every professional health operator from pharmacy to hospital environments.

e-Health allows doctors to deal with patients' medical records anywhere and anytime, with access to test results produced by medical instruments. This prompts a new generation of network connected medical devices. This trend is already in place in telecardiology, where patients with heart problems may be continuously monitored by appropriate medical devices that may remotely raise alarms in critical conditions, thus giving patients an improved quality of life. e-Health will be a significant driver in the coming decades because its widespread diffusion gives real benefits. It is estimated that e-Health patient monitoring applications increase survival rated by 15%, with a reduction in hospital days of 26% and saving of 10% in costs. Electronic prescriptions are estimated to give significant benefits reducing errors by 15% (EC, 2012b).

Unfortunately, these drivers are strongly impacted by the significant **constraints** of the health market sector. A product must be

- safe
- clinically effective
- compliant to the applicable regulatory frameworks
- suitable for the medical organization and its workflow.

The first three factors will be examined in Chapter 8. **Safety** is related to physical injuries or damage to the health of people or animals, or damage to property or the environment. **Effectiveness** means that the device operates as declared by the manufacturer in the intended use/purpose. Safety and effectiveness constraints are ruled by the national or regional regulatory framework which, in the European Union, is the Medical Device Directive (EEC, 2007). Chapter 8 also includes information from other geographical areas.

The regulatory framework specifies other requirements that have to be implemented, such as the manufacturer quality system. In addition to regulations, there are conventional clinical **procedures** for delivering the medical services that impact on the products/services especially when considering e-Health which often encompasses and changes significantly the conventional medical workflow.

The **business context** has been already addressed in Chapter 1. The output is an economic analysis that evidences expected revenues, costs and margins. These economic indicators are derived by specific analyses of the market context that encompass

- the market size and trends
- the economic evaluation of competitors and products
- the distinctive features of competing products

Other elements are related to the customer:

- the customer's expectations of the product in terms of mandatory and optional features
- the price target that customers are willing to pay
- the preferred channel for purchase
- the expected method of promotion.

Regarding customer-related factors, the health market has specific rules and dynamics. The patients and the operators usually have a key role in accepting innovative products and services.

Patient demand is mediated by physicians who perform both the diagnosis and the associated treatment. Telemedicine and e-Health products have therefore to be accepted by all the stakeholders who provide the service. This implies economic convenience and proper usability of the service by operators. In particular, considering for example telemedicine services, the health operators have to have confidence in the effectiveness of the service and, more specifically, in the associated technology.

ICT based technologies have often been perceived as unsuitable for medical applications because of insufficient usability or reliability for a safety-critical context. At this aim, the European Commission recommends focusing on ICT interoperability issues for the integration of e-Health medical devices. Safety and effectiveness may be critical especially for personal health systems and mobile health systems named 'm-Health' (CEU, 2011).

Economic convenience for operators has also to be analyzed. The failure of innovative services may result from non-acceptance by operators owing to insufficient economic benefits compared to conventional medical approaches. For example, telecardiology with devices widely spread over the territory (e.g., pharmacies, surgeries etc.) may be a significant approach to reducing cardiovascular diseases effects through prevention. On the other hand, this may impact on the income and the clinical approach of the specialized operators who may therefore not be willing to adapt such new services.

The patients' perception and acceptability is another key element for the success of e-Health and innovative products. In the area of telemedicine, for example, patients and relatives have to perceive the effectiveness of telemedicine services and have confidence in a system that may reduce the face-to-face interaction with operators. On the other hand, telemedicine may improve the patient quality of life, reducing the number of hospital days. Cultural aspects have to be taken into account. Older patients who could benefit from e-Health are often change and technology adverse.

At this aim, the European Commission stresses that innovation must be patient/user centered and demand driven. This implies an increased involvement of patients, relatives and operators in the early stages of product development (CEU, 2011).

The **benefits** of innovative solutions have to be assessed with respect to conventional approaches. On one side, there is an economic evaluation that involves all the stakeholders who supply and gain benefits from the innovative products or services. On the other side, the benefits on the quality of life has to be clearly evidenced. This benefit analysis may overcome the potential risks and costs associated with the innovative approach.

This section has aimed to offer a holistic approach to innovative products, showing various key factors – as outlined in Table 7.1 – necessary to ensure market success. This is particularly true in the health sector, where specific market dynamics and commercial–technological barriers apply. Technology is only one enabling element that can facilitate the conception of a new product. The other factors listed in Table 7.1 are also not sufficient, alone, to ensure success. These concepts help in considering the right role for technological innovation in medical devices. For example, the emerging cloud technology and e-Health services, presented in the following sections, are important, but will not represent a 'disruptive innovation' in the medical sector if all the success factors listed in Table 7.1 are not be positively addressed. (A disruptive innovation may be defined as a technology that may create a new market improving the perceived customer value of the existing technology or product.).

7.2 The Cloud Computing Model

ICT has an essential role in improving effectiveness and safety and, at the same time, to address the challenge of cost reductions in the health systems. In e-Health there is a growing diffusion of mature information technologies with distributed software applications for the workflow management as well as for core components of medical instruments. At the same time, such IT components are now integrated into wired and wireless communication technologies. These conventional technologies (e.g., networking, mobile communications) are well known to users and engineers. On the other side, cloud technology is now entering the health sector by replacing conventional non-communicating legacy systems and offering access to health data more simply and quickly. Cloud computing is expected to have a deep impact on the e-Health market once issues related to regulation, procedures and security are overcome. Cloud computing for Health is expected to grow to an estimated $5.4 billion by 2017 with an average growth rate of 20.5% per year from 2012 (Marketsandmarkets, 2012). For this reason, we introduce here the term **c-Health** as '*the set of healthcare services supported by cloud computing*'.

In the following, models, virtual machines, services, architectures and platforms related to the cloud, and in particular to c-Health, will be discussed. From Section 7.3, cloud computing will be applied to specific e-Health products and services.

7.2.1 Basics of Cloud Computing

Cloud computing is defined as a model where hardware and software resources are delivered as a service on a public, private or hybrid network. According to the National Institute of Standards and Technology (NIST), cloud computing enables ubiquitous, convenient and on-demand access to a shared pool of configurable resources (computing, storage and services) on the network, changing the resources provision rapidly and with minimal management effort (NIST, 2011).

The concept of **utility** can be extended also to ICT such as cloud computing, considering that ICT resources can be distributed as a service as is the case with electricity. The utility concept is associated with a **business model** where users pay the providers on the basis of usage ('pay-as-you-go'), which is what you do in order to use services such as water, electricity or telephony. Using this approach, information processing can be delivered as a utility and can be defined as 'on-demand delivery of infrastructures, applications, and business processes in a security-rich, shared, scalable and computer based environment, over the Internet, for a fee' (Rappa, 2004). In this model, users access services that they require when they need, regardless of where the service is hosted and how it is delivered. Several computing paradigms have promised to deliver this utility computing vision and these include cluster computing, grid computing and, more recently, cloud computing (Buyya, 2009). Cloud computing is a term that denotes a model where ICT is viewed as a 'cloud' offering mainly computing power and storage. The model may be viewed as 'software as a service' accessible from anywhere, anytime, worldwide. Although there are many similar definitions of 'cloud', there is an agreement on a few basic common **characteristics** that a cloud should have:

- pay-per-use
- elastic capacity, giving the feeling of infinite resources available
- self-service
- virtualization.

In addition, a common feature of the cloud providers is the availability of application programing interfaces (APIs) and development tools to create third-party applications and to enable the deployment of new services on the customer cloud infrastructure.

The current more conventional model is the generation of computing power in-house in places close to where the ICT services have to be supplied. This approach is mainly justified by the large diffusion of PCs and the familiar unavailability of efficient high bandwidth telecommunication networks. At least in the past years, PCs have allowed us to overcome the limits of the centralized systems in use in the 1970s when computing resources were concentrated in data centers and their use was shared between dozens or hundreds of users who could access applications through simple alphanumeric interfaces. The mainframes were the key elements of the IT systems, and were characterized by high computational power, high investment and high maintenance costs. Adoption of the mainframe approach decreased with the advent of microprocessors. The general unavailability of efficient computer networks forced the information processing capability to be generated and consumed near the demand side.

Using the cloud approach, ICT services can be supplied by large-scale data centers with thousands of computers hosting data and applications. In this context, the hardware virtualization can be considered as the basic element to overcome most of the operational burden of deploying and maintaining data centers.

Virtualization of computer resources, including all their components (memory, I/O devices etc.) aims to share and optimize the computation and the storage capacity of a single physical platform that hosts several operating systems and software stacks. Moreover, the market availability of multicore CPUs has increased the virtualization capability of

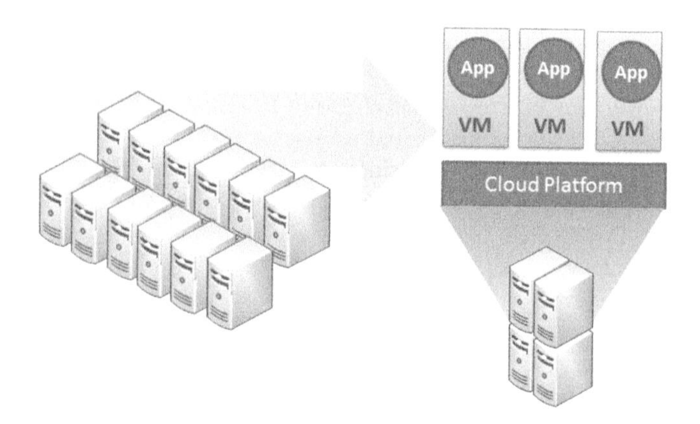

Figure 7.3 Virtualization of resources.

servers, improving their management and reliability. The hardware virtualization allows the introduction of a new one-to-many model with advantages in terms of hardware administration and software update. A virtual server is a **virtual machine** (VM) with an operating system (OS) hosted in a remote data center and some user applications running over this OS. The user experience is similar to what would be experienced by means of a physical PC. As shown in Figure 7.3, this cloud architecture may replace the PC based approach, thus opening a new ICT era (Oswald, 2011).

Virtual servers are only accessible by means of remote desktop programs hosted on a local device. The remote desktop is a software application or an OS facility that allows applications to be used on a local device while running remotely.

The local device can be either a standard PC or a 'thin client', which is a computer with low computational power and a limited operating system and a remote desktop capability.

The **benefits** in the adoption of these scenarios based on ICT virtualization are

- end users work on environments similar to conventional PCs or servers
- applications are identical to those running on PC based local platforms
- administrators can customize and optimize the work environment
- software operations and maintenance activities are simplified.

7.2.2 Cloud Platforms

As shown in Figure 7.4, a cloud hardware virtualization platform consists of a set of virtual machines (VMs), each emulating a single physical computer running applications under its own operating system, hosted by a single computer whose pool of physical resources is dynamically shared between VMs.

Access to the physical resources by individual virtual machines is mediated by a software layer, called a **virtual machine monitor** (VMM) or hypervisor. Each VM is completely isolated from the others, so that errors and exceptions inside a single VM do not affect other VMs. Each VM is fully characterized by its state, which includes a full disk or partition image, configuration files and an image of its RAM. Various configurations may be adopted to implement virtual servers equivalent to the common physical servers such as those

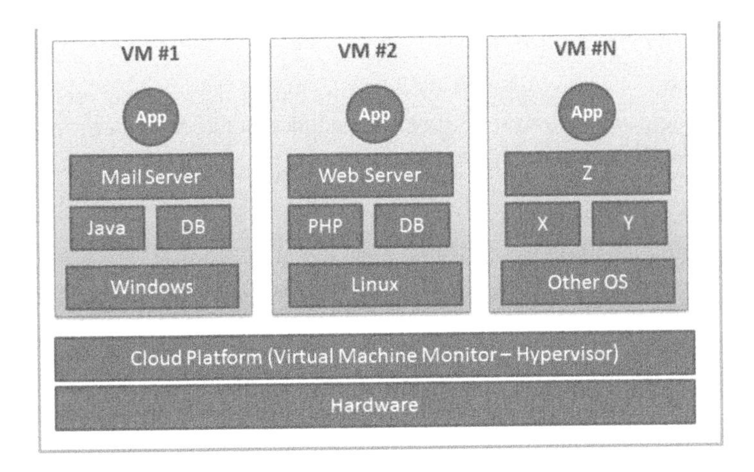

Figure 7.4 Virtualized server hosted on a cloud platform.

employed in data farms. The main VMM open source platforms for cloud computing environments are Xen and KVM. Xen is a open-source hypervisor that can be used as an entry point both for commercial and open-source hypervisors. KVM (kernel-based virtual machine) is a Linux virtualization hypervisor (Buyya, 2011). The main VMM commercial alternatives are VMware vSphere, Microsoft Hyper-V and Citrix XenServer which are briefly described in Table 7.2.

Table 7.2 VMM commercial products

Product	Description
VMware vSphere	VMware introduced its first x86 server virtualization products in 2001, and the related product, vSphere, is one of the more widely used in the market. ESX(i) is a component of a VMware vSphere which offers many advanced features for management and administration, such as vMotion (live migration), high availability, fault tolerance and storage motion. VMware has a powerful infrastructure thanks to the wide choice of third-party tools.
Microsoft ® Hyper-V Server	Hyper-V is the server virtualization platform developed by Microsoft. It has features such as live migration, quick migration and dynamic memory. Microsoft offers its hypervisor in two versions: stand-alone software and a component included in the Windows Server 2008 OS.
Citrix XenServer	Citrix XenServer is a 64-bit virtualization platform with the scalability required by business-critical applications. XenServer provides CPU and memory resource management for the servers that host virtual servers. Citrix XenServer is based on the open source framework XenSource which was acquired in 2007 by Citrix Systems.

7.2.3 Services in the Cloud

The cloud is composed of different IT resources including network, servers, applications, services and storage components. Services can be divided in three main classes according to the associated capabilities and service models:

- infrastructure as a service (IaaS)
- platform as a service (PaaS)
- software as a service (SaaS).

IaaS is the remote usage of hardware resources that are deployed on demand at customer request. IaaS includes virtual resources: virtual machines, servers, storage and network resources. IaaS is the bottom layer of cloud computing systems; users can perform several activities on the virtual resources: machine start/stop, installation of software, virtual disk management and network security management (e.g., security policies, firewall/router configurations etc.).

PaaS is a distributed environment for application development, testing and deployment. PaaS includes virtual resources such as databases, web servers and development tools. These resources are remotely available runtime 'as services', without having to install any software. PaaS offers an environment where developers can create and deploy applications without the knowledge of the physical resources such as processors or memory. In the Paas framework, multiple programing models and specialized services can be offered as building blocks to new applications. For example, Google App Engine is a PaaS offering a scalable environment for developing and hosting web applications.

SaaS is a set of applications exposed to the end customer as a service and deployed on the infrastructure of the service provider that is responsible of the provisioning of all hardware and software resources. SaaS includes virtual resources such as: content relationship management (CRM), e-mail servers, word processors,and so on. Through the SaaS model, the users can access applications by means of web portals, avoiding the locally installed computer programs such as word processing and spreadsheets, thus eliminating the software maintenance burden.

These three levels (IaaS, PaaS, SaaS) can be viewed as a layered architecture where the services on the higher levels can include the underlying layers. On the lower layer, there is the virtual machine monitor that manages the physical resources and the virtual machines deployed on top of the stack and provides additional features such as accounting or billing.

7.2.3.1 Cloud Enabling Platforms

A cloud platform is mainly a hardware-software environment offering cloud abstract services to run software applications. As shown in Figure 7.4, the cloud platform includes virtual machines with their applications software and operating systems. These virtual machines are installed over a virtual machine monitor that offers an abstract hardware layer through which to access the underlying hardware servers. In the cloud scenario, there are many platforms, both in the commercial and in the open source market.

Table 7.3 Commercial cloud platforms

Product	Description
Amazon Web Services[1]	**Amazon Web Services** provides, as a basic service, the Amazon Elastic Computer Cloud (EC2) that is a web service that can assign an application to as many 'computing units'. The company also offers its Simple Storage Solution for storing data. On top of this stack, Amazon has added a whole range of services, from its Simple DB database to a relational database service (RDS) that includes notification and queue service features. On top of the EC2 platform, the customer can run a Linux, Unix or Windows Server with several development tools.
Microsoft Azure[2]	**Microsoft Azure** is a solution aiming to provide an integrated hosting and control cloud computing environment so that software developers are able to create, manage and scale both web and non-web applications by means of the Microsoft data centers. For this, Microsoft Azure also supports a large collection of development tools and protocols such as Live Services, Microsoft.NET Services, Microsoft SQL Services, Microsoft SharePoint Services and Microsoft Dynamics CRM Services. Microsoft Azure also supports Web APIs (such as SOAP and REST) in order to allow software developers to make the interfaces between Microsoft and non-Microsoft tools and technologies.
Google® App Engine[3]	**Google App Engine** allows a user to run web applications written in Python[4]. In addition to Google App Engine, it also supports APIs for the datastore, Google Accounts, URL fetch, image manipulation and e-mail services. Google App Engine also provides for the customer a web-based administration console in order to easily manage the applications running on the web.

Regarding the commercial products, several companies provide solutions starting from EC2 from Amazon[5] up to Azure[6] from Microsoft, and AppEngine from Google.[7] Table 7.3 shows their main characteristics.

Other relevant platforms are the IBM Smart Cloud[8] and Citrix.[9] Open-source software may be of interest in cloud computing to reduce costs avoid single-vendor risk and to customize applications more easily. The two main open-source cloud frameworks are Eucalyptus and OpenNebula, whose features are described in Table 7.4.

[1] Amazon Elastic Compute Cloud (EC2): http://www.amazon.com/ec2/

[2] Microsoft Azure: http://www.microsoft.com/azure/

[3] https://appengine.google.com/start

[4] Pyton is a programing language: http://www.python.org/

[5] Elastic Compute Cloud Amazon Web Services: http://aws.amazon.com/ec2.

[6] Microsoft Azure: http://www.microsoft.com/azure.

[7] Google App Engine: http://code.google.com/appengine.

[8] http://www.ibm.com/cloud-computing/us/en/

[9] http://www.citrix.com/lang/English/home.asp

Table 7.4 Open source cloud platforms

Product	Description
Eucalyptus	**Eucalyptus** has been released as an open-source (under a FreeBSD-style license) infrastructure for cloud computing on clusters that duplicates the functionality of Amazon's EC2 and uses the Amazon command-line tools. Startup Eucalyptus Systems were launched at the University of California, Santa Barbara, and the staff includes the original software architects from the Eucalyptus research project.
OpenNebula	**OpenNebula** VM Manager is an open-source system able to manage many virtual machines. OpenNebula has a virtual infrastructure engine that enables the dynamic deployment and replacement of virtual machines on a pool of physical resources.

7.2.4 *The Cloud Shape*

As shown in Figure 7.5, cloud computing architecture may be implemented in four main deployment models, according to the security requirements and the characteristics of the data (NIST, 2011): public cloud, private cloud, hybrid cloud, community cloud.

In the **public cloud**, the infrastructure, hosted at the vendor's premises, is available to end-user organizations. The single customers, usually small office/home office (SOHO) or small/medium enterprise (SME), have no control of the physical location of the computing infrastructure and the sharing of resources between customers.

A public cloud environment is suited when there are applications shared between many users, as in collaborative projects, or when there are SaaS applications supplied with an acceptable security level.

In the **private cloud**, shown in Figure 7.6, the infrastructure deployed for a single end-user organization is not shared with external entities. The administrators have full control of the

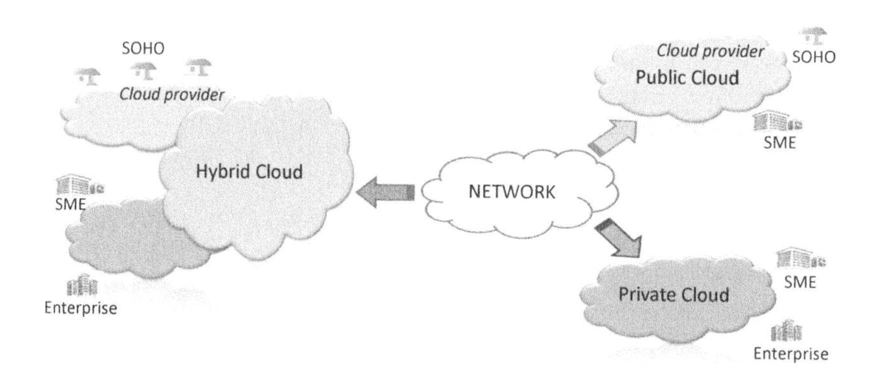

Figure 7.5 The cloud shape.

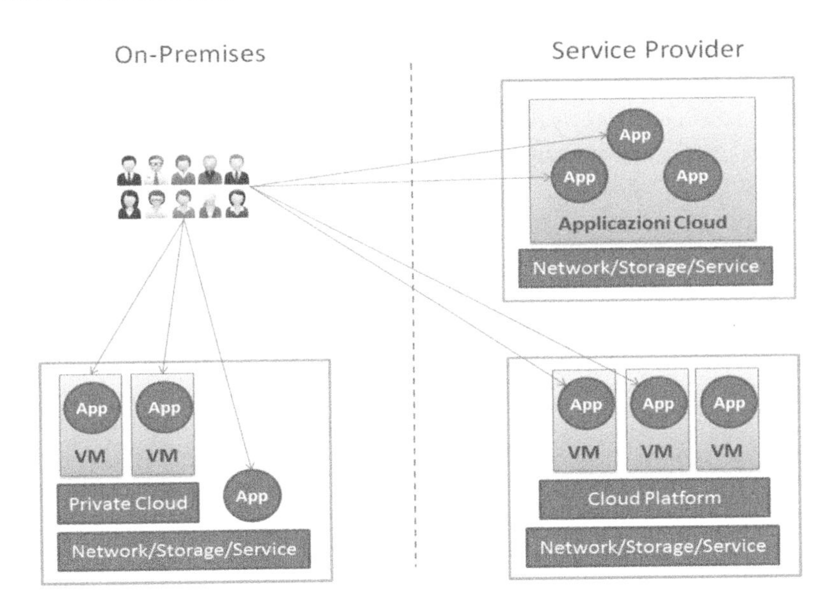

Figure 7.6 Private cloud model.

resources and the security can be managed at the highest levels. The private clouds are mainly of two types:

- on-premises private clouds
- externally hosted private clouds.

In the latter case, the cloud platforms are hosted by a third party specialized in computing infrastructures. An on-premises private cloud is usually selected when the primary business of the company is data and applications, and strict security requirements are mandatory.

The third cloud deployment model (i.e., **hybrid model**) encompasses public and private clouds. Organizations host critical applications on private clouds and applications with limited security requirements on the public cloud. The administrator of a hybrid cloud has to manage multiple different security platforms to ensure that applications can communicate cross different clouds. The hybrid environment is applicable when the company wants to use SaaS applications, applying specific security policies, or when the cloud is used to interact with customers, keeping the data of the organization secure within a private cloud.

The final model is the **community cloud**, where the infrastructure is available to a specific community that shares activities and processes for ICT usage.

Cloud computing also includes storage as a service model. Data is remotely stored in a database or file directory structure. The benefit of data redundancy is counterbalanced by data security and data availability concerns. Data availability risks are addressed by implementing redundant schemes. The security is highly dependent on the security means implemented against attacks and intrusion from outsiders. The confidentiality and integrity of data is mostly achieved using robust cryptographic schemes.

It should be noted that e-Health applications may have strong security requirements. Personal health data must be accessible according to a strict policy to comply with privacy regulations. Service outage may generate unacceptable safety risks by being unable to provide the right patient treatment on time. Corrupted data (i.e., data integrity issues) can also be a source of unacceptable risks.

7.2.5 Features of the Clouds

The cloud model is characterized by five essential characteristics that should be guaranteed by the service providers (Mell and Grance, 2011; NIST, 2011). These key elements have to be implemented according to specific service level agreements (SLA) that formalize the contractual level of the service between the customer and the provider:

On demand self-service
Users of cloud services expect to access resources on-demand, using, customizing and paying for services without the intervention of human operators. The user must be able to manage the services, such as the configuration of security policies or software upgrading through simple interactions with the system.

Metering and measuring
Usage of cloud resources must be automatically monitored and controlled to optimize the load. The user can transparently measure and control resource usage. Metering of several types of services must be allowed, as for example,

- storage
- processing
- bandwidth
- active user accounts.

Elasticity
Elasticity is the capability of managing, in real time, the demand for computing resources coming from applications on the basis of a combination of local and remote capabilities available through an IT infrastructure. And these available capabilities may be rapidly changed, in some cases automatically, in order to scale up and down resource usage. From the user's point of view, the computational resources have to appear unlimited and they can be purchased in any quantity at any time.

Broadband network access
The capabilities of the network in terms of bandwidth and latency must be suitable to allow access from services and resources to a heterogeneous set of devices from mobile equipment to smart phones, tablets and workstations. For example, in a cloud service scenario, mobile devices may be employed to monitor patient health, while smart phones may be useful to send alarms to healthcare operators.

Resource pooling
The computing and storage resources can be pooled to serve multiple users using a 'multi-tenant' model, where the physical and virtual resources are managed dynamically. The

customer generally has no control or knowledge of the location of resources but they can specify high-level requirements on data location determining at least the country or the geographical area where data is stored. This feature may be required by national regulations on digital data management.

The cloud service can be considered as a set of unified computing resources based on SLAs established through negotiation between the service provider and the consumers (Mell and Grance, 2011). The SLA is proposed by the service provider to formalize the delivery of a desired level of quality of service (QoS), to establish the terms of service such as availability or performance and to agree on metrics in order to settle penalties for violating the contract conditions. Through the SLA, the services can be priced on a usage basis, for example, considering the user's activity time (e.g., hours). This model is also referred to as usage-based pricing by the telecoms service providers.

If a user only needs computing resources at specific time periods – say once a week – then there are significant savings when the cloud model is embraced. Let us consider an extreme example: suppose that a customer needs a peak computational power of, say, 100 servers but that all this computational power is needed only for one hour a week. Without the cloud model, the customer would be obliged to pay for and maintain 100 servers that, for most of the week, would be left unused.

Using a cloud business model, the customers can pay for what they actually use, that is, 100 server-hours per week. We now suppose that the cloud supplier charges pro rata, that is, he sells 168 ($=7 \times 24$) server-hours at the same cost as owning and maintaining one server for one week. So, on that pro rata basis, those 100 server-hours will only cost the user 100/168 ($\sim$60%) of the cost of one server and the total cost will be 0.6% ($=1/168$) of what he would have paid for 100 servers.

7.3 e-Health

Adapting from (EC, 2009a), e-Health can be defined as: *the set of ICT based tools and services used in healthcare for prevention, diagnosis, treatment, monitoring and management of health condition.*

The definition encompasses various products that may be grouped into three main areas:

- Healthcare information systems (HIS):

  - *electronic health records (EHR)*
  - *picture archiving and communication system (PACS)*
  - *electronic prescription systems (e-prescription)*

- Telemedicine systems:

  - *telemonitoring: telehomecare, mobile-health devices, etc.*
  - *remote reporting: telecardiology, teleradiology, teledermatology, etc.*
  - *teleconsultation, etc.*

- Specialized devices for healthcare:

– *robotics, advanced systems for surgery, tools for training, grid computing for massive health research, etc.*

In healthcare information systems (**HIS**), there are information systems that manage clinical and administrative data of patients related to diagnosis, prevention, treatment and rehabilitation. These systems may also embed medical workflow through a collaborative model that allows information to be exchanged between healthcare professionals and patients. These systems may be implemented at local level (hospital, clinics, surgeries), at national or at regional level (e.g., European Union) by government authorities that manage healthcare in a specific geographical area.

In the HIS group:

- Electronic health records (**EHR**) are devoted to collecting and storing individual health information in a digital record that can be managed by health professionals. These systems are discussed in Section 7.4.
- In the field of diagnostic imaging, digital image files are replacing the physical X-ray films and paper images. Current healthcare IT systems usually store and manage clinical data archives onsite using internal hospital networks. These systems encompass IT tools, hardware and software applications that manage, collect, store and deliver data. Picture archiving and communication systems (**PACS**) manage the digital storage, transfer and consultation of pictures from image-generating medical instruments. A PACS system consists of a multitude of devices such as servers, storage, networking and workstations, that can host and support several diagnostic applications. The basic element in PACS is the device generating images referred to as a 'modality'. A modality may be a scanner using computed tomography (CT), ultrasound (US), positron emission tomography (PET) or magnetic resonance imaging (MRI). The PACS is currently helping to replace hardcopy images by their digital counterparts that can also be accessed remotely, implementing the telemedicine model. The PACS may also be a component of a HIS in a hospital or healthcare facility interoperating with e-Health systems such as the electronic medical record. PACS provides the storage of electronic images that can be transmitted digitally, eliminating the need to manually transfer files and films. Digital images include additional data, such as scanned documents, that can be incorporated using industry standard formats such as Adobe® portable document format (PDF).
- **Electronic prescription systems** digitally manage the prescriptions from physicians to a pharmacy, possibly embedding all the workflow required by local authority regulation.

Telemedicine is the provision of ICT based healthcare services where actors (patients or operators) are not in the same physical location (EC, 2012b). In this area, **telemonitoring** is a promising area where the patient is connected to medical devices at home, or more recently with mobile technologies (i.e., and m-Health device that relies on wireless mobile technology to transmit the data (Chen, 2010)) that collect and transmit clinical data to HISs, accessible in real time by healthcare operators. In this case, the interaction is between the patient and healthcare professional.

In remote reporting, a specialist sends to patients, or to non-specialist healthcare professionals, reports of clinical data.

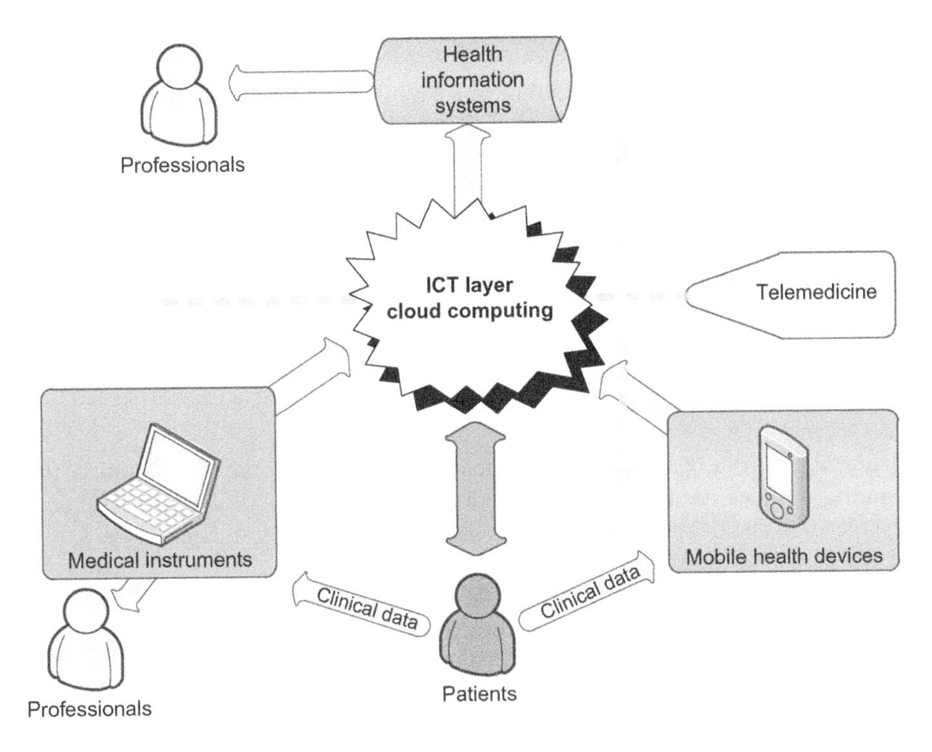

Figure 7.7 e-health architecture.

In **teleconsultation** a healthcare professional interacts with a specialist about a specific diagnostic issue.

The implementation section of this chapter provides additional details with particular regard to telecardiology applications.

7.3.1 Interoperability in e-Health

HISs, telemedicine systems and m-Health devices are experiencing a convergent trend; these components are expected to share data and processes through an ICT layer such as cloud computing. As shown in the conceptual scheme of Figure 7.7, it is expected that medical instruments will transmit clinical data from patients to the HIS through the ICT layer as a preferred operative approach. The same applies for the m-Health devices that transmit data to the HIS. Both these transmission processes come within the concept of telemedicine, where patients and healthcare professionals are in different locations. In this approach, professionals may manage the clinical data through the health information systems 'anywhere, anytime'.

This e-Health model allows the sharing of patients' historical clinical data, thus enabling collaboration between all healthcare stakeholders. Up-to-date clinical history of patients is a key issue in supporting diagnosis and treatment for professionals. This may also be important when patients need healthcare services far from their native geographical area. The model of Figure 7.7 has been proved to significantly improve safety, effectiveness and quality of patient healthcare services (EC, 2009a), and therefore regulatory bodies are strongly supporting this model worldwide. For example, in the directive on patients' rights in cross-border healthcare (EC, 2011c), the European Commission is encouraging the implementation of widespread access to on-line medical records and to telemedicine services by 2020.

This implies distributed interoperable HISs or a single centralized European HIS exchanging data through shared protocols with a variety of sources: administrative national data banks, medical instruments and even m-Health devices.

Following the EU action plan, this implies the following main actions:

1. implement secure on-line access to national health data systems
2. define a shared minimum set of patient data to be accessed/exchanged across countries
3. identify widespread standards, interoperability testing and certification of e-Health systems.

A major issue from the medical instrument point of view is the set of **standards for exchanging clinical data**. Instruments should be able to exchange data on common, widely accepted standards, hopefully in a seamless flow of information among the different devices. This vision generates interoperability concerns that have an impact on safety and effectiveness, especially when devices monitor critical vital parameters. For this reason, regulatory and standards bodies worldwide are recommending focusing on interoperability issues related to the integration of medical devices in e-Health systems (CEU, 2011; NIST, 2011).

Interoperability is a major challenge for medical instruments and in general for the whole healthcare delivery chain.

In general, interoperability should be analyzed at four levels (EC, 2012c):

- *Technical:* technical components (HIS, devices etc.) must be able to exchange health data. This implies standards for protocols message exchange, data formats, security, open interfaces and so on.
- *Semantic:* the exchanged data must have the same meaning for health stakeholders across the different professional background, culture and languages.
- *Organizational:* interoperable workflows and common goals must be shared across cooperating organizations.
- *Legal:* legal frameworks must be compatible across different geographical areas.

In general, interoperability is expected to reduce the cost of the healthcare systems through the replacement of many vendor-specific non-communicating systems to standard systems and to reduce development cost to manufacturers.

In this section, we will focus on the **technical interoperability** and communication standards that are of major concern in the area of medical instruments. We define healthcare interoperability as the capability for heterogeneous healthcare components (healthcare information systems, medical devices etc.) to share data. This means that a seamless flow of information between multivendor, heterogeneous plug-and-play devices and systems over wireless or wired networks is exchanged for use by healthcare operators and patients.

Figure 7.8 helps in clarifying what interoperability means in practical terms. Two healthcare entities such as HISs or medical devices exchange data. Data may span from vital parameters (heart rate, blood pressure, ECG, patient's oxygenation) to PACS images, EHR

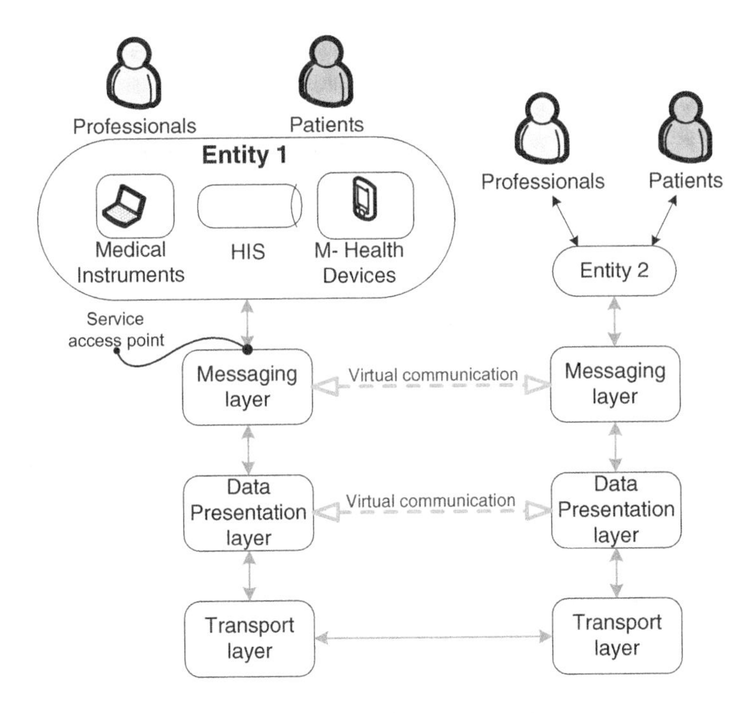

Figure 7.8 Interoperability conceptual model.

contents or even administrative data. First, data has to be exchange following a common messaging protocol specifying procedures to exchange information. Database queries and HTTP web standards are widespread examples. We can consider that any entity has internally an interface to a protocol messaging layer. The interface, usually referred to as a service access point (SAP), contains functions that can be invoked to send and receive data to the other entity. The messaging layer performs all the activities for exchanging data with the message layer of the counterpart.

This is the classic protocol layering model used to design and describe communication protocols and models such as TCP-IP or ISO-OSI described in Chapter 3 (ITU, 2012). To exchange data, the two messaging layers use other lower layers. It can be considered that any layer has a 'virtual communication' with the corresponding layer of the counterpart using its lower layers. In our simplified model, we consider a specific layer that handles the data presentation format. This layer encapsulates and adapts received data according to a common scheme of packet format that may be proprietary or compatible with standards such as the ISO/EN 11073, suitable for specific medical instruments (pulse oximeter, blood pressure monitor etc.).

An additional layer, named 'transport', is charged with sending packets to the other entities using the available communication channel. For example, the transport protocol may rely on serial interface RS-232, USB, Bluetooth, Wi-Fi or TCP/IP standards.

To exchange a piece of information, Entity 1 calls a specific function available in the messaging layer interface (e.g., send (heart rate = 60)). The messaging layer implements all the associated procedures to exchange messages, for example, waiting for an acknowledge from the counterpart entity before confirming that the message has been sent. The message layer uses the functions available at the data presentation layer that formats the packet according to the type of information. The data presentation layer in turn sends the formatted information to the transport layer to be sent to the counterpart.

Interoperability means that each layer in the layer model of Figure 7.8 is compatible with the counterpart's layer, that is:

1. the transports layers allow exchange of raw data between entities in a reliable way
2. the data presentation layers adapt the data to a common format
3. the messaging layers have common procedure for exchanging messages.

The implementation of widely accepted standards with stringent interoperability tests is a key element in enabling interoperability at each level. Table 7.5 shows some relevant standards related to medical devices (Moorman, 2010).

The left column shows the layer names according to the simplified model of Figure 7.8, while the right column shows the corresponding ISO/OSI layer model (ITU, 2012). Transport protocols in the bottom rows of Table 7.5 have to be distinguished in four cases according to the communication channel type. A device may be physically connected to another device usually through serial lines or USB. Devices may also be connected by wireless to a personal area network (PAN). In this case, Bluetooth and ZigBee technologies are preeminent. Device and entities may also be physically connected to a local or wide area network (LAN, WAN) such as in the case of HISs, or they may rely on Wi-Fi or mobile derived standards for a wireless LAN, WAN connection.

Table 7.5 Standards for health communication

<table>
<tr><th>Layers in the simplified model</th><th colspan="4">Standards</th><th>ISO OSI reference</th></tr>
<tr><td>Messaging</td><td colspan="4">Hl7, IHE XDR, DICOM, IEEE 11073- 20601,</td><td>Application</td></tr>
<tr><td>Data presentation</td><td colspan="4">Hl7, DICOM, IEEE 11073-104XX, CCR</td><td>Presentation Session</td></tr>
<tr><td rowspan="4">Transport</td><td rowspan="3"></td><td rowspan="3"></td><td>TCP</td><td>TCP</td><td>Transport</td></tr>
<tr><td>IP</td><td>IP</td><td>Network</td></tr>
<tr><td>PPP, IEEE 802.2</td><td>IEEE 802.2</td><td>Data</td></tr>
<tr><td>RS-232, Serial line, USB,</td><td>Bluetooth, IrDA, IEEE 802.15.4 (ZigBee)</td><td>IEEE 802.3 (wired LAN)</td><td>IEEE 802.11 (Wi-Fi), Telephone protocols</td><td>Physical</td></tr>
<tr><td></td><td>Wired</td><td>Wireless (PAN)</td><td>Wired (LAN/Wan)</td><td>Wireless (LAN/Wan)</td><td></td></tr>
</table>

Regarding messaging and data protocols, **DICOM** (*Digital Imaging and Communications in Medicine,* standard ISO 12052) is a widely used standard for managing, storing, printing and transmitting data for medical imaging. The standard, conceived by the US National Electronic Manufacturers' Association (NEMA, 2012), includes a data model, communication protocols and workflow. The DICOM standard was created to enable communication between all radiology imaging devices in order to facilitate their seamless integration into filmless radiology. It is useful for telemedicine and for teleradiology, where medical images are exchanged between remote locations. The DICOM data model includes additional clinical information; beyond the binary images, there are many attributes that describe imaging parameters and patient conditions (e.g., imaging-specific treatments and preparations, as for example the injection of contrast agents that the patient has gone through). DICOM also provides space to include manufacturer-specific attributes for backward compatibility.

DICOM, used in most hospitals worldwide, is part of the Integrating the Healthcare Enterprise (IHE, 2012) initiative aimed at helping vendors and users to integrate established standards of different imaging and information systems. HL7 and XDS standards are also widely used in healthcare facilities to exchange clinical data. Health Level Seven International (**HL7**, 2012) is a set of ANSI-accredited standards addressed at providing a common framework for electronic health information exchange. For improved interoperability, DICOM has been combined with HL7 within the standard XDS.

The cross-enterprise document sharing (**XDS**) is the integration profile defined by the IHE in order to provide a standard specification for health data sharing between healthcare entities (IHE, 2012). The XDS-I profile extends the XDS to the imaging domain.

Another emerging ICT family of standards is the EN ISO/**IEEE 11073** related to 'Health informatics – Medical/health device communication standards'. The standard addresses real time plug-and-play interoperability between medical, healthcare and wellness devices focusing from the physical layer to the abstract representation of the information.

Cloud computing is also expected to help e-Health interoperability by fostering a global integration of medical resources. Current local non-connected e-Health systems may be integrated in one or many federated health clouds thus evolving into **c-Health** systems. The realization of c-Health systems requires the implementation of interoperability among all associated components. For this, standards such as the Continuity of Care Record (CCR) may help. CCR is a standard used to exchange patient data between carers and to keep updated patient history within an electronic health record (ASTM, 2005). This standard has been used in c-Health and web based solutions since data description is based on XML; this improves the data exchange capability.

The cloud should not to be considered as a simple technology but as a new paradigm for conceiving ICT health systems. Through **c-Health solutions**, the health organizations is able to transform their processes in order to serve patients and partners, while reducing costs, simplifying management and improving the quality of services. This approach will change the operative processes modifying the management of the information and the resources and combining security and reliability. The operative models can be changed, adopting a flexible range of options starting from on-line health and wellness tools, for example for remote monitoring, to data and image storage and sharing in order to address the need of long-term digital preservation. C-Health is expected to foster interoperability at all levels from technology to semantic, organization and legal level.

7.4 Electronic Health Record (EHR)

Many organizations have provided **definitions** of EHR. The international Organization for Standards (ISO) defines a basic generic EHR as 'a repository of information regarding the health status of a subject of care, in computer processable form'. One of the key features is the ability to share patient health information between authorized users of the EHR supporting continuing, efficient and quality integrated healthcare (ISO, 2005). The EHR can also be defined as a 'longitudinal collection of electronic health information about individual patients and populations'. EHR can be 'the central element of an integrated information framework, collecting data coming from different sources including personal, public and private institutions for improving the quality of the care' (Gunter, 2005). According to the US Department of Health and Human Services (HHS), EHR is a 'Health-related information on an individual that conforms to nationally recognized interoperability standards and that can be created, managed, and consulted by authorized clinicians and staff across more than one healthcare organization.'

EHR is also referred to with similar meaning as electronic medical record (EMR) with a focus on clinical data, electronic patient record (EPR) more focused on hospitals' acute care and computerized patient record (CPR) (ISO, 2005).

In general, the EHR has the **purpose** to provide clinical information to support healthcare services of a specific patient by the involved operators. EHR can be considered as a collaboration tool for all health service suppliers and the specific patient.

The capability to **share clinical patient data** across different healthcare contexts is the key driver for the deployment of EHR systems, and a main benefit of EHR is the access to a patient's medical records by different health entities: hospitals, doctors and the patients themselves. Additional benefits are expected in the improvement of the health-related processes and the telemedicine services.

The **contents** of EHR differ between countries, vendors and target users. In general, based on ISO 18308 standard and other sources, the main patient data to be shared are outlined in Table 7.6.

Figure 7.9 shows the **external entities** that have to be interoperable with an EHR to share data. Referring to Table 7.6, administrative data is usually included in public information systems that should also provide some security authentication mechanism. Security has to be considered with great care since health related data is sensitive and is protected by privacy legislation. Clinical data is supplied by physicians with the help of the patients. Nurses and physicians usually need to record annotations in patients' health records. This means that health operators must have access to the EHR of the individual patients through proper security mechanisms.

EHR must include requests and orders by physicians that are useful for recording patient clinical histories. This data may also be useful for automating workflow with pharmacies. Diagnostic test results are a main component of EHR that has to be compatible with the specific data formats to store results (ECGs, PACS images etc.). In addition, EHRs have to be connected to PACS systems, laboratory information systems and medical instrument to share results. The connection with pharmacy network information systems is helpful to implement e-Prescription procedures to manage prescriptions from physicians to pharmacy. EHR may also implement the automatic sharing of monitored vital parameters transmitted by telemedicine medical instruments.

Table 7.6 EHR data contents

1. administrative data
2. clinical information
   - *patient history*
   - *physical examination*
   - *psychological, social, environmental, family and self-care information*
   - *allergies and other therapeutic precautions*
   - *preventative and wellness measures (vaccinations etc.)*
   - *therapeutic interventions (medications, procedures etc.)*
   - *clinical observations, interpretations, decisions and clinical reasoning*
   - *problems, diagnoses, issues, conditions, preferences and expectations*
   - *healthcare plans, health and functional status and health summaries*
3. healthcare operators' annotations
4. physician requests/orders
   - *prescriptions, treatments, other therapies and tests*
5. diagnostic tests results
   - *imaging, ECGs, laboratory analysis etc.*
6. pharmacy related information
7. data from remote monitoring telemedicine systems
8. alerts and reminders

This will allow for a deep patient control with the implementation of alerts based on real-time clinical data. EHR may also include reminders for dates related to health events. This function requires interoperability with scheduling systems. Referring to the left side of Figure 7.9, an EHR must also exchange data with healthcare facilities and more importantly, with emergency systems to provide proper care treatment based on patient history.

EHR systems must be able to provide access to a huge amount of clinical data for patients and health facilities. EHR data exchange implies interoperability based on specific **standards** as discussed in Section 7.3.1 and Table 7.5 with the possibility of importing and exporting data using major standards such as HL7 and DICOM. A seamless EHR patient data exchange between healthcare facilities reduces errors, avoids duplicates in diagnostic testing and improves safety and effectiveness especially in chronic disease management.

The model outlined in Figure 7.9 is suitable at a national level. Less complete models may be tailored to suit smaller entities. In general, EHR may be implemented at local level by care communities (departments, hospitals, clinics etc.), at regional or national level and at transnational level as fostered by the European Commission for more effective cross-border assistance.The adoption of a centralized EHR approach at transnational level can bring **benefits** both for healthcare professionals and for patients in order to access reports, scans, EMRs, prescriptions, patient information, clinical history, insurance claims, prescriptions and lab reports from anywhere (EC, 2010). The deployment of a central repository with a digital copy of the citizen health record mitigates the risk of misdiagnosis or wrong prescription/medication and eliminates the risk of conflicting treatments where multiple healthcare professionals are involved.

The EHR systems can be categorized mainly into **local EHR systems** and **shared EHR systems**. In many healthcare systems, health facilities or providers maintain their own local

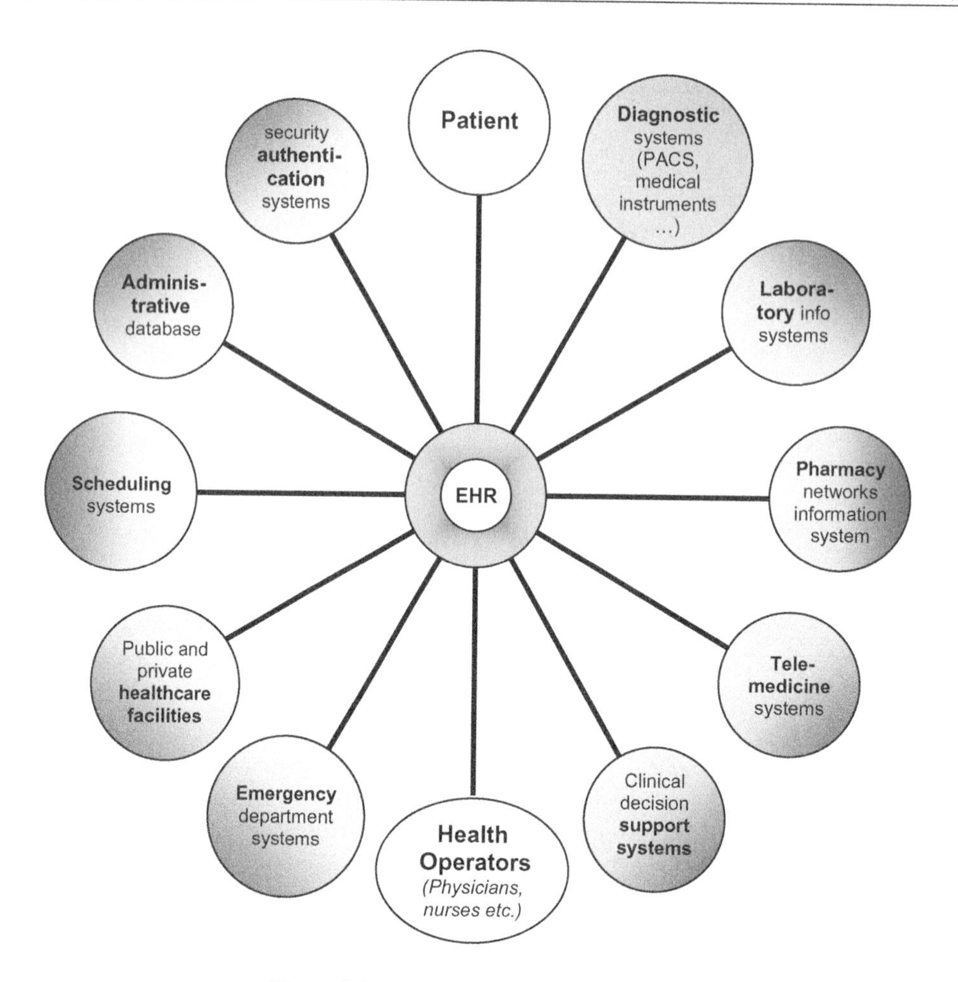

Figure 7.9 EHR system interconnections.

patient data, whether manual or electronic or a combination of both, using a classic client-server approach. These systems store data in-house over servers that use specific hardware and software to be installed in-house. In local EHR systems, access to the information is usually restricted to authorized health professionals within the particular facility. However, in shared EHR systems, data is stored on external servers and can be accessed by any authorized user via a web interface, requiring only a computer with an Internet connection. Shared EHRs are expected to evolve into cloud based c-Health systems that will increasingly replace traditional in-house solutions.

EHR may also be supplied **as a service** according to a c-Health (cloud) approach. There are already companies providing medical solutions as a service and there are web based solutions targeting medical and healthcare services. EMR, PHR and HIS may be currently implemented with cloud approaches and some large corporations (e.g., Microsoft with the HealthVault™ ecosystem) are getting into the market providing their services and solutions.

In the e-Health environment, the main improvements that may be addressed by service based EHR are

- consolidation of document assets towards an interoperable EHR
- management of digital signatures and secure access to data
- image and document archiving with legal validity and long-term preservation.

The main benefits for a health facility may be the opportunity to pay only for the used services. This solution is more scalable and reliable and it ensures some long-term benefits:

- increased IT responsiveness and efficiency
- reduced CAPEX and operational overhead for maintenance
- greater flexibility through an on-demand, pay-as-you-go model that scales with the healthcare facility size
- improved interoperability that ensures better collaboration between health professionals worldwide.

Moreover, with the increased exchange of diagnostic imaging, there is the need to simplify **image-sharing** processes across healthcare organizations.

Cloud computing can offer a cost-effective and scalable solution to all the previous items, for small-medium size medical centers through web-based EHR systems. Indeed, in the **c-Health approach**, the implementation of EHR is simpler because client software runs on web browsers without specific hardware or software installations. The deployment of cloud based EHR systems has benefits both in terms of reduction of time for the activation of services and in terms of cost savings. This is a benefit for small-medium facilities where a local EHR may be too capital intensive (Mosquera, 2011). Since cloud based EHR requires no hardware installation and a different approach for software licenses, the cost of implementation can be greatly reduced. The health facility can pay a monthly fee as per any utility bill. The IT resources can be significantly reduced if medical applications move to the cloud. Instead of requiring equipment purchase and professional activity to install, configure, test, run, secure and update hardware and software, cloud based EHR is managed by the service provider.

The web based interface and the ability to access the system from anywhere, allows physicians, staff and patients to collaborate more effectively with a better service continuity. So national policies are fostering interoperability for PACS and EHR. For example, the US Center for Medicare and Medicaid Services (CMS) and the Office of the National Coordinator (ONC) for Health Information Technology are supporting EHR deployment envisaging that more than 40% of all scans and tests ordered by eligible providers or hospitals will become digitally accessible through certified EHR technology systems during 2013. As pointed out in Table 7.1, these significant changes may be implemented on a large scale only if the stakeholders are given strong reasons to accept the new approaches and the risks are properly addressed.

7.5 c-Health

In ICT for healthcare, also referred to as health informatics, the growing need of shared data, in particular for medical images, is leading to scalability and maintenance issues. Within c-Health (i.e., cloud enabled e-Health) computing and storage in the cloud can be an effective solution to achieve high data availability and reliability performance and disaster recovery facilities.

Most health services and applications can be offered as SaaS (Software as a Service see Section 7.2.3 for details). Cloud computing providers may supply open-source web based c-Health solutions or vendor-specific applications.

Cloud computing enables the access to applications and documents from anywhere, by means of an Internet connection, overcoming traditionally heavy approaches such as the delivery of physical support in PACS. Current technologies are now suitable for storing and also transfering the heavy formats used for imaging, and healthcare facilities are expected to switch to fully digital virtualized solutions. In the c-Health SaaS approach, the user does not have to manage infrastructure and software applications, so the **key benefits** are

- cost effectiveness in setup, maintenance, update and upgrade
- safety
- security
- broadband connectivity
- improved collaboration and sharing
- service reliability and availability.

In this context, we can define the c-Health service as the set of e-Health applications accessed via web interfaces to support all the healthcare processes at an appropriate level of safety, effectiveness and quality. With this approach, patients and operators will be able to access healthcare applications via a web portal without relying on specific computer programs or local computer storage.

We recall that when new services are deployed on the cloud, some **preliminary conditions** should be satisfied, that is, the availability of

- sufficient broadband network capability for all connections for hospitals, physicians, patients and service providers
- a private/public cloud architecture
- a roadmap for the migration of services.

When data is stored on the cloud, main concern for patients is **privacy** and **security**. Patients may fear unwanted access to their personal health information. Systems must usually satisfy severe requirements including strong authentication, logging of all transactions, strict security policies and data encryption. Security of health systems is addressed by the ISO 27799 standard on health informatics that shows best practices to protect health related IT systems. The standard includes controls required for managing health related security, and guidelines and best practices.

The **regulation** framework usually imposes additional **requirements** on digital contents:

- data integrity ensured by digital signature
- timestamp certification
- legal validity for archive procedures and long-term preservation
- strong authentication mechanisms to access documents.

Table 7.7 c-Health risk analysis

ID	Risk description	Availability	Integrity	Confidentiality
1	Compliance risks (standard and regulation compliance may be impacted when migrating to cloud services	✓	✓	✓
2	Loss of governance from the customer in many areas such as security service level agreements (SLAs)	✓		✓
3	Impossible migration across providers	✓		
4	Isolation failure between separated cloud services			✓
5	Management interface compromise: due to the deficiencies of the used web browsing interfaces	✓		
6	Data protection		✓	✓
7	Insecure or incomplete data deletion			✓
8	Malicious insider			✓
9	Service unavailability	✓		

The design of a cloud solution has to consider possible additional requirements from the applicable legislation. For example, some countries do not allow privacy-sensitive data to be physically stored outside the national borders. This implies additional constraints on the service provider. Other requirements may concern: interoperability among clouds, system portability between providers, service level agreements to ensure continuity and accessibility, specific requirements on data storage and backup for the provider. In considering migration to a cloud approach, healthcare customers also have to evaluate the typical risks of cloud based solutions. Table 7.7 shows typical risks, adapted from (ENISA, 2009a).

Following (ENISA, 2009b), we define '**availability**' as the 'degree to which a system or equipment is operable and in a committable state'. **Integrity** has to ensure that 'data is whole or complete; data must be identically maintained during any operation (such as transfer, storage or retrieval)'. **Confidentiality** guarantees that information is accessible only to authorized users. Referring to Table 7.7, the main risk is the compliance to regulations that may be more easily obtained when using an in-house solution, but it may be cumbersome when outsourcing. The other risks may have a deep impact on confidentiality and availability. For example, a SLA that ensures 99.5% availability – as in some standard public sector on-line services – is critical since it is expected that unscheduled downtime may occur for around 44 hours per year. A 99.9% availability, ensuring a maximum of 9 hours unavailability per year is usually considered sufficient for critical public services, but critical health services may experience unacceptable risks with such a service level. Notably, 99.999% availability is probably sufficient since downtime is reduced to 5.3 minutes per year. Table 7.7 clearly indicates that protective measures have to be implemented when migrating to cloud solutions to reduce risks to acceptable levels.

c-Health systems are expected to have a wide diffusion in the coming years with an estimated growth to $5.4 billion by 2017 (Marketsandmarkets, 2012). On one side they offer clear benefits, but their realization is conditioned by the elimination of unacceptable risks and the satisfaction of all the stakeholders that have to find strong perceived advantages in embracing these new solutions.

Part II: Implementation

In the theory part, we have discussed models, technologies and solutions for c-Health and e-Health. A main e-Health sector is telemedicine which, according to research institutes, is expected to have a high growth in the coming decades thanks to

- advances in technology
- improved patient perception
- pressure on cost reduction for the public health sector
- shortage of professionals.

Telemedicine is defined as the provision of healthcare services through the use of ICT when the actors (patient and professionals) are not in the same physical location (EC, 2012b). Telemedicine addresses specific needs in the health sector:

- treatment of chronic patients or individual patients in the usual patient environment
- health related knowledge distribution
- improvement of the quality of life of older people
- patient empowerment
- more efficient distribution of highly specialized healthcare professionals.

The main associated areas are

- teleconsultation and second opinion
- telecardiology
- teleradiology
- telepathology, teledermatology
- telemonitoring and home care
- training.

The implementation part of this chapter will focus on telemonitoring and telecardiology which address problems (i.e. cardiovascular diseases) accounting for the main causes for death worldwide. Telecardiology is the most consolidated and widespread service within telemedicine, with an increasing trend. In the area of telecardiology there are numerous scientific clinical studies analyzing outcomes and many suitable medical instruments. Organizational processes are also diffused; telecardiology may be implemented according to various business models such as: business to business (B2B), business to business to consumer (B2B2C) and business to consumer (B2C). After a brief description on telecardiology in Section 7.6, the medical device technology is addressed in Section 7.7; the associated organizational processed are focused on in Section 7.8 and risks are outlined in Section 7.9.

7.6 Telecardiology

Telecardiology may be defined as an application of telemedicine that enables the delivery and management of clinical cardiology data through the use of ICT. The main objectives of telecardiology are to

- record and share ECGs and vital signs without the patient and the specialist having to be at the same premises
- summon the emergency services in case of detection of critical conditions
- do follow-up treatment with a 'stress free' approach for the patients
- optimize therapy through ongoing monitoring
- improve care and quality of life of patients in their home
- reduce patient transfers to the health facilities
- reduce inappropriate hospitalizations
- reduce unnecessary emergency room visits.

Telecardiology is one of the most consolidated medical disciplines in the area of telemedicine and e-Health (Birati and Roth, 2011) with services supplied at the patient residence; these services are referred to as 'home telehealth' (Pecina, 2011).

In addition to the provision of care for patients with heart disease, telecardiology helps patients to improve their compliance to the therapy and to support them in a more healthy life-style, interacting day by day with a platform that can have educational content. Moreover, unnecessary transportation of patients can be avoided by using remote expert counseling. Finally, patients can receive second opinions, and physicians can consult other experts; such possibilities have proven to have a beneficial effect on both patient survival and recovery. Some studies on telemonitoring in congestive heart failure and other chronic diseases have been demonstrated to decrease hospital readmission rates (Antonicelli, 2008), the number of bed-days of care in hospital and emergency department visits (Pecina, 2011) and, last but not least, they reduce mortality (WSD, 2011).

Among the most common applications of telecardiology there is the possibility of recording an electrocardiogram, through mobile devices, and transmitting it in real time to a service centre for specialist reporting, storage and subsequent analysis.

The transmission of the clinical data take place by means of traditional analog trans-telephonic systems with the public switched telephone network (PSTN) or through digital communication with the usual IP networks (xDSL, GPRS, UMTS, satellite networks etc.).

The main instrument for the treatment of cardiac patients is the ECG with transmission capability. As shown in Figure 7.10, for multiparameter monitoring, the ECG may be used with other telemedicine instruments.

In addition, the ECG and the blood pressure can also be detected over a period of several hours by using Holter recorders.

7.6.1 *Application Scenario*

Telemedicine has been achieving increased success, mainly due to the following factors:

- the increased aging of the population
- the budget constraints of national health systems

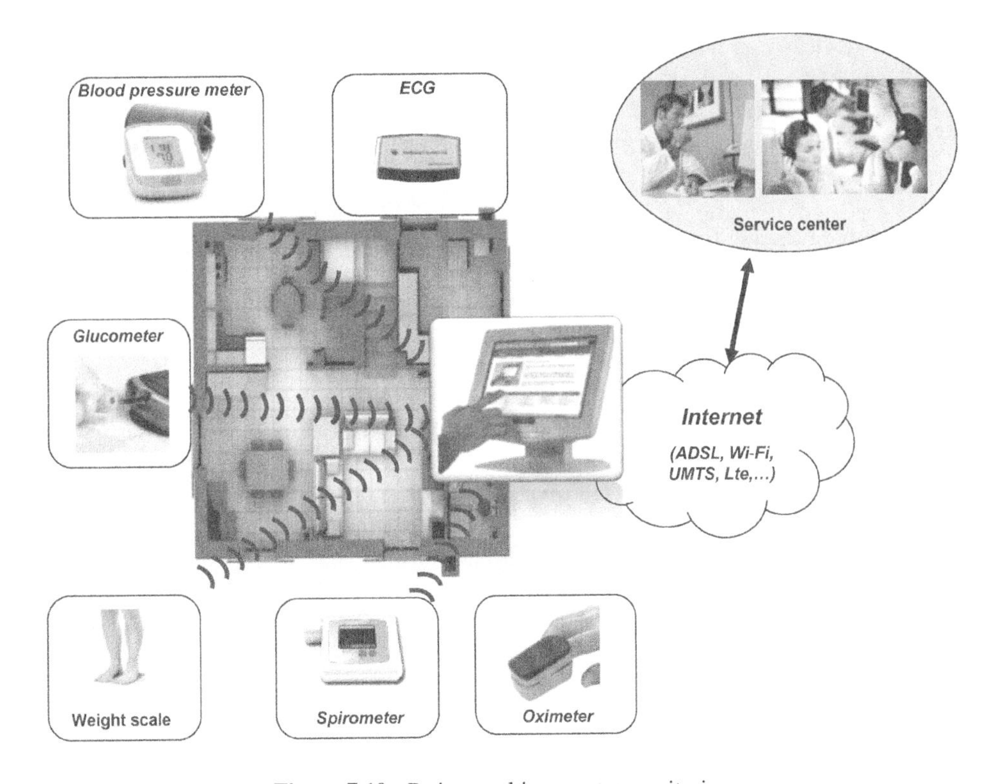

Figure 7.10 Patient multiparameter monitoring.

- the reorganization of healthcare services away from hospitals by delegating them either to local health facilities or to the patients themselves, who are trained to manage their own disease (patient empowerment)
- the need to provide a uniform level of care for people living in remote areas.

In this context, telecardiology and telemonitoring could provide a major contribution, but effective telemedicine services require specific ingredients:

1. suitable technology
2. committed operators involved in the service (cardiologists, health operators etc.)
3. in place procedures to supply the service
4. proper organization with a service center
5. a business model to satisfy all suppliers.

Some of these key factors will be analyzed in the following sections.

7.7 Telecardiology Technology

There are many telecardiology devices on the market that can be grouped into

- home devices
- professional use devices.

Both the devices can transmit ECGs to a service center for archiving, displaying and reporting, but home devices can be used by the patient themself or their relatives/carers, after a short and simple training. These devices must be certified for an intended use that specifically includes non-specialized or non-health personnel. This means that a specific design with proper labeling and instructions has to be performed by the manufacturer so that the intended users can manage the product without generating unacceptable risks in terms of safety and effectiveness. These certification issues will be addressed in Chapter 8. In home use devices, users may only record and possibly transmit the ECG since only cardiologists can write an ECG report. Professional devices have to be used by health operators, and they include tools for doing ECG reports.

Professional devices may have the following features:

- real time ECG display or print to check correct electrode positioning
- international standard (e.g. SCP, XML FDA) for ECG transmission and data exchange to help integration with clinical information systems
- certified ECG quality according to standards used in hospital devices
- storage of multiple ECGs recorded from different patients
- transmission capability of ECGs through wireline or wireless networks
- pacemaker detection and defibrillator proof (this latter compulsory for in emergency environments)
- automatic or assisted interpretation.

Home devices may miss some of the previous features considering that they are often associated to a unique user, that is the patient. In turn, personal devices have to provide improved capability in lead positioning so that users are able to position electrodes correctly. Recent devices have both the professional features and the usability required for non-trained personnel or for personal use. For example, Gamma Cardio CG has two types of graphical user interfaces, one for cardiologist resembling to the classical ECG devices and one for patients or non-trained operators with a step by step straightforward procedure to record and transmit the ECG signal.

Holters are another class of devices that has a reduced quality with respect to diagnostic ECGs since they usually have a frequency response from 0.5 to 40 Hz compared to diagnostic ECGs with 0.05–150 Hz bandwidth. On the other hand, Holters have the capability to monitor electrical heart activity for at least 24 hours. Several devices for cardiac Holter monitoring can transmit data to a service center with appropriate receiving stations This allows remote cardiologists to monitor patients and write the associated reports.

Technology is improving the devices: patient burden is now limited and the level of monitoring is increasing through real time transmission of a greater number of parameters. The advances in mobile technologies allow the design of mobile devices that can monitor patients in real time in daily life, using mobile cellular networks. These devices, while interesting from a technology point of view, have to be accepted at clinical level by all stakeholders following the criteria included in Table 7.1. This is currently limiting the diffusion of such devices.

Wearable devices represent a frontier for non-invasive heart activity monitoring. T-shirts with built-in sensors can measure the ECG signal and other parameters such as

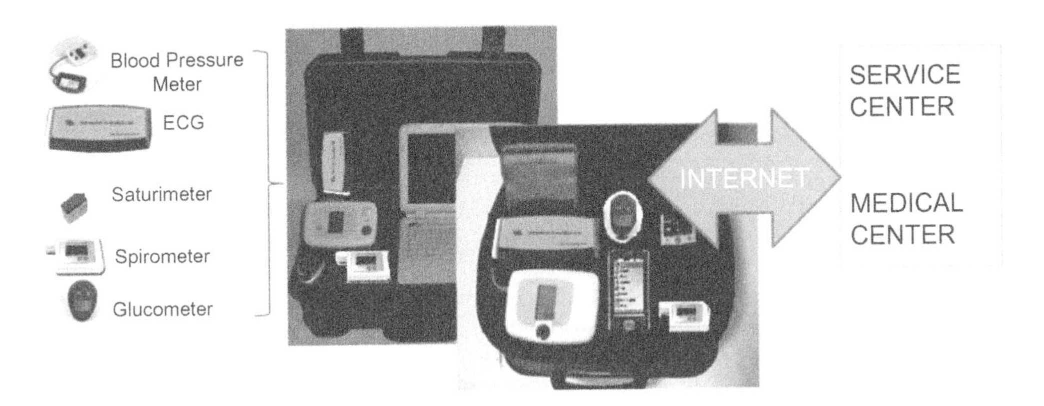

Figure 7.11 Telemedicine briefcases.

respiratory rate, etc. (Patel, 2012). The T-shirt with sensors can be washed, and the data can be transmitted to the receiving station by a mobile phone. Among wearable devices, **wrist devices** with sensors and possibly mobile connectivity are increasing their presence. These devices look similar to watches with the capability to monitor parameters such as heart rate, blood pressure and oximetry and to transmit their values. Although appealing from the technological point of view, they have to be well integrated within usual clinical practice. The factors of Table 7.1 may help in showing the weaknesses of such products.

Health telemedicine units, as shown in Figure 7.11, are organized to contain a set of medical instruments that record the main diagnostic parameters and transmit the data to service centers. These units are useful for providing medical services remotely, using central healthcare facilities. Data can be transmitted to the service centers where specialists manage diagnosis, treatment and rehabilitation. Portable units usually contain medical instruments (ECG, blood pressure, saturimeter, glucometer etc.) connected to a PC/device that transmits data to the remote facilities. The PC has a wireless connection to the Internet via mobile, wired or satellite links, etc.

These units may be helpful for emergency situations in remote areas, especially when including an automated external defibrillator. In Europe, there are annually 700,000 cardiac arrests with a survival rate of only 5–10%. Within 5 minutes of a cardiac arrest, the first brain injuries occur. In a heart attack or acute myocardial infarction (AMI) – the survival rate is inversely proportional to the time before medical treatment starts: 50% of deaths happen within the first hours after an AMI. For this reason, guidelines on AMIs suggest that treatment has to start a soon as possible. Unfortunately, the choice of treatment depends on the type of AMI, that can be detected through a diagnostic ECG. A health telemedicine unit, managed by a general health professional, could then be used for an ECG, transmitting the result to a cardiologist, who could recommend the proper treatment. In addition, he automated external defibrillator (AED) is helpful in the case of potentially life-threatening cardiac arrhythmias or cardiac arrest.

The **service center** is the technical and organizational structure that receives the data transmitted by the devices, manages patient data and stores ECGs in an EHR. The reports of the specialists are also included in the EHR which implements security policies to grant controlled access to front-end operators, physicians, specialists, nurses, patients and carers.

A service center is made up of a receiving system used by front-end operators and specialists. The receiving system usually performs the following tasks:

- receive data from devices
- identify device and patient
- store the data in an EHR
- generate automatic alarms triggered when parameters are outside predefined limits
- provide secure access to data to authorized health specialists
- display data with specific viewers and tools for reporting
- show historical clinical patient data
- include tools for report generation.

A front-end operator usually performs the following activities:

- manage alarms
- contact patient or carers
- check the technical quality of the received data and
  - if the data cannot be used because of the bad quality of the ECG record, ask the patient (or carers) to retransmit the ECG
  - if the data are of good quality, transmit them to the specialists for analysis and reporting.
- connect the patient (or carers) and the specialist telephonically
- send the report, by e-mail, fax etc., to the patient or to the sender of data.

A cardiology specialist will then

- customize the monitoring protocol and the threshold values for each physiological parameter
- monitor the health status of patients
- report measures of the received physiological parameters
- interact with the patient through text messages, conference calls or videoconferences
- evaluate the patient's responses to customized questionnaires.

If data are directly accessible to the patient or carers, the EHR connected to the receiving system must be designed properly to avoid unacceptable risks for patients. The EHR must include a user-friendly man–machine interface designed for the target users who include sick people and the elderly (Pecina, 2011). The user interface must be extremely intuitive so that users can easily follow the prescribed medical advice and protocols. Particular care has to be addressed to

- graphical layout
- accessibility of the information
- data presentation (clear, but scientifically correct)

- guidelines for the main user activities
- contextual pathology information that has to be focused on the patient's specific disease.

The service center and the associated EHR can be internal to a specific public health institution or it can be managed by a private company, providing services to health institutions according to a B2B model or directly to the patients as customers (B2C model).

7.8 Workflow in Telecardiology

Technology is only one factor related to the service provision in healthcare. As shown in Table 7.1, other factors have to be considered to implement a successful service. In this section, we will look at the usual workflows implemented in telecardiology services. These workflows have been shown to be

1. in line with usual clinical practice
2. perceived as favorable by patients and healthcare operators
3. economically sustainable.

7.8.1 *Basic Workflows*

Figure 7.12 shows a generic workflow for a telecardiology service aimed at monitoring arrhythmias. The workflow shows the interactions and the data exchange between the patient, the service center and the specialist.

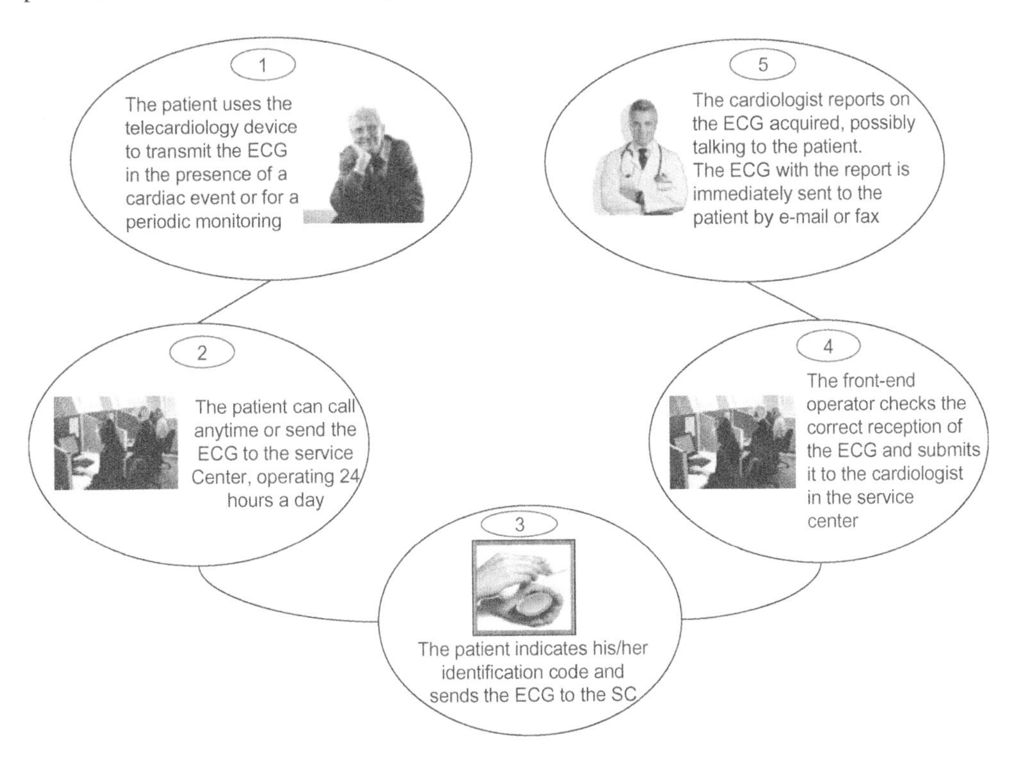

Figure 7.12 Telecardiology basic workflow.

At the beginning, the patient (or the relatives/carers) receives the ECG transmitting device and the ID (identification code) after medical prescription. The device is used to record ECGs for periodic monitoring or in the case of cardiac events. The patient can contact the service center anytime (i.e., 24/7) to have assistance from a front-end operator. The patient indicates the ID to obtain authorization to transmit the ECG. The receiving station in the service center automatically records the data; then, the front-end operator checks the received ECG. If the quality is acceptable, the ECG is transmitted to the cardiologist for reporting; otherwise, the operator asks the patient to record a new ECG or to transmit it again. After receiving the ECG online, the cardiologist can talk to the patient, compare the ECG with previous results stored in the database (if any), and report the ECG. The ECG with the report is immediately sent to the patient (or caregiver) by e-mail or fax.

In more automated solutions, or using m-Health, the ECG device automatically transmits through a gateway to the service center. The service center operator does scheduled contacts with the patient or when there is an abnormal situation.

If monitoring is focused on specific diseases such as Arrhythmias, simpler and smaller devices can be used. Figure 7.13 shows the conceptual scheme of a suitable instrument (e.g., the CG1 from the Gamma Cardio Soft).

The device is a single-lead ECG that can record the rhythm of the electrical heart activity and transmit it through acoustic coupling using wired telephones or mobile phones. For this purpose, the I lead is recorded by measuring electrical activity on the two thumbs that are in contact with two sensors. As shown in Figure 7.13, sensors are connected to a preamplifier.

The preamplifier stage contains all the protective filters that avoid electromagnetic interference, as explained in Chapters 5 and 8.

The output of the preamplifier is a high pass signal filtered at 0.5 Hz to eliminate respiratory and other unwanted signals. The amplifiers increase the signal dynamics so that the signal is suitable for ADC conversion. Two low pass filters avoid aliasing and eliminate unwanted high frequency signals such as noise and muscle-derived signals. The microprocessor elaborates and encodes the digital signal so that it is suitable for the wired telephone or a mobile acoustic channel. The power module controlled by the microprocessor has to manage energy to reduce battery consumption. Table 7.8 shows the main features of the monitoring device.

ECGs with performance like the CG1 are well suited to specifically monitor arrhythmias. For diagnosis purposes, ECGs must have minimum safety and performance requirements

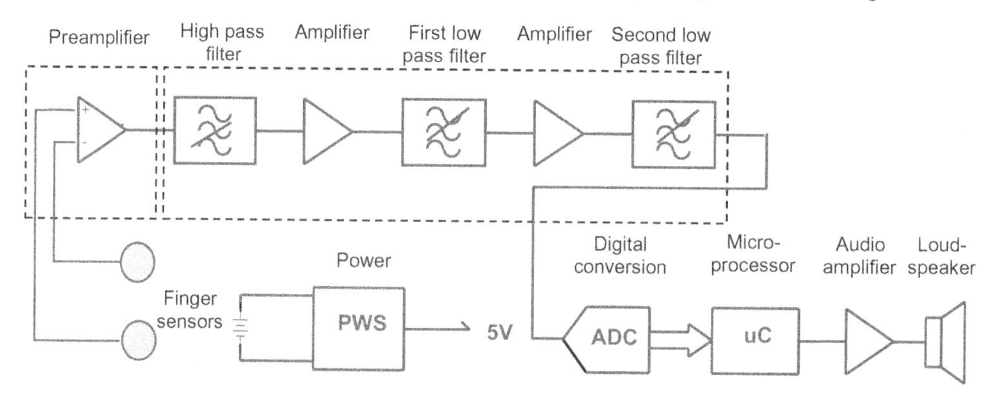

Figure 7.13 CG1 conceptual scheme from Gamma Cardio Soft.

Table 7.8 CG1 technical specification

Feature	CG 1 Specifications
ECG channels	I lead single channel recording
Input source	Ag/AgCl integrated electrodes on the thumbs
Input dynamic range	± 2.5 mV
Input impedance	$> 20\,M\Omega$
Frequency response	$0.5 \sim 30$ Hz
CMRR	> 95 dB
Input maximum offset	DC ± 200 mV
Output signal encoding	Digital for wired telephone lines, FM for GSM channels
FM central frequency	1800 Hz
Frequency deviation	200 Hz/mV
Internal memory	Maximum four events each of 120 seconds
Resolution	130 Hz sampling, 19.5 μV LSB
User interface	Two LEDs, two buttons (one to start recording, one to start transmission)
Power	single 1.5 V AA cell

that will help ensure a reasonable level of clinical efficacy and patient safety. Compliance to standards such as AAMI EC 11 (diagnostic electrocardiographic devices) ensures that these minimum safety and performance requirements are met. Devices that do not have this level of performance may not be able to detect specific heart diseases. Chapters 2 and 5 have additional details on the performance and the associated standards. More simply, telemedicine devices may be intended to

1. detect and monitor arrhythmias
2. diagnose general heart diseases.

In the first case, devices with performance as in CG1 are suitable, but, in the latter case, telemedicine devices have to have minimum essential requirements to achieve sufficient diagnostic capability; compliance to a performance standard may ensure this condition. Next sections introduces alternative workflows for telemedicine that again requires devices with minimum diagnostic capabilities to be sufficiently safe and effective.

7.8.2 *Alternative Workflows*

In telecardiology other workflows may be implemented according to the nature of the services and the proposed business model. Such additional workflows can be grouped into four areas:

- patient early or assisted discharge
- follow-up and home treatment of patients with chronic heart disease

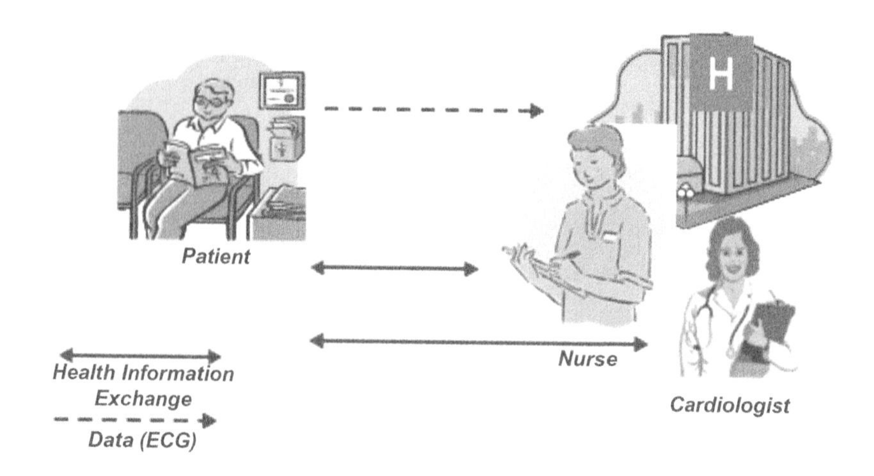

Figure 7.14 Home hospitalization workflow.

- primary care of cardiac patients by general practitioners
- periodical check.

Early or assisted discharge is one of the most common alternative workflows used in post-cardio-surgery rehabilitation after interventions such as coronary artery bypass graft and cardiac valve replacement, or after an ablation for supraventricular and ventricular arrhythmias. Telecardiology could be used to monitor patients at home with the so-called medically managed **home hospitalization** (Aimonino, 2008; Scalvini *et al.* 2011). The typical workflow outlined in Figure 7.14 is

- the hospital cardiologist discharges the patient under specific conditions with a customized monitoring protocol; the institution assigns a nurse 'tutor' and telemedicine devices to a patient; a nurse could go to the patient's home to activate the devices and do the training
- the 'tutor' contacts the patient according to a scheduled plan
- the patient transmits ECGs to the hospital service center at the planned time intervals or on a voluntary basis
- the hospital cardiologist reports on the ECG and interacts with the patient.

This approach can be classified as a B2C model since there is a direct interaction between the healthcare facility and the patient.

Follow-up and home treatment of patients with **chronic heart disease** is a common service model in many countries. This model has proven to be effective in reducing unnecessary hospitalizations or unnecessary emergency department attendances of patients with chronic heart failure (Parati, 2009). Examples of diseases that can be treated are

- chronic heart failure
- heart rhythm disorders
- hypertension

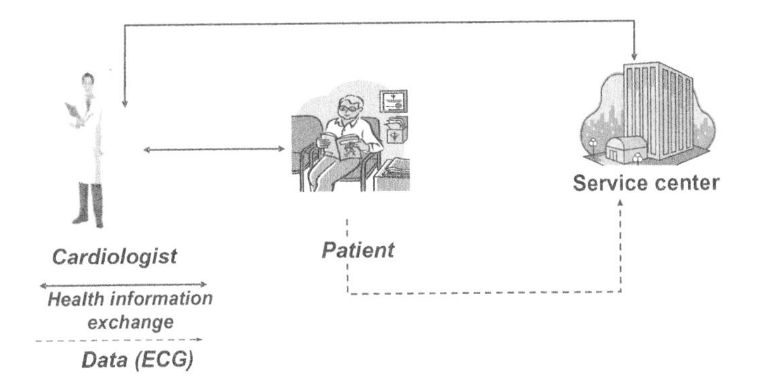

Figure 7.15 Follow-up workflow.

- chronic ischemic heart disease
- presence of pacemaker and defibrillator
- cardiovascular risk factors (obesity, diabetes etc.)
- respiratory failure
- other diseases for which the physician needs cardiologic home monitoring.

The workflow depicted in Figure 7.15 may be

1. the cardiologist recommends cardiac home monitoring under specific patient conditions
2. the patient is equipped with an ECG device at home
3. the patient transmits ECGs to a service center (periodically or upon specific events)
4. the service center operators provides front-end services, checking the quality of the data (i.e., if the ECG is good enough to be reported); if necessary, the operators ask patients for a new transmission or a new recording, explaining to them the correct electrode placement; the service center stores all the data
5. the cardiologist accesses the service center data via the Internet and reports on the ECG, referring to previous data if available, and interacts with the patient.

Since there is a direct relationship between the service center, the cardiologist and the patient, this model can be defined as a B2B2C (Business to Business to Customer).

Primary care of cardiac patients may be performed **by general practitioners (GPs)** supported through teleconsultation with cardiologists (NHS, 2005; Rice, 2011). In this case, the workflow shown in Figure 7.16 is defined as

- a GP (or a primary care physician, family physician or pediatrician) records the ECG and transmits the signal to a service center which provides the service of ECG reporting by a cardiologist; the service center could be hosted in a public hospital or in a private institution/company
- the cardiologist reports on the ECG
- the GP/pediatrician can interact with the cardiologist after receiving the report
- the GP/pediatrician interacts with the patient, or the parents in the case of children.

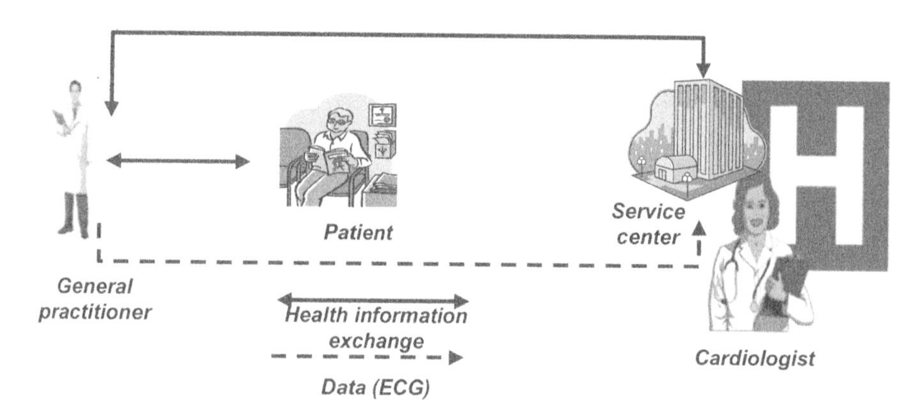

Figure 7.16 'Primary care' workflow.

An alternative workflow, as in Figure 7.17, may be

- the patient goes to the pharmacy periodically
- the pharmacist records the ECG and transmits it to a service center
- the cardiologists in the service center check the ECG and report on it
- the general practitioner who manages the patient can access the data and read the reports
- the general practitioner uses the data and the reports in his routine work.

Since there is a relationship between the service center, the physician, the pharmacy and the patient, this model can be considered as B2B2C or B2B model according to the specific contract.

7.8.3 Where and When Telecardiology Can Be Used

Telecardiology is suitable for many environments such as

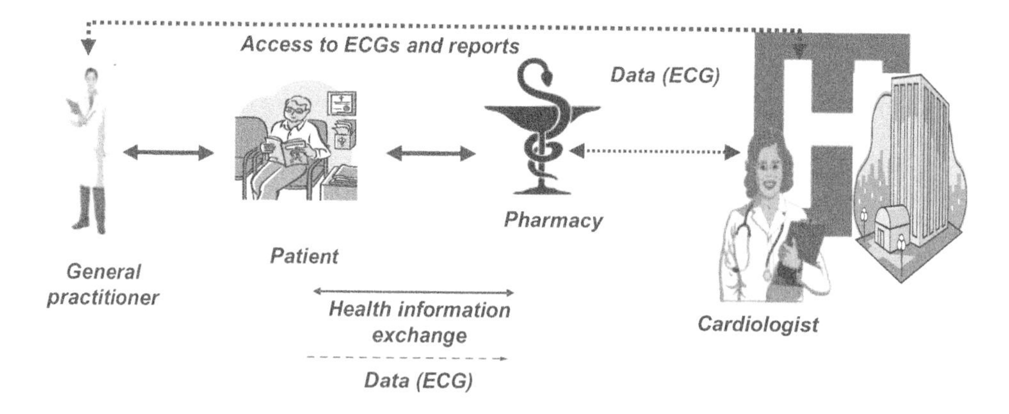

Figure 7.17 'Periodical check' workflow.

• patient's home	integrated homecare services
• health facilities	pharmacies, GP facilities, nursing homes, small clinics
• critical environments	shelters of the army, prisons, small islands, refugee camps, engineering camps in remote areas
• transport	ships, cruise ships, recreational boats, fishing vessels, trains, aircraft
• emergency vehicles	ambulances, mobile vehicles, helicopters.

In the following sections, we will show telecardiology examples in emergencies, prisons and for the integrated homecare services.

7.8.3.1 Telecardiology in Emergency

For various acute cardiac diseases immediate treatment is crucial for the reduction of damage, the best follow-up and the probability of patient survival.

The primary objective of telecardiology in emergencies (Birati and Roth, 2011) is to record a diagnostic ECG as soon as possible in the place where the patient is located and to transmit it to a specialized center, in order to diagnose as quickly as possible the type of heart disease. This allows for a more timely treatment, and directing the patient to the center that is most appropriate for the specific treatment.

When no disease is reported on the ECG, patient can be quickly reassured, thereby avoiding inappropriate treatments and unnecessary hospital visits. This approach allows for a significant saving in resources with a more effective use of emergency departments.

An application of the telecardiology in emergency is equipping ambulances with electrocardiographs so that the ECGs can be sent to the emergency department (ED). ED specialists can read the ECG and give appropriate directions to the ambulance crew for

- immediate therapy
- transportation to the most suitable center for the particular pathology and for the right therapy (e.g., angioplasty).

As discussed at the beginning of this section, ECG devices must have diagnostic performance in compliance with the ANSI AAMI EC 11 standard to ensure that heart diseases are not neglected on ECG records.

It is interesting to give some numbers relating to real experiences based on this workflow. In a specific project accounting for 219,000 ECGs recorded in ambulances or in first-aid facilities (Puglia, 2010), 45% of patients was pathological. Of the pathological results, 41% evidenced life-threatening disease, more specifically 4% was related to acute myocardial infarction, 14% to cardiac ischemia and 19% to serious arrhythmias. These numbers show the significant results of a telecardiology service in term of life saving.

7.8.3.2 Telemedicine in Prisons

One specific area of telemedicine is aimed at the prevention, diagnosis and continuity of care in prisons. Telemedicine technologies and services are particularly advantageous for prisoners within a special prison regime in which prisoner contact with the outside world is particularly complex, even for healthcare reasons.

Telemedicine, and specifically telecardiology, has the following objectives in prisons:

- provide tools to improve care and lifestyle of prisoners through the primary prevention of chronic diseases
- ensure continuity of care throughout the period of imprisonment
- reduce prisoner transport to hospitals
- reduce the time of hospitalization.

7.8.3.3 Integrated Homecare

When target patients are unable to move to health facilities, integrated homecare assistance may replace the many actors usually present in specialist centers. In this case, patients are in nursing homes or assisted in their home by a team that periodically visits them and monitors their health. Figure 7.18 shows a workflow for the management of patients attended in the integrated homecare approach. The data are transmitted via mobile networks to the receiving platform in the service centre. The ECGs can be read 24/7 by cardiologists. In this case, the associated health facility may immediately assess vital signs and ECGs so that prompt treatments may be performed in the case of emergency.

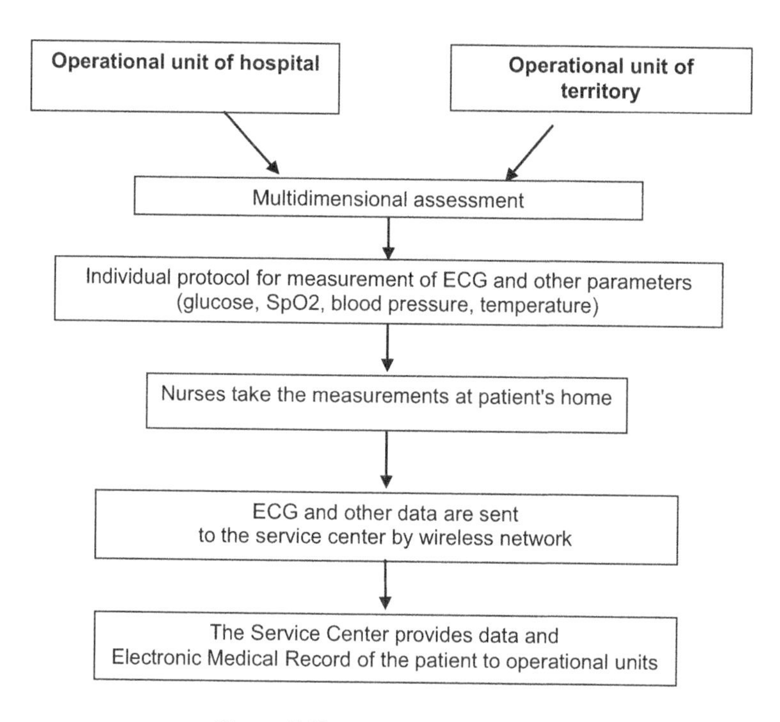

Figure 7.18 Integrated home care.

7.9 Risks of Telecardiology

The previous sections have outlined the possible workflow of telemedicine services. These workflows contain many risks, since telemedicine shifts part of the healthcare service activity onto patients and non-specialized health operators. This has a possible impact on the safety and/or on the effectiveness of the service itself. This issue may be particularly serious in the case of safety-critical areas such as telecardiology. In addition, the shift of the activity onto patients and non-specialized operators is also perceived as dangerous by specialized operators and by the patients themselves which may therefore hinder the take-up of telemedicine projects. On the other hand, the potential benefits of telemedicine services have fostered trials and clinical evaluations to make risks acceptable and include telemedicine in common clinical practice (EC, 2009a). Telemedicine trials and studies such as (ENISA, 2009b) have noted many risks that reduce the take-up of telemedicine in practice. Table 7.9 shows some risks focused on telecardiology services.

With respect to the usual hazards experienced in medical instrument practice, we have additional risks due to

1. the **technology** when managed by the intended non-expert telemedicine users
2. the **users** themselves who have reduced training
3. the **service** which is not part of the routine healthcare workflow.

Table 7.9 Potential risks in telecardiology

ID	Harm area	Risk description
1	**Technology** failure	System malfunctions and breakdowns
2	Technology usability	Usage problems of technologies by elderly, sick people or by inexperienced carers
3	Technology data protection	Data breaches with impacts on security and privacy
4	**User** problems	Instructions and procedures not attended by users
5	User problems	Equipment damaged or stolen
6	User problems	Telecardiology service perceived as alternative to emergency services and therefore accessed by users in life-threatening situations, delaying access to the emergency department
7	**Service**: inadequate on privacy	Non-compliance with data protection legislation
8	Inadequate service	Inadequate provision or availability of telemedicine service
9	Inadequate service	Human error in emergencies
10	Inadequate service	Inadequate feedback to users after service, user without instructions in the after-care period.

Regarding technology, it is expected that the users (elderly and ill people) may have different instruction levels, with an anxious attitude to the use of telecardiology devices. These patients may also use devices without reading the instructions even if strongly recommended. This implies that devices have to supply proper immediate feedback for any operation and that procedures all have to be intuitive for any class of user. This requires a proper design and certification of the mobile devices. Telemedicine services imply also that the patient's sensible data are distributed over on-line databases. This generates risks of non-compliance with privacy legislation and security since sensible health data may be fraudulently accessed an modified to damage patients.

A telecardiology service may also be used in the case of emergency event diagnosis. The patient may perceive the telecardiology service as an alternative to the emergency services with the risk of activating this service in a life-threatening situation, delaying access to the emergency department (Pecina, 2011). For example, in the case of chest pain, this service may be used to evaluate the cardiac origin that may also imply life-threatening causes (myocardial infarction, angina pectoris). Clinical and legal issues have to be considered in this case, since a remote positive diagnosis may save the life by early detection of acute heart disease, but on the other side a false negative diagnosis is (e.g. infarction not reported) would be an unacceptable risk.

It is evident that the telecardiology service has to be designed according to different criteria than for conventional medical services, because of the non-professional users and because the service uses processes that are different from healthcare facility assistance. This may generate a lower priority in telecardiology service implementation and also a reduced attention to service follow-up. For example, a negative ECG report may remain without consequences and the patient may be confused, wondering about the next steps of the services.

In recent years, government health authorities have appreciated the significant benefits of a widespread telemedicine implementation and they are therefore fostering its diffusion in the face of the obstacles to its implementation (WHO, 2009). At the European level, the following actions have been considered (EC, 2009b):

1. solve the **technical problems** such as the lack of adequate community-wide broadband infrastructure and interoperability of telemedicine devices
2. increase confidence and acceptance of telemedicine services among **users** by encouraging provision and dissemination of scientific evidence of its effectiveness and cost effectiveness
3. change the **legislation** to support telemedicine services.

Telemedicine is expected to change medical instrument usage in the coming decades since it improves service quality while reducing cost and increasing accessibility.

Diagnostic instruments may overcome geographical barriers with a wider access to healthcare services making use of ICT. This is particularly significant in rural and underserved areas characterized by inadequate healthcare facilities.

Telemedicine, e-Health and c-Health promise healthcare service access anywhere and anytime. On the other hand, to overcome barriers to these emerging approaches, practical services have to be implemented so that all the risks and critical issues listed in Table 7.1. This implies t minimum to address: business sustainability, positive perception, elimination of legal regulation barriers and risk acceptability, with a safe and effective service ensured in all conditions.

References

Aimonino R. *et al.*, (2008), 'Substitutive "Hospital at Home" Versus Inpatient Care for Elderly Patients with Exacerbations of Chronic Obstructive Pulmonary Disease: A Prospective Randomized, Controlled Trial' Journal of the American Geriatrics Society, Wiley.

Antonicelli R. *et al.* (2008), Impact of telemonitoring at home on the management of elderly patients with congestive heart failure. *J Telemed Telecare.*

ASTM (2005), American Society for Testing and Materials, ASTM E2369–05e2 Standard Specification for Continuity of Care Record (CCR).

Birati, Y., and Roth, A., (2011), Telecardiology, IMAJ VOL 13.

Buyya, R., *et al.*, (2009), Cloud Computing and Emerging IT Platforms: Vision, Hype, and Reality for Delivering Computing as the 5th Utility – Future Generation Computer Systems.

Buyya, R., *et al.*, (2011), Cloud Computing – Principles and Paradigms, John Wiley & Sons.

CEU (2011), Council of The European Union: 'Council conclusions on innovation in the medical device sector'

Chen, M., (2010), Body Area Networks: A Survey, Springer Science+Business Media, LLC.

Della Mea, V., (2001), What is e-Health (2): The death of Telemedicine? J. Med. Internet Res.

EC (2009a), eHealth in Europe, The European Files, European Commission.

EC (2009b), European Commission, On Telemedicine for the benefit of patients, healthcare systems and society, Communication from the Commission to the European Parliament, the Council, the European Economic and Social Committee and the Committee of the Regions, COM(2008)689 final.

EC (2010), European Commission, Study Report: Interoperable eHealth is worth it.

EC (2011a), European Commission recommendation 2011/413/EU: 'More years, better lives the potential and challenges of demographic change'

EC (2011b), MethoTelemed Final Study Report on Methodology to assess Telemedicine applications commissioned by the European Commission.

EC (2011c), Directive 2011/24/Eu On Patients' Rights In Cross-Border Healthcare.

EC (2012a), European Commission, Research and Innovation Health: http://ec.europa.eu/research/health/medical-research/index_en.html

EC (2012b), European Commission, Information Society, eHealth, ICT for better healthcare in Europe: http://ec.europa.eu/information_society/activities/health/index_en.htm.

EC (2012c), European Commission: Information Society, Interoperability: connecting eHealth services: http://ec.europa.eu/information_society/activities/health/policy/interoperability_and_standardization/index_en.htm

EEC (2007), Council Directive 93/42/EEC of 14 June 1993 concerning medical devices: http://ec.europa.eu/health/medical-devices/index_en.htm amended by 2007/47/ECin force on 21 March 2010.

ENISA (2009a), European Network and Information Security Agency, Cloud Computing: Benefits, risks and recommendations for information security.

ENISA (2009b) , European Network and Information Security Agency, 'Being diabetic in 2011': Identifying emerging and future risks in remote health monitoring and treatment.

FDA, (2011), Center for Devices and Radiological Health, CDRH Innovation Initiative.

Gillespie, A., (2007), Foundations of Economics, 'PESTEL analysis of the macro-environment'. Oxford University Press.

Gunter, T., *et al.*, (2005), LLM – The Emergence of National Electronic Health Record Architectures in the United States and Australia: Models, Costs, and Questions – J. Med. Internet Res.

HL7 (2011), HL7 ANSI Standards: http://www.hl7.org

IHE (2012), Integrating the Health Enterprise (IHE): http://www.ihe.net/

ISO (2005), Health informatics – Electronic health record – Definition, scope and context – ISO/DTR 20514.

ITU, (2012), ITU series recommendations: http://www.itu.int/rec/T-REC-X/en

Kotler, P., *et al.*, (2011), Principles of Marketing, Prentice Hall.

Marketsandmarkets (2012), Cloud Computing (Clinical, EMR, SaaS, Private, Public, Hybrid) Market – Global Trends, Challenges, Opportunities & Forecasts (2012–2017).

Mell, P., and Grance, T., (2011). The NIST Definition of Cloud Computing Special Publication

Moorman, O B., (2010), Medical Device Interoperability: Standards Overview, IT World.

Mosquera, M., (2011), Despite incentives, cost is a barrier to small provider EHR use, GovernmentHealthit: http://www.govhealthit.com/news/congressional-panel-explores-barriers-small-provider-ehr-use

NEMA (2012), Digital Imaging and Communication in Medicine (DICOM), National Electrical Manufacturers Association: http://medical.nema.org/dicom/

NHS, (2005), NHS-North West – Cardiac Telemedicine in Primary Care, Delivering Benefits for Patients and the NHS in Lancashire & Cumbria, report available at: http://www.northwest.nhs.uk/

NIST (2011), National Institute of Standards and Technology, Medical Device Interoperability.

Oswald, E., (2011), IBM Declares the End of the PC Era – Today@PCWorld – by Ed Oswald, PCWorld.

Patel, *et al.*, 2012, A review of wearable sensors and systems with application in rehabilitation, Journal of Neuro-Engineering and Rehabilitation.

Parati, G., *et al.*, (2009), Home blood pressure telemonitoring improves hypertension control in general practice, J. Hypertens.

Pecina, J. L., Takashai, P. Y., and Hanson, G. J., (2011), Current Status of Home Telemonitoring for Older Patients: A Brief Review for Healthcare Providers, Clinical Geriatrics, 2011.

Puglia, (2010), Regione Puglia (Italy), Telecardiology for emergency medical service in the Puglia area, report on data from 11/10/2004 to 31/1/2010.

Rappa, M. A., (2004), The utility business model and the future of computing systems, IBM Systems Journal.

Rice, P., (2011), Teleconsultation for Healthcare Services, Health Innovation and Education Cluster (HIEC) Yorkshire and the Humber.

Scalvini, S., Giordano, A., and Glisenti, F., (2011) Ten Years Experience of the 'Fondazione Maugeri' Network for Cardiovascular Diseases: http://www.mespe.net/website_objects/Articles/MESPE_Scalvini.pdf

WHO (2009), World Health Organization, Telemedicine: opportunities and developments in Member States: report on the second global survey on eHealth.

WSD (2011), Department of health – UK, Whole System Demonstrator programme: Headline findings: http://www.dh.gov.uk/health/2011/12/wsd-headline-findings/

8

Certification Process

Ensure safety and effectiveness for every manufactured device

Chapter Organization

The certification process aims to ensure that a medical device is safe and effective for all manufactured samples. This suggests that certification is less involving when addressed at the earliest stages of product design. In previous chapters we discussed specific certification issues in detail, in this chapter, we will focus on additional topics of certification focusing on design verification and testing. The material of this chapter is organized following a simplified certification process shown Figure 8.1, each topic is addressed in the specific section

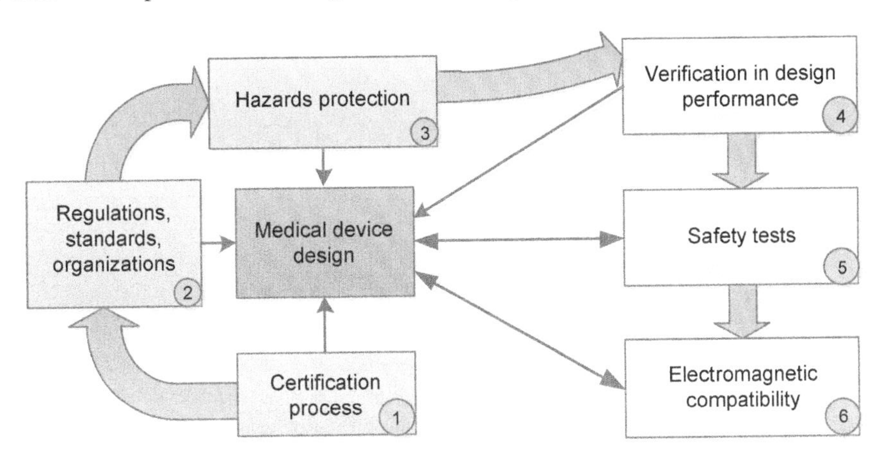

Figure 8.1 Chapter structure.

Medical Instrument Design and Development: from Requirements to Market Placements, First Edition.
Claudio Becchetti and Alessandro Neri.

indicated by the circled number (from 8.1 to 8.6) in the theory section. The implementation issues related to the certification are discussed in the sections with the corresponding unit numbers (from 8.11 to 8.16) using the real example of the Gamma Cardio CG device disclosed by the Gamma Cardio Soft (www.gammacardiosoft.it).

Referring to Figure 8.1, first the processes are outlined. In section 8.1 standards are introduced in Section 8.2, a summary of regulations and standards is included at the end of the chapter. Basic concepts on protection from main risks are discussed in Section 8.3. The following sections address verifications for design performance, safety tests and electromagnetic compatibility (EMC) certification. Through the comparison of the theory and implementation corresponding sections, the reader will be able to understand how certification is applied to a typical medical device. At the end of this chapter, the reader will understand the main hazards and the associated protections that have to be implemented at the design stage and tested on the final device. The reader will find how these protections are implemented in a real product (the Gamma Cardio CG) in the implementation part. This chapter may guide for device certification but these contents do not replace the careful reading of the updated versions of the referenced standards and regulations. The 'Summary of Regulations and Standards' section at the end of this chapter indicate the document versions considered in this book.

Part I: Theory

8.1 Certification Objectives and Processes

A medical product must have an approval from a national or regional regulatory body before being marketed. This process is called **certification** in areas like the European market or **market approval/clearance** in other regulatory frameworks such the USA.

The certification process aims to ensure that a medical device is:

1. safe
   and
2. effective
3. for all the manufactured items.

Definitions and processes may change across standards and regulatory frameworks, but the previous three pillars remain the essential objective of the certification.

Regarding these three concepts, the next step is to clarify: *what* they mean, *why* they have to be accomplished and *how* they are achieved.

A product is **safe** if patients, users and third parties do not run unacceptable risks of physical hazards (death, injuries etc.) in its intended use. This is achieved by introducing protective measures on the devices that make all the foreseeable risks acceptable. The residual risk must be evaluated as acceptable when compared with the benefit derived from the use of the device.

A product is **effective** if it performs as declared by the manufacturer in the intended use. More specifically, products must attain the **performance** levels attributed to them by the manufacturer. This may be achieved by performing clinical evaluation, by complying with applicable performance standards or by demonstrating the substantial equivalence with an already marketed device of the same type.

The previous features have to be ensured **for all the manufactured items** of the medical device. This requires that the organizations that manage the medical device must have in place a quality system that ensures quality for all the critical processes such as design, manufacturing.

Figure 8.2 summarizes the previous concepts. The outer circle includes the risks in the use of a medical device that justify *why* the certification process is in place. The middle circle shows *how* to mitigate risks so that they become acceptable. The inner circle includes the three pillars achieved after mitigating all the foreseeable risks so that the medical device can be marketed. All the other topics related to certification are mainly consequences of these three pillars.

Following the (WHO, 2003) perspective, the previous discussion suggests that absolute (i.e., 100% deterministic) safety cannot be ensured for a medical device. Safety is ensured in terms of residual risks in the usage of the device that are acceptable relative to the benefit of the device itself. Safety is therefore assessed through a risk management approach. Safety is not the only element, clinical effectiveness and performance also have to be aligned to what is declared by the manufacturer. These elements have to be guaranteed for all the manufactured devices throughout the product life span until disposal.

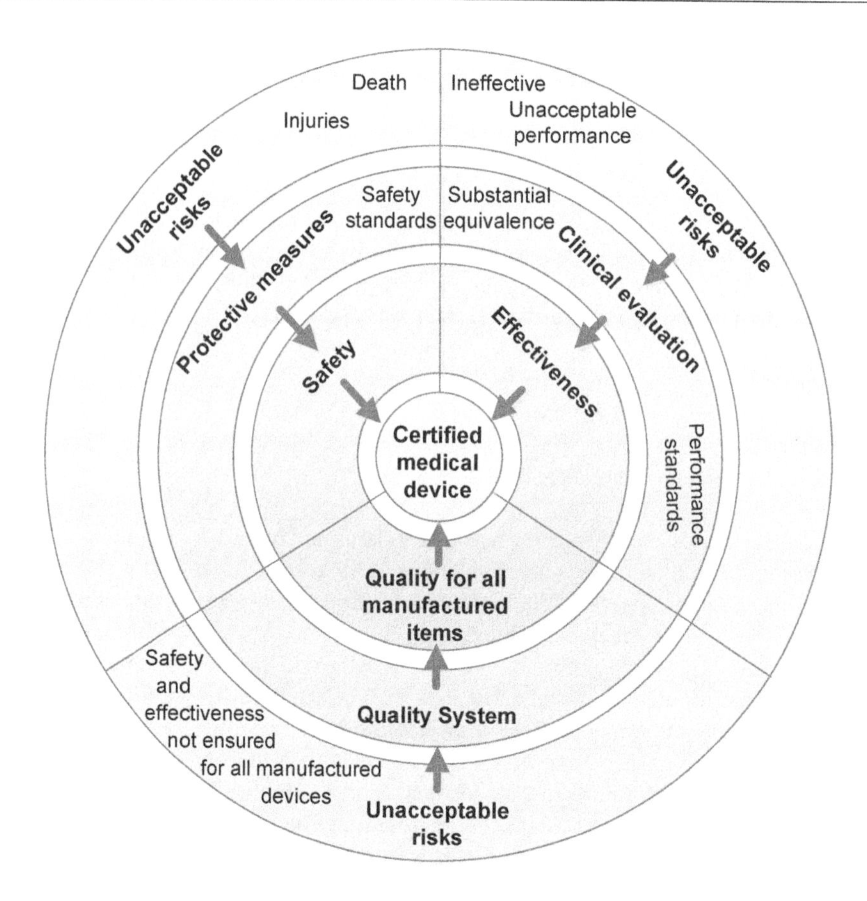

Figure 8.2 Certification model.

This requirement can be guaranteed by a proper quality system focusing on all the critical issues of the device. The life span of the product encompasses several stakeholders that have to share **responsibility**, so that each device maintains its safety and effectiveness.

We refer first to the manufacturer who has a main role up to product delivery. Then there are the vendors who sell to the users. They have the responsibility of advertising, sales, storage, device tracking and 'postmarketing surveillance'. (Postmarketing surveillance is the activity of monitoring the safety and effectiveness of a medical device after it has been released on the market.) Users have the responsibility to use the device according to the specifications, ensuring proper training for the operators. Patients may help in providing feedback and, in the coming years, will have an increased role in homecare, assuming activities previously performed by trained operators. Finally, the regulatory bodies have the responsibility of monitoring the process.

8.1.1 *Certification, Standards and Definitions*

Many hazards have been experienced in the use of electronic equipment. Electric shock, as discussed in Chapter 5, but also mechanic and other hazards may produce significant harm.

Medical equipment adds extra risks that threaten the safety of people's health. These additional risks may be generated by medical-specific hazards (e.g., from ineffective sterilization or disinfection) or from unexpected behaviors of the device's basic performance that affect the device's effectiveness. National or regional regulatory bodies have implemented processes to ensure that risks from medical device are reduced to acceptable levels.

In the past, device performance was not considered as critical in terms of safety. This concept was summarized by the sentence: 'The *ability of an electric kettle to boil water is not critical for its safe use*' (IEC/EN 60601-1).

This meant that for an ECG, safety was focused on physical injury or damage to the health of people, such as from an electric shock. A distorted ECG signal that could lead to a wrong medical treatment was not primarily addressed in the scope of certification in some market areas. As discussed in Chapter 2, an ECG device that hides a signal elevation of 1 mm, may not detect a myocardial infarction. Other examples of safety-critical performances are related to devices that deliver energy or therapeutic substances to the patient.

For the benefit of the patient, it is now widely accepted that certification has to ensure that the device is not only intrinsically **safe in usage** but also that its **performance does not generate unacceptable risks** for the patient. The performance concept used in the EU region is substantially similar to the 'effectiveness' term used in the USA.

Some risks (e.g., from electric shocks) were addressed in previous chapters focused on design. This is in line with the recommended strategy: '**design for certification**' that suggests addressing certification at the earliest stages of development. Addressing certification after device development is a great risk for the manufacturer due to the unexpected increase in cost and marketing time. In addition, there are technologies, such as safety-critical software, that may be considered safe only when the design process is managed and controlled.

These considerations have also suggested that a **risk management** process must be an integral part of the design and in general of all the device life cycle stages.

Certification requirements and associated tests must be considered at the earliest stages of the design. These requirements have to address all the expected hazards such as electric shock, fire, overheating, mechanical risks, explosion and dangerous radiation. The best approach for this is to rely on **standards** that have already identified hazards and associated protective countermeasures to reduce risks to acceptable levels. In addition, in the EU and in other countries, if a device is compliant to specific standards, there is the presumption that the device is also compliant to requirements and performance prescribed by the regulation framework which, in the EU, is the Medical Device Directive MDD (EEC, 2007).

IEC/EN 60601-1 third edition (IEC/EN 60601, 2005) approved without modification at European level as EN 60601-1:2006 (EN 60601, 2006) is a good starting point because it identifies hazards of medical electrical equipment.

The IEC/EN 60601-1 has the double approach to ensuring basic safety (from usage) and the essential performance of the device through general requirements to be imposed on the device.

The following **terms** and their relations are crucial to an understanding of standards and, in general, certification. The standard defines '**basic safety**' as: '*the freedom from unacceptable risk directly caused by physical hazards . . . under normal condition and single fault condition*'.

Essential performance is a performance required to avoid unacceptable risk. When the essential performance is absent or degraded, there would be an unacceptable risk. The

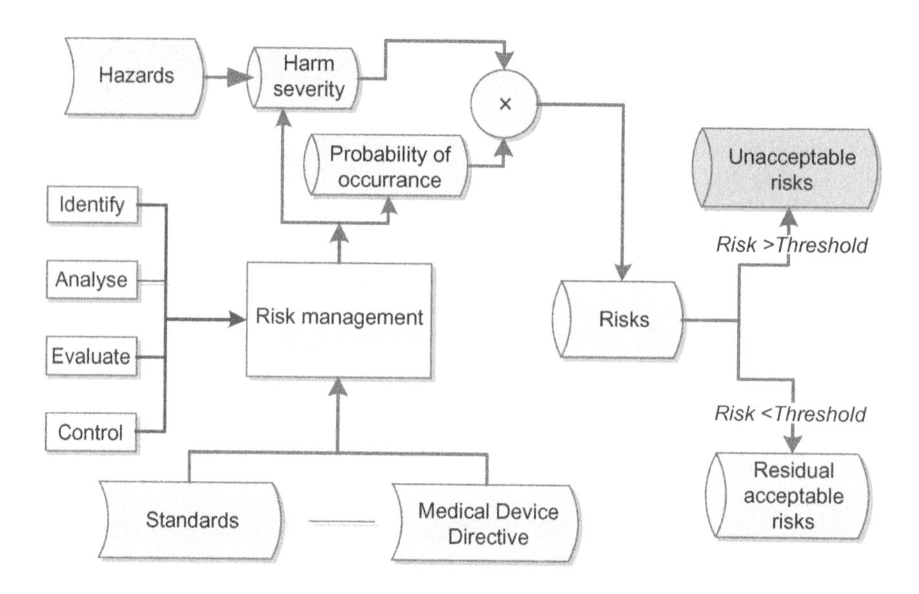

Figure 8.3 Risk management: unacceptable risks.

standard covers all relevant essential requirements as given in Annex I of the European MDD which includes over a hundred essential performance items. These items have been outlined in Chapter 2. When medical electric equipment is compliant with this standard, then there is the presumption of conformity to the MDD. A similar approach holds in the USA and other countries, as outlined in Section 8.2.

The standard also introduces general requirements for the risk management processes that have to be performed for the whole life cycle of the product (i.e., also after release of the product onto the market). As shown in Figure 8.3, derived from Figure 6.9, risk management has to identify, analyze, evaluate and control risks.

A **risk** is defined as: ‘a combination of the probability of occurrence of harm and the severity of that harm’ (IEC/EN 60601-1), where **harm** is a ‘physical injury or damage to the health of people or animals, or damage to property or the environment’ (ISO/EN 14971). **Hazard** is a potential source of harm. (IEC/EN 60601-1).

As shown in Figure 8.4, **unacceptable risks** are to be managed introducing **protective measures** that make the **residual risks** acceptable. The residual risk is the risk that remains after the protective measures are in place. The manufacturer has to define the policy for determining which risk is acceptable and the evaluation of acceptability of the residual risks. Unacceptable risks are eliminated by introducing protective measures, complying with the associated basic safety and the essential performance requirements. When basic safety or essential performance is not satisfied, there will be an unacceptable risk. When complying with a requirement of a standard that addresses specific hazards (say IEC 60601-1, or its associated ‘collateral’ or ‘particular’ standards), the residual risk is presumed to be acceptable unless there is evidence to the contrary.

(Note that **IEC 60601** is a family of standards where the general standard is labelled as IEC 60601-1 – Medical equipment/medical electrical equipment – Part 1: General requirements for basic safety and essential performance.)

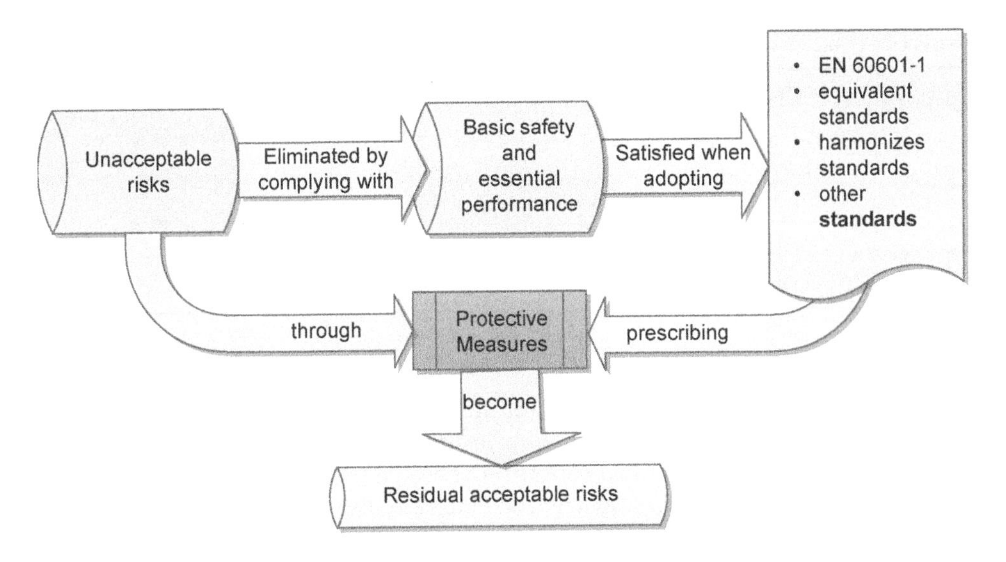

Figure 8.4 Risk management: protective measures.

This standard includes general requirements that may be overridden by specific requirements contained in other standards of the same family. IEC 60601 **'collateral standards'** (identified as IEC 60601-1-x) specify the requirements for specific aspects of safety and performance (for example, electromagnetic compatibility). **Particular standards** (identified as IEC 60601-2-x) include requirements for a specific class of products, for example ECGs.

In the European market, compliance with IEC/EN 60601-1 (identical to EN 60601-1) provides one means of conformity with the specified essential requirements of the medical device directive.

The main hazards and associated protections focused on in IEC/EN 60601-1 are:

- electrical hazards
- mechanical hazards, such as moving parts, surface-corners-edges, instability, expelled parts, acoustic energy, pressure and vessels, support systems
- unwanted and excessive radiation hazards
- excessive temperatures, fire
- overflow, spillage, leakage of liquids
- hazards from insufficient disinfection, sterilization
- hazards from biocompatibility
- hazardous outputs (accuracy of control and instruments).

Main specific hazards will be discussed through this chapter. Table 8.1 summarizes the first steps of the certification process. The corresponding implementation Section 8.11 shows the implementation of the process. Readers interested in understanding the overall process from theory to implementation would do well to read Section 8.11. after this section. The next theory sections will describe in more detail the specific certification steps. In particular, Section 8.2, will show some details on standardization organizations. This provides a better

Table 8.1 Certification: the first steps

1.	Identify requirements from the regional or national legal framework
2.	Select applicable standards
3.	Implement risk management process with risk analysis
4.	Design with certification in mind: consider the certification requirements and tests at the earliest stages of development

understanding of the standards themselves. Section 8.3 introduces some basic concepts of protection provided by insulation, the enclosure and by other means to avoid electric shocks. Section 8.4 outlines protection to be performed at design time and Section 8.5 addresses the main safety tests. EMC (Electromagnetic compatibility) compliance is addresses in Section 8.6.

8.2 Regulations, Standards and Organizations

Medical devices have a crucial impact on people's health. A defective device may result in misdiagnosis, delays in treatment, injuries or even death. An accurate analysis by the medical device throughout its product life cycle is therefore crucial to reducing risks. Countries worldwide have implemented specific approaches to regulating market access for medical devices.

In the more general terms, the scheme is: a country or a set of countries at a regional level (e.g., the EU) defines **a legal framework** (e.g., the Medical Device Directives in Europe) where regulatory requirements are set on medical devices and on the associated management processes. Requirements have to be preferably satisfied by standards recognized at international, regional or national level. When requirements are satisfied by applicable standards accepted by local authorities, this generally implies that there is a presumption that such requirements have been fulfilled. In the USA the accepted standards are referred as '**recognized consensus standards**', and in Europe, they are referred as '**harmonized standards**'. **Guidelines** published by the authorities are helpful for having a common interpretative approach to regulation although they are not legally binding. The section 'Summary of Regulations and Standards' at the end of the chapter brings together this information for the USA and the EU.

Unfortunately, legal frameworks and regulatory requirements are different across country systems worldwide although, in principle, the safety level of devices should not differ. This creates a heavy additional cost for industry in certifying its products since manufacturers have to perform different procedures possibly with different regulatory requirements in each country where they plan to market. There is also a negative impact on patients who may not have the availability of specific medical products, since manufacturers tend to concentrate on the main markets so as to reduce development effort. In general, the main addressed market areas tend to be the EU, the USA and specific Asian countries.

Some effort has been taken to harmonize the different systems. In particular, the **Global Harmonization Task Force** (GHTF) was created in 1992 to respond to the growing need for international harmonization in the regulation of medical devices. GHTF is a voluntary group of representatives from national medical device regulatory authorities and from industry to develop uniformity among systems so that harmonization can be achieved by selected countries in the regulation of medical devices. GHTF includes the USA, the EU, Australia, Japan, Canada, the Asian Harmonization Working Party (AHWP), the International Organization for Standardization (ISO) and the International Electrotechnical Commission (IEC).

Despite the GHTF's efforts, compliance to standards with a worldwide recognition remains currently the best strategy for improving efficiency in medical device certification.

In more details, according to European Committee for Standardization (CEN), a **standard** is defined as 'a publication that provides rules, guidelines or characteristics for activities or their results, for common and repeated use. Standards are created by bringing together all interested parties including manufacturers, users, consumers and regulators of a particular material, product, process or service'.

Standards do mainly include requirements, recommendations for products, systems, processes or services. Standards may also include measurements and test methods and may determine terms and definitions for a specific sector.

Standards are voluntary: usually there is no legal obligation to apply them unless specific laws and regulations may refer to them and even require compulsory compliance. In any case, standards compliance is a main and preferred approach for demonstrating conformity to medical device safety and performance requirements.

In the EU, the compliance to 'harmonized standards' gives presumption of conformity with the essential requirements specified in the Medical Device Directive.

A **harmonized standard** is defined as: 'a European standard elaborated on the basis of a request from the European Commission to a recognised European Standards Organisation to develop a European standard that provides solutions for compliance with a legal provision. Such a request provides guidelines which requested standards must respect to meet the essential requirements or other provisions of relevant European Union harmonisation legislation'

In the USA, '**recognized consensus standards**' are used for market access of medical devices. The Food and Drug Administration (FDA) and in particular its Center for Devices and Radiological Health (CDRH) is responsible for regulating firms who manufacture, repackage, relabel and/or import medical devices sold in the Country. The FDA is also responsible for the recognition of national and international medical device consensus standards and the maintenance of the associated public database. **Consensus standard** is defined as: 'A standard developed by technical or professional societies or by national and international standards-setting organizations *according to a well-defined procedure for consensus agreement among representatives* of various interested or affected individuals, companies, organizations, and countries.'

Standards are developed by standards organizations. Here, we consider organizations involved in technical standards and organizations focusing on standards in electrical and medical equipment. Table 8.2 summarizes some standardization bodies that address general technical standards.

International Organization for Standardization (ISO) is the world's largest developer of voluntary international standards, composed of more than 160 representatives from various national and regional standards bodies. The organization promulgates worldwide proprietary,

Table 8.2 Standard organizations

Area	Organization	Standard code	Description
International	ISO	ISO	International Standards Organization
Europe	CEN	EN	European Committee for Standardization
German	DIN	DIN	German Institute for Standardization
UK	BSI	BSI	British Standards Institution
France	AFNOR	AFNOR	Association Française de Normalisation
Italy	UNI	UNI	Italian Organization for Standardization
US-International	ANSI	ANSI	American National Standards Institute

industrial and commercial standards. ISO standards may be recognized without modification at national or regional level.

For example, about 30% of the European EN standards from the CEN committee are identical to ISO standards. These EN ISO standards have the dual benefits of automatic and identical implementation in all CEN European member countries, and global applicability.

CEN is the European Committee for Standardization that contributes voluntary technical standards which promote free trade, the safety of workers and consumers, environmental protection, exploitation of research and development programs and public procurement. CEN with **CENELEC** (European Committee for Electrotechnical Standardization) and ETSI ratifies the European EN standards.

European Standards are generally developed in a global perspective for recognition. CEN has signed the Wien Agreement with the ISO, through which European and international standards can be developed in parallel.

In CEN, national standardization bodies such as **DIN** (the German Institute for Standardization), **BSI** – British Standards Institution, **AFNOR** (Association Française de Normalisation), **UNI** (Italian Organization for Standardization) are represented.

Outside the European areas, the American National Standards Institute (ANSI) has issued many standards of medical device interest. The Institute oversees creation and use of thousands of norms and guidelines that directly impact businesses in nearly every sector.

The previous organizations address standards on general sectors. Historically, there are specific standard organizations focused on electrotechnical systems as outlined in the Table 8.3.

The International Electrotechnical Commission (**IEC**) is the main international organization that prepares and publishes standards for all electrical, electronic and related technologies. The equivalent European organization is the **CENELEC** (European Committee for Electrotechnical Standardization). These two organizations have a close cooperation: around 76% of all European standards adopted by CENELEC are either identical to or based on IEC standards.

Table 8.3 Electrotechnical standard organizations

Area	Organization	Standard code	Description
International	IEC	IEC	International Electrotechnical Commission
Europe	CENELEC	EN	European Committee for Electrotechnical Standardization
German	DKE	DIN/VDE	German Commission for Electrical, Electronic & Information Technologies
UK	BSI	BS	British Standards Institution
France	UTE	UTE	Union Technique de l'Electricité et de la Communication
Italy	CEI	CEI	Italian Electrotechnical Commission
US-International	AAMI	AAMI	Association for the Advancement of Medical Instrumentation
US-International	AHA	AHA	American Heart Association

National standards bodies or national electrotechnical committees of the 32 European member countries are represented in CEN and CENELEC respectively. Some of them are listed in Tables 8.2 and 8.3.

An EN (European Standard) 'carries with it the obligation to be implemented at national level by being given the status of a national standard and by withdrawal of any conflicting national standard'. For this reason, a European Standard (EN) automatically becomes a national standard in each of the 32 CEN-CENELEC member countries.

This and the other considerations covered in this section suggest that:

1. standards compliance is the preferred approach to demonstrate safety and performance requirements of medical devices
2. compliance to standards with a worldwide recognition (e.g., ISO/IEC standards with US and EU recognition) is the recommended strategy for improving efficiency in certification of world marketed medical devices.

8.2.1 Technical Standards for Medical Devices

There are many standards related to specific medical products or to classes of products such as the ISO/EN 10993 series of standards on biocompatibility or IEC 60878 standards that deal with symbols for medical device. In this section, we will focus on standards related to electrical and electromechanical aspects. In particular, in this book, we have focused on the standards related to quality systems, risk management, safety of electrical equipment, usability, EMC, alarms and software. These standards, referenced at the end of this chapter in 'Summary of Regulations and Standards' section, are referred to as '**horizontal standards**', that is, they are suitable for a wide range of different medical devices. There are also '**vertical**' standards that are specific to a single medical device class: IEC/EN 60601-2-2 for high frequency surgical equipment, IEC/EN 60601-2-5 for ultrasonic physiotherapy equipment, IEC/EN 60601-2-25 for ECG etc. The section 'Summary of Regulations and Standards' at the end of the chapter contains further details. These standards are structured to include additional requirements or tests focused on the specific equipment starting from the general requirements included in the IEC/EN 60601-1 family.

It should be noted that there are additional standards related to specific risks such as the ISO/EN 10993-1 for biocompatibility. These standards as well as the standards focused on specific devices are not discussed in this chapter.

Standard compliance is demonstrated through test reports, which describe the verification activity. A test report is a single integrated document that contains verification for both the general requirements of IEC/EN 60601 Part 1 and device-specific requirements of Part 2. For other risks not strictly related to electrical hazard (e.g., biocompatibility), a separate test report is suggested.

Note that the previous standards are related to design and prototype verification. There are other security standards related to factory acceptance tests and to extraordinary or periodic maintenance. More recently, such issues have been ruled by a harmonized standard that encompasses the existing national standards: Medical electrical equipment – Recurrent test and test after repair of medical electrical equipment (IEC 62353).

8.2.2 European Context

In this section, an outline of the European context is introduced, in order to give a better understanding of the regulatory issues in medical devices.

The regulation of trade in medical equipment in all European Union countries has changed significantly, with the publication of the directive 'Medical Devices' 93/42/EEC of 14 June 1993 (and subsequent amendments to the EC 2007/47) which represents the application of a new approach established by the European Union (former European Community) for technical harmonization and standardization.

The EU aimed at achieving the single market and free trade, breaking down customs, technical and legal barriers between the member countries. To implement these goals effectively, the EU needed to:

- define a unified **new approach** for the assessment of safety performance
- **harmonize procedures** for the import and approval of sales for products that were regulated by national laws and classified as 'dangerous' in their design performance or target use – this is the case for medical devices
- **harmonize the technical standards** used in different countries to avoid the need for manufacturers to build different products depending on the country of destination.

The overall EU regulatory scheme is depicted in Figure 8.5.

In June 1984, a directive (84/539/EEC) related to the devices used in human and veterinary medicine was issued. In practice, this directive was never fully implemented since:

- the national laws governing the approval for marketing medical devices continued to be in place, and these laws implied country-specific procedures with significant differences in the certification authorities and the technical requirements of devices

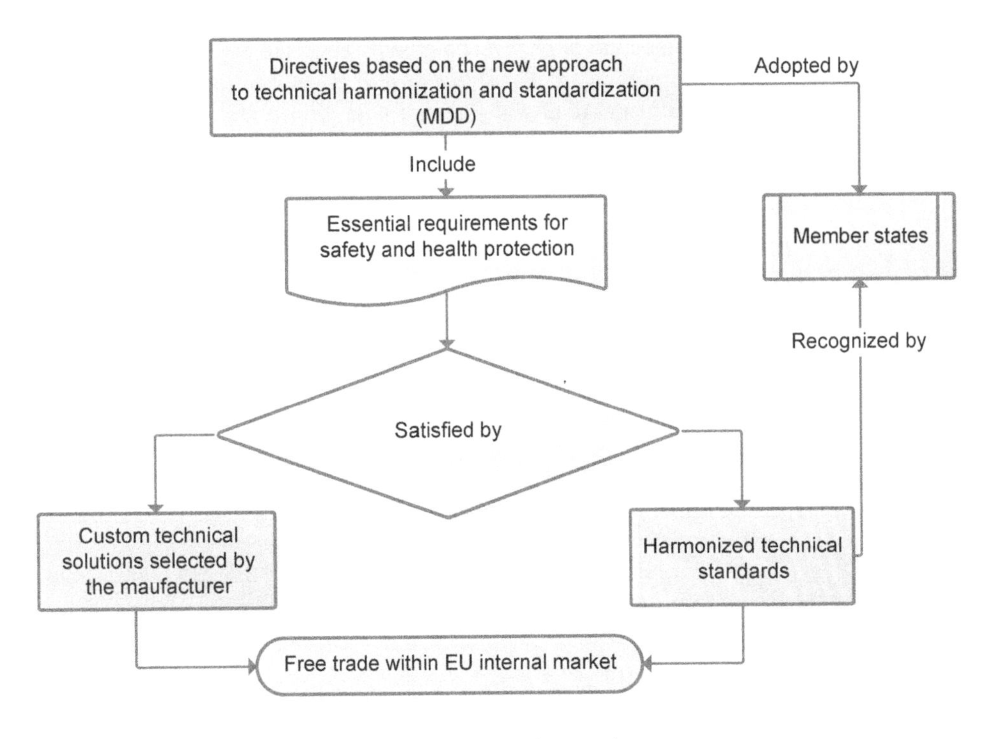

Figure 8.5 EU regulatory scheme.

- the directive was 'static': it included a reference to a specific technical standard without considering the natural evolution of technological and scientific requirements, so unfortunately, with this approach, requirements quickly became obsolete
- the directive did not imply a certification procedure, but was based only on self-certification whatever the device and its associated hazard.

To overcome such problems, common also to other areas, the European Union (previously the European Community) has systematically addressed the issue of standardization and certification of products with the Council Resolution related to a new approach to technical harmonization and standards (85/C136/01). This resolution was then addressed more organically through the new Council Resolution (1989) concerning a global approach to conformity assessment. The basic idea is that *the directives were to address the minimum essential safety requirements, and that the harmonized standards contained the specific requirements associated with the essential requirements*. From a practical point of view, the harmonized standards are not mandatory, but their application ensures the fulfillment of the essential safety requirements. When harmonized standards are not used, the manufacturer has to prove that the device has a level of security similar to that guaranteed by the relevant harmonized standards. In general, *products and systems have to be aligned at the security level achieved at the* ***state of the art***. Design engineers are also obliged to follow an approach oriented to prudence, diligence and skill. In practice, it is not that easy to assess the state-of-the-art level, and consequently, when a risk can be considered acceptable. Broadly speaking, the state of the art reflects the medium–high level commonly accepted by industry experts. In this sense, a new invention applied in an experimental form is not the state of the art, but it may indicate just one evolutionary trend. Only after a long consolidated experience and application can an innovation actually become the state of the art. The documents issued by the technical committees of standard bodies express this level of security, jointly agreed and shared by all stakeholders. In practice, the safety level implied within standards represents the minimum level under which it is not allowed to go, even if these standards are not legally binding. In general, we may say that there is no responsibility for injuries caused by products or equipment compliant with relevant standards.

8.3 Basic Protection Concepts

The following sections will address some main protection concepts that help in understanding the requirements specified by the standards. Before going into the discussion, Table 8.4 lists some useful terms used for test descriptions.

We also introduce the different types of **insulation** that are required to separate parts at dangerous voltages. In general, we can consider the types of insulation shown in Table 8.5.

For more details on the terminology, please refer to clause 3 of the IEC/EN 60601-1 standard, which includes the formal definitions of all the required terms.

8.3.1 Protection Against Electric Shock

By electric shock, we mean the physiopathological effect resulting from an electric current flowing within the human body. This situation can occur when a person comes in direct contact with two parts at different voltage. Protection against electric shock includes the means of avoiding hazards both through direct and indirect contact via the operators.

The IEC/EN 60601-1 standard prescribes the use of **two separate means of protection** for accessible parts; this arises from the assumption that one of the two means may possibly

Table 8.4 Main terms for safety tests

Term	Explanation
Enclosure	outer surface of an electrical device
Functional earth	earth of a device not used for safety purposes
Protective earth	earth conductor in the power supply cord or earth cable
Mains supply	source of electrical energy with voltage supply of 115 V or 230 V or 400 V according to the power distribution network
Single fault	condition that leads to the occurrence of one single safety-critical problem at a time, such as the opening or interruption of the earth conductor
Applied part	any part that is intentionally in contact with the patient
Accessible part	any part of the device that can be touched without the use of a tool
SIP/SOP	signal input/output part: connectors for signal input/output

Table 8.5 Type of insulation

Type of insulation	Definition
Functional	insulation between live parts, and between live parts and ground, for operating purposes only
Basic	insulation applied to live parts to provide a layer of protection against electric shock (in Class I equipment)
Supplementary	independent insulation applied in addition to the basic insulation to provide protection against electric shock in the case of failure of the basic insulation
Double	insulation that includes both a basic and a supplementary insulation (e.g. insulation made up of two separate layers of insulation completely coating the active parts)
Reinforced	single insulation protection of live parts that provides a degree of protection against electric shock equivalent to a double insulation

be compromised during the product lifetime. The first means of protection is performed by the isolation between the case and the circuitry. For equipment with a metal enclosure, one additional means of protection is provided by connecting all the accessible metal surfaces to the protective earth ground. In the event of a fault in the basic insulation of the enclosure, the connection of the ground protection will cause the fault to generate a short-circuit that triggers the short-circuit protection devices. This action removes power from the device and eliminates the risk of electric shock. The different types of insulation are listed in Table 8.5. When a device has a plastic enclosure not connected to ground, the standard allows for the use of the **double insulation** approach: the main insulation of the electronic circuits from the enclosure and an additional insulation composed of the insulation material of the enclosure itself. In case of failure of the basic insulation, the plastic enclosure is a second means of protection for the operator.

Another method of protection against direct and indirect contact consists of using only low voltage supply within the equipment. This reduces the risk of electric shock by eliminating the source of danger (high voltage).

Medical electrical devices are classified according to the approach used to protect against direct and indirect contacts: Class I, Class II equipment and devices with internal power source.

8.3.1.1 Class I Equipment

Class I equipment relies on basic insulation and an additional safety precaution that is constituted by a connection of all the accessible metal parts to protective earth. The accessible parts of the equipment are often constituted mainly by the metal enclosure. If the enclosure can carry voltage in the case of insulation failure and the enclosure is *accessible* in normal condition (i.e., it can be touched), the enclosure is called the '**earth**'.

The most usual means of protection against indirect contact is to connect the earth of the unit to ground, through a conductor, called the 'protective earth conductor'.

The requirement of protection depends on the electric power system. The system has to be designed to ensure an automatic circuit break in the event of electric shock hazard for the users. More specifically, circuit breakers have to trigger with a speed proportional to the voltage level on the enclosure. The trigger time must be compatible with the voltage–time curve limit that is safe for human body protection.

Note that if there are conductive parts inside the equipment that are accessible only when the fixed enclosure is removed using a specific tool, these parts can become live by means of a connection to the earthed part. In general, these parts must not be considered earth, and therefore they must not be connected to the circuit ground. However, if the enclosure can be removed under normal operations, for example, to carry out operations of tuning, fuse replacement, or restoration of thermal overload relays, the related enclosure parts are to be considered accessible and therefore they have to be connected to the ground. A conductive part separated from live parts by double or reinforced insulation is not to be considered as earth, because it does not become live in the case of single fault condition. Similarly, a conductive part that becomes live during an insulation fault, because it is in contact with earth, can not be classified as earth.

8.3.1.2 Class II Equipment

Class II equipment is protected with basic insulation and with an additional double or reinforced insulation. In this case, there is no need for protective earthing, and the equipment does not rely upon correct earth system installation conditions to ensure safety requirements. When the basic insulation fails, operators and patients are protected by the supplementary insulation or reinforced insulation. **Reinforced insulation** is defined as a single insulation system with two means of protection. Note that Class I devices rely on the integrity of the safety protection circuit of the electrical environment. This means that protection may be at risk even if the Class I equipment is working properly. On the contrary, Class II devices are intrinsically safe since they don't rely on the integrity of external means of protection (the earth connection and the earth system). Class II equipment is extremely widespread for small appliances, especially because of their light plastic enclosure and the agreeable aesthetic design. Typically, Class II equipment is safer than Class I because of the

- protection against indirect contacts (e.g., via medical staff), which is a *safer solution from a statistical point of view*
- protection against direct contact, involving *improved insulation compared to the basic insulation class equipment.*

8.3.1.3 Devices with Internal Power Source

These devices are powered by an internal source whose voltage is lower than the safety limit. The equipment design must guarantee that these limits are not exceeded during normal and fault conditions. The circuits at low voltage SELV (safety extra low voltage) must be safe in normal and fault conditions, so the voltage source and the wiring must be adequately separated from hazardous voltage circuits (e.g., using a certified safety transformer). IEC 60364-4-41 considers extra low voltage limits as 25 V RMS (35 V peak) for AC and 60 V for DC for general appliance.

In medical devices, the approach to **protection against direct and indirect contacts** is very different from other equipment, because ventricular fibrillation may be triggered by much lower values of current (even 1000 times lower than the safe limits required in other contexts). In addition,

- the patient is often unconscious and can not react to electric shock risks
- the patient is in poor health, and the effect of a given leakage current is greater
- applied parts may be directly in contact with the cardiovascular system through catheters or invasive parts, which means that all the current can flow through the heart, and the current not only affects the heart muscle surface but also flows within the heart mass, with higher risks of ventricular fibrillation.

For these reasons, the standard does not refer to safe voltages limits, as is the case for general electrical equipment (household appliances, office equipment, etc.). The standard prescribes safety limits in terms of the maximum current that can flow between the circuits and the body. This may happen in case of indirect/direct contact or under 'double-indirect contact' condition, that is, when in contact with a different person (e.g., a nurses) who is in contact with the device.

The standard introduces an additional classification based on the leakage current that can flow to ground, in normal/fault condition, from the enclosure or from the parts applied to the patient. Chapter 5 details the different types of current. As outlined in Table 8.6 first column, there are three types of medical electrical equipment (MEE) according to the level of leakage currents.

The equipment type must be clearly distinguished on the device through the associated symbols shown in Table 8.6.

To verify whether an equipment part can be in direct contact with patients or operators, suitable tools have been standardized. For example, the standard 'test finger' is used as a model for the human finger. Since equipment may be indirectly touched through accessories and tools in normal usage, there is also a standard test plug to verify whether a specific device part can come indirectly in contact with the patients or the operators.

Table 8.6 Classifications for applied parts

Type	Symbol	Description
B		MEE having reduced leakage currents; designed with a specific patient safety approach; suitable for direct application to the patient
BF		MEE having the patient applied part insulated from ground (floating); has a greater security level than type B
CF		MEE with highest safety level, suitable for direct application to the patient's heart

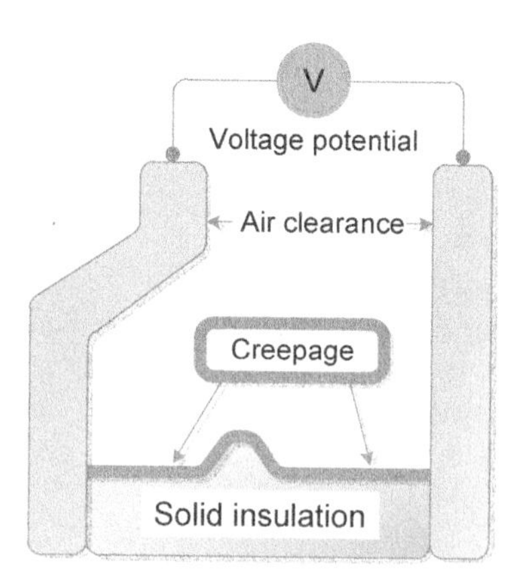

Figure 8.6 Different insulations on devices.

8.3.2 *Insulation*

The separation of parts from dangerous voltage is obtained by appropriate insulation, as defined in Table 8.5.

The materials used to make the insulation must be suitable for the intended use, with specific requirements for flammability, and additional mechanical and thermal properties of non-deformability. For example, the materials to be used for the enclosure must have more stringent flammability characteristics than those that cover the live parts. In addition, the insulation must not be hygroscopic (i.e., the material must not attract or hold water from the surrounding environment) since humidity jeopardizes the insulation and therefore the equipment safety. As show in Figure 8.6, the insulation can be made by

- the solid insulation
- superficial insulation from the creepage distance
- air clearance insulation.

Creepage is defined as the shortest path between two conductive parts (or between a conductive part and the bounding surface of the equipment) measured along the surface of the insulation. **Air clearance** is defined as the shortest distance through the air between two conductive elements. The requirements that must be met are related to

- dielectric strength
- insulation resistance
- size
- flammability
- resistance to surface currents.

Regarding the resistance to surface currents, it should be noted that between two live parts, separated by a solid insulation, a conductive layer of electrolytic nature can be generated on the surface of the insulator. The origin may be pollution on the surface or condensation of atmospheric moisture. The flow of a small current produces a progressive alteration of the insulating surface, which becomes more conductive. This degradation process is called tracking.

The **comparative tracking index (CTI)** is used to measure the electrical breakdown (tracking) properties of an insulating material. The tracking index (CTI, also known as PTI, proof tracking index) is dependent on factors such as pollution, humidity, type of material and the time for which the leakage current has been flowing. The applied voltage increases this effect, so the insulation between an active part and the insulating enclosure is less susceptible to tracking in that the leakage current flows only when the enclosure comes into contact with a person or with metal parts connected to ground.

The tracking index is expressed as the value of voltage between two electrodes, which must be withstood without tracking (i.e. breakdown) by the material with 50 drops of a conductive solution dropped between the electrodes. Other specific test contamination conditions are also specified. The normalized values standardized in IEC/EN 60601-1 are: Group I, $600\,V \leq CTI$; Group II, $400\,V \leq CTI < 600\,V$; Group IIIa, $175\,V \leq CTI < 400\,V$; Group IIIb, $100\,V < CTI < 175\,V$.

Among the insulating materials, phenols and polyesters are very sensitive to tracking, while ceramic materials do not exhibit tracking effects. Ribs and ridges improve tracking resistance since they create a discontinuity of the conducting layer.

Insulation material is critical to guaranteeing safety. Insufficient CTI also increases the risk of fire. The supplier of the insulating material has to be able to provide specifications on the flammability characteristics and tracking index of the material. The equipment designer must select the appropriate material for the intended use.

8.3.3 Degree of Protection Provided by Enclosures

In addition to protection against electric shock, the enclosure provides protection from foreign bodies and external weather conditions. This degree of protection has been encoded by the letters **IP** (ingress protection rating or international protection rating) followed by two digits and optional letters, according to standard IEC 60529:

1. The first digit in the range 0–6 indicates the degree of protection:
   a. protection of the enclosure against the access or contact by persons to hazardous parts (moving parts, conductors etc.)
   b. protection of the enclosure against penetration of external material.
2. The second digit in the range 0–8 indicates the protection against the penetration of harmful water.
3. The additional letters (optional), A–D, are used to indicate a level of protection for people against access to dangerous parts greater than indicated by the first digit. This additional indication gives the protection against solid bodies as
   A – protection against access by the back of the hand
   B – protection against access of a finger
   C – protection against access by a tool
   D – protection against access by a wire.

Table 8.7 shows the specific meaning of the first digit. The second digit's meaning is outlined in Table 8.8.

Table 8.7 IP Protection against access to hazardous parts (first digit)

First digit	Degree of protection	
	Protection for the persons	Protection against access to external objects
0	unprotected	unprotected
1	effective against the back of the hand	protected against solid external objects of diameter ≥50 mm
2	effective against fingers	protected against solid external objects of diameter ≥2.5 mm
3	effective against tools such as a screwdriver	protected against solid external objects of diameter ≥2.5 mm
4	effective against access by a wire	protected against solid external objects of diameter ≥1.0 mm
5	effective against access by a wire	protected against dust – the penetration of dust is not totally excluded, but the dust must not enter in such quantities as to affect the normal operation of the equipment or impair its safety
6	effective against access by a wire	totally protected against dust

8.4 Verification of Constructional Requirements

8.4.1 Choice of Safety Critical Materials and Components

Medical devices may contain components that have an impact on safety. These components must be classified as *critical*. As a general rule, any components that may provoke harm are to be considered as critical. At the beginning of the design stage, the manufacturer should identify the critical components (e.g., optocouplers for galvanic isolation of floating parts) so that their selection on the market may be performed with due care (e.g., selecting models whose pins withstand the necessary creepage). Technical documentation of critical components (e.g., datasheets) is needed to demonstrate compliance to applicable standards since non-compliance of a single critical component may compromise safety of the whole device. Table 8.9 outlines the main components that are usually considered to have impact on safety. The table also list the applicable standards.

The safety standards require that critical components comply with the specific IEC standards. Alternatively, the manufacturer must test all the components that are not compliant with IEC standards. This approach may be suitable for specific components such as resistors, transformers, capacitors. In any case, it is recommended that components such as fuses, relays, plugs, connectors and cables are already compliant to IEC standards. Compliance for these components does not increase their price significantly, and tests on non-compliant types is time- and cost-consuming. In addition, the selection of certified components increases the quality of the final product since these components are usually also tested on other relevant performances such as strength and reliability.

Table 8.8 IP Protection against access to hazardous parts (second digit)

Second digit	Degree of protection
0	unprotected
1	protected against vertically falling water drops; the drops of water falling vertically must not cause harmful effects
2	protected against vertically falling water drops when the enclosure is tilted up by 15°; the drops of water falling vertically must not cause harmful effects when the enclosure is tilted up by 15°
3	protected against rain; the water sprayed in a direction forming an angle with the vertical up to 60° must not cause harmful effects
4	protected against splashing water; the water sprayed against the enclosure from any direction must not cause harmful effects
5	protected against water jets; the water jet sprayed against enclosure from any direction must not cause harmful effects
6	protected against powerful water jets; water projected in powerful jets against the enclosure from any direction must not cause harmful effects
7	protected against the effects of temporary immersion; water must not have harmful effects when the enclosure is temporarily immersed in water under specified conditions of pressure and duration
8	protected against the effects of continuous immersion; continuous immersion must be possible under conditions agreed between the manufacturer and the user; these conditions are more stringent than those specified in the previous case (IPx7)

The designer who uses IEC compliant components must verify that

- the applicable standards for the components are suitable for the device
- the selected component is suitable for the intended use in the target environment
- the component has nominal performance values suitable for the intended installation (e.g., temperature, voltage, frequency, power).

Transformers have to be analyzed carefully in medical device design when they are part of the safety insulation. Transformers may be used for signal transfer or for power supply.

Signal transformers suitable for high frequencies or low voltages may be employed to transfer signals through a basic or reinforced insulation. These transformers have to be designed to comply with the safety insulation requirements. Therefore, they are often encapsulated in resin to improve reliability even with reduced distances between the parts to be insulated. This is possible when the air and surface spaces are completely filled with insulating material that is resistant to heating and fire.

Power supply transformers are the most safety-critical elements. Usually, the most demanding requirements are set on the minimum distances on air, on surfaces through the

Table 8.9 Main critical components and applicable standards

Component	Standard
Plug and socket connector	EN 60320
Switches	EN 60898, EN 60934, EN 61008-1/-2-1, EN 61009-1/-2-1
Appliance couplers	Series EN 60320-1
Enclosures	UL94, IEC 60695-11-10
Fuses and fuseholders	EN 60127-1/2/3/6 for miniaturized fuses EN 60269-1/2/3 for the other fuses
Y capacitors	IEC 384-14
Other capacitors	EN 60143, EN 60252, EN 60931-1/2 EN 61048, HD 207 S1 S1/597
X2 capacitors	IEC 384-14
Integrated protectors, fuse resistors	IEC 127/BS 4265
Switchgear and control gear	EN60947
Plugs and sockets	IEC 60598-1, EN 60320, EN 50075
Industrial plugs and sockets	EN 60309-1/2
Adapters	IEC 60884-2-5
Power modules	IEC/EN 60950, EN 61204
Power cables	IEC 245 227/IEC
Optocouplers	EN 60950-1
Relay	IEC 328
Transformers	EN 60742, EN 61050, EN 61558-1, EN 61558-2
Thermal-protectors	EN 60730-1, EN 60730-2-9
Connecting devices, etc.	EN 60998-1, EN 60999, EN 61210
Control switches, switches	EN 61058-1
Contactors	EN 61095

insulation and on heating and insulation. When the secondary circuits (i.e., a circuit not directly connected to primary power. Secondary circuit is powered by a battery, or an isolation device such as a transformer or a converter.) have to be powered at SELV (Safety Extra Low Voltage) voltage or they are directly connected to applied parts, a safety transformer compliant to IEC/EN 61558-1 has to be used.

The transformer must be:

1. protected against short circuits by design or by additional suitable devices
2. protected against overload or by design by suitable devices, for which a test is performed at the maximum current before the protection device triggers
3. compliant to the required distances in air, surface and through the insulation:
   - 2 mm through the insulation between primary and secondary winding throughout all the winding (in general, compliance with this requirement is critical at the transformer's edges)
   - 25 mm between the primary and secondary terminals
   - 8 mm between primary and secondary surface (insulating barriers may be used if necessary)

   these safety distances must be maintained throughout the component life time to avoid risks; this is done by fixing the windings using tape or some other suitable method.
4. manufactured with insulating material (insulation of wires, coil, strip and insulating barriers) classified according to IEC 85 'Methods of determination of thermal class of the electrical insulation'
5. designed to prevent
   - conductor and terminal displacement
   - the risk that any conductive part can short-circuit insulation in case of failure
6. designed so that temperature does not to exceed the values allowed for the insulating materials used in ordinary conditions (with test voltage equal to 110% of the supply voltage) and in fault conditions.

In order to meet requirements, fuses or thermal protection devices have to be used. Usually a safety transformer may be equipped with

- fuses on the primary winding
- fuses on the secondary winding or a thermal device on the primary winding
- alternatively, a thermal switch on the primary.

When using the combined protection of fuses and thermal switches, fuse provides short-circuit protection, the thermal device prevents overload.

8.4.2 Creepage Distances and Air Clearances

Creepage distances and air clearances are related to the paths of voltage 'breakdown' between active parts or between an active part and other parts separated by a basic, supplementary or reinforced insulation. Non-active parts may be ground-connected elements of circuits at SELV. In particular, the **air clearance** between two parts is the shortest path of separation measured, in air, between:

- two conductive parts
- a conductive part and a metal surface connected to the protection circuit.

We define a circuit as '**primary**' when it is directly connected to the external supply mains. A **secondary circuit** is a circuit that has no direct connection to the primary power and derives its power from a battery, a transformer, a converter or an isolation power device. The minimum air clearance that avoids air dielectric breakdown is dependent on the

- type of circuit (primary or secondary)
- degree of pollution (1, 2 or 3, where 2 is the value usually selected)
- working voltage and the possible presence of transient voltages
- type of isolation (basic, supplementary, reinforced)
- material group classification (I, II, IIIa or IIIb) as in Section 8.3.2.

The **creepage** distance is the smallest distance measured along a surface of separation between:

- two conductive parts
- a conductive part and a metal surface connected to the protection circuit.

Creepage distances shorter than the air clearance are not allowed since it is preferred that breakdown occurs in air rather than on surfaces, as the latter event may permanently damage the insulation layer.

Not all the live parts inside the equipment have to be isolated. Live parts that are properly fastened and sufficiently far from other parts may be insulated by air.

The minimum distance that must be met, for example, are those between:

- live parts
  - of opposite polarity
  - and parts connected to earth
  - and parts at SELV
  - and applied parts
- floating applied parts
  - and enclosure or ground
  - and SELV parts (e.g., signal I/O circuits).

During the project design, manufacturing tolerances due to welding, milling and finishing tooling etc. have to be taken into account when establishing minimum distances. The minimum distances are therefore to be exceeded to include such tolerances. Compliance of insulation is verified by inspection, by measurement of thickness and by testing the dielectric strength where the insulation has to withstand the appropriate test voltages.

8.4.3 Markings

The IEC/EN 60601-1 standard prescribes specific labels, for warning and danger. Labels must not be placed on removable parts and they must pass a durability test to ensure that they remain readable throughout the expected product life. Labels and warnings do not replace safety requirements; they must be considered as an additional means of integrating the safety requirements of the equipment. Medical equipment must show

1. the manufacturer's name (logo or trademark)
2. the model reference

3. the serial number
4. an indication for the year of manufacture
5. the nominal power (voltage, frequency and current consumption, the symbol for the AC/DC voltage)
6. the symbol for Class II (if applicable) ◻
7. the degree of IP protection (if different from IP20 – see Tables 8.7 and Table 8.8)
8. the classification of applied parts (B, BF or CF, with associated symbols shown in Table 8.6).
9. the symbol that warns to read the user manual  .

Software also has to be identified with the product identification, revision and release date. Specific warnings are also required

- to identify hot surfaces that may come into contact with the operator or service personnel
- for accessory units required but not supplied with the equipment
- for capacitors that can produce electric shock to service personnel
- for fuses in the neutral conductor
- for risks of exposure of service personnel to X-rays
- for devices that intentionally use radio frequency radiation.

Below, are some general rules to be considered in designing the product packaging:

- The symbol  has to be placed on the enclosure or near parts with dangerous electric voltage.
- The means of disconnecting must show the indication of the position (open/closed: O/I).
- The control devices must show the indication of the direction of adjustment (e.g., + and –).
- Equipment containing liquid must indicate the maximum allowed level.
- Rated voltage must be indicated close to the associated lamps.
- The ground terminal must be marked with associated symbol.
- The Class I must have a label on the power cord: 'WARNING: This equipment must be grounded.' Class I instructions must include: '*WARNING: To avoid the risk of electric shock, this equipment must only be connected to a supply mains with protective earth.*'
- All instructions must be durable: they must withstand a rubbing test with specific gasoline, water and alcohol.
- The components must be clearly identifiable. For components with limited size, an acronym is sufficient.
- For replaceable batteries, the type and model must be indicated.
- If the device is equipped with lithium batteries a warning must be shown close to the batteries or on the instruction manual: '*WARNING, Danger of explosion if battery is replaced incorrectly. Replace only with the same type or with an equivalent type recommended by the manufacturer. Discard used batteries according to manufacturer's instructions.*'

The fuses used must be indicated close to the fuse position: the rated voltage, rated current, breaking capacity and breaking speed. Warnings labels are also generally required where

- chemicals are used
- mechanical hazards are possible.

Furthermore, it is good practice to alert service personnel concerning

- the need to isolate the unit from the mains before removing the enclosure
- indication of parts that remain live.

Service personnel generally assumes that

- non-marked metal parts and components are safe
- fans inside the unit are equipped with protective gratings
- any lifting operation on the device requires only one person
- the mechanical parts will not move, causing a hazard
- laser or other radiations are not present.

If any of these conditions is not satisfied, a warning has to be shown; the general warning symbol is ⚠. The warnings must be provided in the language of the destination country. Warning labels must withstand durability test with water and alcohol. At the end of the test, they must remain readable and they must not be detached from the equipment.

8.4.4 Conductors

Internal and external conductors must be chosen considering the required insulation that has to be suitable for the voltage rating and the temperature of use. Table 8.10 shows some codes that identify cables according to (CENELEC HD 1999). The cable codes specify the main cable features.

The choice of the installation approach impacts on heat dissipation and therefore it determines the type and the section of the cable. The choice of the conductors of electric equipment is influenced by

- rated working voltage
- maximum expected current
- ambient temperature
- requirements for electromagnetic interference (EMI) emission or immunity
- installation environment conditions (e.g., presence of water, corrosive substances etc.)
- mechanical stress.

The power plugs must have a nominal value greater than 125% of the nominal value of equipment consumption. Plugs and power cords must be certified; for example, it is possible to use the HAR scheme (HAR, 2002) which is the 'agreement on the use of a Commonly Agreed Marking for Cables and Cords complying with Harmonized Specifications'. The power supply cords must comply with IEC 227 or IEC 245 and have an adequate section to avoid overheating and ensure that short-circuit protection devices can switch before cable damage occurs.

The wires connected to the power supply network and for external wiring must be harmonized according to EN 50525-2-21. These cables are identified by the HAR marking. The colors of the wires must be

- brown for phase
- blue for neutral
- yellow-green for ground.

Table 8.10 Cable classification adapted from (CENELEC HD 361)

	Assignment to standard	
	Cable as per harmonized standard	H
	Nationally recognized cable type	A
	Other type of national type	N
	Complies to IEC standard	J
A.	**Rated voltage**	
	≤100/100 V	00
	100/100 V ≤ voltage ≤ 300/300V	01
	300/300 V	03
	300/500 V	05
	450/750 V	07
	600/1000 V	1
	Insulation material	
	Special PVC compounds for T > 70°C	V2/V3
	Natural and/or synthetic rubber	R
	Silicone rubber	S
	Vinylacetate ethylene	G
	Sheath material (if any)	
	PVC	V
	Natural or synthetic rubber	R
	Polychloroprene	N
	Braid of glass fiber	J
	Textile braid	T
B		
	Construction details (if any)	
	Flat cable cores separable	H
	Flat cable, inseparable cores	H2
	Type of conductor	
	Single wire	U
	Stranded	R
	Stranded for fixed	K
	Stranded for mobile application	F
	Highly flexible rope	H
	Tinsel wire	Y
	Number of cores	—
C	**Protective conductor**	
	Without yellow-green protective conductor	X
	With yellow-green protective conductor	G

A cable with a suitable number of conductors has to be used: for example, Class II equipment must have a cable with two conductors while Class I equipment needs three wires. The yellow-green wire with insulation must be used only for the protective conductor.

All the internal conductors must be maintained in a fixed position so that they do not become dangerous if they move and get in contact with another circuit or with the metal case. This is especially important if there are SELV circuits or circuits connected to the patient (applied part). When cables at SELV and cables at dangerous voltage are located in the same cable duct, the SELV cables must be insulated at maximum voltage. For example, in the presence of 230/400 V cables, SELV cables have to ensure insulation at 300/500 V to avoid that an accidentally detached phase conductor creating a danger on the SELV cables.

When SELV circuits are not properly isolated, it is always necessary to physically separate the wiring of the circuits at low and very low voltage. The cable routing must be such that the insulation cannot be compromised by sharp edges or protrusions. The internal conductors must not be subjected to traction stresses by opening the removable parts. If in specific equipment parts, temperature may rise above 70°C, conductors with silicone insulation or for high temperature are to be used. In this case, PVC is unsuitable.

The internal conductors must be fixed both electrically and mechanically, for example by sheaths, thread guides and clamps. The soldered conductors must be mechanically fixed before soldering. Conversely, wires to be tightened into screw terminals should not be tinned.

8.4.5 *Connections to the Power Supply*

The connection to the mains supply must be designed to ensure

- an easy connection to the power distribution system
- a reduced risk of failure of the power supply cord due to mechanical stress
- a reduced risk of short circuits or incorrect ground connections due to disconnection of the wires from their terminals.

From a practical point of view, the connection to the power supply can be made using the following approaches:

- an appliance (i.e., male plug) that is connected to the equipment through a power supply cord anchored to the equipment
- terminals for detachable cords
- a male plug incorporated into the equipment enclosure.

Connectors and terminals used in the equipment must be certified, because the insulating materials that support live parts have to withstand specific tests to determine their resistance to high temperature and their flammability characteristics. It should be noted that the usual PVC clamps do not pass these tests.

If the anchoring system uses screws, it must be guaranteed that screws do not cut the wires when they are tight. The end of the screw must therefore be rounded or provided with a clamping plate.

If the power cord is replaceable (without special tools) conductors have to be distinguished as: L for the phase, N for the neutral and ⏚ for protective ground terminal.

When connections are soldered, the wire has to be fastened by some other means (gland or clamping). It is generally forbidden to solder power cord wires directly onto the PCB. A conductor at dangerous voltage, if it becomes disconnected, must not touch low voltage parts in any circumstance. Insulation barriers are often necessary to ensure this requirement.

8.4.6 Fire Enclosure

The IEC/EN 60601-1 standard has introduced specific requirements on the enclosures and related materials to be used. This ensures that even in the case of fire inside the unit, flames do not propagate outside. The requirements, derived from IEC/EN 60950-1 for information technology equipment, specify that

- the enclosure must not have holes or openings on the upper part such that an object can fall within the enclosure
- the openings on the lateral sides must be designed so that any projection of internal objects or parts outside can not find an opening in its path
- the openings on the bottom of the enclosure must be reduced (e.g., mesh holes of 2 mm diameter interspaced by a distance 1 mm) or protected by baffle plates.

Plastic enclosure must have a flammability classification equivalent to FV-2 (according to IEC 60695 family standards) or better for portable units and equivalent to FV-1 or better for fixed appliances.

8.5 Medical Equipment Safety Tests

The basic purpose of safety tests on electrical equipment and medical systems is to ensure that the device is safe for both the patient and the medical and paramedical staff. This safety level must be attended to in the design and production and must be maintained in normal use. Standard IEC/EN 60601-1 has detailed the design requirements that must be satisfied for patient and operator safety. This approach is in line with other safety standards for operator safety (e.g., IEC/EN 60950-1 for information technology equipment safety), but it implies more stringent requirements for patients considering their vulnerability. The greater vulnerability of a patient compared to a healthy person is mainly due to

- the inability to move away from live parts (i.e., at high voltage) when the patient is under anesthetic
- the danger of microshocks due to the increased sensitivity to currents when a patient has no protection of the skin and for the possible direct electrical connection to internal vital parts and to heart.

The safety tests are particularly demanding in the design phase since safety must be integrated into the design and not just verified after production. The knowledge of the appropriate standards can drive the designer to a safe implementation rather than having to solve problems on the final product with additional costs.

When the design project has been verified as safe, it will be necessary for the manufacturer to carry out tests on the final product. A theoretically safe device may become dangerous because of an unexpected short-circuit on the printed circuit board, an incorrectly connected socket or other problems that may jeopardize the protective earthing system. Only tests on the final product can guarantee a safe product for the customer.

Compliance with IEC/EN 60601-1 implies testing and construction verification on ten main areas that are source of hazards:

1. **Leakage currents:** which are classified into (see also Section 5.9 for the impact of this hazard on medical devices hardware design):

Earth leakage current	mainly the current flowing through the protective conductor in the supply cable to ground
Enclosure or touch leakage current	the current that may flow from any accessible part of the enclosure when touched by a human being
Patient leakage current (or leakage current in the applied parts)	the current flowing to or from an applied part through the patient.
Patient auxiliary current	current due to potentials on the patient applied parts, present in normal use for bias or for desired signals, as for example the right leg applied signal generated by the ECG to reduce common mode interference.

2. **Dielectric strength:** the ability of the insulation to withstand high voltages (the test involves applying a high voltage between the separate insulated parts under test)
3. **Heating:** the temperature of various surfaces and accessible components of the product
4. **Mechanical risks:** stability, mechanical strength of enclosures: test on the capability of the enclosures to provide protection from electrical, mechanical and fire hazards; test on the moving parts and the associated risk of truncation, mutilation, puncture or abrasion
5. **Abnormal operations and fault conditions:** test on the preservation of the target safety level, when device is in fault condition or in a predictable malfunction
6. **Continuity of protective earthing:** the capability of the connection protection system to withstand a current of 25 A flowing in the protection circuit in order to verify its continuity
7. **Residual voltage:** the voltage at the pins of the plug or between the plug pins and the enclosure, one second after the power plug is disconnected (also referred to as the discharge test); additionally, the voltage that remains inside the device (e.g., in the capacitors) once the enclosure is removed
8. **Voltages on accessible parts:** the voltage on any accessible part, including the parts protected for service
9. **Stored energy:** the energy stored in any accessible part or that can become accessible under fault conditions, including hydraulic and pneumatic energy or energy accumulated in the pipes and tanks

10. **Current and power consumption:** the power and the power consumption absorbed by the device from the power network.

Note that these ten areas are usually focused on by all standards, regardless of their regional origin or enforcing authority. We will now give details on tests associated with these ten areas.

8.5.1 Leakage Current

Leakage current was introduced in Section 5.9. Here it is of interest to discuss some additional details on the associated tests. During the leakage current tests, normal conditions and single fault conditions are simulated by reproducing all the events that may occur from an electric point of view. The reproduced single faults are conditions that may generate potential safety issues. Since two or more faults are unlikely to occur simultaneously, faults are simulated one at a time. These are the simulated single fault conditions:

1. interruption of the protective conductor
2. interruption of the neutral conductor
3. interruption of the external protective conductor
4. interruption of the external neutral conductor
5. mains supply on the signal connections (SIP/SOP)
6. mains supply on the applied parts
7. mains supply on the unearthed enclosure
8. short circuit on the insulation.

Normal conditions are defined as electrical situations that occur daily and do not imply safety problems. In normal conditions, all the means of protection against hazards are working properly. For example, the grounding of an applied part, which occurs whenever the electrodes of an ECG come in contact with a metal grounded enclosure. The normal conditions used in the IEC/EN 60601-1 standard are

1. inverted power line
2. functional earth connected to protection ground
3. applied parts connected to ground
4. isolated metal parts connected to ground
5. incorrect polarity in SIP/SOP connections
6. incorrect polarity in applied part connections
7. functional earth of the external supply connected to protective ground.

Voltage rate and frequency of the mains supply is another source of safety hazards. To certify a product worldwide, voltage and frequency ranges have to be tested in the worst case (usually at 240 V, 50 Hz). Whatever the selected values of frequency and voltage, IEC/EN 60601-1 requires testing with an additional margin of 10% (e.g., for 240 V voltage rate, tests are performed at 264 V). The final and most critical issue is related to the instrumentation that is used for safety tests. The standard requires the use of instrumentation for the leakage currents that shows the true RMS value. For this purpose, it is suggested to use an

Table 8.11 IEC/EN 60601-1 limits for leakage current (values in μA)

	Normal condition		Single fault condition	
Earth leakage:	5,000		10,000	
Touch current (i.e., Enclosure leakage current):	100		500	
	Type B-BF	**Type CF**	**Type B-BF**	**Type CF**
Auxiliary leakage (between on patient connection and all the other): Patient leakage (from patient connection via patient to earth or from external source via patient to a patient connection)	DC < 10 AC < 100	DC < 10 AC < 10	DC < 50 AC < 500	DC < 50 AC < 50

instrument that can measure leakage currents with a crest factor (i.e. the ratio between the peak amplitude of a waveform and its root mean square) of 10 and an accuracy better than 10%. The instrument must also have enough passband capability (1 MHz) and a frequency-compensated filter in order to give more weight to low frequency components of leakage current, which are more dangerous.

8.5.1.1 Earth Leakage Current

The earth leakage current test is critical for safety. The test measures the sum of all the leakage currents that flow into the protective earth conductor. This test is performed both under normal and single fault conditions. IEC/EN 60601-1 specifies that the test must be done with neutral connection opened and closed in all possible combinations of the polarity of the power supply. The measured values of the leakage current must not exceed the IEC/EN 60601-1 limits outlined in Table 8.11. These limits were discussed in Section 5.9.

8.5.1.2 Touch Current

The measurement of the touch current on the enclosure is performed by assessing the current that flows from a person who touches the enclosure of the unit under test. The leakage current must be measured between earth and a conductive accessible part of the enclosure, such as connectors, external metal parts, knobs and dials or, if enclosure is insulated, between a metal foil (of area of 200 cm^2) that covers the instrument under test.

8.5.1.3 Patient Leakage Current

The measurement of the patient current on the applied parts (i.e., on the device parts normally in contact with the patient) is the most critical of all the safety tests. In the case of invasive devices, applied parts are under the skin of the patient where the resistance is the lowest. In this case, even a 15 μA leakage current can be fatal. The measurement of these low and critical values of current requires very sensitive and accurate instrumentation. This test is complex also because of all the possible combinations have to be measured under all possible conditions. When, ten (e.g., the ECG) or more conductors (e.g., EEG) are connected to the patient and the patient can touch the other objects near to test environment, the number of tests may be very high. The IEC/EN 60601-1 specifies detailed measurement methods. Basically tests are performed as follows:

- Type B and BF applied part: from and to every patient connection related to every single function of the applied parts. Connections are connected together or loaded as in normal operating conditions.
- Type CF applied part: from and to, every patient connection, one at a time.

In addition, for all the equipment, the auxiliary leakage current is measured between each patient connection and all the other patient connections connected together or loaded as in normal operating conditions.

8.5.2 *Heating*

Excessive heating can cause fires, damage to the materials used to isolate dangerous voltage and deformation of the enclosure. Overheating may be expected on enclosures, on parts normally accessible (knobs, switches, handles, applied parts) and in inner parts such as magnetic components (transformers, motors, relays), electrolytic capacitors, switches, fuses, PCB areas under heating components.

The maximum allowed overheating should not be exceeded under normal conditions (i.e., 110% of the rated power supply) and under single fault conditions. It should be noted that the limits of the standards are referred to an ambient temperature 25°C. When the device is intended to be used at higher temperatures (e.g., to 40°C) the limits reduce by the difference between standard ambient temperature and the expected ambient temperature (i.e., 40°C – 25°C = 15°C). Temperature must be measured with sensors (thermocouples) on components and surfaces. Alternatively, for transformers and motors, the change in resistance with regard to the windings can be measured. In this case, the increase in temperature is given by:

$$\Delta t = \frac{R_2 - R_1}{R_1}(234.5 + T_1) - (T_2 - T_1) \tag{8.1}$$

where R_2 and R_1 indicate the resistance value at the end and beginning of the test and T_2 and T_1 represent the ambient temperature at the end and the beginning of the test and Δt is the temperature increase. Δt must not exceed the limits specified by the respective manufacturers for the class of materials used, for PSBs and for windings; the monitoring probes will have to be positioned on the parts at higher temperatures such as

- the top of cases (usually above the power supply)
- the inner surface of the enclosure
- between a switching power supply transformer and its associated core
- the deflection coils (if used)
- the voltage regulators
- the power circuits
- the insulation.

The temperature limit for the applied parts depend on the intended of use of the device; if the device transfers heat to the patient, a higher temperature limit is allowed if this has been evaluated in the risk analysis.

8.5.3 Dielectric Strength

Dielectric strength indicates the resistance to insulation breakdown. The dielectric strength test is always preceded by a preconditioning in a climatic chamber with humidity of $93 \pm 3\%$, and temperature between 20°C and 32°C for a period of 48 hours (7 days for appliances with protection from IPx1 to IPx8). These test conditions create condensation on the device, which must therefore not have hygroscopic insulation. The test has to be repeated after the heating test, that is when the device has reached operating temperature during normal operating conditions. This is because the thermal effect may have changed the dielectric property of the insulating materials. When the device contains vessels for liquids, vessels have to be filled with a quantity equal to 110% of the capacity to pass the dielectric strength test.

Areas where fluids are contained have to be separated from areas with electric parts to avoid electrical sparks; this effect may occur particularly when fluids are volatile.

The dielectric strength tests have to be performed on all devices in order to assess the effectiveness of the insulation. The test voltages are related to the values of the working voltages and the type of insulation. During the test, which lasts 1 minute, no discharges or insulation breakdown has to occur within the device. The test voltage is usually applied between

- the primary circuits with dangerous voltage and ground (basic insulation) (according to IEC 60950, a **primary circuit** is an internal circuit directly connected to the external supply mains or other equivalent source, such as a motor-generator set)
- the primary circuits with dangerous voltage and the applied parts (with reinforced or double insulation)
- the primary circuits and the plastic enclosure (reinforced/double insulation)
- the primary circuits and the signal input/output (double or reinforced insulation or basic insulation and shield connected to earth)
- the applied parts and enclosure for class BF and CF devices.

8.5.4 Stability and Mechanical Strength

8.5.4.1 Strain Relief

Equipment where power supply cords can not be separated (i.e., the usage of a tool or the unit's destruction is required for separation) must be equipped with an adequate mean to ensure a proper cord anchorage (normally a gland). The anchorage must be effective when the cord is pulled, pushed and twisted. The standard specifies various test conditions. For example, the cord must be torqued and pulled for 1 minute at specific strength values according to the mass of the equipment itself.

The anchorage must avoid introducing the cord inside the unit and damaging the cord itself, its conductors and insulation, or the internal parts of the units. After the tests performed 25 times, the cord must not have moved more than 2 mm.

8.5.4.2 Stability

Medical equipment may be tilted and may fall on the operator. This is a mechanical risk that has to be minimized. The standard specifies that all the equipment must be able to return to the equilibrium position once placed on a plane inclined by 10°. For this test, all doors not fixed by a tool must be placed in the most unfavorable condition, as well as the electrical connections and mechanical adjustments. Stability must also be tested in transport conditions, for example, for devices equipped with casters wheel, simulating the inclination and the overcoming of a step 20 mm high.

8.5.4.3 Mechanical Strength of the Enclosures

Enclosures must have sufficient mechanical strength and hardness. This is verified by the following tests:

- drop test up to a maximum of 1 m from the ground
- pressure test of 45 N on the enclosure with a circular surface with diameter of 30 mm
- shock-test: a sphere of diameter (50 mm 500 g) dropped onto the surface from a height of 1.3 m
- simulation tests for the rough handling of equipment with wheels; the test must verify the effect produced on the equipment after rapid ascent or descent of steps or obstacles.

Knobs, levers and control elements are tested in torsion and traction. The metal structure of the external and internal case, as well as the supporting parts or near the cables must not have sharp edges; the edges must be rounded and smooth.

8.5.5 Abnormal Operating and Fault Conditions

The measurement of the temperature must be made also in reasonable abnormal operating conditions and in fault conditions. The applicable fault conditions that have to be tested, always one at time (because only a single fault is considered probable), are

- locked rotor of motors or fans (the temperature of the winding coils must be measured, and the value must be included from 150°C to 210°C according to the class of the winding)
- thermal devices in short circuit
- sensors used for adjustments (e.g., NTC, Negative Temperature Coefficient sensors) replaced so that the control conditions are never reached
- short-circuit and overload of transformers
- failure (in case of a short-circuit or open-circuit) of each component of the electronic circuitry that can generate a power greater than 15 W
- overloading of equipment containing heating elements (tested with an absorption 1.27 times the nominal value)
- short-circuit of capacitor connected between the supply conductors and ground
- open-circuit of one of the parallel branches of power resistors
- short-circuit of the output signal and the applied parts
- short-circuit of the deflection coils
- short-circuit of any two junctions of power semiconductors

- short-circuit of two pins of a rectifier
- short-circuit of the output of a power supply
- short-circuit of the output of a transformer
- jamming of moving parts or moving material handled by the equipment.

The risk analysis must guide the choice of the fault conditions that needs to be simulated. Under fault conditions, there are temperature limits that must not be exceeded for parts that may cause fire, emit gas or degrade insulation. For example, PCBs must not exceed a temperature of 105°C above room temperature.

Sockets, plugs, switches and subsystems have their temperature limits, indicated on the component or in its specifications.

During each condition of breakdown or abnormal status, the equipment under test must be safe in terms of the standard (i.e., the risk of mechanical, fire and electrical shock must not increase). The standard does not require that the equipment is working properly after or during this test, but requirements for dielectric strength, high temperature and installation guidelines must hold. These requirements must be checked at the end of the tests. During the tests simulating faults, the following means of protection can be added:

- a fuse
- a thermal switch
- a pico fuse.

8.5.6 Continuity of Protective Earthing

The safety performance in Class I devices is based on the integrity of the connection to the earth system: this is a necessary condition. In these products, live parts are protected by the basic insulation, while the accessible conductive parts have to be protected through a reliable connection to the protective earthing system. The reliability of this connection must be experimentally verified by measuring the resistance. In addition, the following construction conditions have to be satisfied:

- all accessible metallic parts must be grounded
- the protection circuit (i.e., the circuit that establishes the connection to the earth) must not contain fuses or protection devices
- the protection circuit must be bare or with yellow and green insulation
- the protective conductor must be secured to the chassis with a closed ring terminal attached to a welded nut that does not perform any other mechanical functions.

The IEC/EN 60601-1 standard specifies that the safety of the device ground system must be tested to ensure that the earthed device parts are safely connected to circuit ground through a protective earth connection. The enclosure and all accessible parts must also be connected properly to the ground connection of the product. The test is performed as follows. A current equal to 25 A, or greater than 1.5 times the current absorbed by the product, is introduced into the circuit under test from a voltage source of maximum 6 V. The measured resistance must be less than 0.1 Ω for equipment with detachable power supply cord or 0.2 Ω for equipment with non-detachable power supply cord.

8.5.7 *Residual Voltage*

The IEC/EN 60601-1 standard requires that the residual voltages measured between the terminals of the power supply cable or between any power supply terminal and the enclosure, is less than 60 V after 1 second from the disconnection to the mains supply.

If this limit is exceeded, the total energy must be less than 2 mJ (J = joule = watt × second). These tests have been introduced to ensure that when a device is disconnected, the user cannot experience an electric shock by unintentionally touching the pins of the power plug after plug disconnection. This voltage is often caused by an excessive capacitance to ground, which slowly discharges when power is interrupted.

This test is quite complex, since it requires a measure of the instantaneous voltage 1 second after the disconnection, without applying any load in the range of 1 second. Attention must be paid to the use of an oscilloscope which must not load the circuit for a period of 1 second (for this purpose it is suggested to use a probe with impedance of at least 100 MΩ).

Alternatively, the measuring instrument can be connected exactly 1 second after the disconnection and then the maximum peak voltage can be considered as a correct measure.

The test must be performed over the three configurations:

- between line and neutral conductors
- between line and earth conductors
- between neutral and earth conductors.

Similarly, the discharge time of the internal parts that are able to maintain the voltage after equipment disconnection has to be measured (e.g., capacitor banks). The discharge time of a specific electrical part must be less than the time required to gain physical access to that part even with a tool; this ensures safety protection for maintenance operators.

8.5.8 *Voltage on the Accessible Parts*

IEC/EN 60601-1 specifies that the voltage between the accessible parts and ground must not exceed 42.4 V AC peak-to-peak or 60 V DC in normal and single fault condition. This ensures that there are no excessive voltage differentials between the accessible parts and ground during normal operation or after the equipment has been powered down and the enclosure removed. If the equipment is marked with a label that specifies the delay time before the enclosure can be touched or opened, this test can be used to verify that the label information is correct. An additional test is required if the measured voltage on the accessible parts exceeds the limits of 60 V DC or 42.5 V AC peak-to-peak. In this case, it must be verified that there is less than 2 mJ of energy stored in the accessible parts. To conduct this test, an energy meter able to detect one millijoule has to be used.

8.5.9 *Energy Stored – Pressurized Part*

Hydraulic or pneumatic circuits, and, more specifically vessels, must be equipped with a safety device that prevents high pressures. Valves or other pressure-limiting safety devices may be avoided if the circuit is intrinsically safe, that is, no dangerous pressures can be generated even in fault condition (e.g., due to the limited capacity of a pump). However, it is prescribed that vessels and the associated circuits must be tested in overpressure conditions whenever the pressure is higher than 50 kPa or the value of energy (energy = pressure × volume) is greater than 200 kPa-litre. The test pressure to be used is given by the maximum permissible working pressure times a multiplication factor in the range 1.3 – 3, according to the maximum working pressure.

8.5.10 Current and Power Consumption

The standard requires that the maximum current absorbed by the equipment is compared with that reported on the markings; the measured current consumption must be less than 110% of the rated current. The current value must be measured through the supply main power cord during normal operation, under all voltage and frequency rated values.

The standard requires that the measured value of the power consumption does not exceed 110% of the rated value. Some care has to be considered in the measurement. The active power is to be detected, since phase shift can produce measurement errors of up to 15%.

8.6 Electromagnetic Compatibility

Electromagnetic interferences (EMI) has gained growing attention in recent years in medical device regulation, due to the increase of electrical equipment systems in the patient environment with potentially harmful interactions. In specific cases, radiation interference emitted by earlier devices has produced negative effects causing hazards and significant documented incidents (Silberberg, 2001; FDA, 2000a). For example, as documented in (Banana skins, 1998), a patient connected to an ambulance's monitor-defibrillator died because of defibrillator failure due to interference from the ambulance's radio. Another fatal incident happened to a patient with a pacemaker who went into ventricular fibrillation after passing through a metal detector. EMI also caused failure in patient monitoring alarms system, and two patients died. The US Food & Drug Administration (FDA) has warned of more than 100 reports of EMI-related incidents between 1979 and 1993 caused by common medical elements such as electrosurgery, fluorescent lights and radio transmitters (Silberberg, 1993).

Specific requirements have been introduced in the regulatory frameworks to reduce EMI risks. Annex I of the MD directive 93/42/EEC (EEC, 2007) include requirements for both electromagnetic compatibility (EMC) **immunity** and **emission**:

> *'9.2. Devices must be designed and manufactured in such a way as to remove or minimize as far as is possible:*
>
> *. . .*
>
> *– risks connected with reasonably foreseeable environmental conditions, such as magnetic fields, external electrical influences, electrostatic discharge . . . ,*
>
> *– the risks of reciprocal interference with other devices normally used in the investigations or for the treatment given,*
>
> *. . .*
>
> *12.5 Devices must be designed and manufactured in such a way as to minimize the risks of creating electromagnetic fields which could impair the operation of other devices or equipment in the usual environment.'*

Similar requirements are stated at FDA level (FDA, 2007). As a general rule, the application of recognized standards is the quickest way to achieve compliance.

Table 8.12 IEC/EN 60601-1-2 emission requirements for conducted/radiated EMI

Test	Reference standard	Description
Radiated E-field emissions	EN 55011 (CISPR 11) EN 55014 (CISPR 14)	Product emission constraints on the 30–1000 MHz frequency band
Conducted emissions	EN 55011 (CISPR 11)	Product emission constraints on the 0.15–30 MHz frequency band
Harmonic emissions	IEC/EN 61000-3-2	Limits for harmonic frequency emissions on public mains network for devices with input current lower than 16 A per phase
Flicker emissions	IEC/EN 61000-3-3	Limits for voltage changes, fluctuations and flicker injected into the public mains supply network for devices with input current lower than 16 A per phase

Unfortunately, these standards are not 'design standards', that is, they do not contain requirements that can be used in design. The standards simply prescribe limits on emissions and immunity levels and they define the associated test methods. Standard compliance can be obtained only through testing of a final sample of the equipment. This approach provides only an a posteriori validation of the criteria adopted in the design stage.

The standards for EMC are grouped into three classes:

- general standards applicable to all products operating in a given environment
- basic standards that define, for each class of EMC interference, methodology, tools and the configuration for the associated test
- product-specific standards.

These standards may be published in the European Official Journal as harmonized EMC standards, and when recognized as harmonized standards, they can be applied to guarantee presumption of compliance to the EU directive essential requirements already cited. See the end of the chapter for a 'Summary of Regulations and Standards'.

IEC/EN 60601-1-2 is a starting point for complying with the essential requirements of the regulatory framework on EMC. The standard contains

- the reference to the basic standards and the related electromagnetic tests to be considered
- the level of severity for tests
- the acceptance criteria under which tests will be considered as having passed.

The standard contains requirements for emissions and immunity by referring to basic standards listed in Table 8.12 and Table 8.13. These topics were discussed in Chapter 5.

In addition to compliance to applicable standards, medical devices must also have specific supplementary documentation that specifies

- guidelines to prevent, identify and solve unwanted electromagnetic effects
- constraints related to the use of equipment and/or systems

Table 8.13 EMC IEC/EN 60601-1-2 immunity requirements for conducted/radiated EMI

Test	Reference standard	Description	Compliance level
Electrostatic discharge (ESD)	IEC/EN 61000-4-2	Product immunity to ESD: contact, coupling plane, air discharge	±6 kV contact, coupling plane ±8 kV air
RF radiated immunity	IEC/EN 61000-4-3	Product immunity to electric fields generated by intentional transmitters (e.g. mobiles)	3 V/m from 80 MHz – 2.5 GHz.
Electrical fast transient/burst	IEC/EN 61000-4-4	Product immunity to switching and transient noise tested on AC input and I/O cabling located outside the building	±2 kV
Surge immunity	IEC/EN 61000-4-5	Product immunity to lightning strike, tested on AC input and I/O cabling	±1 kV differential mode ±2 kV common mode
RF conducted immunity	IEC/EN 61000-4-6	Product immunity to low frequency fields generated by intentional transmitters through cabling (mobiles phones, radios, TV etc.), tested on AC input and I/O cabling	3 V_{eff}/m from 150 kHz to 80 MHz
Power frequency H-field immunity	IEC/EN 61000-4-8	Product immunity to low frequency magnetic fields	3 A/m at 50/60 Hz
Variations on voltage supply power lines	IEC/EN 61000-4-11	Voltage dips, and voltage variations Short interruptions	$V_{rms} < 5\%$ for 0.5 cycle $V_{rms} = 40\%$ for 5 cycles $V_{rms} = 70\%$ for 25 cycles Interruptions $V_{rms} < 5\%$ for 5 s

- description of the levels of immunity for equipment and/or systems for patient use
- rationale for the application of possibly lower levels of immunity and actions to be taken by the installer and/or the user.

EMC was introduced in Chapter 5; in this section, we will focus on the associated tests.

8.6.1 Emissions

EN 5501 is the basic referenced standard for tests according IEC/EN 60601-1-2. EN 5501 is a standard on ‘radio-frequency industrial, scientific and medical (ISM) radio disturbance characteristics. Limits and methods of measurement’.

Table 8.14 EN 5501 classification of radio equipment

Group/Class	Description
ISM Group 1	Equipment in which there is **intentionally generated** or used **conductively coupled** RF energy that is necessary for the internal functioning of the equipment itself, so its RF emissions are very low and not likely to cause any interference in nearby electronic equipment
ISM Group 2	Equipment in which **RF energyis intentionally generated** or used in the form of electromagnetic radiation for the treatment of material and spark erosion equipment
Class A	Equipment suitable for use in all establishments **other than domestic** and those connected to a **low voltage power supply** network which supplies buildings used for domestic purposes
Class B	Equipment suitable for **use in all establishments** including domestic establishments and those directly connected to the public low voltage power supply network that supplies buildings used for domestic purposes

This standard provides a classification into groups and classes of equipment. As shown in Table 8.14 there are two types of emission tests to be performed:

- **conducted emissions:** tests measure the voltage interference generated on the supply lines
- **radiated emissions:** tests to measure the intensity of the noise field radiated from the equipment under test.

8.6.1.1 Conducted Emissions

The purpose of the test is to measure the amplitude of the interference voltage (RF conducted emission) on the supply line. An elevated noise may interfere with other devices connected to the same power supply network. The conducted interference is propagated through the power supply cables or other connection cables between the different systems. Due to the high capacity, usually connected on power line terminals of the powered device, the cutoff frequencies of these lines are relatively low. This means that only low frequency interference may propagate at longer distances. For this reason, only noise interference whose spectrum is under 30 MHz is measured.

The conducted emission test has to be performed on equipment supplied by the power network. The values are measured with mean (average) value and quasi-peak detectors. The mean value detector is used for the measurement of the narrow-band noise, the quasi-peak detector assesses the broadband noise, more specifically:

- **Average value detector:** takes the output from an IF (i.e., intermediate frequency bandpass) filter and the signal is averaged with an RC network. The output measures the average noise level. This network reduces the effects of an interference of high intensity and short repetition period even if such interference is significant (e.g., a few millisecond duration and repetition period of 10 ms with a duty cycle of 0.1%).

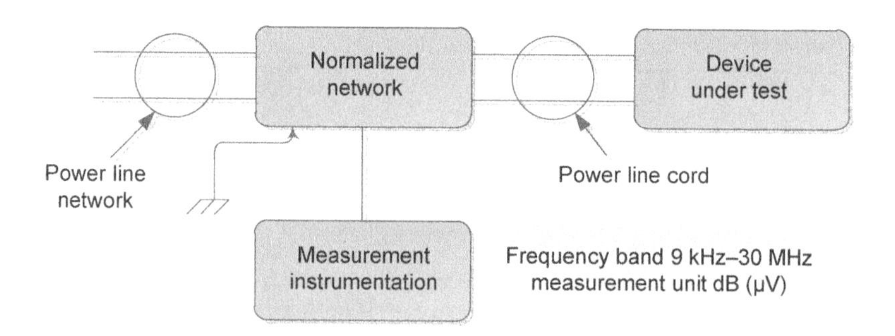

Figure 8.7 Block diagram of the emission test conducted.

- **Peak detector:** the output corresponds to the maximum level, regardless of the pulse repetition frequency. This detector is sensitive to sporadic random interference of high amplitude, even if the repetition rate is low and an additional lower level interference with a 100 Hz repetition rate is present.
- **Quasi-peak detector:** a compromise between peak and mean value detector. At the hardware level, it corresponds to a peak detector with a charge-discharge network with charge times of 1 ms and discharge time of 160 ms.

The measurement of RF conducted emissions is performed by inserting an additional network – the line impedance stabilization network (LISN) – between the power supply network and the device under test as shown in Figure 8.7. For higher currents, a voltage probe is used. The measurement performed though an LISN provides a better reproducibility and accuracy with respect to the standard limits.

The purpose of the LISN is twofold:

- ensures a constant impedance (50 Ω) between the phase conductor and the earth wire and between the neutral conductor and earth conductor over the whole measurement frequency range
- avoids measurement of interferences conducted by external devices connected to the power supply network.

The equivalent circuit of the stabilization network is simply given by a resistor of 50 Ω between the phase wires and the ground and a resistor of 50 Ω between the neutral and ground wires. Tables 8.15 and 8.16 outline the applicable limits for emission.

8.6.1.2 Radiated Emissions

This test aims to measure the amplitude of the RF noise electric field intensity (in V/m) emitted by the equipment and its connected cables. An elevated radiated noise in the frequency range of 30 MHz–1 GHz may interfere with other devices. Emitted radiations may generate improper behavior on closed telecommunication receivers or other electronic devices that may capture noise interferences through their cables and circuits.

Table 8.15 Limits for conducted emissions of Group 1 and 2 Class B equipment

Frequency range (MHz)	Limits (dB μV) input terminals	
	Quasi-peak	Medium
0.15–0.50	66: decreasing linearly with the log of the frequency to 56	56: decreasing linearly with the log of the frequency to 46
0.50–5	56	46
5–30	60	50

The test environment should be performed in an open-area test site (**OATS**) that must be situated on flat terrain, free from obstructions (e.g., buildings and fences). The OATS has the ground screen, which acts as a standardized reflective surface, to simulate the typical condition of propagation in a terrestrial environment. Unfortunately, because of open-air EMI pollution, this test site cannot be done in practice, since the measured frequency spectrum will include radiation from telecommunications devices.

An **anechoic shielded chamber** is a suitable alternative. This chamber allows the measurement to be isolated from the external environment with propagation conditions typical of a 'clean' terrestrial environment. The anechoic chamber is a shielded room in which walls and ceiling are coated with a special material that absorbs electromagnetic waves, and floor that is a perfect reflecting plane.

The measurement is made with broadband antennas that cover the entire frequency range of 30 MHz–1 GHz (with bilog antennas) or two sub-bands: 30MHz–200MHz (with a biconical antenna) and 200 MHz–1 GHz (with a log-periodic antenna). Tables 8.17 and 8.18 show the applicable limits for the emission for the groups defined in Table 8.14.

When measurements are performed at a lesser distance (i.e.$<$3 m), the limits are increased by 10 dB. The Figure 8.8 shows a typical measurement of radiated emission where x axis shows frequency on a log scale up to 1 GHz and the y axis shows the measured values in μV/m and continuous horizontal lines indicate specific limit values. For medical electrical equipment with motors and no electronic circuits emitting at frequencies above 9 kHz, the test may be replaced by the measurement of the radiated power from the power cord in

Table 8.16 Limits for conducted emissions of Group 1 and 2 Class A equipment

Frequency Range (MHz)	Group 1 Input terminals (dB μV)		Group 2 Input terminals (dB μV)	
	Quasi-Peak	Medium	Quasi-Peak	Medium
0.15–0.50	79	66	100	90
0.50–5	73	60	86	76
5–30	73	60	115	105

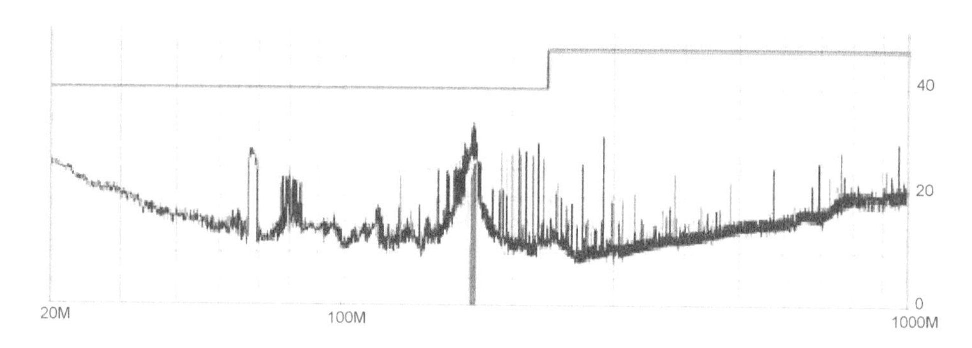

Figure 8.8 Class B radiated emissions measured at 3 m (dB μV/m).

accordance with EN 55014. The measurement is performed with a clamp set around the power cable measuring the noise power detected in the 30 MHz–200 MHz frequency range.

8.6.1.3 Harmonic Current Emissions

The purpose of the test is to measure the current amplitude emitted from the unit under test at even and odd harmonic multiples of the power supply frequency. The generation of harmonic currents is due to the non-linear behavior of components. Limited harmonic current emission avoids an unacceptable distortion of the sinusoidal waveform of the power supply line. The test is performed on the power plug, simulating the product's full cycle of operations.

8.6.1.4 Flicker

The purpose of the test is to measure the amplitude of the voltage variations, which can produce flicker in lighting systems due to fluctuations in voltage produced by variations in load on the network. The problem generally arises in equipment with sudden changes in current absorption. This has an impact on the supply voltage that produces flickers on the lighting. The measurement is performed on the power plug, simulating one or more complete

Table 8.17 Limits for radiated emissions of Group 1 equipment

Frequency range (MHz)	Measured in test site		Measured in situ
	Class A 30 m distance dB (μV/m)	Class B 10 m distance dB (μV/m)	Class A 30 m away from the walls of the building dB (μV/m)
0.15–30	Under study	Under study	Under study
30–230	30	30	30
230–1000	37	37	37

Table 8.18 Limits for radiated emissions of Group 2 equipment

Frequency range (MHz)	Quasi-peak measurement of the electric field at 10 m distance	Quasi-peak measurement of the magnetic field at 3 m distance
0.15–30	—	39: decreasing linearly with the logarithm of the frequency to 3
30–80.872	30	—
80.872–81.848	50	—
81.848–134.786	30	—
134.786–136.414	50	—
136.414–230	30	—
230–1000	37	—

cycles of operation of the equipment under test to cover a 10-minute interval. For equipment that does not have cyclic operations, the test time is increased to two hours.

8.6.2 Immunity

Immunity tests ensure that the equipment does not become dangerous or unsafe when noise is radiated or conducted into the equipment from the environment. In the presence of interference, the equipment must maintain a proper level of performance, depending on the type of applied noise, and must not suffer from failure or, worse, create hazards. For this, the acceptance criteria differ for life-support devices. The test result must be compared with those reported in the risk analysis according to the essential requirements and the intended use of the device. In general, for impulse shaped noise, performance degradation is allowed if it is in line with the risk analysis and the performance required for the device (e.g., performance degradation is not allowed for a life-support device). First, the essential performance and allowable lack of performances must be stated, according to the risk analysis, before performing the tests.

8.6.2.1 Electrostatic Discharge (ESD)

This test, derived from EN 61000-4-2, aims to verify that the equipment works properly in the presence of electrostatic discharge applied directly on the enclosure (i.e., parts accessible by the operator) or applied on a reference ground plane.

The test simulates the interference generated by human contact ESD or by electrostatically charged insulating materials. The test is performed by applying interference with a generator (gun) with different probes to simulate air discharge and contact discharge. The generated noise waveform has a very steep rising edge that corresponds to high frequency spectral contents, which may interfere with the electronic control circuits. Interference values are 6 kV for contact, 8 kV in air (IEC/EN 60601-1-2). The contact discharges are applied to the conductive surfaces; air discharges are applied to insulating surfaces. The contact discharge has a greater repeatability and so it is always to be preferred; The air discharge depends on the approach speed of the probe. Both discharges are strongly influenced by environmental parameters and in particular by moisture.

8.6.2.2 Immunity to Fast Transients and Bursts

This test, derived from EN 61000-4-4, aims to verify that the equipment works properly in the presence of fast and repetitive transients on the power supply line, on signal lines and on other lines.

The test, which simulates the effects of noise generated by switching of inductive loads, contacts relays and so on, is performed by a noise generator and a coupling network to induce interference in the power cords. Interference is induced on the signal lines by a capacitive clamp. The interference is a burst of 300 ms, repeated at 2.5 kHz–5 kHz frequency. The rising edge of the single pulse is 5 ns and the falling edge is 50 ns. The noise levels limits defined in IEC/EN 60601-1-2 are

- 2 kV/5 kHz for the test performed on the power supply cord
- 1 kV/5k Hz for the signal cables, if conductors have a length greater than 3 m.

8.6.2.3 Immunity to Pulses

This test, derived from EN 61000-4-5, aims to verify that the equipment works properly in presence of voltage transients caused by switching capacitive electrical loads, by SCR-based switching devices, by ground circuit faults or by lightning (excluding direct lightning strike). The test is performed using a noise generator which adds the surge pulse (with rising edge of 1.2 μs, and 50 μs time at the half value), to the power cord of the equipment under test. The test is performed in common mode (i.e., between the phases and ground) and in differential mode (i.e., between phase and neutral). The noise levels are provided by EN 60601-1-2: 1 kV for differential mode noise and 2 kV for common mode noise.

8.6.2.4 Immunity Voltage Dips and Variations

The test, derived from EN 61000-4-11, aims to assess the immunity of electrical and electronic equipment when subjected to dips, short interruptions and voltage variations, due to faults in the power supply networks or installations, or to sudden load changes. The voltage variations are produced by continuously varying loads connected to the network. The test is performed with different values of time length and voltage variations as indicated in Table 8.13.

8.6.2.5 Immunity to Mains Frequency Magnetic Fields

The test, derived from EN 61000-4-8, aims to assess the immunity of electrical and electronic equipment when subjected to external magnetic fields at network frequency, generated by power supply lines and electrical equipment with electromagnetic radiating components (transformers, electric motors, etc.). The noise current may be generated by long-term magnetic fields in normal operating conditions, or by short-term (<milliseconds) magnetic fields produced under fault conditions. The test is performed on equipment that may be sensitive to low frequency magnetic fields as for examples cathode ray tubes, Hall-effect sensors and coils. The minimum level of immunity required for the test is 3 A/m. The device under test is radiated by the field produced by a coil connected to the noise generator.

8.6.2.6 Immunity to RF Currents

This test, derived from EN 61000-4-6, aims to verify that the equipment works properly in the presence of RF interference (frequency swept electromagnetic fields with amplitude modulation (AM)) injected on power cords and signal conductors. This test verifies the proper equipment operation with radio transmitters, when the RF interferences are injected into supply or signal conductors. The test is performed by using a sinusoidal signal swept between 150 kHz and 80 MHz, 80% AM modulated with 1 kHz modulation frequency. The interference signal has an amplitude level of 3 V, or 10 V for devices with vital functions. The interference is applied using a capacitive coupling network, or a clamp by injection.

8.6.2.7 Immunity to Electromagnetic Fields

The test, derived from EN 61000-4-3, aims to verify that the equipment works properly in the presence of RF interference (electromagnetic field sweep frequency amplitude modulated). The test verifies the proper equipment operation with radio transmitters when the EM field couples with the enclosure and the conductors.

The interference is 80% AM at 1 kHz modulation frequency. The frequency ranges from 80 MHz to 2.5 GHz. The test is performed in a semi-anechoic or anechoic chamber, which ensures the requirement of adequate field uniformity. The typical level of interference according to EN 60601-1-2 is 3 V/m, or 10 V/m for devices with vital functions. During the test, all the lateral faces of the unit are radiated by the interference field.

8.6.3 The Test Report

The test report must provide evidence of the tests and, at the same time, it must provide all the required information for test reproducibility. The document will include

- name and address of the laboratory
- unique identification of the report by a serial number on each page
- name and address of the applicant
- description and identification of the device under test
- essential performance considered for the immunity test
- test execution date and date of issue of the report
- identification and description of the test methods, and deviations from the referenced procedures
- signature and title of the person who has the technical responsibility for the report
- a statement indicating that tests are related only to the sample product under test.

For each performed test, the following information will be detailed:

- level of severity
- acceptance criteria
- cycle of operation during the test
- diagrams and photographs
- test conditions (load, regulations, environmental conditions)
- instrumentation and associated deadlines of calibration
- test results.

The customer equipment documentation must include the EMC tables that identify the proper use and limitations with respect to other devices and the equipment environment. An example of such documentation is included in the implementation part.

Part II: Implementation

This section shows how certification concepts are applied for the certification of a commercial medical device sold in the market. This section has a broader perspective, since certification steps are described in terms applicable to any device and then applied to the Gamma Cardio CG's specific certification.

This is done by disclosing the certification details of the product **Gamma Cardio CG**. The implementation details are provided for this book by the Gamma Cardio Soft Company (www.gammacardiosoft.it), who aim to supply low-cost high-performance devices and to open the associated know-how.

For a better understanding, we will first outline the general process of European CE certification. However, US certification, although different in procedures, has similar goals as in Figure 8.2.

In general, certification aims to ensure that all the products are safe for all users-patients according to the intended use as declared by the manufacturer. In this sense, all country-specific certification processes are similar and the European-US process may be an indicative case for all.

8.11 The Process

A simplified certification process with the associated deliverables is outlined in Figure 8.9.

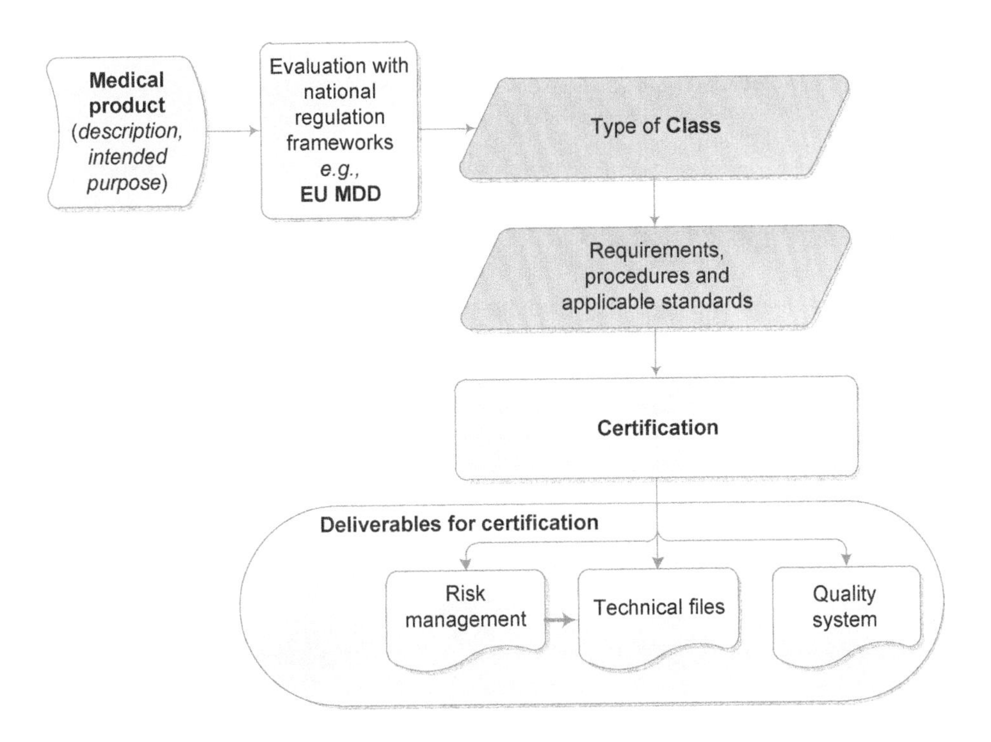

Figure 8.9 The certification process.

First, the product has to be described and its intended use, purpose, performance and safety requirements have to be declared. This information is used to define the class of the medical device. The **class** determines the **level of control** for the device, for example the applicable requirements, procedures and standards to be used for certification. Classes and the associated levels of control are determined by the regulatory framework in use in the specific area where the device is to be placed on the market. For example, in the European Union, the Medical Device Directive (EEC, 2007) is the reference regulation framework for medical devices. The certification consists of fulfilling the prescribed level of control that is certified through the production of deliverables: usually the risk management files, the technical file containing elements that ensure that the product is safe and effective, and the quality system. In the following sections, these steps will be addressed in more detail.

8.11.1 Device Description

The device description produced by the manufacturer is the first step that determines all the successive procedures. Basically, risks can only be assessed when it is clear what is the use and the purpose of the device. The certification process is then dependent on the identified risks. The higher the risks, the more rigorous the certification process. The intended use/purpose is defined as 'the use of a product . . . in accordance with the specifications, instructions and information provided by the manufacturer' (IEC/EN 60601-1). For more consolidated devices such as an ECG, the description and the intended purpose is simpler.

For the **Gamma Cardio CG** we have

> *The Gamma Cardio CG is a professional electrocardiographic system that can be used in hospital, home, ambulance and out-patients' environment. For safety, the system is compliant to IEC/EN 60601-1, IEC/EN 60601-2 and EN 60601-2-25. Regarding performance, the system is designed according to AAMI EC 11.*

The previous description allows us to derive the medical class according to MDD.

8.11.2 Medical Device Classes

Medical devices are grouped into classes to have different level of control (i.e., requirements, assessment procedures, applicable standards) according to their potential risks. The description and the intended purpose are useful for determining the device **class**, since they suggest the potential level of hazard of the device.

This class-based system allows for an economically feasible certification since devices are certified with an appropriate level of rigorousness associated to their class. More simply, it would be economically and operationally unfeasible to certify all medical devices with the most demanding procedures.

The EU classification is a system '*based on the vulnerability of the human body taking account of the potential risks associated with the technical design and manufacture of the devices*' (EEC, 2007). The EU's Medical Device Directive (MDD) defines a set of rules that determine the proper classification according to the device features.

Rules are based on criteria such as: '*the duration of contact with the patient, the degree of invasiveness and the part of the body affected by the use of the device*'.

These rules, described in Annex IX of MDD, are similar to the classification rules defined by the Global Harmonization Task Force (GHTF) in their guidance document (GHTF, 2006).

Table 8.19 EU medical device classes

Class	Risk	Level of control
Class I	low	sole responsibility of the manufacturer (apart from devices of class Is and Im)
Class IIa	medium risk	the intervention of a **notified body** is compulsory at the **production stage**
Class IIb	high risk potential	inspection by a **notified body** is required with regard to the **design and manufacture** of the devices
Class III	high risk potential	inspection by a notified body is required with regard to the design and manufacture of the devices **explicit prior authorization** with regard to conformity is required for them to be placed on the market

Medical device classes vary according to the applicable regulatory framework defined by national or regional regulatory authorities. The European Commission – the regulatory body of the EU – has defined the classes included in Table 8.19.

Class I has two subgroups: the sterile (Class Is) and the measuring devices (Class Im). For all classes, the manufacturer has to produce a declaration of conformity. For devices not belonging to Classes Is, Im, IIa, IIb or III, a simple self-certification is required to place device on the market.

For classes Is, Im, IIa, IIb or III, the declaration of conformity has to be verified by a notified body through a certificate of conformity. A notified body is a public or private organization that has been accredited in the EU to validate the compliance of a medical device with respect to the MDD.

In the USA, there are three '*regulatory classes based on the level of control necessary to assure the safety and effectiveness of the device*'. Classes are described in Table 8.20.

In the USA, the medical device class may be found by searching for similar products in the FDA classification database or finding the medical specialty of the device itself (FDA, 2013a). This resource also shows the corresponding regulation. ECG is found as a Class II device.

Table 8.20 US medical device classes

Class	Risk/usage	Level of control
Class I	Not intended for use in supporting or sustaining life, or to be of substantial importance in preventing impairment to human health, a potential unreasonable risk of illness or injury may not be present	General control
Class II	Held to a higher level of assurance than Class I devices, and designed to perform as indicated without causing injury or harm to patient or user	General controls and special controls
Class III	Usually those that support or sustain human life, are of substantial importance in preventing impairment of human health, or which present a potential, unreasonable risk of illness or injury	General controls and premarket approval (PMA)

In the EU, the rules in Annex IX of MDD or, more simply, their graphical representation in (MEDDEV, 2010) allow for device classification. For classification purposes, three groups of devices are identified:

- Non invasive devices rules 1, 2, 3, 4
- Invasive devices rules 5, 6, 7, 8
- Active devices rules 9, 10, 11, 12

These three groups have the specific rules of Annex IX indicated above. In addition, specific devices, such as those including medicinal products and human blood derivatives are covered by rules 13–18.

We will focus on **active medical devices** that include most of medical instruments. Such devices are defined as

> *'Any medical device operation of which depends on a source of electrical energy or any source of power other than that directly generated by the human body or gravity and which acts by converting this energy. Medical devices intended to transmit energy, substances or other elements between an active medical device and the patient, without any significant change, are not considered to be active medical devices.* ***Stand alone software*** *is considered to be an active medical device.'*

The electrocardiograph is clearly an active device. Figure 8.10 shows the applicable rules for active devices (i.e., rules 9–12).

Rule 10 applies to active devices intended for diagnosis. In rule 10, shown in Figure 8.10: 'vital physiological processes and parameters' also embraces 'respiration, heart rate, cerebral functions, blood gases, blood pressure and body temperature'. Rule 10 associated devices form a broad group including medical instruments (ultrasound diagnosis, therapeutic and

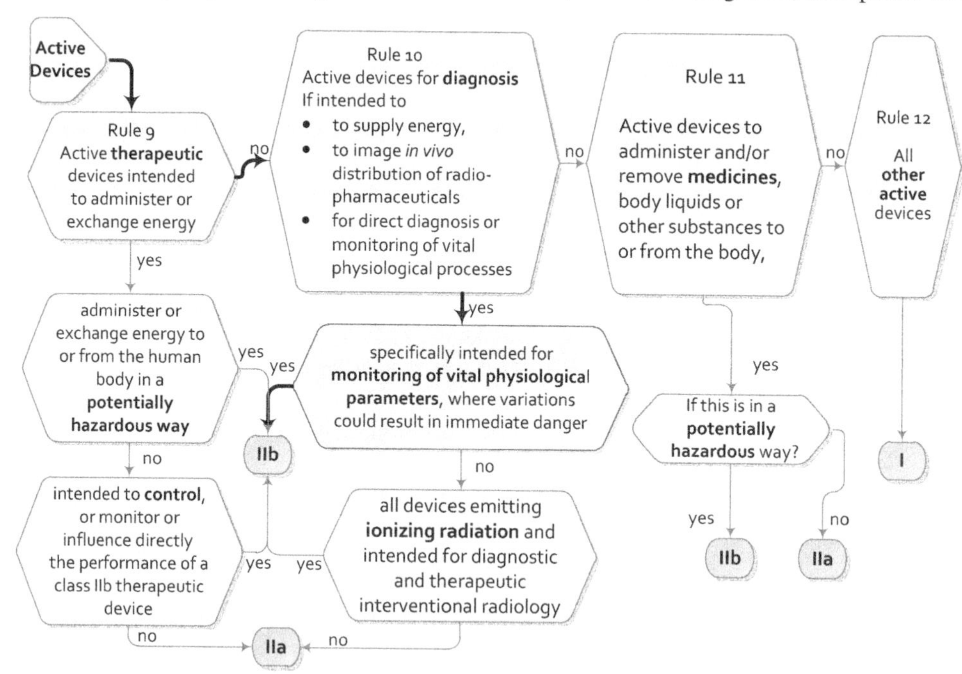

Figure 8.10 EU classification for active devices.

diagnostic radiology, ECG, EEG etc.). These active diagnosis devices may be classified as IIa or IIb. More specifically, ECGs are classified as IIb when they monitor vital physiological parameters where variations could result in immediate danger.

This latter case implies higher risks and therefore a higher level of control. If an ECG is IIa classified, the intended use and the instructions should show that the equipment must not be used to 'monitor vital physiological parameters where variations could result in immediate danger'.

Specifically, (MEDDEV, 2010) clarifies that 'Medical devices intended to be used for continuous surveillance of vital physiological processes in anesthesia, intensive care or emergency care are in Class IIb, while medical devices intended to be used to obtain readings of vital physiological signals in routine check-ups and in self-monitoring are in Class Ilia'. Since the Gamma Cardio CG is certified for use in ambulances, ECG parameter variations in an ambulance may result in immediate danger. This requires that Gamma Cardio CG class must be set to the higher level IIb. Many ECGs are IIa classified since they are intended to be used only for routine check-ups.

8.11.3 EU Conformity Assessment

The medical device class determines the level of control that the device must comply with to be placed on the market. This level of control translates into a specific set of methods depending on the device's class. In EU market, the manufacturer can choose different certification methods within a specific class as detailed in the MDD Article 11.

The **Conformity assessment** is defined as 'the method by which a manufacturer demonstrates that its device complies with the requirements of Directive 93/42/EEC'.

The method of certification implies the use of different procedures that are detailed in the Annexes of the MDD. Figure 8.11 shows the different procedures included in the MDD

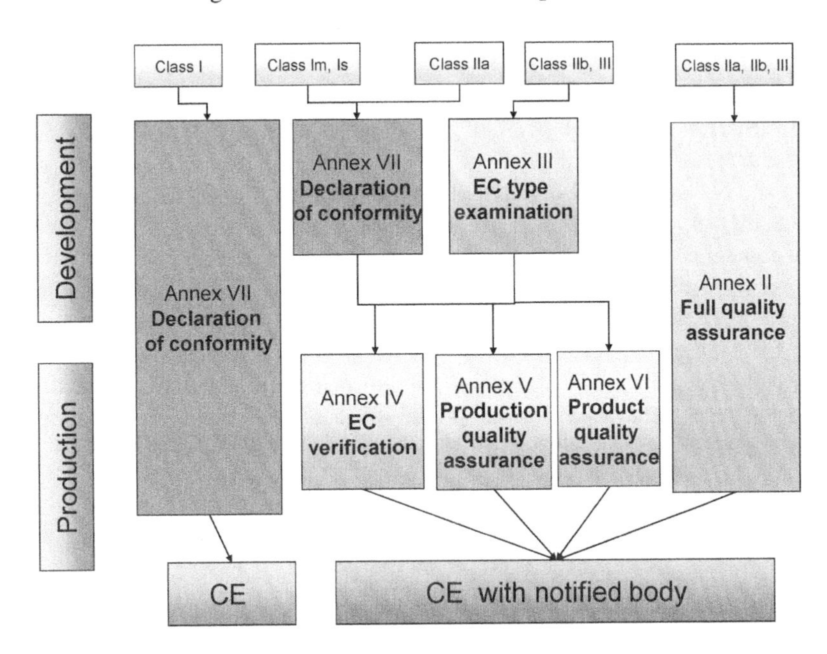

Figure 8.11 EU conformity assessment procedures included into Annexes.

Annexes that are applicable according to the class (MEDDEV, 2010). Class I devices are subdivided into general Class I, sterile (Is) and measure devices (Im). For example, Class IIb devices that are not custom-made or intended for clinical investigations, may perform conformity assessment using

1. procedure of Annex II: 'full quality assurance' (excluding point 4 Annex II)
   alternatively
2. procedure of Annex III: 'EC Type examination'
   and
   a. Annex IV: 'EC verification'
      or
   b. Annex V: 'Production quality assurance'
      or
   c. Annex VI: 'Product quality assurance'

Details of these procedures can be found in the MDD. Basically, using these procedures implies tests and verifications that are summarized in specific deliveries: risk management, technical file and quality system deliverables. The MDD procedures are mainly part of an ISO quality certification process. The Annex II procedure, used for the Gamma Cardio CG, is more onerous in the short term with respect to other certification approaches. On the other hand, this procedure is more effective in the long term since it implies a full quality assurance that can be used for any other device.

8.11.4 Risk Management Deliverable

As shown in Figure 8.9, the **risk management** file is the first deliverable that drives subsequent design and manufacturing. *Medical device certification aims to reduce risks to an acceptable level with respect to the benefits for the patients.* This is done by designing and manufacturing the device in an appropriate way, introducing protection methods to reduce critical risks to an acceptable level. Regarding the risk management framework, we will refer to **ISO/EN 14971** (Application of risk management to medical devices) that is the de facto standard for medical device worldwide: ISO/EN 14971 is a US FDA consensus standard and an EU harmonized standard. In addition, the ISO 13485 quality system standard recommends ISO 14971 for risk management. The US FDA guidance (FDA, 2000b) is also a valuable document addressing risk management for medical devices.

Risks have to be assessed according to the **intended use** declared by the manufacturer. According to ISO/EN 14971 and IEC/EN 60601-1, intended use (also referred to as intended purpose in EU MDD) is defined as: 'use for which a product, process or service is intended according to the specifications, instructions and information provided by the manufacturer'.

The same device may have a different risk analysis, depending on the intended use specified by the manufacturer. For example, a device that is constantly supervised by a trained operator has different risks from the same device left unattended for long periods. A device exclusively used in anesthesia has to be considered different from the same device also used to support patients in homecare. So the ISO/EN 14971 standard includes, in Appendix C, many questions to help identifying the performance and the intended use of a medical device.

Medical device risks encompass toxicity, flammability, contamination, EMC, measurement error, radiation, risks related to electrically powered and software managed equipment.

In the EU market, items 7 to 12 of the MDD Annex I contain the essential requirements that have to be satisfied in addressing risks. These requirements are the base for the related risk management file included in the technical file of the medical device.

Risks have already been addressed in the previous chapters. Here we focus on the final deliverable that contains risk management. We recall that a **risk** is defined as: 'a combination of the probability of occurrence of harm and the severity of that harm' (IEC/EN 60601-1, 2005) where **harm** is a 'physical injury or damage to the health of people or animals, or damage to property or the environment' (ISO/EN 14971, 2009). The **hazard** is 'a potential source of harm'. (IEC/EN 60601-1, 2005).

ISO/EN 14971 suggests a model of risk that the manufacturer may adapt to its own case. Figure 8.12, derived from Figure 6.10, shows an adapted model suitable for medical devices. This model has been used to define the Gamma Cardio CG deliverable of the risk management whose sample columns are reported in Table 8.21. In Figure 8.12, the risk management input are hazards that may stem from experience from similar devices, from scientific literature, from standards (e.g., ISO/EN 14971, IEC/EN 60601-1) or from the regulatory framework.

ISO/EN 14971 includes some examples related to energy, biological/chemical, environment, operational and information hazards accounting for improper labeling, warning and operating instruction. The EU risk hazards included in items 7–12 of the MDD Annex I may also be helpful for any other regulation framework.

Referring to Figure 8.12, derived from Figure 6.10, from a generic hazard (e.g., high temperature and flammability), a specific hazard scenario may be derived. The scenario must contain a sequence of events or circumstances that cause the hazardous situation. Note that a hazard does occur only when triggered by a sequence of events. A hazardous situation may occur in normal operation or in a fault condition.

For example, let us consider a specific hazard (flammability), a failure or shortcut (sequence of events) may create excessive current and overheating on specific components (hazardous scenario). This hazard scenario has a specific probability of occurrence. Overheating does not always generate a harm, which, in this specific case, may be an internal fire. This suggests estimating an additional probability that the harm occurs given that the hazardous situation has occurred. The probability of harm $Pr\,(\text{Harm})$ is given by the probability of harm conditioned to the event of the hazardous situation times the probability of the hazardous situation:

$$Pr\,(\text{Harm}) = Pr\,(\text{Harm/hazardous situation}) \times Pr\,(\text{hazardous situation}) \tag{8.2}$$

The Pr (Harm) is usually cited as the probability level. The severity level of harm is a qualitative or quantitative estimation of the harm seriousness assuming that the harm occurs.

Assuming a severity quantitative index, the risk level is given by the product of the probability and the severity level. Note that the analysis performed up to now has not considered any technical solution. This analysis is simply based on the intended use of the device. The steps up to now are referred to as **risk analysis**, described under clause 4 of ISO/EN 14971.

The risk level may be acceptable as it is or it may require one or more control (protective) measures to reduce risk to an acceptable level. This step is referred to as **risk evaluation** described under clause 5 of ISO/EN 1491. The two steps (risk analysis and risk evaluation) are also called **risk assessment**.

If the risk level is considered as acceptable, a completeness check closes the specific risk management, otherwise the additional step of **risk control** is performed (clause 6 of ISO/EN

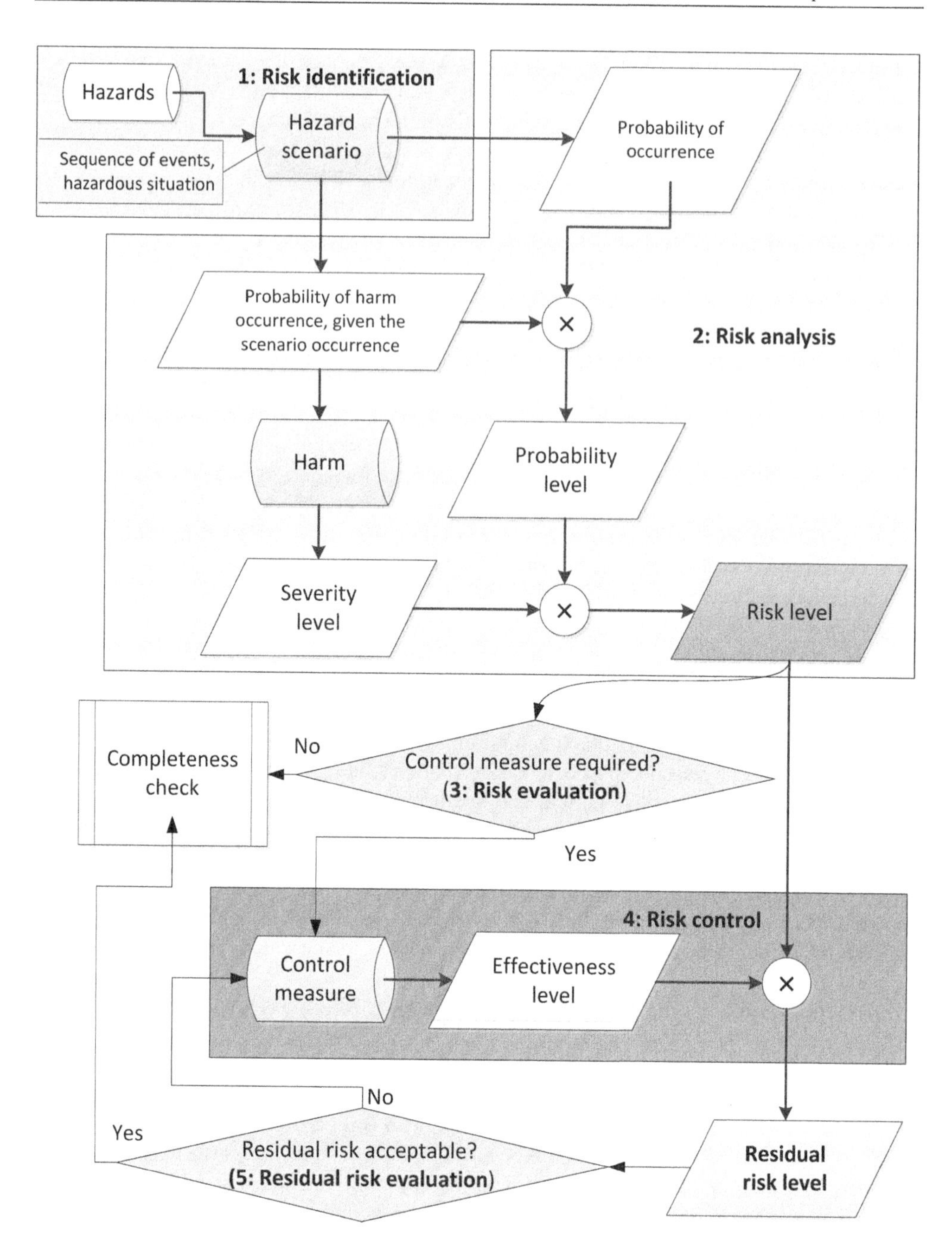

Figure 8.12 Risk management model.

Table 8.21 Risk analysis example

ID	MDD 7.1	MDD 10.1	MDD 12.7.4	MDD 12.09
Safety requirement	**Internal flammability**	**Measurement errors** due to accuracy and stability	Minimize risks due to **electric energy**	Understandable **controls and indicators**
Hazard scenario (sequence of events, hazardous situation)	Excessive internal temperature due to failure of non-fireproof components: transformers, wires, enclosure	Due to bed calibration, operational amplifier drift, design or manufacturing problems, device shows inaccurate time or signal amplitude	Unacceptable leakage currents due to faults or design/ manufacturing problems	Wrong selection of commands and filters
Probability level (PL)	1 (almost impossible)	5 (remote)	3 (improbable)	5 (remote)
Harm description	Device fire triggered by internal device components	Wrong diagnosis due to wrong ECG signal display	Death due to electric shock	Wrong diagnosis
Severity level (SL)	6 (serious)	10 (catastrophic)	10 (catastrophic)	8 (critical)
Risk level (= SL × PL)	6	50	30	40
Control measure	UL94V0 electronic board, fireproof cables, enclosure material resistant to maximum internal temperature that is limited by the reduced dissipated energy	The device is compliant to AAMI EC 11 that specifies the required minimum signal performance	Designed according to Class II CF (IEC/EN 60601), protected against defibrillator discharge; leakage currents are well below CF safety limits	Visual system reproduces the panel of a conventional ECG
Effectiveness level (EL)	0.1	0.3	0.1	0.3

(continued)

Table 8.21 (*Continued*)

ID	MDD 7.1	MDD 10.1	MDD 12.7.4	MDD 12.09
Additional control measure	none	Devices are individually tested, all tests related to signal accuracy are automatically performed and controlled by the device itself through a built-in automatic test system	Cables, system connections and power supply compliant to safety standards, safety tests performed for each manufactured device	User manual includes associated instructions, warnings and explanations
Additional effectiveness level (AEL)	1	0.3	0.3	0.6
Residual risk level (= SL × PL × EL × AEL)	0.6	4.5	0.9	7.2
Supporting documents	Electrical safety report, enclosure datasheet	Compliance to EC 11 tests, technical product description, device final test	Electrical scheme, components datasheets, electrical safety report, test procedure	User manual
Risk Area	Acceptable	ALARP	Acceptable	Acceptable

1491). This step aims to introduce control measures to reduce the residual risk to an acceptable level. A control measure is associated with a particular technical solution that in the specific case may be the adoption of fireproof components with current reduction mechanisms in case of short-circuit. The effectiveness level of Figure 8.12 accounts for the reduction in risk when a control measure is introduced. The residual risk level is then given by:

$$\text{Residual risk level} = \text{Probability level} \times \text{Severity level} \times \text{Effectiveness level} \quad (8.3)$$

This final step is called residual risk evaluation, described under clause 7 of ISO/EN 1491. Table 8.21 shows how the risk management model is translated into a risk management file.

These file is well described using a spreadsheet format. Each column of Table 8.21 refers to a specific hazard situation of the Gamma Cardio CG. The whole spreadsheet accounts for around 120 identified risks. The manufacturer has to define one column with specific data for each possible hazard. The first row (ID) of the table contains the identification of the hazards. For the ID row, we may use a reference to the MDD or other conventions. The second row includes a short description of the hazard that is derived from the MDD requirement. The third row describes the specific hazard situation with possible sequences of events.

For each hazard situation, there may be many triggering sequences of events that have specific probability of occurrence. The **probability level** (PL) given by (8.2) must account for all possible different triggering sequences. This probability is assessed based on experience, know-how about the specific device, the intended use, the environment and the operators who will use the device. The probability may also be estimated using specific analytical or simulation techniques. The probability must take into account the whole product life that has to be specified in the technical file. Table 8.22 shows an example of how quantitative levels may be defined. This value is used in the formula (8.3) to compute the risk level.

The harm description outlines the specific harm that may happen if the hazard scenario takes place. The **severity level** (SL) is the measure of seriousness of the harm. Table 8.23 shows typical levels, adapted from ISO 14971.

The severity level and the probability level are combined to obtain the risk level. In this specific case, the two indexes are simply multiplied. Since risk level may be unacceptable,

Table 8.22 Probability levels for risk analysis

Qualitative probability estimation	Quantitative probability estimation range	Probability level (PL)	
		Index value	**Probabilistic value**
almost impossible	Probability $\leq 10^{-6}$	1	10^{-6}
improbable	$10^{-6} <$ Probability $\leq 10^{-5}$	3	10^{-5}
remote	$10^{-5} <$ Probability $\leq 10^{-4}$	5	10^{-4}
occasional	$10^{-4} <$ Probability $\leq 10^{-3}$	7	10^{-3}
probable	$10^{-3} <$ Probability $\leq 10^{-2}$	9	10^{-2}
frequent	$10^{-2} <$ Probability	10	10^{-1}

Table 8.23 Severity levels for risk analysis

Severity level name	Possible result	Severity level	
		Generic value	**Economic value**
catastrophic	patient/operator death	10	10^{10}
critical	permanent impairment or life-threatening injury	8	10^{9}
serious	major injury or impairment requiring professional medical intervention	6	10^{7}
minor	minor injury or impairment not requiring professional medical intervention	4	10^{5}
negligible	inconvenience or temporary discomfort	2	10^{3}

Table 8.24 Effectiveness levels for risk control

Effectiveness of the control measure	Effectiveness level (EL)	
	Generic value	Probabilistic value
useless	1	10^{0}
scarce	0.8	10^{-1}
moderate	0.5	10^{-2}
good or effective	0.3	10^{-3}
Safe or compliant to harmonized standard	0.1	10^{-4}

the use of a control measure reduces the risk level to a lower residual risk level through the quantitative estimation of the **effectiveness level** (EL) as in formula (8.3). Table 8.24 shows possible quantitative levels. An additional control measure with associated additional effective level (AEL) may be required.

The acceptability of risks may be assessed, correlating the severity level and the probability of occurrence in the single matrix shown in Table 8.25. This table shows three areas of acceptability that may be associated to the ALARP (as low as reasonably practicable) approach (ISO 14971).

The three areas are defined as follows:

Area A: residual risk acceptable or negligible compared to other risks or to the expected benefits; no additional control measures are required

Area B: (ALARP) residual risk is accepted considering that further risk reductions may be impractical for technical/economic reasons and that the acceptance of such risks implies clear benefits that outperform expected risks

Area C: unacceptable risk

Table 8.25 Risk acceptability matrix

Frequency of occurrence	Severity				
	Negligible *Temporary discomfort*	**Minor** *Impairment*	**Serious** *Non-permanent impairment*	**Critical** *Permanent impairment*	**Catastrophic** *Death*
frequent	B	C	C	C	C
probable	A	B	B	C	C
occasional	A	B	B	C	C
remote	A	A	A	B	C
improbable	A	A	A	A	B
almost impossible	A	A	A	A	A

Table 8.26 Risk management checklist

1. Identify hazardous scenarios (i.e. the sequence of events and the consequent situation) from standards, regulatory frameworks or other sources
2. Identify the associated harm
3. Estimate probability of occurrence and harm severity level
4. If risk (probability, severity) is unacceptable:
5. Introduce appropriate and effective control measures until residual risk is acceptable
6. Perform the previous steps in a continuous process that reduces risks for possible new hazards throughout the product life cycle

More specifically, the control measures and the associated effectiveness levels, shift the residual risks position to the lower left area in the acceptability matrix (Table 8.25). These control measures may impact either on the probability of occurrence either on the severity so that the risk position is located in Area A (acceptable risk) or Area B (ALARP). In this latter case, risk has to be clearly justified. The main impacting risks have to be highlighted in the user manual. The management risk result must show that all foreseeable risks are acceptable. So risks addressed by conformance to an applicable standard are to be considered acceptable. When a risk is not acceptable in fault conditions, the probability of occurrence of the fault may be reduced through fault monitoring measures, or periodic maintenance controls.

Risk management is a key element ing ensuring that medical devices are safe. Since the manufacturer is responsible for the safety of its devices by law, it is worth the manufacturer producing an extremely accurate risk management with all the supporting documentation and verifications on the control measures and associated technical solutions. This guarantees that, in the case of an accident due to the device, the manufacturer has performed all possible activities to reduce risks to the minimum possible level. Although the risk management activity may be quite complex, its main underling concepts are simple; Table 8.26 summarizes the process of the risk management.

8.11.5 The Technical File

The documentation, data and records for the application for medical device certification are usually collected in one main document called the technical file.

This document has to demonstrate the conformity of the device to the applicable regulatory framework and, more simply, that the device is safe and effective. Other documentation has to be provided to demonstrate that a suitable quality system is in place for all stages affecting the quality of the device. This ensures that all the marketed devices are safe and effective. Technical files are similar across different regulatory frameworks. We will base this section on the EU framework while Section 8.12 will outline the US case. Table 8.27 shows the general structure of the file with all the associated contents prescribed by MDD Annex II 3.2 for a declaration of conformity that implies a full quality assurance

Table 8.27 EU generic technical file structure

ID	Section title	MDD Annex II section 3.2 requirements
1	General description and intended use	A general description of the product, including any variants planned, and its intended use(s)
2	Referenced documents.	The **design specifications**, including the **standards** which will be applied
3	Labeling plan	The draft label
4	Risk analysis	**Results of the risk analysis** The solutions adopted as referred to in Annex I, Chapter I, Section 2: – eliminate or reduce risks as far as possible (inherently safe design and construction) – where appropriate, take adequate **protection measures** including alarms if necessary, in relation to risks that cannot be eliminated – inform users of the **residual risks** due to any shortcomings of the protection measures adopted.
5	Instructions for use	Where appropriate, instructions for use
6	Product technical description	Also a **description of the solutions** adopted to fulfill the essential requirements which apply to the products if the standards referred to in Article 5 are not applied in full
7	Design verifications	The techniques used to control and **verify the design and the processes** and systematic measures which will be used when the products are being designed If the device is to be **connected to other device(s)** in order to operate as intended, proof must be provided that it conforms to the essential requirements when connected to any such device(s) having the characteristics specified by the manufacturer
8	Manufacturing and final inspection	the inspection and **quality assurance** techniques at the manufacturing stage and in particular – the processes and procedures which will be used, particularly as regards sterilization, purchasing and the relevant documents – the product identification *procedures* drawn up and kept up to date from drawings, specifications or other relevant documents at every stage of manufacture the appropriate **tests and trials** which will be carried out before, during and after manufacture, the frequency with which they will take place and the test equipment used; it must be possible to trace back the calibration of the test equipment adequately

Table 8.27 (*Continued*)

ID	Section title	MDD Annex II section 3.2 requirements
9	Clinical Evaluation	The preclinical evaluation
		The **clinical evaluation** referred to in Annex X
	Not applicable	A statement indicating whether or not the device incorporates, as an integral part, **a substance or a human blood derivative** referred to in Section 7.4 of Annex I and the data on the tests conducted in this connection required to assess the safety, quality and usefulness of that substance or human blood derivative, taking account of the intended purpose of the device
	Not applicable	A statement indicating whether or not the device is manufactured **utilizing tissues of animal origin** as referred to in Commission Directive 2003/32/EC (1)

system. The requirements of the MDD placed in the third column of Table 8.27 are useful to prepare a technical file with all the contents required by the regulation.

The structure of the technical file is also useful, to understand the overall process of the certification, since its sections are organized in the logical sequential order of the development process.

8.11.5.1 Intended Use

Referring to Table 8.27, the first section includes the **description of the product** and **intended use** (also referred to as the intended purpose in EU MDD). This part is a crucial initial step for understanding the device accurately and completely. All the certification activities and considerations stem from the intended use specified by the manufacturer. A device cannot be put into service outside of its intended use because this may create unexpected and unacceptable risks. For example, medical instruments are classified as IIa devices unless they are used for monitoring vital physiological parameters where variations could result in immediate danger. In that case, they are classified as IIb devices with more stringent controls. More specifically, due to the high risk potential of class IIb, a notified body also has to inspect the design of the device not assessed for Class IIa devices. Class IIa devices may have design errors (not inspected by the notified body) that may create unacceptable risks if misused for monitoring parameters whose variations may result in immediate danger. The intended use and the description should include the following information (adapted from FDA, 2000a):

- overall device operation
- general use scenarios that describe how the device will be used
- needs of users for safe and effective device use, and how they are met by the device
- characteristics of the intended user population
- expected use environments
- expiry date and periodic controls
- accessories.

Section 1 of the technical file must also include the class of the device, accessories and possible restrictions on product's shelf life.

8.11.5.2 Referenced Documents

The **referenced document** section contains the applicable standards that are accepted within the specific market area (e.g., US recognized or EU harmonized standards); these give the presumption of conformity to the requirements of the regulatory framework.

8.11.5.3 Labeling

Section 3 of the technical file (Table 8.27) refers to **labeling**, also referred to as 'marking' discussed in section 8.4.2. Labeling is defined as 'written, printed or graphic matter affixed to a medical device or any of its containers or wrappers, or accompanying a medical device, related to identification, technical description, and use of the medical device, but excluding shipping documents' (ISO 13485). This definition also includes the instruction manual. It should be noted that in other contexts, such as IEC/EN 60601-1, the labeling term refers only to information affixed to the device and packages, so it does not include information within the accompanying documents, such as the instruction manual.

Inadequate labeling is a common source of hazards that can lead to the patient's death or serious deterioration in their state of health (EEC, 2007). Hazards may be generated by poor description of performance, intended use or existing limitations. In the EU MDD context, any restrictions on use must be indicated on the label or in the instructions for use. For high-risk/high-benefit devices, labeling should also include adequate information to ensure that appropriate risk/benefit decisions are made by individuals.

Labeling is a typical control measure to reduce risks. For example, medical instruments should have labeling specifying the calibration interval to avoid hazards due to non-calibrated instruments. Labels that clearly specify the measurement unit (e.g., mm/sec) are another usual control measure to reduce risks of misinterpreting device data.

IEC/EN 60601-1 includes a guide for requirements on marking and labeling. Other applicable standards are the EN 1041 (Information supplied by the manufacturer of medical devices) and IEC 60878 or EN 15986 grouping medical devices symbols.

The contents prescribed by EU MDD are included in Table 8.28. Similar requirements hold for the US case, as explained in the FDA documentation. Requirements in Table 8.28 are useful to validate labeling of a specific device. The third column of table 8.28 includes the checklist with respect to the Gamma Cardio CG device and includes how the requirements are implemented (N/A means 'not applicable').

Figure 8.13 shows the label present on the back of the ECG device. Regarding warnings (requirement (k) of Table 8.28), the following **symbols** are included: ⚠ Caution, Class II equipment ⧈, Defibrillator-proof type CF equipment from standards ♥.

Symbols are often present on medical devices since they replace text more effectively, especially when space is limited. The main requirement is then to use symbols consistently. Inconsistent use of symbols may generate errors and hazards.

This suggests using only standardized symbols which, for medical devices, are grouped into the IEC 60878 or EN 15986.

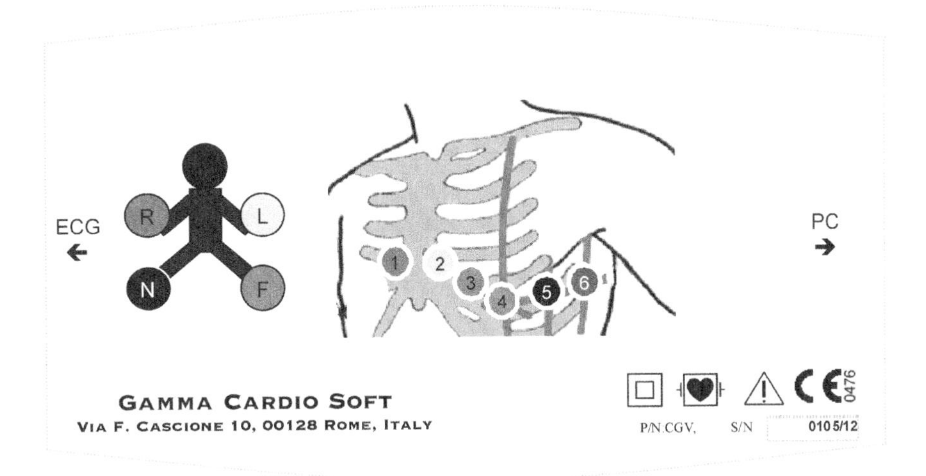

Figure 8.13 Gamma Cardio CG equipment back label.

8.11.5.4 Risk Analysis

The risk analysis and management, to be included in the technical file section 4 (see Table 8.27) of the technical file, was addressed in Section 8.11.4.

8.11.5.5 Instructions for Use

Section 5 of the technical file is devoted to the instructions for use of the medical device. Instructions are another critical component because, as with labels, they are a source of errors, and therefore they may generate unacceptable risks. The EU MDD prescribes specific requirements for the instructions for use. Right column of Table 8.29 shows all the MDD requirements (Annex I, clause 13.6.) that are sorted according to a possible chapter structure whose headings are included in the left column. For the instruction manual see also the valuable recommendations from (FDA, 1993).

8.11.5.6 Product Technical Description

The **product technical description** included in Section 6 of the technical file must describe the mechanical, electronic and software structure of the device with the aim of demonstrating the conformity to regulatory requirements.

The electronic part is usually discussed including all the electrical diagrams. These diagrams have already been included in the hardware chapter of this book for the Gamma Cardio CG devices. The critical components' datasheets, discussed in Section 8.4.1, have also to be included as an annex of the technical file.

The **mechanical structure** has to be described through proper drawings to demonstrate conformity to requirements. For example, for Class II devices it must be shown that there are no accessible conductive parts. Possible screws have to be positioned so that they can be touched only through the use of a tool. The **layout** of the components is also useful to show the implementation of physical means of protections (air clearance, creepage etc.).

Table 8.28 MDD label requirements checklist

MDD topic	MDD label requirement	CG-ECG validation
Manufacturer identification	the name or trade name and address of the manufacturer – for devices imported into the Community, in view of their distribution in the Community, the label or the outer packaging or instructions for use, must contain, in addition, the name and address of the authorised representative where the manufacturer does not have a registered place of business in the Community	OK (name and address)
Device identification	the details strictly necessary to identify the device and the contents of the packaging especially for the users	OK (device part number)
Sterile	where appropriate, the word 'STERILE'	N/A
Serial number	where appropriate, the batch code, preceded by the word 'LOT', or the serial number	OK (S/N)
Expiring date	where appropriate, an indication of the date by which the device should be used, in safety, expressed as the year and month	
Single use	where appropriate, an indication that the device is for single use – a manufacturer's indication of single use must be consistent across the Community	N/A
Custom-made	if the device is custom-made, the words 'custom-made device'	N/A
Clinical investigations	if the device is intended for clinical investigations, the words 'exclusively for clinical investigations'	N/A
Special storage/ handling conditions	any special storage and/or handling conditions	N/A
Special operating instructions	any special operating instructions	OK Electrodes application and colors
Warnings	any warnings and/or precautions to take	OK See below
Year of manufacture	year of manufacture for active devices other than those covered by (e). This indication may be included in the batch or serial number	OK included in S/N
Method of sterilization	where applicable, method of sterilization	N/A
Human blood derivative	in the case of a device within the meaning of Article 1(4a), an indication that the device contains a human blood derivative	N/A

Table 8.29 Instructions for use and MDD requirements

Section	MDD instructions for use requirements
Front page	(a) the details of Table 8.28 (Manufacturer, device identification etc.) (q) **date of issue** or the latest revision of the instructions for use.
	Table of contents, introduction
1: *Physical description of the device and accessories*	
	(o) **medicinal substances** or human blood derivatives incorporated into the device as an integral part in accordance with Section 7.4
2: General warnings and precaution	(e) where appropriate, information to avoid certain risks in connection with **implantation of the device** (k) **precautions** to be taken in the event of **changes in the performance** of the device (f) information regarding the risks of **reciprocal interference** posed by the presence of the device during specific investigations or treatment
3: Intended use	(b) the **performances** referred to in Section 8.3 (intended purpose) and any undesirable **side effects**
4: Setup	(c) if the device has to be installed with or **connected to other medical devices** or equipment, in order to operate as required for its intended purpose, sufficient details of its characteristics to identify the correct devices or equipment to use, in order to obtain a safe combination (d) all the information needed to verify whether the device is properly **installed** and can operate correctly and safely
5: Before each use	(g) the necessary instructions in the event of damage to the sterile packaging and, where appropriate, details of appropriate methods of **resterilization** (h) if the device is reusable, information on the appropriate processes to allow **reuse**, including cleaning, disinfection, packaging and, where appropriate, the method of sterilization of the device to be resterilized, and any restriction on the number of reuses; where devices are supplied with the intention that they be sterilized before use, the instructions for cleaning etc. (i) details of any further treatment or handling needed before the device can be used (e.g. **sterilization**, final assembly, etc.)
6: During the use (operating instructions)	The instructions for use must also include details allowing the medical staff to brief the patient on any **contra-indications** and any precautions to be taken; these details should cover in particular: (k) **precautions** to be taken in the event of **changes in the performance** of the device; (l) **precautions** to be taken as regards **exposure**, in reasonably foreseeable environmental conditions, to magnetic fields, external

(*continued*)

Table 8.29 (*Continued*)

Section	MDD instructions for use requirements
	electrical influences, electrostatic discharge, pressure or variations in pressure, acceleration, thermal ignition sources, etc.
	(m) adequate information regarding the **medicinal product** or products which the device in question is designed to administer, including any limitations in the choice of substances to be delivered
7. Maintenance	(d) plus details of the nature and frequency of the **maintenance** and **calibration** needed to ensure that the device operates properly and safely at all times; ((a) expiry date)
	(n) precautions to be taken against any special, **unusual risks** related to the disposal of the device
8. Technical performance	(j) in the case of devices **emitting radiation** for medical purposes, details of the nature, type, intensity and distribution of this radiation
	(p) degree of **accuracy** claimed for devices with a **measuring function**

Software (firmware and/or application layers) is another critical component to be certified. In the latest MDD version, stand-alone software is considered to be an active medical device and therefore it inherits all the requirements associated with these devices. If a harmonized standard such as IEC/EN 62304 has not been applied, the manufacturer must provide a description of the solutions adopted to fulfill the essential requirements. For this, (MDD Annex I 12.1a) specifies 'software must be validated according to the state of the art taking into account the principles of development life cycle, risk management, validation and verification.' Requirements from IEC/EN 60601-1 related to programmable electrical medical systems are also to be taken into account.

The process and the associated structure of a typical software validation document may be as follows:

1. Describe the **solution** and the intended use (use case, software environment etc.)
2. Select applicable **standards**
3. Describe the **development life cycle** process if not covered by an applicable standard
4. Define a **risk management file** (ISO/EN 14971)
5. Set **software requirements** from
    a. regulatory requirements
    b. safety requirements
    c. requirements from standards
    d. requirements from risk analysis
    e. functional and performance requirements
6. **Design** the system architecture and the associated subsystems (modules) showing the associated requirements
7. Provide **test descriptions** for validation and verification
8. Give a **traceability matrix** that associates all the requirements to defined tests.

Another document included in the technical file has to report the test results of the specific software version. All the software modifications have to be recorded through change reports, and the new versions have to be validated and verified taking particular care with the safety requirements.

8.11.5.7 Design Verification

Section 7 of the technical file is related to the **design verifications** and should include the techniques used to control and verify the design, the processes and the systematic measures used when the product is designed.

A medical device has to be verified according to the applicable harmonized standards, as specified by the manufacturer. In addition to software validation and verification already included in Section 6 of the technical file, a medical device usually has to be tested for conformity to the main horizontal standards:

- IEC/EN 60601-1 (Medical electrical equipment Part 1: General requirements for basic safety and essential performance)
- IEC/EN 60601-1-2 (Electromagnetic compatibility).

Other vertical standards (i.e., devoted to a specific device) may apply. For example, for the Gamma Cardio CG, EN 60601-2-25 adds additional specific safety requirements for the ECG. Requirements are usually tested by specialized laboratories that produce a conformity test report to the specific standard whose structure is outlined in Section 8.6.3.

8.11.5.8 Manufacturing and Final Inspection

Section 8, related to **manufacturing and final inspection**, has to ensure that all the produced devices are at the expected safety and effectiveness level. This is guaranteed by a quality system that encompasses all the manufacturing stages. Appropriate tests and trials have to be considered throughout the product life in order to ensure safety, until the product is disposed of.

The first step is a manufacturing process, which has to be described within the technical file. The manufacturing process includes an electronic and mechanical assembly. Electronic assembly (i.e., automatic assembly of surface mount devices) implies the use of specialized machines that are often owned only by specific companies.

Specific device subcomponents, such as electronic boards, may then be supplied by third parties. Mechanical assembly may also be performed by third parties. So suppliers of safety-critical components have to ensure the full quality of subcomponents through a proper company certification.

The technical file has to specify the general manufacturing process and the specific procedures. A simple manufacturing process is outlined in Figure 8.14. Although designed for the Gamma Cardio CG it has a general validity for most of the devices.

The next step consists of detailing the procedures to allow any trained operator to manufacture the specific device. The detailed procedure may be included in a specific document named: ‘Instructions for manufacturing and final inspection’. The beginning

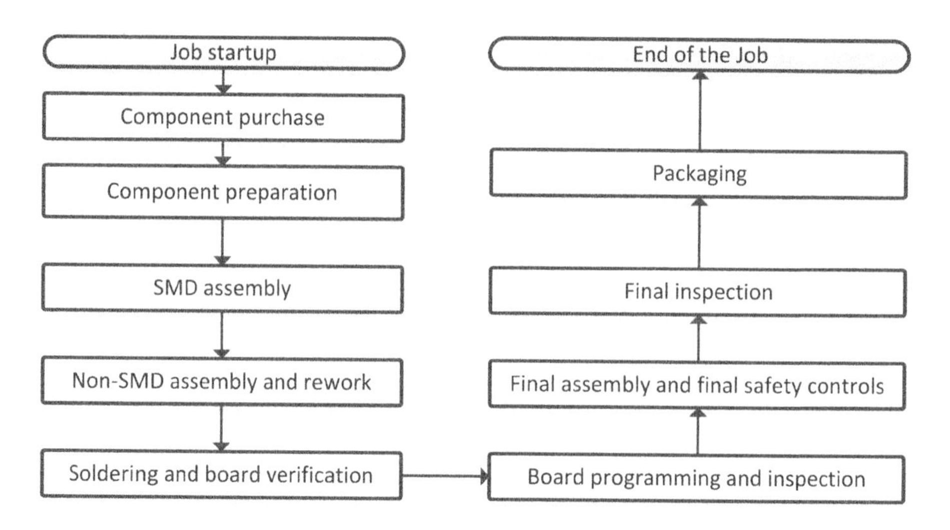

Figure 8.14 Outline of a manufacturing process.

of this document must include the prerequisites to start the production stage. These may be

1. device manufacturing documents
   a. bill of material
   b. electrical, mechanical assembly diagrams
2. tools for programming and final inspection
   a. equipment
   b. software
   c. firmware
3. activity documents for the specific manufacturing job
   a. production plan
   b. job instructions for SMD assembly machines
   c. tracing documents to link orders, manufacturing lots, components, serial numbers
   d. non-compliance register.
4. For each item, it is compulsory to specify the applicable version.

8.11.5.9 Clinical Evaluation

The **clinical evaluation** is the final item (Section 9) of the suggested technical file template. Clinical evaluation must demonstrate

1. safety use for patients, users and third parties; risks expected in the intended use must be acceptable with respect to the benefits to the patients
2. performance specified by manufacturers.

According to MDD (Annex X), the **clinical evaluation** must be generally based on clinical data, taking into account of any relevant harmonized standards. Clinical evaluation must be based on one of the three following methods:

1. evaluation of the relevant scientific literature related to a predicate device (i.e., a device already in the market that is similar to the one to be evaluated) that demonstrates
   a. equivalence of the predicate device with the device to be certified regarding safety, performance, design characteristics and intended use
   b. compliance with the relevant essential requirements
2. evaluation of the results of all clinical investigations made
3. a combination of the two previous methods.

Note that the first approach is close to the US FDA approach for certification, where a device must be substantially equivalent to a predicate device (see Section 8.12).

When clinical evaluation is not based on clinical data, it may be performed through performance evaluation, bench testing and preclinical evaluation. For this, the manufacturer must provide appropriate justification: the use of performance evaluation, bench testing and preclinical evaluation has to be duly substantiated. For example, the ECG has been in clinical use for many decades and the underlying mechanisms and the required essential diagnostic and safety performance are well known. In the USA, from 1998, the FDA has been recommending that the approach to demonstrating the substantial equivalence of an ECG is conformity to the AAMI EC 11 standard (FDA, 1998). Clinical evaluation of an ECG may then be based on the conformity to the EC 11 **standard**, as in the Gamma Cardio CG, apart from performances that are already covered by other applicable standards, such as IEC/EN 60601-1 and IEC/EN 60601-2-25.

In general, standards that define the minimum performance of a device are recommended because it is expected that a standard ensures a complete test coverage over the most relevant diagnostic cases from a risk point of view. When clinical evaluation is based on the comparison with a predicate already marketed device, *clinical data may not cover all the significant space of the clinical observations.*

8.11.5.10 Declaration of Conformity

For the EU market case, an additional final section has to be present. This section contains the declaration of conformity that the manufacturer can deliver to the customer. This section includes the manufacturer's and device's identification, serial number, applicable class, associated certification procedure annexes, identification of the notified body and the harmonized standards to which the device is compliant.

8.12 Regulatory Approaches to Medical Device Market Placement

In this section we will outline a comparison of the regulatory approaches focusing on the US and EU markets. As discussed in Section 8.1, the goals of a medical device certification are similar for any regulatory system:

1. demonstrate the **safety** of the device
2. demonstrate the **effectiveness** of the device
3. demonstrate that all the marketed devices have the same **quality** level in terms of safety and effectiveness.

The third goal ensures that any produced device has the same effectiveness and safety as the specific device used in certification. This is achieved through a quality system applied to the organizations manufacturing and managing the devices.

The three elements (safety, effectiveness and quality) are intrinsic features of the device and of the organizations that manage the device itself. This suggests that the country-specific regulation frameworks should not differ in goals and contents although they may employ different procedures for achieving the three goals. Since country-specific regulations have been developed separately, contents may also differ. As discussed in the theory section (Section 8.2), these differences are going to decrease through the efforts of the Global Harmonization Task Force, and the use of international standards that are widely accepted at country level.

The US approach for medical device 'approval-clearance' is similar in contents to EU certification although different in its processes. Approval and clearance are procedures that the FDA requires when a review is needed before marketing a medical device. FDA will either

- '**clear**' the device after reviewing a premarket notification, (510(k) procedure)
- '**approve**' the device after reviewing a premarket approval (PMA) application.

The 510(k) or PMA application procedures are used based on the classification of the device. Market clearance using the 510(k) procedure is used when the device is 'substantially equivalent' to a device that is already legally marketed in the USA for the same use.

Device approval through a PMA application is obtained by demonstrating that the device is reasonably safe and effective. Clear explanations are present on the FDA website for interested readers (FDA, 2013b).

A **510(k) is a premarket submission** procedure that aims to demonstrate that the device is at least as safe and effective, that is, '**substantially equivalent**', to a US legally marketed device that is not subject to premarket approval.

The FDA will assess the provided information and will give clearance to market the device, if such device is considered substantially equivalent to another device that is marketed in the USA. The FDA will then be entitled to perform a quality system inspection after 510(k) clearance. We have again to demonstrate the three features: safety, effectiveness and quality which guarantees safety/effectiveness for all the marketed devices.

In the US approach, a device is substantially equivalent if, in comparison to an already marketed device, it has:

1. the same intended use and technological characteristics

   or

2. the same intended use and different technological characteristics but
   a. no new questions on safety and effectiveness arise

      and

   b. the device is demonstrated to be at least as safe and effective as the legally marketed device.

According to the FDA, substantial equivalence is based on 'intended use, design, energy used or delivered, materials, chemical composition, manufacturing process, performance, safety, effectiveness, labeling, biocompatibility, standards, and other characteristics, as applicable'. These elements are close to the EU MDD requirements (safety, performance, design characteristics and intended use).

Safety and effectiveness is demonstrated through the contents of a **technical file**. The technical files submitted for US approval (510(k) procedure) and for EU certification (CE Marking) are similar, and most of the specific contents are the same, as shown in Table 8.30,

Table 8.30 US FDA technical file structure

FDA Sect.	510(k) section title	MDD section	Topic
1	MDUFMA cover sheet		
2	CDRH premarket review submission cover sheet		
3	510(k) cover letter		
4	Indications for use statement	1	General description and intended use
5	510(k) summary or 510(k) statement		
6	Truthful and accuracy statement		
7	Class III summary and certification		
8	Financial certification or disclosure statement		
9	Declarations of conformity and summary reports	2	Referenced documents.
		10	Declaration of conformity
10	Executive summary		
11	Device description	6	Product technical description
12	Substantial equivalence discussion		
13	Proposed labeling	3	Labeling plan
		5	Instructions for use
14	Sterilization/shelf life	7	Design verifications
15	Biocompatibility		
16	Software		
17	Electromagnetic compatibility/electrical safety		
18	Performance testing – bench	9	Clinical evaluation
19	Performance testing – animal		
20	Performance testing – clinical		
21		4	Risk analysis
22		8	Manufacturing and final inspection

where a simple comparison of the two structures is outlined. In addition, the adoption of standards or, better, US-EU accepted **standards**, reduces the certification effort between these two systems. For example, the Gamma Cardio CG is compliant to AAMI ANSI EC 11, 'Diagnostic Electrocardiographic Devices' that is a recognized consensus standard by the US FDA. This guarantees safety and effectiveness (i.e., substantial equivalence) for the US 510(k) approval procedure. More specifically, AAMI ANSI EC 11 conformance is a recommended approach by the FDA. Other possible approaches for an ECG device to obtain FDA clearance may be based on: (FDA, 1998):

- sufficient comparison testing with a legally marketed predicate device
- conformance to any other standard which meets or exceeds the requirements of the EC11 standard.

Regarding the EU market, since EC 11 guarantees the minimum safety and performance requirements for an ECG device, the conformity to this standard can be used to demonstrate compliance to the EU essential requirements related to clinical validation as required by Annex X of MDD.

Regarding the **quality system**, the use of the ISO 13485 is a good strategy to reduce certification efforts when addressing different markets. For example, this standard allows for a substantial overlap of requirements and activities when placing devices in the USA within its *Quality Systems* Regulation under 21CFR 820 and EU market within the MDD framework.

Table 8.30 shows the FDA recommended sections for a traditional or abbreviated 510(k) submission that is the process to obtain FDA clearance to market a medical device. In the **traditional 510(k)**, 'the submitter provides descriptive information about the indications for use and technology and, if not identical to the predicate, results of performance testing to demonstrate substantial equivalence.' An **abbreviated 510(k)** relies on one or more:

- FDA-recognized consensus standards
- special controls established by regulation
- FDA guidance documents.

In the abbreviated 510(k) approach, full or partial declarations of conformity to an FDA-recognized consensus standard are usually the main part of the technical file.

8.13 Basic Concepts in Device Implementation

In this section, we show the implementation of the concepts outlined in Section 8.3, taking into account the example of the Gamma Cardio CG. The main topics have already been discussed in Sections 5.9 and 5.29. Here we will summarize results related to the certification deliverables and in particular to the conformity to the IEC/EN 60601-1.

The Gamma Cardio CG shown in Figure 8.15 is connected on the left side to the PC through a USB cable. Through this connection, data are exchanged and the USB 5V power is supplied to the acquisition board. On the right side, the Gamma Cardio CG is connected to the patient cable that is, in turn, connected to the 10 electrodes.

The Gamma Cardio CG, shown in Figure 8.16, is safety extra low voltage (**SELV**) equipment since the power voltage is 5 V and no high voltage is included in the enclosure. The power is supplied by a PC USB port which must be certified to the specific standard. This is

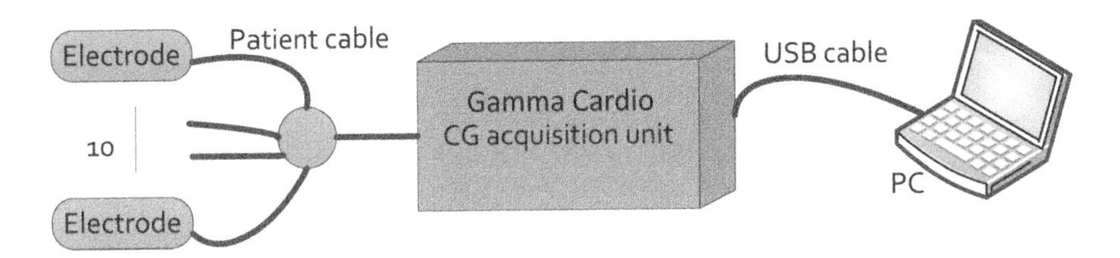

Figure 8.15 Gamma Cardio CG connection scheme.

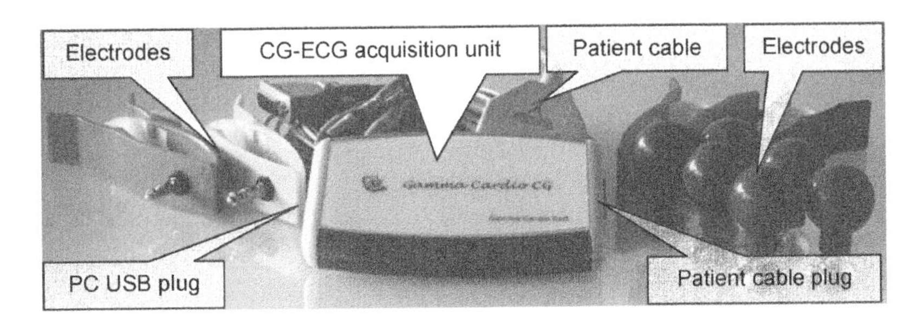

Figure 8.16 Gamma Cardio CG components.

a major simplification for the Gamma Cardio CG's certification since it avoids many requirements associated to high voltage live parts and mains supply within a device.

Another simplification is the use of double insulation which classifies the equipment as class II.

This equipment is protected with basic insulation and an additional double insulation. When basic insulation fails, operators and patients are protected by supplementary insulation or by reinforced insulation. In this case, there is no need to rely on the protective earth and the correct earth system installation conditions. Table 8.31 lists the applicable terms used for test descriptions and the Gamma Cardio CG associated components. Definitions of these terms are given in Table 8.4 and Table 8.5.

8.13.1 Protection Against Electric Shock

Electric shock scenarios might occur when using an ECG in normal use or when in presence of a single fault condition. These scenarios have to be tested so that the residual risk is acceptable in the marketed device. The main risks envisaged by MDD are reported in Table 8.32.

The table includes the main control measure set by the Gamma Cardio CG. Additional control measures, not included in this table, are added when risk is still unacceptable. Table 8.32 shows that the choice for a Class II SELV equipment reduces electric shock risks.

8.13.2 Insulation

Insulation ensures separation of the device parts from dangerous voltages. Considering Figures 8.15 and 8.16, this voltage may be present on the patient connection due to

Table 8.31 Applicable safety tests for the Gamma Cardio CG example

Term	Explanation
Enclosure	outer surface of the Gamma Cardio CG
Applied parts	10 electrodes and the patient cable (in normal use they are necessarily in physical contact with the patient for the ECG to perform its function)
Accessible parts	enclosure, cable, electrodes
Accessories	electrodes, patient cable, USB cable
SIP/SOP	signal input/output parts: USB power-data connection, patient cable connection
Single fault	safety critical problems, such as mains supply on applied parts or on SIP/SOP, failure-short-circuit on one electrical component, short-circuit on specific electronic components, failure of mechanical parts, shorting of basic or supplementary insulation
Double insulation	insulation that includes both basic and supplementary insulation – in addition to the basic insulation, the plastic enclosure of the Gamma Cardio CG is a supplementary insulation that ensures protection against electric shock in case of failure of the basic insulation, and when live parts are in contact with the plastic enclosure, the plastic material insulates the live parts

defibrillation (normal intended use) or due to an unwanted contact of the electrodes with the mains supply (single fault of some other equipment). On the USB side, a 5 V level is expected. Dangerous voltage may occur only when the PC is in fault condition due to insulation problems. Finally, the enclosure may be in contact with dangerous voltages. In all these conditions, the device has to be safe and therefore proper control measures have to be in place.

The defibrillator has a specific test with 5 kV voltage. During this test, the device must not generate insulation problems (e.g., arcs on the PCB, insulation breakdown, air-superficial discharges) and must recover the ECG signal display within 10 seconds. The test results show a complete signal recovery within 5 seconds. Chapter 5 has additional details of this test. The requirements also prescribe that no dangerous voltage is generated at the device terminals. This test is performed by measuring the voltage at the terminals with the device switched off. The test results show a voltage lower than 0.5 V at the USB connector.

A similar test is performed to ensure **dielectric strength** between the patient connection and either the enclosure or the USB connection. During this test (1500 V AC for 1 minute), no on-air or superficial electric discharges have to be generated on the device.

This requirement can be ensured by proper design of the PCB (i.e., increasing insulation distances) and proper selection of the enclosure material. For example, an ABS plastic enclosure has an improved dielectric strength, that is, the material can withstand higher test voltagea for a given thickness.

Figure 8.17 shows the inside of the Gamma Cardio CG PCB superimposed over the drawing of the enclosure. At the bottom left, there is the minimum distance between the outer

Table 8.32 Main electric shock risks from Gamma Cardio CG usage

ID (MDD ref.)	MDD hazard	Hazard scenario	Harm description	Main control measure
7.6	Unintentional **ingress of substances** into the device taking into account the device and the nature of the environment in which it is intended to be used	Shortcut of internal parts connected to the PC due to liquid/dust penetration.	Electric shock due to voltage transferred by PC	The device is not connected to the mains supply network: a double fault is then required on the PC and on the device in order to have a dangerous voltage on applied parts; the device is contained in an enclosure that is closed and protected against possible liquid penetration even in the operating room or in an out-patient environment
12.6	**Accidental electric shocks** during normal use and in single fault condition, with devices installed correctly	Higher leakage currents	Probable death due to electric shock	Designed according to IEC/EN 60601-1 class II CF
12.7.1	**Mechanical risks** connected with resistance, stability, etc.	Enclosure breaking due to accidental collision or crash, and consequent accessibility to electrical parts connected to mains supply network	Possible death due to electric shock	Risk not applicable since device is powered by SELV PC USB power; mechanical components are compliant with IEC/EN 60601-1.
12.7.4	Risks from **terminals and connectors** to the electricity, gas or hydraulic and pneumatic energy supplies which the user has to handle	Higher leakage currents due to problems on terminals	Probable death due to electric shock	Designed according to IEC/EN 60601-1 Class II CF with defibrillator-proof applied parts
12.8.1	Energy supply: accuracy of flow rate within specified limits	Higher leakage currents due to non-accurate flow rate	Probable death due to electric shock	Designed according to IEC/EN 60601-1 Class II CF with defibrillator-proof applied parts

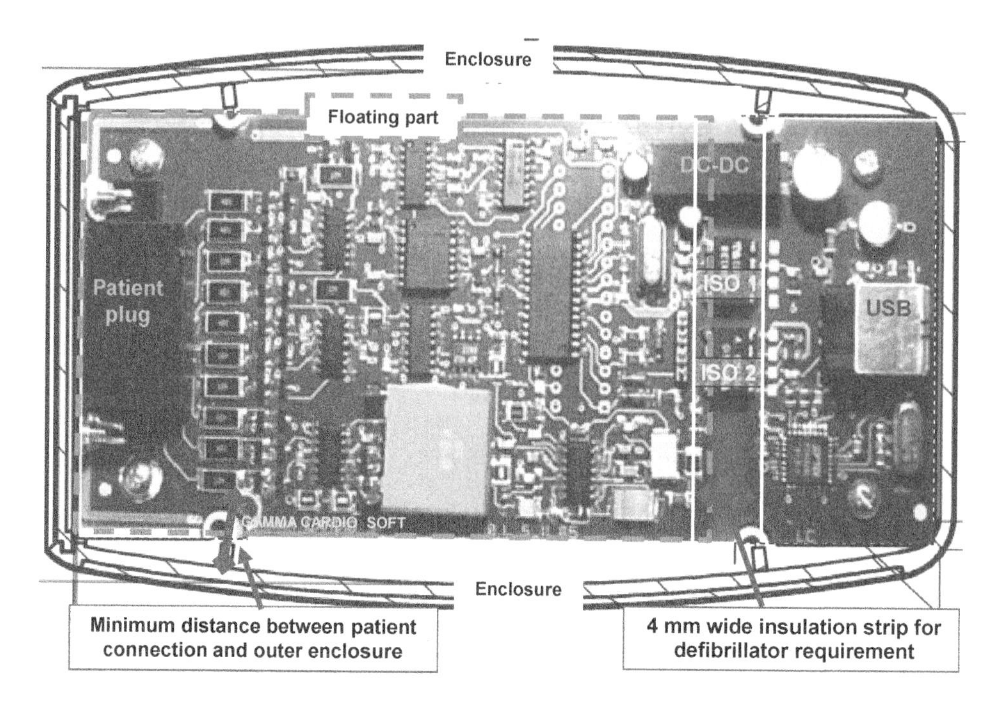

Figure 8.17 Gamma Cardio CG PCB.

enclosure and the PCB tracks that may have a dangerous voltage. This possible **dielectric breakdown** path (thick line) includes a PCB surface, an air distance and the insulation material of the ABS enclosure. The overall distance can withstand higher voltages than required by the IEC/EN 60601-1. A similar consideration holds when 1500 V is applied between the patient connection and the USB plug. In this case, there is also the 4 mm insulation (bottom right of Figure 8.17) that is required to withstand the defibrillator's 5 kV short duration impulse.

8.13.3 Enclosure IP Protection

Regarding IP protection, as discussed in 8.3.3, the enclosure of a medical device has to provide protection from external bodies and external weather conditions. The protection level is encoded by the IP code. The Gamma Cardio CG is IP 20, that is, it is effective against access to hazardous parts from a finger, but it is not protected against ingress of water.

8.14 Verification on Design Performance

8.14.1 Safety-critical Materials

As discussed in Sections 5.29 and 5.30, the optocouplers and the DC–DC converter are the two safety-critical components since their performance ensures the safety to the device. These components cross the vertical 4 mm wide insulation strip shown in the right side of Figure 8.17, where at the top there is the DC–DC converter and at the bottom the two optocouplers indicated with the ISO 1 and ISO 2 label. Datasheets of the critical components have to be provided to demonstrate the required safety performance. First, the distance between pins crossing the insulation strip has to be more than 4 mm. The other performances impacting safety are

1. resistance and capacitance isolation to ensure limited leakage currents
2. rated working voltage to avoid electric breakdowns
3. short-circuit protection for DC–DC to avoid dangerous overheating.

Regarding the leakage current, as discussed in Sections 5.29 and 5.30, these components ensure the galvanic insulation between the floating and the non-floating parts, that is, there is no substantial physical connection that allows (leakage) current to pass between floating and non-floating parts of the Gamma Cardio CG. The only connection that allows some leakage current to pass is represented by the limited insulation capacitance and resistance of these components which, in this case, are extremely high. The higher their value, the lower the leakage current. For example, the selected optocouplers have insulation resistance $>10^{12}\,\Omega$ and insulation capacitance <1 pF. The DC–DC converter has 20 GΩ minimum insulation resistance and insulation capacitance of 40–72 pF. These reduced values are the origin of the limited leakage current, which is $\leq 0.1\,\mu$A, apart from current leakage caused by mains supply on applied parts, which is $\leq 1.1\,\mu$A (requirement $\leq 50\,\mu$A).

8.14.2 Creepage and Air Clearance

Creepage distance and air clearance discussed in Section 8.2.2 have to be proportional to the voltage differential present on the board to avoid dielectric breakdown. The Gamma Cardio CG is powered by low voltage and so such requirements are less stringent. Nevertheless, a defibrillator may be used when the ECG is connected to the patient. This requires a specific insulation distance to avoid dielectric breakdown during defibrillator discharge. According to IEC/EN 60601-1 and IEC/EN 60601-2-25, 4 mm is considered sufficient for the 5 kV discharge. In the bottom right of Figure 8.17, this 4 mm distance is indicated, separating the floating applied part from the PC connected part. In this 4 mm-wide strip, there must not be any conductive PCB path, so the strip includes only PCB insulation material, and critical components cross the floating and non-floating areas. These components have to be selected so that they maintain insulation at the prescribed working voltage and they withstand defibrillator discharge. It is recommended to increase these distance so that requirement is maintained even with the limited precision of PCB manufacturing, so the Gamma Cardio CG has a measured 5 mm insulation distance.

8.14.3 Other Verifications

Regarding the other sections discussed in the theory part, marking (also referred to as labeling) was extensively discussed in Section 8.11.5.3. External wiring discussed in Section 8.4.4 is a simple task, since the USB cable must be compliant with the relevant standard, and the patient cable is compliant to the MDD. The connection to the power supply discussed in Section 8.4.5 is not applicable for the Gamma Cardio CG. Regarding the mechanical strength and the protection against fire, the enclosure has to be verified with specific tests according to IEC/EN 60601-1 (push, drop etc.). After tests, no damage or deformation or break was experienced.

8.15 Safety Tests

8.15.1 Leakage Current

Leakage currents have been extensively covered in Section 5.9, 5.29 and in 8.5.1. Here we will focus on the specific test results. Tests require that leakage current has to be measured for any possible combination of input/output with ground, mains supply and the enclosure. Leakage currents have a catastrophic impact on safety, since a higher value of these currents may lead to death from electric shock. This suggests performing extensive testing at the final inspection of each device to be delivered. Unfortunately, this is the only test that cannot be performed through the automatic control system of the Gamma Cardio CG device. This means that all the tests have to be performed manually by an operator with risks of errors in the measurement. To reduce these risks, the Gamma Cardio CG has been designed to achieve safety limits that are much better than the safest limits required for the direct application to the heart (type CF). This also allows the tests to be simplified because a number of tests can be combined in a single test. For example, earth leakage may be performed by measuring the current between earth and all the 10 patient connections together instead of performing 10 measurements, one with each patient connection. The higher aggregate current value is still much lower the required maximum value for an individual connection. The highest value of leakage current (11 μA) is obtained in the single fault condition when one patient terminal is connected to the mains supply. This patient leakage from patient connection to mains supply must not exceed the 50 μA according to the requirements. The other values are at around 1 μA or lower, while the maximum requirement values are shown in Table 8.33 for IEC/EN 60601-1 3rd edition. Note that other standards and the previous version IEC/EN 60601-1 have different limits.

8.15.2 Heating

Excessive heating is a possible cause of fires, damage to the materials used to isolate dangerous voltage, and deformation of the enclosures. Overheating may be expected on enclosures, on parts normally accessible (knobs, switches, handles, applied parts) and in the inner parts, such as magnetic components (transformers, motors, relays), electrolytic capacitors, switches, fuses and PCB areas under heating components.

Table 8.33 IEC/EN 60601-1 leakage current limits for CF devices (values in μA)

Current leakage type	Normal condition	Single fault condition
Earth leakage	5000	10,000
Enclosure leakage current (touch)	100	500
Auxiliary leakage (either from external voltage to patient or from patient to ground)	DC < 10	DC < 50
Patient leakage (from patient connection to earth or from external voltage to SIP/SOP)	AC < 10	AC < 50

Table 8.34 Gamma Cardio CG heating test results

Device part	Measured value (°C)	Limits
Test chamber environment	19.0	90
External conductors insulation	23.0	70
Accessible parts without a tool	21.0	85
Patient accessible (surfaces held for short time)	21.0	50
Terminals for external conductors	23.0	85
Thermoplastic materials	21.0	85
Electrolytic capacitors	23.0	85
Test chamber environment	19.0	90
Printed circuit board	24.0	105
Applied parts	19.0	40

The maximum allowed overheating should not be exceeded under normal conditions (i.e., 110% of the rated power supply) and under single fault conditions such as short-circuit of electrodes and short-circuit of internal components. In this case, temperatures must not rise significantly. This can be achieved by selecting active components, and in particular the DC–DC converters, protected by short-circuits. Table 8.34 shows the temperature values of various Gamma Cardio CG parts after 2 hours continuous operation, with an environment temperature of 19°C. Values are well below prescribed limits. This ensures that the device can work safely, continuously for long periods.

8.15.3 Other Safety Tests

The other safety tests discussed in the theory section (Section 8.5) are already covered or not applicable. **Dielectric strength** has been discussed in the section on insulation (Section 8.13.2).

Mechanical strength tests, introduced in Section, 8.5.4, have to demonstrate that expected mechanical stresses do not create unacceptable risk. This requires a proper choice of enclosure and external connectors. In addition, mechanical design of the internal connections must be robust to avoid stresses creating damage and disconnections on the electronic circuits. The applicable tests for the Gamma Cardio CG are

- drop test up to a maximum of 1 m from the ground
- pressure test of 45 N on the enclosure
- shock-test: a sphere of diameter (50 mm 500 g) dropped onto the surface from a height of 1.3 m.

Figure 8.17 shows the inside of the Gamma Cardio CG. Components are all soldered and tightly fixed to the PCB. Thus, disconnections or damage are unlikely to occur with foreseeable stresses. This is confirmed by the tests, since no damage, deformation or breakage was experienced.

Abnormal operating and fault conditions (Section 8.5.5) have an impact on heating, and they were discussed in the corresponding Section 8.15.2.

Since the Gamma Cardio CG is a low voltage, Class II CF, USB powered portable device, the tests discussed from Sections 8.5.6–8.5.10 are not applicable (continuity of protective earthing, the residual voltage, voltage on the accessible parts, the energy stored and the current and power consumption tests).

8.16 Electromagnetic Compatibility

Electromagnetic compatibility (EMC) means that an electrical device is able to operate in its electromagnetic environment without unacceptable risks affecting safety or performance of the device or of the surrounding devices.

The **applicable standards** are the IEC/EN 60601-1 and the IEC/EN 60601-1-2 on electromagnetic compatibility, as detailed in Section 8.6.

The standard contains requirements and tests for emission and immunity. Table 8.12 and Table 8.13, already discussed in Chapter 5, contain the specific tests and the associated limits. Some tests are not applicable to devices that are not powered by mains supply, such as the Gamma Cardio CG. In particular, conducted, harmonic and flicker emissions of Table 8.12 do not need to be included. Variations in voltage supply power lines and surge immunity tests of Table 8.13 can also be omitted.

Each **EMC Test report** usually contains the following information: scope, ports where values are measured (connectors, enclosure etc.), applicable standards, limits, instruments used for the test (name, manufacturer, model, calibration expiry date), test procedure, uncertainty in the measurement, results. An example of the test report is given in Table 8.35.

The **accompanying documents** of medical devices compliant to IEC/EN 60601-1 must include information on EMI performance and suggestions on methods for minimizing this interferences. More specifically, the following information is to be provided:

- tables regarding the emissions and immunity of the medical device
- a table with safety separation distances for portable and mobile RF transmitters
- a list of the product's essential performance.

If applicable, the following information is also to be specified:

- a list of all the parts and associated performances that may affect EMC compliance (e.g., cables with admissible length – cables act as passive-active antennas)
- a warning that the use of non-recommended accessories may result in increased emissions and/or decreased immunity
- a warning that the medical device should not be used close to or over other equipment
- any justifications for lower immunity levels.

With this information in mind, the user/customer can organize an EMC environment that is suitable for the specific device. The user has the responsibility of ensuring that the device is used within the intended EMC environment. A specific device and its surrounding devices are ensured to operate without unacceptable risks if hosted in the specified EMC

Table 8.35 Gamma Cardio CG emission test report

Scope	This test aims to measure the amplitude of the RF noise electric field intensity (in V/m) emitted by the device and its connected cables. An elevated radiated noise in the frequency range of 30MHz–1GHz may interfere with other devices. Emitted radiation may generate improper behavior on nearby telecommunication receivers or other electronic devices that may capture noise interference through their cables and circuits.		
Ports	Enclosure		
Applicable standards	EN 55022 Class B, EN 60601-1-2		
Limits	30–230 MHz: <30 dBμV/m quasi-peak value 230–1000 MHz: <37 dBμV/m quasi-peak value		
Instruments used for the test	**Instrument**	**Manufacturer, model**	**Calibration expiring date**
	Anechoic chamber Receiver RF amplifier Antenna bilog	*to be specified*	*to be specified*
Test procedure	Radiated emissions have been measured in accordance with methods and limits of EN 55022. The distance between the equipment under test and the antenna was 3 meters. The equipment was connected to the PC through the USB port. During the test, the device side with the maximum emission has been identified (i.e. the frontal plane).		
Uncertainty in the measure (2σ)	+6/−6.7 dB between 30 MHz and 230 MHz +4.6/−5.2 dB between 230 MHz and 1 GHz		
Results	Compliant to EN 55022 Class B, since the measured emission values are within the limits for the quasi-peak detector in both the horizontal and vertical polarization (see Figure 8.18).		

environment. In the following sections, we will outline examples of test results based on the Gamma Cardio CG.

8.16.1 Emission

The emission test is performed to assess possible interference with adjacent devices. Table 8.35 shows the associated test report.

Figure 8.18 shows the RF radiated emissions of the Gamma Cardio CG on the horizontal and vertical polarization. The x axis shows the frequencies in log scales from 30 MHz to 1 GHz. The vertical axis shows the emission expressed in dBμV/m. The emissions shown in Figure 8.18 are mainly due to the DC–DC converter that has an internal switching oscillator generating significant harmonics. Emission limits may be satisfied following the manufacturer's recommendations on EMI and, in particular, providing the electronic board with proper EMI filters. Alternatively, cables may be wound onto external ferrite coils.

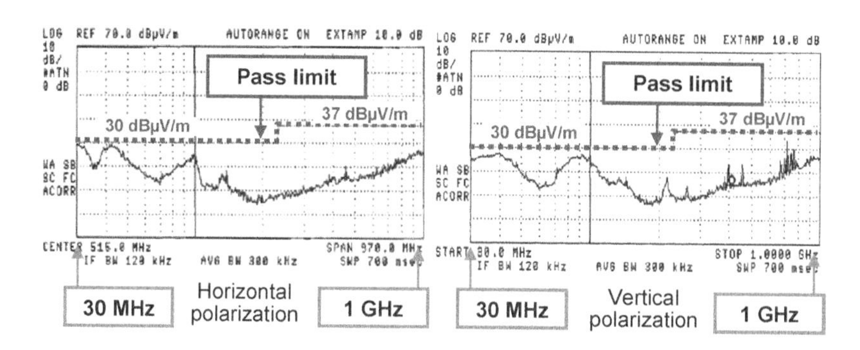

Figure 8.18 Gamma Cardio CG emission results.

Table 8.36 Gamma Cardio CG Emission test information

	Aspects of emission	
Emission test	**Compliance**	**Electromagnetic environment – guide**
RF emissions CISPR 11	Group 1	Equipment in which there is **intentionally generated** or used **conductively coupled** RF energy that is necessary for the internal functioning of the equipment itself. Therefore, its RF emissions are very low and are not likely to cause any interference in nearby electronic equipment.
RF emissions CISPR 11	Class B	Equipment suitable for **use in all domestic establishments** and those directly connected to the public low-voltage power supply network that supplies buildings used for domestic purposes.

When the device is compliant to the test, the information of Table 8.36 is to be included in the documents for use.

Devices that are powered by the mains supply have to be compliant with the limits of the harmonic emissions and emissions of fluctuations of voltage/flicker as shown in Table 8.12. Table 8.37 shows typical information to be included in the documents for use.

8.16.2 Immunity

Immunity tests ensure that the equipment does not become dangerous or unsafe when noise is radiated or conducted into the equipment by the environment. In the presence of interference, the equipment must maintain a proper level of performance depending on the type of applied noise and must not suffer from failure, or, worse, originate hazards. For the Gamma Cardio CG, the following criteria have been adopted to assess conformity:

1. the device must not evidence
   - component failures
   - parameter variations

Table 8.37 Emission test information for device powered by mains supply

Aspects of emission		
Emission test	**Compliance**	**Electromagnetic environment – guide**
Harmonic emissions IEC 61000-3-2	Class A Compliant	The device is suitable for use in all establishments, including domestic establishments and those directly connected to the public low-voltage power supply network that supplies buildings used for domestic purposes.
Emissions of fluctuations voltage/flicker IEC 61000-3-3	Compliant	

- changes in the intended operating mode
- false alarms
- interruptions or cessation of the expected functions
- beginning of unwanted operations.

2. The safety performance has to be maintained, verifying that the ECG acquisition with the simulated signal is correctly displayed.

Table 8.38 shows a test report related to fast transient immunity. Similar test reports are produced for the other immunity tests.

Table 8.39 shows typical immunity information to be included in the documents for use.

Table 8.38 Gamma Cardio CG fast transient immunity test report

Scope	Verify that the equipment works properly in the presence of **fast and repetitive transients** on the power supply line, on signal lines and on other lines. This test simulates the effects of noise generated by the switching of inductive loads, contacts relays, etc.		
Ports	USB		
Applicable standards	EN 61000-4-4, EN 60601-1-2		
Limits	Level 3 (1 kV peak voltage at 5 kHz repetition frequency on signal data and control ports)		
Instruments used for the test	**Instrument**	**Manufacturer Model**	**Calibration Expiry date**
	Immunity test generator	*to be specified*	*to be specified*
Test procedure	Test performed with the device arranged as in 7.2 of EN 61000-4-4 in accordance to the procedure described in 8.1 EN 61000-4-4, using the capacitive clamp for 2 minutes as a coupling device.		
Results	Compliant (no effect visible on the ECG display)		

Table 8.39 Immunity test information

Aspects of immunity			
The product is intended for use in the electromagnetic environment specified below. The customer or the user should ensure that the product is used in the specified environment			
Immunity Test	**Test level IEC/EN 60601-1-2**	**Level of Compliance**	**Electromagnetic environment guide**
Electrostatic discharge IEC/EN 61000-4-2	±6 kV in contact ±8 kV in air	±6 kV in contact ±8 kV in air	Floors should be wood, concrete or ceramic tile; if floors are covered with synthetic material, the relative humidity should be at least 30%.
Electrical fast transient/burst IEC/EN 61000-4-4	±1 kV signal lines	±2 kV lines signal supply	The quality of the mains voltage should be that of a typical commercial or hospital environment.
Magnetic field IEC/EN 61000-4-8	3 A/m	3 A/m	Magnetic fields at mains frequency should be at values of a typical area in a commercial or hospital environment.

Aspects of immunity to RF			
The ECG product is intended for use in the electromagnetic environment specified below. The customer or the user should ensure that the device is used in such an environment.			
Immunity test	**Test level to EN 60601-1-2**	**Level of compliance**	**Electromagnetic environment – guide**
RF conduct EN 61000-4-6	3 V_{rms} from 150 kHz to 80 MHz	3 V_{rms} from 150 kHz to 80 MHz	The RF communications equipment (portable and mobile) should not be used near to any part of the ECG, including cables, except for equipment that complies with the recommended separation distances calculated according to the equation applicable for the transmitter frequency. Recommended separation distances $d = 1.2 \cdot \sqrt{P}$ from 150 kHz to 80 MHz $d = 1.2 \cdot \sqrt{P}$ from 80 MHz to 800 MHz $d = 2.3 \cdot \sqrt{P}$ from 800 MHz to 2.5 GHz
RF Radiated EN 61000-4-3	3 V_{rms} from 80 MHz to 2.5 GHz	3 V_{rms} from 80 MHz to 2.5 GHz	where P is the maximum rated output power of the transmitter in watts (W) according to the transmitter manufacturer and d is the separation distance in meters (m).
The field strengths from fixed RF transmitters, as determined by an electromagnetic site survey, may be less than the compliance level in each frequency range. Interference may occur in the vicinity of equipment marked with this symbol: ((•))			

Table 8.40 Immunity test information for device powered by mains supply (U_T = Nominal Voltage Power Line)

Immunity test	Test level to IEC/EN 60601-1-2	Level of compliance	Electromagnetic environment – guide
Surge IEC/EN 61000-4-5	±1 kV differential mode	±1 kV differential mode	The quality of the mains voltage should be that of a typical commercial or hospital environment.
Voltage dips, short interruptions and voltage variations on the input lines IEC/EN 61000-4-11	<5% U_T (>95% dip in U_T) for 0.5 cycles 40% U_T (60% dip in U_T) for 5 cycles 70% U_T (30% dip in U_T) for 25 cycles <5% U_T (>95% dip in U_T) for 5 seconds	<5% U_T (>95% dip in U_T) for 0.5 cycles 40% U_T (60% dip in U_T) for 5 cycles 70% U_T (30% dip in U_T) for 25 cycles <5% U_T (>95% dip in U_T) for 5 seconds	The quality of the mains voltage should be that of a typical commercial or hospital environment. If the user requires continued operation during a power supply voltage, it is recommended that the equipment is supplied with an uninterruptible power supply (UPS) or battery.

For devices that are powered by the mains supply, additional tests related to surge and interruption immunity have to be performed. The additional information given in Table 8.40 has to be provided.

Finally, the manufacturer must disclose the safety separation distances between its device and portable and mobile RF transmitters as shown in Table 8.41.

Table 8.41 Recommended separation distance with RF transmitters

Recommended separation distance between portable and mobile radio communication equipment and the ECG device
The ECG product is intended for use in an electromagnetic environment in which the radiated RF interference is under control. The customer or the operator of the device can help prevent electromagnetic interference by maintaining a minimum distance between RF mobile and portable communication devices (transmitters) and the ECG device, as recommended below, based on the maximum output power of the radio equipment.

(*continued*)

Table 8.41 (*Continued*)

Maximum rated power output of the transmitter (W)	Separation distance to the transmitter frequency (m)		
	150 kHz to 80 MHz $d = 1.2 \cdot \sqrt{P}$	**From 80 MHz to 800 MHz** $d = 1.2 \cdot \sqrt{P}$	**800 MHz to 2 GHz** $d = 2.3 \cdot \sqrt{P}$
0.01	0.12	0.12	0.23
0.1	0.38	0.38	0.73
1	1.2	1.2	2.3
10	3.8	3.8	7.3
100	12	12	23

For transmitters rated at a maximum output power not included above, the recommended separation distance d in meters (m) can be calculated using the equation applicable to the frequency of the transmitter, where P is the maximum rated output of the transmitter in watts (W) according to the transmitter manufacturer.

Notes:

(1) At 80 MHz and 800 MHz, the applicable frequency range is the highest
(2) These guidelines may not be applicable in all situations. Electromagnetic propagation is affected by absorption and reflection from structures, objects and people.

References

Banana skins, (1998), UK EMC Journal, vol. 15, p. 8. Compliance Engineering, vol. 10, p. 25.

FDA (1993), Recommendations for Developing User Instruction Manuals for Medical Devices Used in Home Health Care.

FDA (1998), Guidance for Industry Diagnostic ECG Guidance (Including Non-Alarming ST Segment Measurement).

FDA (2000a), Medical Devices and EMI: The FDA Perspective, available at FDA website.

FDA (2000b), Medical Device Use-Safety: Incorporating Human Factors Engineering into Risk Management.

FDA (2007), Draft Guidance for Industry and FDA Staff Radio-Frequency wireless Technology in Medical Devices.

FDA (2013a), Classify your device, available at FDA website: http://www.fda.gov/MedicalDevices/DeviceRegulationandGuidance/Overview/ClassifyYourDevice/default.htm

FDA (2013b), how to market medical device: http://www.fda.gov/MedicalDevices/DeviceRegulationandGuidance/HowtoMarketYourDevice/default.htm

GHTF (2006), http://www.ghtf.org/documents/sg1/SG1-N15-2006-Classification-FINAL.pdf

HAR (2002), Agreement on the use of a Commonly Agreed Marking for Cables and Cords complying with Harmonized Specifications, HAR Group, EEPCA.

MEDDEV (2010), European Commission, DG Health and Consumer, Medical Devices: Guidance document – Classification of medical devices, 2. 4/1 Rev. June 2010.

Silberberg (1993), Performance degradation of electronic medical devices due to electromagnetic interference, Compliance Engineering vol. 10 p. 25.

Silberberg, J., (2001), Achieving Medical Device EMC: The Role of Regulations, IEEE International Symposium on Electromagnetic Compatibility.

WHO (2003), World Health Organization, Medical Device Regulations Global overview and guiding principles.

Summary of Regulations and Standards

Legal frameworks

European Union

EEC [2007], European Council Directive 93/42/EEC of 14 June 1993 concerning medical devices available at http://ec.europa.eu/health/medical-devices/index_en.htm amended by 2007/47/EC into force on 21 March 2010.

The regulatory body is the European Commission. See also: http://ec.europa.eu/health/medical-devices/index_en.htm

United States

US Food and Drug Administration (FDA) is the US legal authority for medical devices, see: http://www.fda.gov/MedicalDevices/default.htm

Most of FDA's medical device and radiation-emitting product regulations are in Title 21 CFR (Code of Federal Regulations) Parts 800-1299. Main applicable Title 21 CFR Parts are:

Part 807	Establishment registration
Part 807	Medical device listing
Part 807 Subpart E	Premarket notification 510(k)
Part 814	Premarket approval (PMA)
Part 812	Investigational device exemption (IDE)
Part 820	Quality system regulation (QS)/ Good manufacturing practices (GMP)
Part 801	Labeling
Part 803	Medical Device Reporting –

See also the FDA 21 CFR searchable database:
http://www.accessdata.fda.gov/scripts/cdrh/cfdocs/cfcfr/CFRSearch.cfm

Accepted standards by the regulatory body

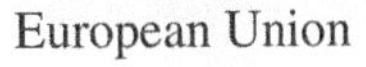

European Union

Harmonized standards for medical devices:
http://ec.europa.eu/enterprise/policies/european-standards/harmonised-standards/medical-devices/index_en.htm

Harmonized standards for Electromagnetic compatibility (EMC):
http://ec.europa.eu/enterprise/policies/european-standards/harmonised-standards/electromagnetic-compatibility/index_en.htm

United States

Recognized Consensus Standards:
http://www.accessdata.fda.gov/scripts/cdrh/cfdocs/cfstandards/search.cfm

(continued)

	Main widely accepted horizontal standards 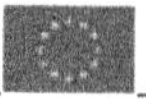 ***indicates EU harmonized or US recognized***
Quality system	ISO EN 13485:2003/AC:2009: Medical devices – Quality management systems – Requirements for regulatory purposes (~)
Risk Management	ISO EN 14971:2007: Medical devices – Application of risk management to medical devices (EN 14971:2009)
Safety and performance for MEE	IEC/EN 60601-1:2005-12: Medical electrical equipment (MEE) Part 1: General requirements for basic safety and essential performance (identical to EN 60601-1:2006)
Usability	IEC EN 60601-1-6:2010: MEE – Part 1-6: Collateral standard: Usability
Alarms	IEC/EN 60601-1-8:2006: MEE – Part 1-8: General requirements for basic safety and essential performance – collateral standard: General requirements, tests and guidance for alarm systems in medical electrical equipment and medical electrical systems (EN 60601-1-8:2007/AC 2010)
EMC	IEC/EN 60601-1-2:2007 Medical electrical equipment – Part 1-2: General requirements for basic safety and essential performance – Collateral standard: Electromagnetic compatibility – Requirements and tests (EN 60601-1-2:2007 Modified)
Software	IEC EN 62304:2006: Medical device software – Software life-cycle processes (EN 62304:2006/AC:2008)

Medical device certification resources

Australia

Therapeutic Goods Administration: http://www.tga.gov.au/

Canada

Health Canada: http://www.hc-sc.gc.ca/dhp-mps/md-im/index-eng.php

China

State Food and Drug Administration P.R. China: http://eng.sfda.gov.cn

European Union

European Commission: DG Health & Consumers http://ec.europa.eu/health/medical-devices/index_en.htm

India

Central Drugs Standard Control Organization: Ministry of Health and Family Welfare, Government of India http://cdsco.nic.in/

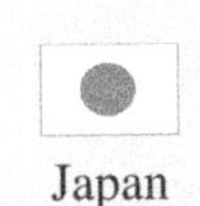
Japan

Ministry of Health, Labor and Welfare: www.mhlw.go.jp/english/index.html Pharmaceutical and Medical Devices Agency http://www.pmda.go.jp/english/index.html

United States

US Food and Drug Administration: http://www.fda.gov/MedicalDevices/default.htmwww.fda.gov/cdrh/index.html.

Guidelines (*non-legally binding*)

European Union

Guidance MEDDEVs: http://ec.europa.eu/health/medical-devices/documents/guidelines/index_en.htm

United States

Good Guidance Practice (GGP): http://www.fda.gov/MedicalDevices/DeviceRegulationandGuidance/GuidanceDocuments/default.htm

---

Other Standards

AAMI (1991), Association for the Advancement of Medical Instrumentation, American National Standard, Diagnostic Electrocardiographic Devices, ANSI/AAMI EC11-1991.

CENELEC HD 361 S3 (1999), System for cable designation.

EEC (2007), Council Directive 93/42/EEC 14 June 1993 concerning medical devices available at http://ec.europa.eu/health/medical-devices/index_en.htm amended by 2007/47/EC 21 March 2010.

EN 980 (2008), Symbols for use in the labeling of medical devices.

EN 1041 (2008), Information supplied by the manufacturer of medical devices.

EN 50525-2-21 (2011), Electric cables – Low voltage energy cables of rated voltages up to and including 450/750 V (Uo/U) – Part 2-21: Cables for general applications – Flexible cables with crosslinked elastomeric insulation.

HL7 (2011), HL7 Ansi Standards see http://www.hl7.org

IEC/EN 14971 (2009), Medical devices. Application of risk management to medical devices.

IEC/EN 60601-1 (2005): Medical electrical equipment (MEE) Part 1: General requirements for basic safety and essential performance (identical to EN 60601-1:2006).

IEC/EN 60601-1-2 (2007), MEE. General requirements for safety. Collateral standard. Electromagnetic compatibility. Requirements and test (IEC Modified).

IEC/EN 60601-1-6 (2010), MEE – Part 1-6: General requirements for basic safety and essential performance – Collateral standard: Usability.

IEC/EN 60601-2-25 (1999) MEE – Part 2-25: Particular Requirements for the Safety of Electrocardiographs.
IEC/EN 60601-2-27 (2006) MEE – Part 2-27: Particular requirements for the safety, including essential performance, of electrocardiographic monitoring equipment.
IEC/EN 60601-2-47 (2001) MEE – Part 2-47: Particular requirements for the safety, including essential performance, of ambulatory electrocardiographic systems.
IEC/EN 60601-2-51 (2003) MEE – Part 2-51: Particular requirements for safety, including essential performance, of recording and analyzing single channel and multichannel electrocardiographs.
IEC/EN 60601-3-2, ed 1. Medical electrical equipment – Part 3-2: Particular requirements for the essential performance of recording and analyzing single channel and multichannel electrocardiograph, 1999.
IEC/EN 60950-1, (2011), Information technology equipment – Safety – Part 1: General requirements.
ISO/IEC/EN 62304, (2006), Medical device software – Software life-cycle processes (EN 62304:2006/AC:2008).
IEEE 14764 (2006), Software Engineering – Software Life Cycle Processes – Maintenance.
ISO (2005), Health informatics – Electronic health record – Definition, scope and context – ISO/DTR 20514.
ISO/EN 10993-1 (2009), Biological evaluation of medical devices. Evaluation and testing within a risk management process.
ISO/EN 14971 (2009), Medical devices. Application of risk management to medical devices.
ISO/DIS 31000 (2009), Risk management – Principles and guidelines on implementation. International Organization for Standardization.
ISO/IEC 25010 (2011), Systems and software engineering – Systems and software Quality Requirements and Evaluation (SQuaRE) – System and software quality models.
MIL-STD-498 (1994a), Software Development and Documentation Standard, AMSC NO. N7069, US Department of Defense.

Index

Medical Instrument Design and Development: from Requirements to Market Placements, First Edition.
Claudio Becchetti and Alessandro Neri.